G. Morrison

AF837482

SYMPOSIUM ON DEEP FOUNDATIONS

Proceedings of a Symposium
Sponsored by the
Committee on Deep Foundations of the
Geotechnical Engineering Division of the
American Society of Civil Engineers

at the
ASCE National Convention
Atlanta, Georgia
October 25, 1979

Frank M. Fuller, Editor

AMERICAN SOCIETY OF CIVIL ENGINEERS FOUNDED 1852

Published by the
American Society of Civil Engineers
345 East 47th Street
New York, New York 10017

Copyright© 1980 by the American Society of Civil Engineers,
All Rights Reserved.
Library of Congress Catalog Card Number 80-69155
ISBN 0-87262-256-8
Manufactured in the United States of America.

FOREWORD

At the April 1978 meeting in Pittsburgh of the Deep Foundations Committee, ASCE Geotechnical Engineering Division, initial plans were layed to sponsor a one day (two session) symposium on deep foundation case histories, to be held during the Atlanta Conference in October 1979. The primary objective of this symposium and these subsequent proceedings was to bring into the public domain some of the interesting and useful field data on the testing and performance of deep foundations. From the 52 papers offered, 38 were selected as complying with the basic objective of the sumposium. From the accepted papers, twenty were selected for presentation at the one-day symposium held October 25, 1979 in Atlanta, Georgia.

Ten papers were presented at each of the two symposium sessions. Following the ten minute presentations by the authors, the balance of each session was devoted to a panel discussion and question and answer period. The morning session was moderated by M. T. Davisson and the afternoon by G. G. Goble. The panels were comprised of the authors making presentations.

It was planned from the beginning that all accepted papers even though not presented at the Symposium, would be published in a subsequent Proceedings. Due to time constraints, only a limited number of papers could be presented but no limit was established for publication. Eventually, out of the 38 accepted papers, 12 were withdrawn by their authors. The remaining papers were passed through a peer review process and authors were given an opportunity to prepare final manuscripts based upon review comments, suggestions and in some cases required revisions. Before being submitted for publication, all papers were edited by the undersigned.

The Deep Foundations Committee trusts that this compilation of case histories will be found useful and informative. The Committee recognizes that there is still a wealth of unpublished data relating to the design, installation and performance of deep foundations. It is the feelings of this Committee that some way must be found to extract and publish these practical and useful data to supplement the theoretical and highly technical papers generally being published. We believe these Proceedings are a step in that direction and recommend future symposiums having practicality as their major theme.

Frank M. Fuller
Symposium Chairman

Houston, May 1980

CONTENTS

HISTORY

DEEP FOUNDATIONS COMMITTEE
GEOTECHNICAL ENGINEERING DIVISION
AMERICAN SOCIETY OF CIVIL ENGINEERS

The Deep Foundations Committee of the ASCE Geotechnical Engineering Division can trace its origin back to 1913 when the first ASCE Foundation Committee was formed "to codify the practice on bearing value of soils." The activities of this original committee have had a profound affect on early development of soil mechanics in the United States.

Over the years, the extent of activities that fell under the jurisdiction of the original foundation committee increased to such a degree that in 1968 a decision was made to sub-divide this committee into three technical committees: soil dynamics, shallow foundations and deep foundations. The original membership of the newly formed Deep Foundations Committee included Elio D'Appolonia (contact member), M. T. Davisson, N. W. Koziakin, Raymond Lundgren, Tonis Raamot, Reuben Samuels, Earl Sibley (secretary), and Aleksander S. Vesić (chairman).

The purpose and scope of the committee was defined as follows:

> To carry out the following general functions as appropriate to deep foundations: (1) evaluation of current research and literature, (2) preparation of state-of-the-art reports, (3) organization of technical programs, (4) liasion with related organizations, (5) preparation of manuals of practice and (6) advising the Society and other agencies in the specialty area of the Committee.

A. S. Vesić served as chairman until October 1974 at which time M. T. Davisson was appointed chairman. In October 1978 Davisson turned the gavel over to Raymond Lundgren who is the current committee chairman. The current committee membership is listed below.

The Deep Foundations Committee has generally held annual meetings usually in conjunction with either the Spring or Fall national ASCE Conventions. These meetings have been devoted to pursuing the various activities and objectives of the committee such as the development of building code provisions relating to deep foundations, the planning and preparation for technical sessions on deep foundations and to keep abreast of developments of standards, codes, manuals of practice and other documents affecting the design and installation of deep foundations.

Apparently there was very little committee activity in 1973, 74, and 75. In October 1974 several new members were appointed to this committee and in February

1976, Chairman Davisson called a special meeting of the committee for the purpose of reorganization and reactivation. Four subcommittees were established to further the work and objectives of this committee as follows:

Technical Sessions on Piles (F. M. Fuller, chairman)

Technical Sessions on Piers (R. Samuels, chairman)

Continuing Education Program (D. York, chairman)

Research Needs (T. Raamot, chairman)

At a subsequent meeting of the committee in Chicago, October 1978, the subcommittees were reorganized as follows:

1. Research: Evaluation and Needs, (T. Raamot, chairman)
2. Technical Programs (F. M. Fuller, chairman)
3. Practice, Codes and Standards - Piers (J. D. Smart, chairman)
4. Practice, Codes and Standards - Piles (F. M. Fuller, chairman)
5. Continuing Education (D. L. York, chairman)
6. Liaison with other Organizations (M. T. Davisson, chairman)

The work of these subcommittees has substantially increased the activities of the Deep Foundations Committee.

Codes and Standards

Since 1969, members of the Deep Foundations Committee (Davisson, Fuller, Liver and Raamot) have served on a special task force to revise and update the ASCE Manual of Engineering Practice, No. 27, "Pile Foundations and Pile Structures." This project is currently in its final stages. Members of this committee are also participating in the development of a new set of foundation and excavation standards which will replace the ASCE Manual of Engineering Practice No. 32, "Building Code Requirements for Excavations and Foundations." Four members of the Deep Foundations Committee (Davisson, Fuller, Hunt and Parola) serve on the special ASCE Committee on Foundation and Excavation Standards. M. T. Davisson is the contract writer for the pile standard, responsible for the preparation of a document for committee action.

In September 1977 by letter to the American Society for Testing and Materials, the committee as a whole, expressed concern over an ASTM Standard, D 2899-74 "Standard Method for Establishing Design Stresses for Round Timber Piles," because of its misleading and potentially dangerous provisions. This Standard prescribes a method for determining what is termed the design or working compressive stress but which actually is a ten-year ultimate stress. Since 1975, members of this committee (Davisson, Fuller and Lundgren) have been working within the ASTM Subcommittee, D 07.07, responsible for D 2899 to obtain the necessary revisions to this Standard. A breakthrough finally came in October 1979 when at a meeting of ASTM Subcommittee D 07.07, the necessary revisions were unanimously approved. However, this matter will not be settled until the necessary revisions are approved within ASTM by Subcommittee, Main Committee and Society ballots. This committee will continue to follow the developments leading toward the reissuance of ASTM D 2899.

Continuing Education

A course entitled "Field Control of Pile Installation and Concreting Operations," has been inaugurated by this committee. The course has been successfully given in Chicago, October 1978 and Atlanta, October 1979 by M. T. Davisson and E. T. Mosley (non-committee member). The course will be offered again in October 1980 at the Hollywood, Fla. ASCE meeting and it is planned to continue offering the course as long as sufficient interest is generated. Other courses are in the planning stage.

Liaison With Other Organizations

Through its various members this Committee maintains a cross-membership and thus an active liaison with such organizations as:

American Concrete Institute - Concrete Pile Committee
- Drilled Pier Committee

American Society for Testing and Materials - Deep Foundations Committee
- Committee on Round Timbers (piles)

International Conference of Building Officials - General Design Subcommittee

National Safety Council - Committee on Safety Code for Pile Driving

Prestressed Concrete Institute - Prestressed Concrete Pile Committee

Southern Building Code Congress International - General Design Subcommittee

Structural Engineers Association - California - Various foundation and code committees.

Transportation Research Board - Committee on Foundations of Bridge and Other Structures.

Through its various members, the committee also maintains indirect liaison with such organizations as the American Railway Engineers Association, American Association of State Highway Transportation Officials, American Iron and Steel Institute, Building Research Advisory Board, American Petroleum Institute, Association of Drilled Shaft Contractors, American Wood Preservers Institute, National Bureau of Standards, Naval Facilities Engineering Command, National Forest Products Association, American Wood Preservers Association, etc.

Many of the organizations with which this Committee maintains either direct or indirect liaison develop and promulgate various types of codes, standards and specifications governing the design and installation of pile and pier foundations. Therefore, a contact with such organizations is necessary to keep this Committee abreast of current developments.

Technical Sessions

The following technical sessions were sponsored, organized and presented by the Deep Foundations Committee:

Drilled Piers in Cohesive Soil, Chicago, October 1969

Use of Wave Equation for Analysis of Pile Driving, Chicago, October 1969

Behavior, Analysis and Design of Piles, Houston, October 1972

Allowable Stresses in Piles, Houston, October 1972

Pile Foundation Case Histories, Dallas, April 1977

Drilled Pier Foundations, Pittsburg, April 1978

Symposium Deep Foundation Case Histories, Atlanta, October 1979

The ASCE Deep Foundations Committee through its varied activities will continue to promote sound engineering principles in the development of codes, standards and recommended practices governing the design and installation of deep foundations. This committee will continue to voice its objections to the promotion of unsound principles by special interest groups and will continue its efforts to improve the state-of-the-art.

Roster - November 1979
*Raymond Lundgren
(chairman)

Bengt B. Broms	Jai B. Kim	*Tonis Raamot
*Robert D. Darragh	Norman L. Liver	*Reuben Samuels
*M. T. Davisson	John E. Lyons	Gregory A. Shekas
Anthony Dover	James E. McNutt	John D. Smart
Frank M. Fuller	Joseph Ray Meyer	Ernest Selig
Robert W. Gillespie	George G. Meyerhoff	William F. Swiger
George G. Goble	Michael W. O'Neill	V. N. Vijayvergiya
Hal W. Hunt	Jerry F. Parola	Donald L York

* Members of original (1968) committee.

(Prepared by Frank M. Fuller, May 1980.)

LOAD TESTS ON PILES DRIVEN INTO CRETACEOUS SAND

By Richard E. Ackley[1] and Richard L. Sanders[2], M. ASCE

SYNOPSIS

The test pile program described herein was conducted to aid in pile-type selection at the Arkansas Power and Light Company Independence Steam Electric Station. The program was designed to develop installation data to evaluate performance of four selected pile configurations.

INTRODUCTION

The Arkansas Power and Light Company Independence Steam Electric Station is under construction at a site located on the White River near Newark, Arkansas. Foundation drawings based on a 100-ton (890 kN) design single-pile capacity were available for a similar 1400 gross megawatt facility. The test pile program was designed to provide an economical 100-ton (890 kN) pile configuration and avoid costly re-design and scheduling difficulties for the new facility.

This paper presents the soil conditions, pile data, pile installation, load test data, details of tests, and summary of test results for four types of piles. Data on the pile driving equipment and pile driving resistance are also presented.

A total of fifty-three piles were driven during the test pile program. The test and exploratory pile locations in relationship to boring and final structure locations are shown on Figure 1. The pile types included composite piles (Pipe/Step-Taper), Raymond Step-Taper piles, Steel Pipe piles and H-piles (HP 14 x 102 and HP 14 x 73). A total of twenty piles were driven in Test Area A, twenty-one in Test Area B and ten in the lateral load test pile group. Two additional piles were driven in outlying areas.

The load test layouts are shown on Figures 2, 3 and 4. Piles C-1 through C-7 were loaded in compression. Pile E-10 was test loaded in tension and Pile P-1 and P-7 were test loaded laterally. All compression test pile locations were pre-drilled with an 18-inch (457 mm)

---

[1]Richard E. Ackley, Project Engineer, McClelland Engineers, Inc. Little Rock, Arkansas.

[2]Richard L. Sanders, Formerly Project Engineer, Geotechnical Group, Raymond Technical Facilities, Inc., Houston, Texas

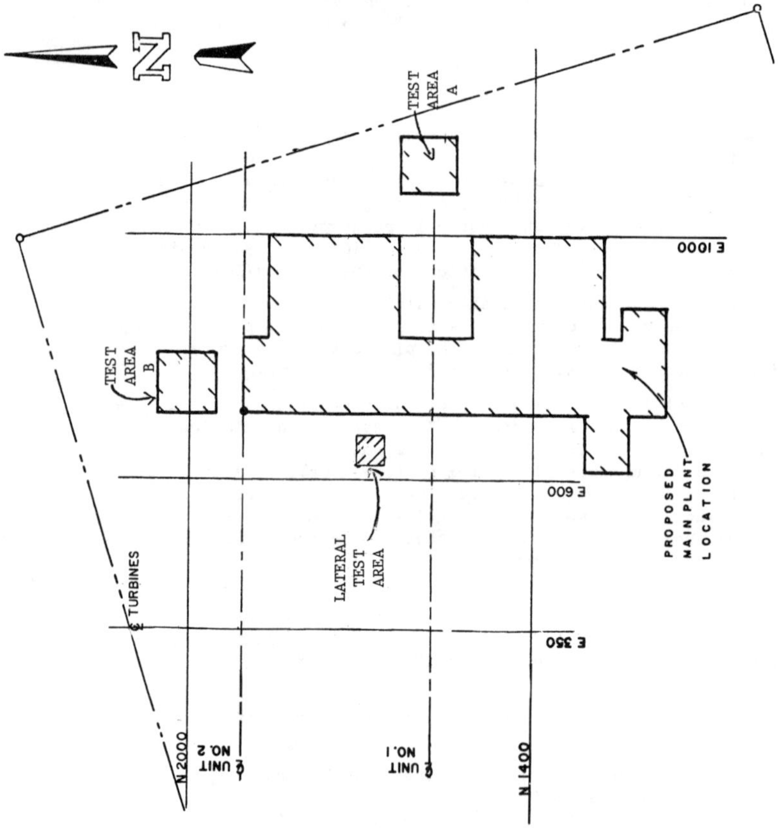

FIG. 1 LOCATIONS OF TEST AREAS

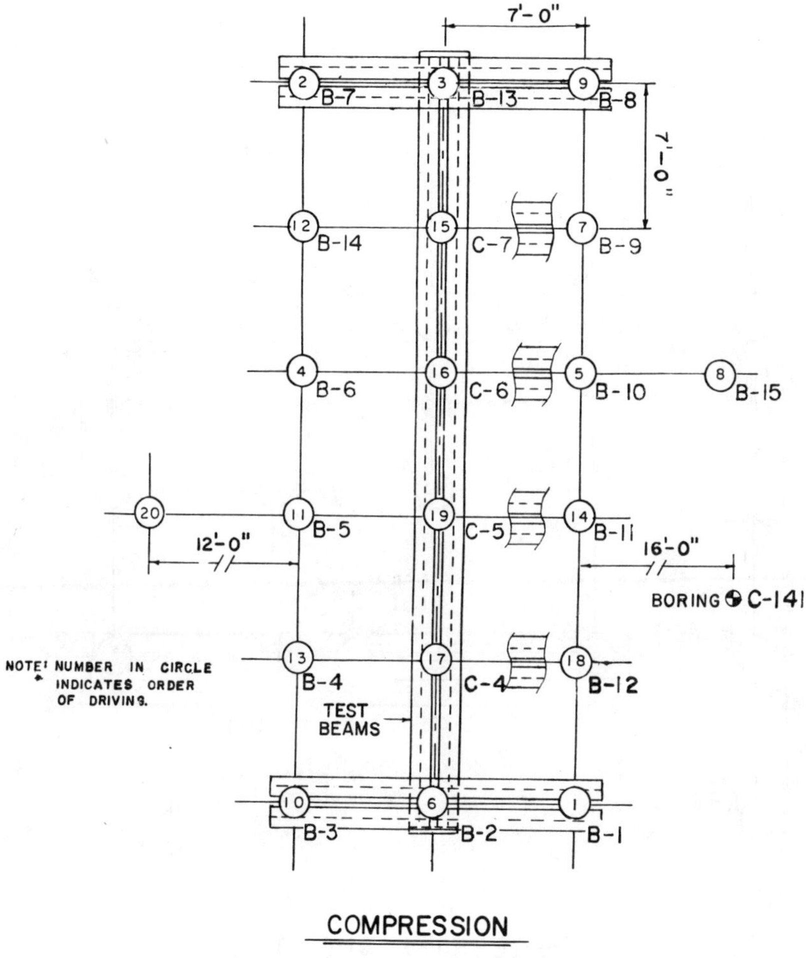

1 ft = 0.305 m   1 in. = 25.4 mm

FIG. 2 TEST PILE LAYOUT - SITE A

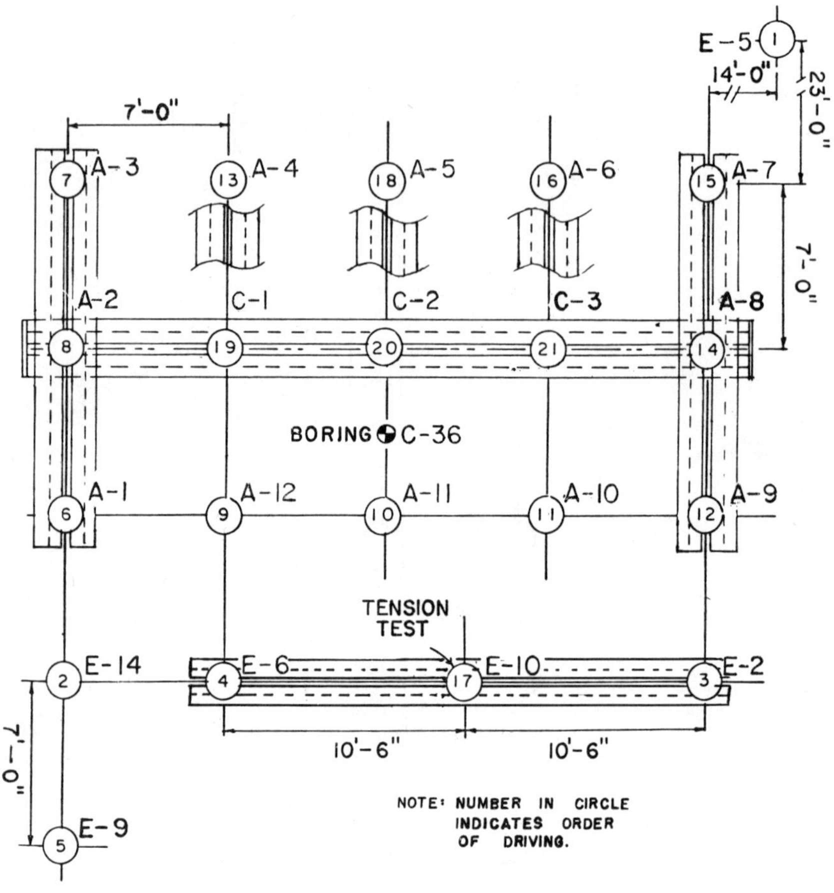

COMPRESSION AND TENSION

1 ft = 0.305 m 1 in. = 25.4 mm

FIG. 3 TEST PILE LAYOUT - SITE B

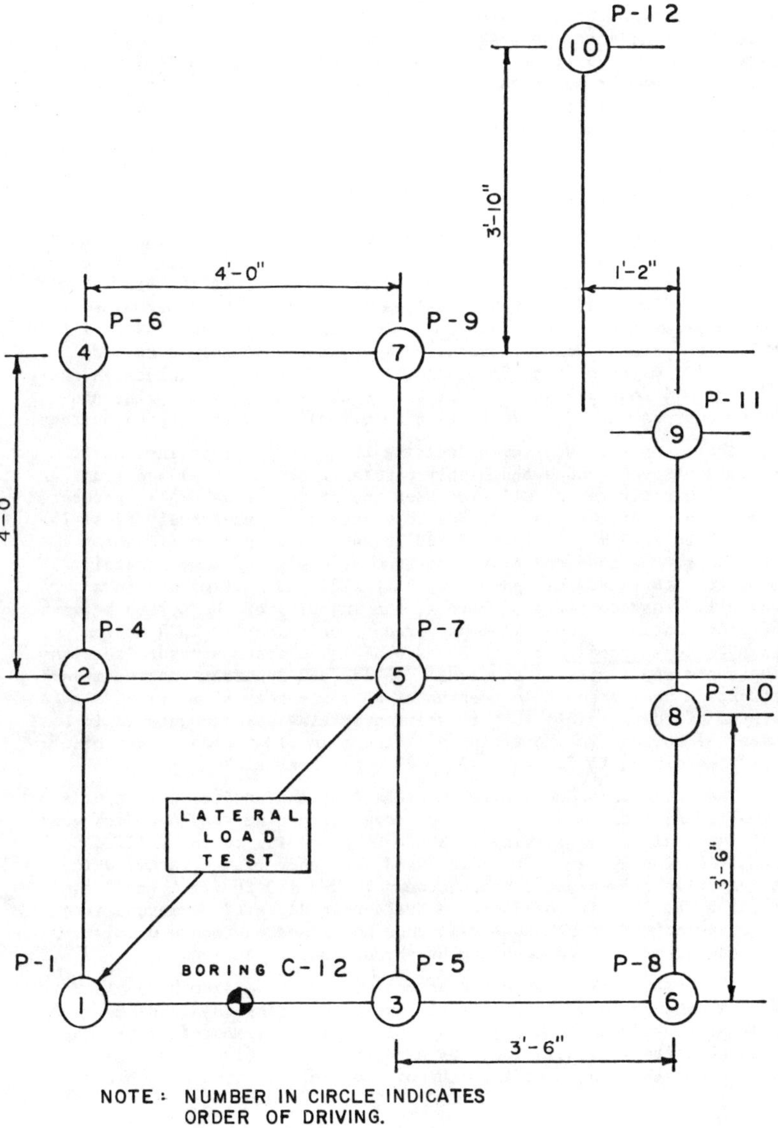

1 ft = 0.305 m 1 in. = 25.4 mm

FIG. 4 TEST PILE LAYOUT - LATERAL TEST

diameter earth auger to a depth of 20 ft (6.1 m). Pre-drilling was conducted to eliminate the skin friction contribution of the stiff silty clay. The uplift test pile was pre-drilled with a 14-in. (35.6 cm) diameter auger to a depth of 20 ft (6.1 m) and the pile tested for lateral load was pre-drilled with a 12-in. (30.5 cm) diameter auger to a depth of 20 ft (6.1 m).

SOIL CONDITIONS

The project site is underlain by four major strata. The upper stratum consists primarily of geologically recent alluvial silty clay and clayey silt to depths of approximately 20 to 25 ft (6.1 to 7.6 m), to near El 280 to 285. Cohesion values obtained in the upper fine-grained stratum vary between 0.5 and 1.2 tons per sq ft (47.9 to 115.0 kPa). In certain isolated areas, however, approximately 2 to 4 ft (0.6 to 1.2 m) of soft to firm sandy clay with cohesion values in the order of 0.4 tons per sq ft (38.3 kPa) underlies the silty clay and was encountered at 18 to 20 ft (5.5 to 6.1 m) below the ground surface.

The major soil stratum underlying the upper fine-grained stratum consists of open-graded and highly permeable chert gravel and sand. The geologically recent alluvial sand and gravel is generally present from the base of the clay stratum to depths of approximately 65 to 75 ft (19.8 to 22.9 m), to near El 230 to 240. The upper and lower portions of the sand and gravel stratum at depths of approximately 25 to 30 ft (7.6 to 9.2 m) and 60 to 70 ft (18.3 to 21.4 m) are in a generally dense condition; whereas, the majority of the strata between the aforementioned zones is considered to be primarily in a medium dense to dense condition with occasional loose zones and soft to firm clay seams and layers. Occasional cobbles and apparent rock seams are present in some areas below depths of approximately 65 ft (19.8 m). A stratum of compressible soft to stiff gray clay was encountered in a number of borings in the vicinity of the main plant area at depths of approximately 68 to 72 ft (20.7 to 22.0 m), near El 240.

The third major soil stratum present at the project site consists primarily of interbedded calcareous gray silty fine sand and dark gray fine sandy clay, classifying as SM to CL according to the Unified Soil Classification System. The silty sand and sandy clay stratum, which is generally present at depths between 70 and 110 ft (21.4 to 33.6 m), El 235 to El 195, is considered a Cretaceous deposit. Numerous irregular and apparently discontinuous sandstone and limestone seams and layers were encountered at various depths throughout the stratum.

The basal stratum consists of medium soft to medium hard black carbonaceous shale with occasional limestone stringers and seams. The shale is considered to be of the Mississippian Age Moorefield Formation, which characteristically consists of dark gray to black platy and calcareous shale interbedded with some impure limestone and chert.

PILE DRIVING

GENERAL VIEW OF TEST SET-UP

The upper surface of the shale stratum is generally soft and apparently weathered, but the effects of weathering and degree of softness decrease with depth.

PILE DRIVING

PILE TYPES

The basic pile types initially considered suitable for support of the structure and the aforementioned soil conditions included: 1) Steel Pipe; 2) Steel H-piles (HP 14 x 102 and HP 14 x 73); 3) Composite piles (Pipe/Step-Taper); and 4) Raymond Step-Taper piles. One of the primary concerns in pile selection was the use of piling which could be driven into or through the gravel stratum without appreciable damage to the piling. Pile lengths were altered from those originally contemplated following the establishment of driving conditions over the site. Pre-cast concrete piling were not included in the test program. Test pile characteristics are summarized on Table I.

PILE DRIVING EQUIPMENT

The pile driving equipment consisted of:

1) A Vulcan 80C differential hammer having a ram weight of 8000 lbs (3064 kg) and a manufacturer's rated energy of 24,450 ft-lbs (33,154 N-m); and

2) A Raymond 3/0 single-acting hammer having a ram weight of 12,500 lbs (5663 kg) and a manufacturer's rated energy of 40,625 ft-lbs (55,088 N-m).

The capblock used for both hammers consisted of 11-14.5-in. (36.8 cm) diameter by 0.5-in. (1.3 cm) thick aluminum plates and 11-14.5-in. (36.8 cm) diameter by 1.0-in. (2.5 cm) thick micarta plates alternately stacked. The Raymond 3/0 hammer was used to drive the H-piles. Both hammers were operated with a 2000 cfm (56.0 m^3/min) air compressor producing 135 psi (931.5 kPa) pressure at the compressor. The Step-Taper and composite Pipe/Step-Taper piles were driven with a Raymond thick-walled Steel Step-Taper mandrel and core head.

PILE DRIVING CRITERIA

All test piles, with the exception of Pile C-1, were driven to a minimum driving resistance based on Wave Equation analyses provided by Raymond Technical Facilities, Inc. The ultimate pile capacities of each of the test piling were also predicted on the basis of the Pacific Coast, Janbu, and the Modified Engineering News Formulae. The

TABLE I

SUMMARY OF TEST PILE DATA

TEST SITE	PILE NO.	PILE DESCRIPTION TYPE	PILE DESCRIPTION SIZE	PREDRILL DIA. IN.	PREDRILL DEPTH FT	FINAL SET BPI	LENGTHS DRIVEN FT	LENGTHS TESTED FT	LENGTHS EMBEDDED FT	EMBEDDED AREA FT^2	TYPE TEST
B	C-1	Pipe	13.375 X 0.380	18	20	2	40	35.2	35	52.5	Compression
B	C-2	Pipe Step Taper	Pipe:10.75 X 0.25 X 24' Shell: 1 to 3 @ 12' 4 to 5 @ 8' Mandrel: 00-3 @ 12' 4-6 @ 8'	18	20	12	84	72.5	72	156.0	Compression
B	C-3	Pipe Step Taper	Same as C-2	18	50	8	84	76	76	170.0	Compression
B	E-10	Pipe Step Taper	Pipe:10.74 X 0.25 X 24' Shell: 1 to 6 @ 8' Mandrel: 00 to 0 @ 12' 1 to 6 @ 8'	14	20	10	72	72	71.2	237.4	Uplift
A	C-4	Pipe	13.375 X 0.380	18	20	50	82	73.8	73.0	185.5	Compression
A	C-5	H	14 HP 102	18	20	80	80.5	78.8	77.0	273.6	Compression
A	C-6	Step Taper	Shell: 0 to 5 @12' Mandrel: 0 to 5 @12'	18	20	10	72	74.7	73.0	156.6	Compression
A	C-7	Pipe Step Taper	Same as C-2	18	20	9	84	73.5	73.2	159.8	Compression
	P-1	Pipe Step Taper	Same as E-10	12	22	8	72	71.3	71.3	--	Lateral
	P-7	Pipe Step Taper	Same as E-10	12	22	8	72	71.3	71.3	- -	Lateral

aformentioned dynamic pile capacity equations were used to determine their applicability in establishing pile driving criteria at the project site. The computations were performed using a 70 percent efficiency for the differential hammer and an 80 percent efficiency for the single-acting hammer. A coefficient of restitution of 0.8, based on dynamic tests performed by Raymond International in 1957, was assumed for the aluminum-micarta capblock assembly. The dynamic capacities were computed using the actual pile dimensions and physical properties, together with the driving resistance.

PILE DRIVING CONDITIONS

Comparative logs for Test Area A are presented on Figs. 5 & 6 for Test Area B on Figs. 7&8 and for the lateral load test piles on Figure 9. The comparative logs show pile driving resistance, basic pile configuration and generalized soil stratigraphy. Also indicated are the final driving resistance and the recorded driving time.

As shown on the comparative logs, the exploratory pile driving generally indicated dense soil conditions at a penetration of approximately 28 ft (8.5 m) below existing grade, but a much looser soil condition between depths of 30 and 60 ft (9.2 and 18.3 m). No difficulties were encountered, however, in penetrating to depths of approximately 65 ft. Driving through the gravel stratum also indicated the presence of very loose gravel and/or soft clay layers with driving resistances in the order of 1 to 4 blows per ft (3 to 13 blows per m), with the Vulcan 80C hammer.

Prior to test driving, some concern existed with regard to possible difficulties in penetration through the dense cobble and gravel strata, occasionally present below depths of 65 ft (19.8 m), and subsequent penetration through the compressible gray clay stratum which underlies the cobble and gravel stratum in certain areas. In areas tested, no major difficulties were experienced in penetration through both the cobble and gravel zone and the underlying soft, compressible clay strata to the dense Cretaceous deposits.

Piles P-1 and P-4 through P-12 were installed at spacings of 3.5 ft (1.1 m) and 4 ft (1.2 m) in an attempt to evaluate driving conditions within a pile group similar to those anticipated in production and to determine the degree of densification at design spacings. The comparative log for the exploratory piles P-1, P-7 and P-11 within the test pile group is shown on Figure 9. As shown, some densification did occur. The degree of densification during driving of the group occurred generally throughout the profile, but not to the extent that would seriously affect driving penetration. Practical refusal for Pile P-11 was obtained at a depth of 66 ft (20.1 m); nevertheless, it should be noted that the inability to obtain the desired penetration is considered to be related to the presence of a hard chert boulder or layer at that depth, rather than densification, as evidenced by the full penetration of Pile P-12.

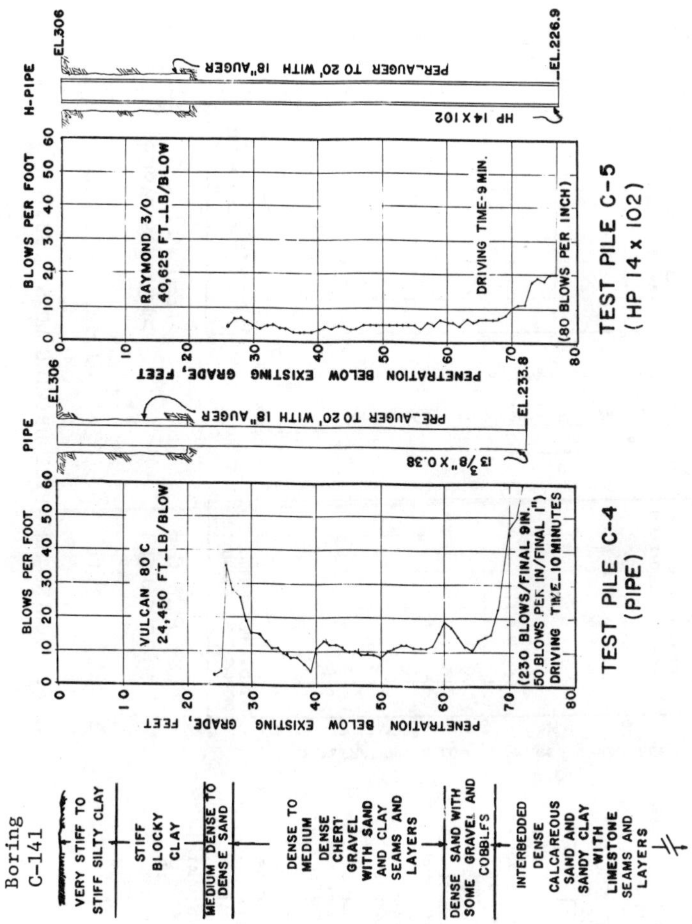

1 ft = 0.305 m 1 in. = 25.4 mm 1 ft-lbf = 1.36 J

FIG. 5. BORING AND PILE DRIVING LOGS - SITE A

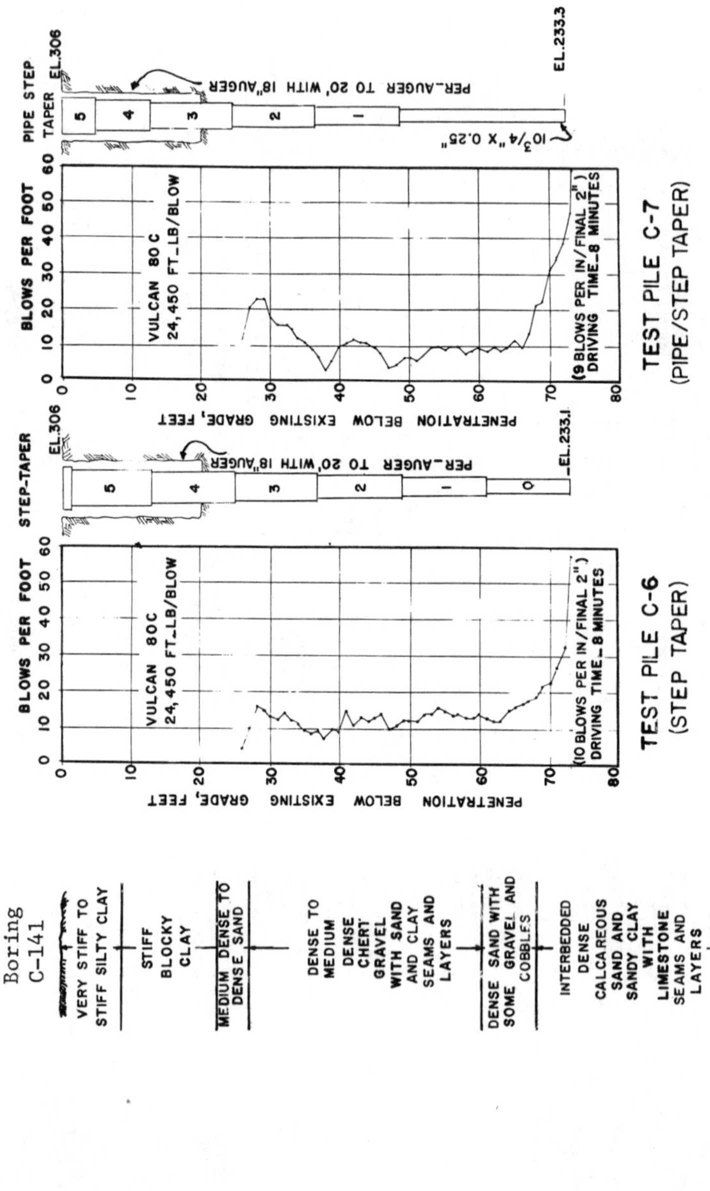

1 ft = 0.305 m 1 in. = 25.4 mm

FIG. 6. BORING AND PILE DRIVING LOGS - SITE A

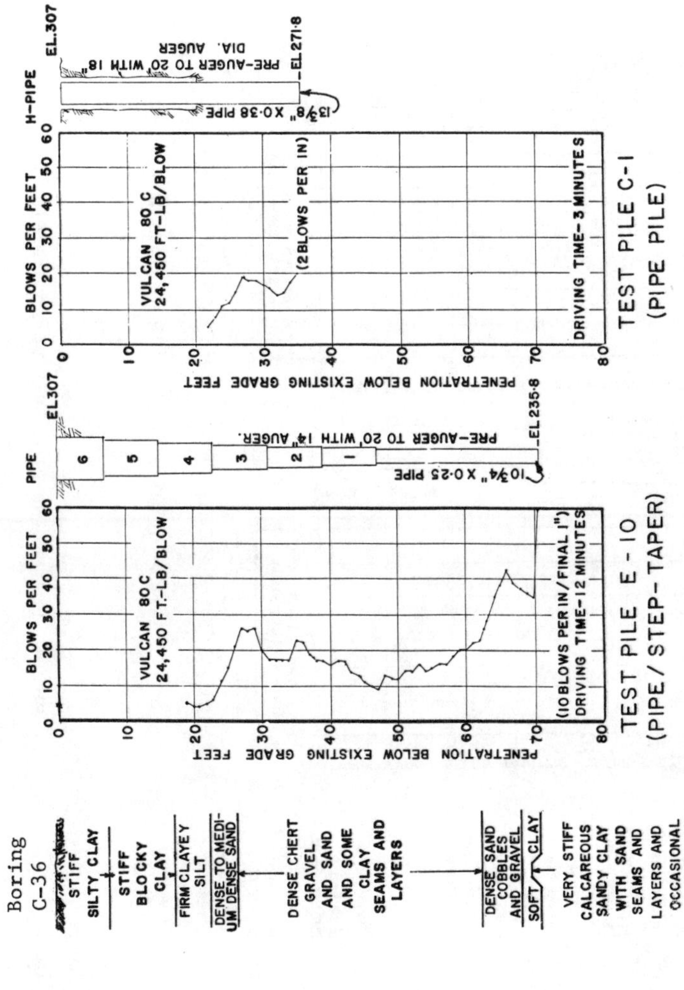

1 ft = 0.305 m   1 in. = 25.4 mm

FIG. 7. BORING AND PILE DRIVING LOGS - SITE B

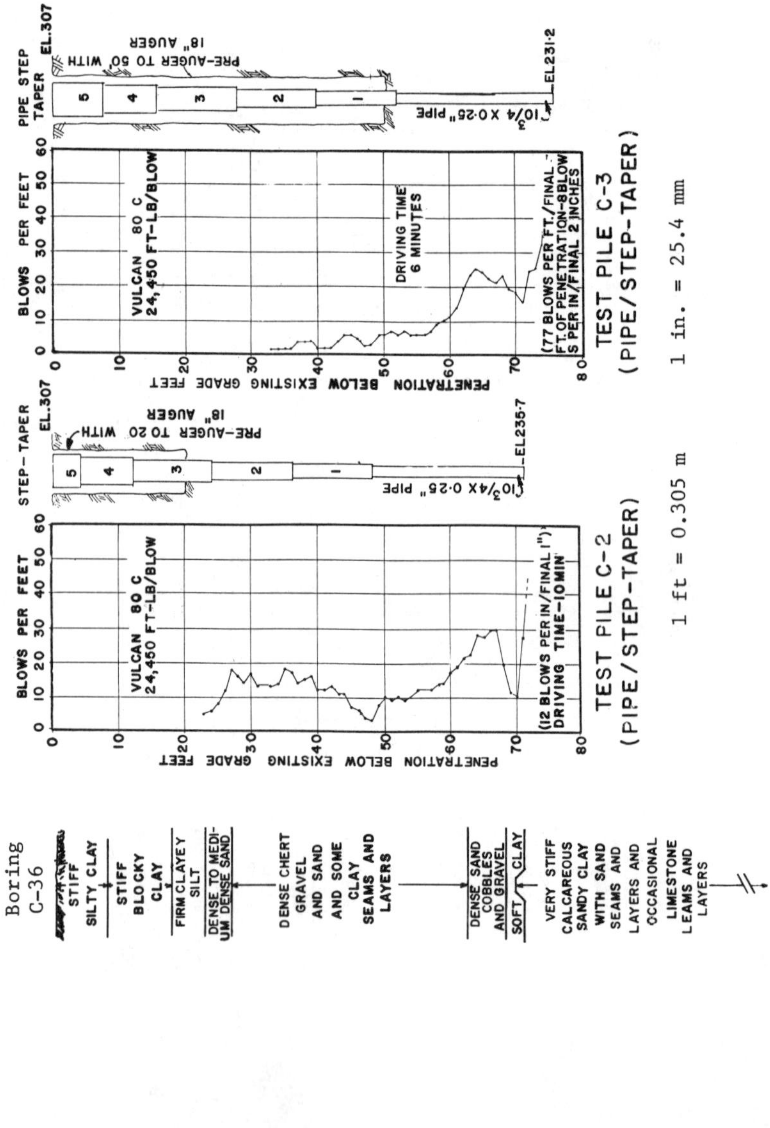

FIG. 8. BORING AND PILE DRIVING LOGS - SITE B

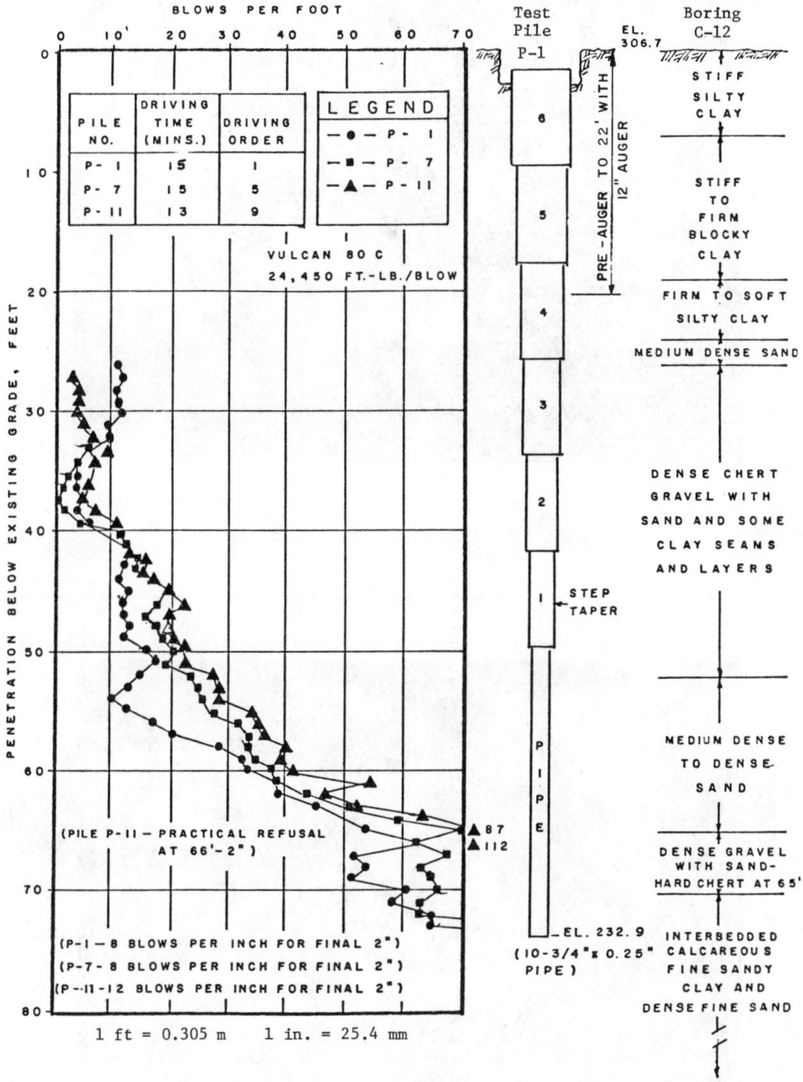

1 ft = 0.305 m   1 in. = 25.4 mm

FIG. 9. BORING AND PILE DRIVING LOGS - LATERAL TEST SITE

PILE BUTT AND TELL-TALE INSTRUMENTATION

COMPRESSION TEST UNDER WAY

To determine if piling could potentially penetrate through the Cretaceous deposits to the Paleozoic shale, H-piling (HP 14 x 102) equipped with a reinforced tip (Associated Pile and Pipe Fitting point BP75600) was driven using the 3/0 hammer in Test Area A. As shown on the comparative log, Figure 5, practical driving refusal was obtained at El 226.9 on what is apparently a limestone layer within the Cretaceous deposits.

LOAD TEST SET-UP AND PROCEDURES

COMPRESSION TEST

Compressive loads were applied to the test piles by either a Simplex hydraulic assembly consisting of four 100-ton (890 kN) jacks or by a Rodgers hydraulic assembly consisting of two 200-ton (1.8 MN) jacks. The compression test reaction frame consisted of twin 36WF230 test beams connected to the anchor pile system by a series of twin 24WF84 tie-down beams. The reaction frame in Test Area A was anchored with total of fourteen piles, which consisted of two Step-Taper, eleven Pipe/Step-Taper and one H-pile. The reaction frame in Test Area B was anchored with a total of eleven piles, which consisted of ten Pipe/Step-Taper piles and one H-pile. The beams were connected to each reaction pile by one 1.25-in. (3.2 cm) diameter Dywidag bar embedded approximately 7 ft (2.1 m) into the pile with an end plate attached to the Dywidag bar. The reaction frames were designed to allow load testing of two test piles simultaneously. The compressive load test piles and anchor piles were installed at 7 ft (2.1 m) center-to-center spacings.

Each of the test piles contained a telltale installed to the tip of the pile. In addition, Test Pile C-6 contained an intermediate telltale installed to a depth of 60 ft (18.3 m) below pile cutoff. The apparatus for measuring butt movement consisted of three dial indicators spaced at 120 degrees around the perimeter and in contact with a separately supported reference frame. A dial indicator was also mounted to each telltale rod, which was placed in contact with the reference frame. Readings were obtained on each dial indicator to the nearest 0.001 inch (0.0025 cm) and each loading increment applied to the pile was measured using a load cell. A typical load test arrangement is shown on Figure 10 and the test layouts are shown on Figures 2 and 3.

The compressive load tests were performed in general accordance with current ASTM D-1143 Procedures with certain modifications consistent with soil conditions to expedite testing and/or obtain additional pertinent data. The compressive test piles, with the exception of Test Piles C-3 and C-5, were loaded in 25-ton (220 kN) increments to 0.5 times design load, design load of 100 tons (890 kN), 1.5 times design load, 2.0 times design load and 2.5 times design load. The test piles were allowed to rebound under zero load from each of

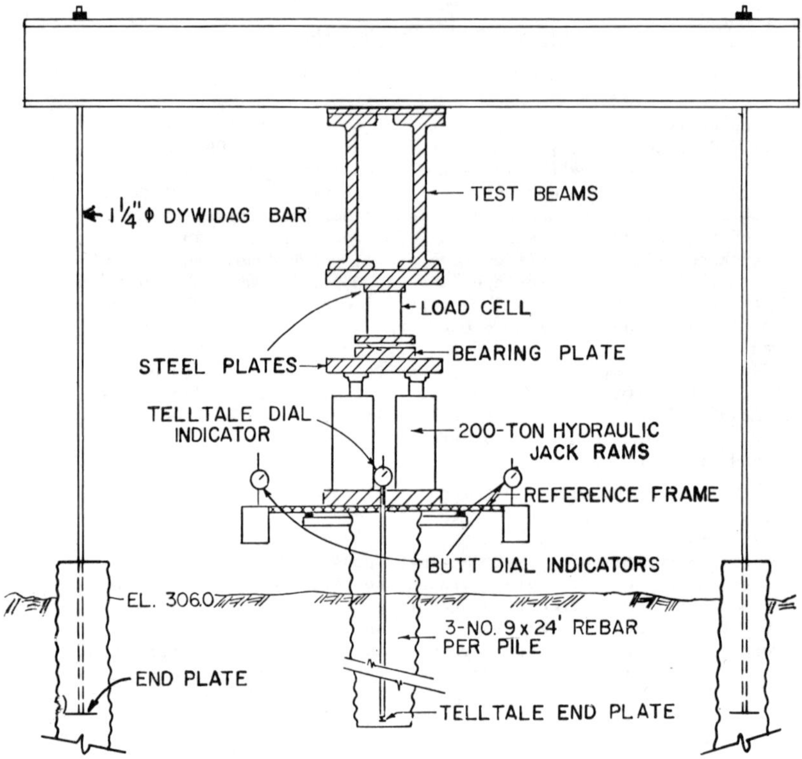

1 ton = 8.9 kN   1 ft = 0.305 m   1 in. = 25.4 mm

FIG. 10. COMPRESSION TEST SET-UP

CRP MONITORING INSTRUMENT

TWO TESTS CONDUCTED AT SAME TIME
NOTE LOAD CELLS

the indicated loads in order to establish the elastic settlement characteristics of the pile. Test Pile C-2 was additionally loaded to loads of 260, 270 and 280 tons (2.3, 2.4, and 2.5 MN) in an effort to fail the pile.

Test Pile C-5 (H-pile) was loaded in 40-ton (360 kN) increments to a total of 320 tons (2.8 MN) using the standard loading procedures; and the pile was allowed to rebound decrementally from loads of 80, 160, 240 and 320 tons (0.7, 1.4, 2.1, and 2.8 MN) to better evaluate the elastic deformation characteristics of the H-pile.

Test Pile C-3 was loaded to failure using the Constant Rate of Penetration (CRP) method rather than the standard loading procedure. The CRP method is an optional procedure in accordance with ASTM D-1143 Standards. The other compressive test piles were also loaded using the CRP method after the standard loading procedure was applied. Test Pile C-5 was terminated prior to failure at a total load of 400 tons (3.6 MN).

Testing of Pile C-6 was terminated initially at a load of approximately 350 tons (3.1 MN) due to insufficient "travel" in the hydraulic jack assembly. The second attempt to fail the pile was terminated upon collapse of the jack assembly.

TENSION TEST

Tensile loads were applied to Test Pile E-10 by a Simplex hydraulic assembly consisting of three 100-ton (890 kN) jacks. Load was transferred to the test pile through three No. 11 reinforcing bars extending 63 ft (19.2 m) below the top of concrete in the pile. The reaction frame consisted of a 24-in. (61.0 cm) twin I-Beam supported by a single pile on each end. Vertical pile movement was measured using three dial indicators spaced at 120 degrees around the pile perimeter and in contact with a separately supported reference system. Each dial indicator reading was obtained to the nearest 0.001 inch (0.0025 cm) and each load increment was measured by use of a load cell mounted above the hydraulic jack assembly. The tensile load arrangement for Test Pile E-10 is shown on Figure 11 and the test layout is shown on Figure 3.

Test Pile E-10 was loaded in 10-ton (90 kN) increments to a total of 80 tons (720 kN) and permitted to rebound under zero load. Following stabilization of pile movement, the test pile was then loaded to failure using the CRU procedure.

LATERAL LOAD TEST

Lateral load was applied to Test Pile P-7 by means of a 100-ton (890 kN) Simplex hydraulic jack reacting against Pile P-1 and a 1800-lb (815 kg) follower block. Lateral movement at the ground line was measured for Pile P-7 using dial indicators in contact with an

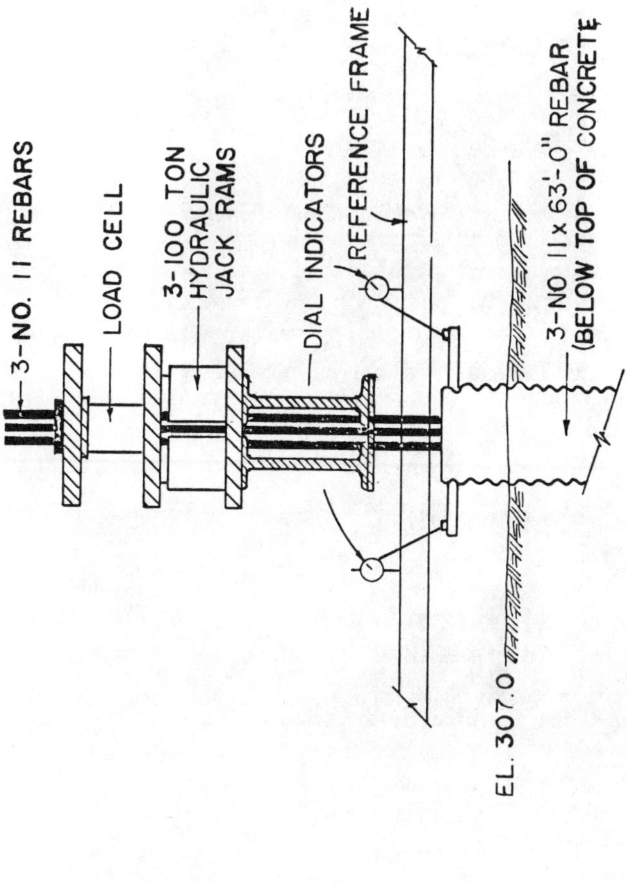

1 ton = 8.9 kN   1 ft = 0.305 m   1 in. = 25.4 mm

FIG. 11. TENSILE TEST SET-UP

UPLIFT TEST

LATERAL TEST

independent reference frame. Dial indicator readings were obtained to the nearest 0.001 inch (0.0025 cm) and loads applied to the test piles were measured using a load cell. The lateral load test arrangement for Piles P-1 and P-7 is illustrated on Figure 12 and the test layout is shown on Figure 4.

Lateral load was applied to Pile P-7 in 2-ton (18 kN) increments to a total of 8 tons (70 kN). The pile was allowed to rebound decrementally at 4 (35 kN) and 8-ton (70 kN) loads. Following the final rebound, the pile was loaded to 8 tons (70 kN) in 100 repetitions. Following the load repetitions, the pile was then loaded in 4-ton (35 kN) increments to failure, 23.5 tons (210 kN).

SPECIAL INSTRUMENTATION

Telltales were placed in all compression test piles. Telltale rods consisting of a 9/16-in. (1.4 cm) O.D. rod, were installed to the test pile tip inside a 1.25-in. (3.2 cm) O.D. pipe casing provided with a plate welded at the bottom. This casing provides for free movement of the telltale rod. The telltale rod was extended out beyond the outside pile wall to a reference beam either above or below the base plate. The telltale casing was attached to the H-pile by intermittent welds spaced about 18 in. (45.7 cm) along the fillet.

The applied loads in all tests were measured using a calibrated load cell. The load cell was connected to an electrical strain indicator which had a readout in micro-inches. The load at each increment was held constant by the use of a nitrogen pressure regulator called an accumulator.

LOAD TEST RESULTS

COMPRESSION TESTS

Results of the compressive load tests are presented graphically on Figures 13 through 19 as Load in Tons versus Vertical Pile Deflection in Inches. The scales were selected so that the elastic line would make approximately an angle at 20 degrees with the horizontal. In addition to the gross pile butt movement and tip movement during the standard test procedures, pile tip movement as recorded by the telltales and pile butt movement recorded during the CRP method of test are shown. The ultimate pile capacities were determined by curve intersection of the gross butt movement curves at standard loading procedure with a failure criteria line drawn below and parallel to the elastic line of the pile as suggested by Davisson [1].

Failure loads are summarized on Table II. Also shown on Table II are comparisons of the test failure loads and the dynamic capacities as predicted by the Pacific Coast, Janbu, Modified Engineering News formulae and the Wave Equation. The ratio of test capacity to predicted

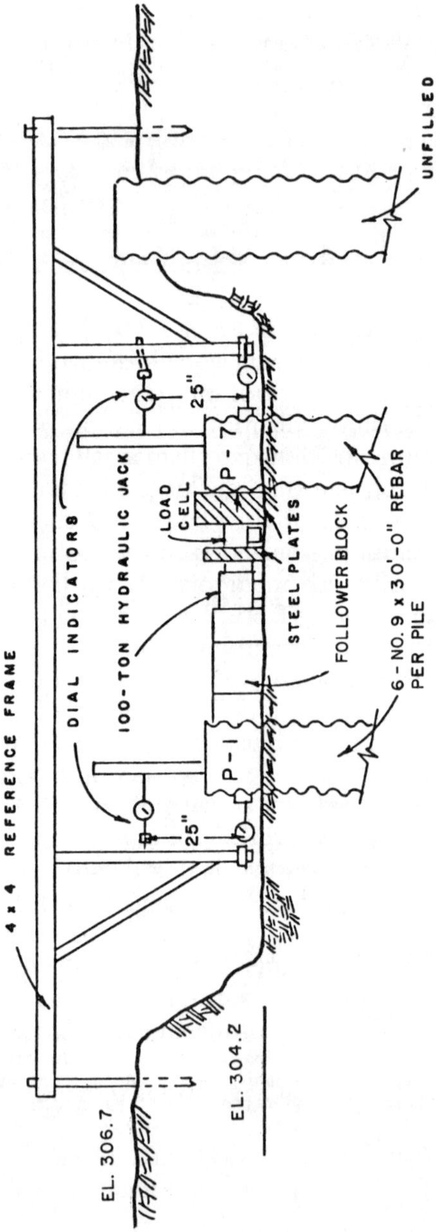

1 ton = 8.9 kN   1 ft = 0.305 m   1 in. = 25.4 mm

FIG. 12. LATERAL TEST SET-UP

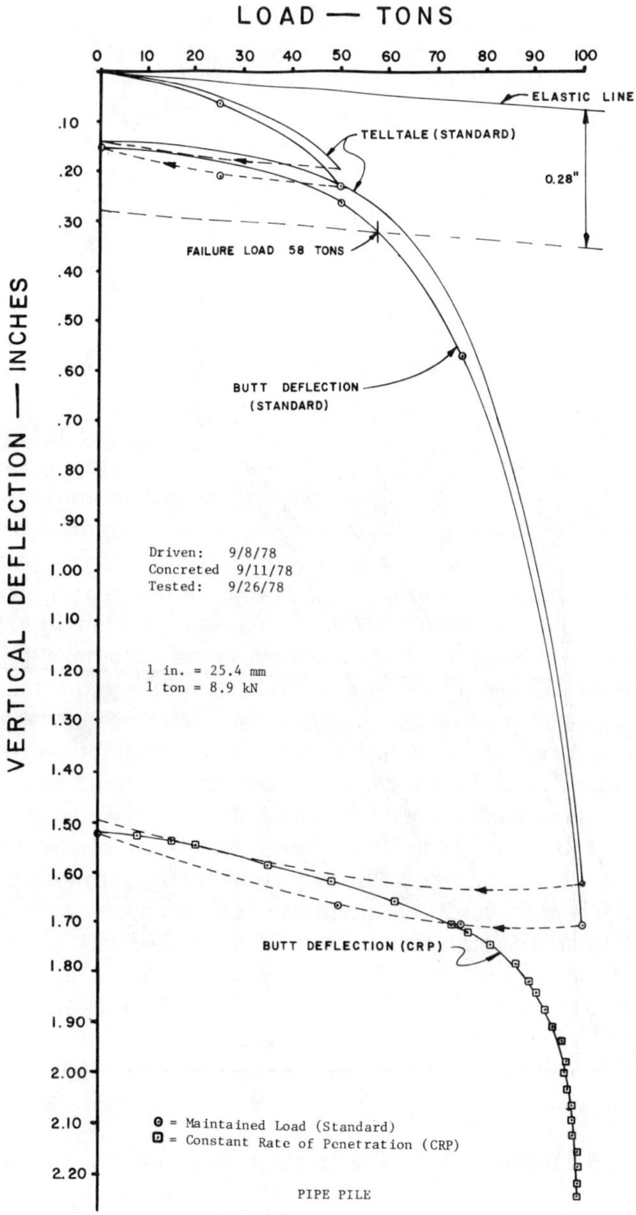

FIG. 13. RESULTS OF COMPRESSION TEST ON PILE C-1 SITE B

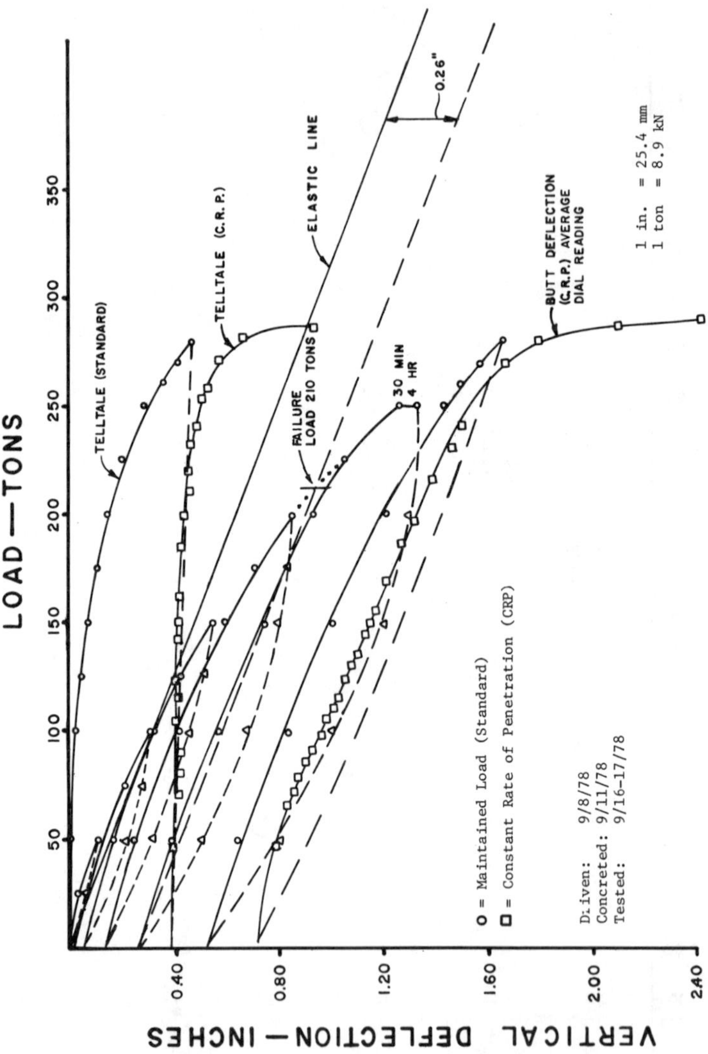

PIPE STEP TAPER PILE

FIG. 14. RESULTS OF COMPRESSION TEST ON PILE C-2 SITE B

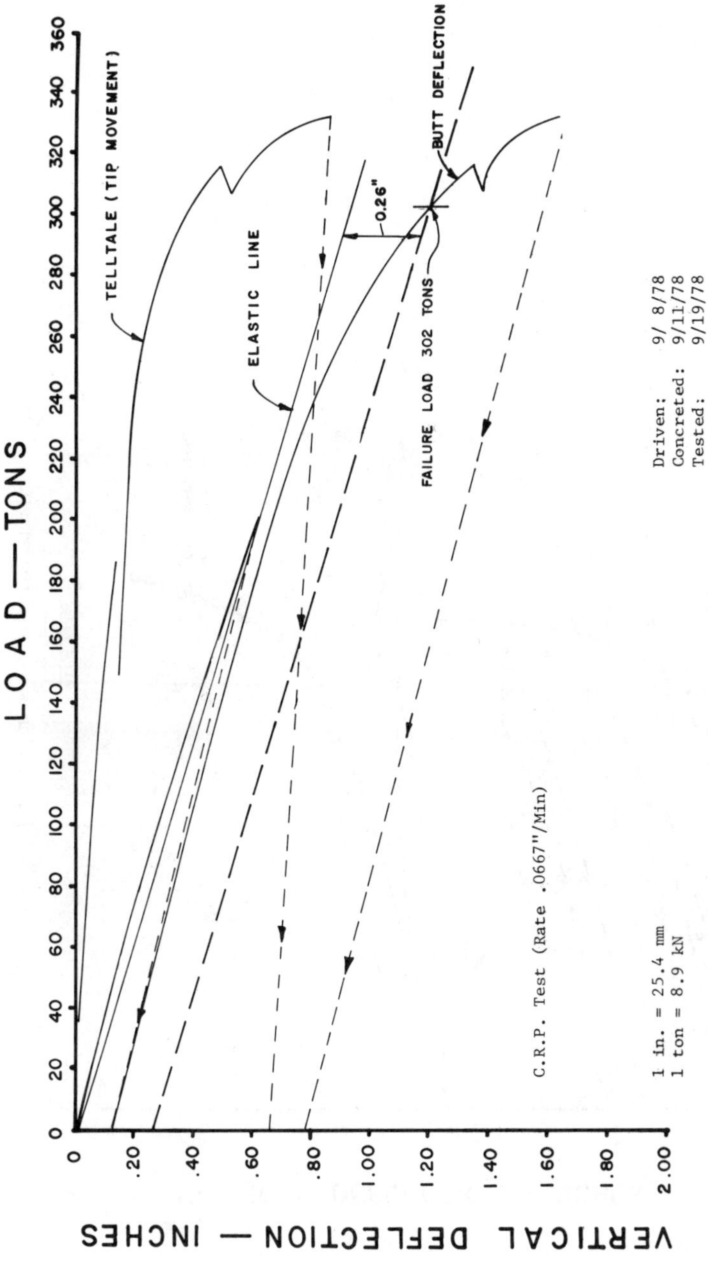

PIPE STEP TAPER PILE

FIG. 15. RESULTS OF COMPRESSION TEST ON PILE C-3 SITE B

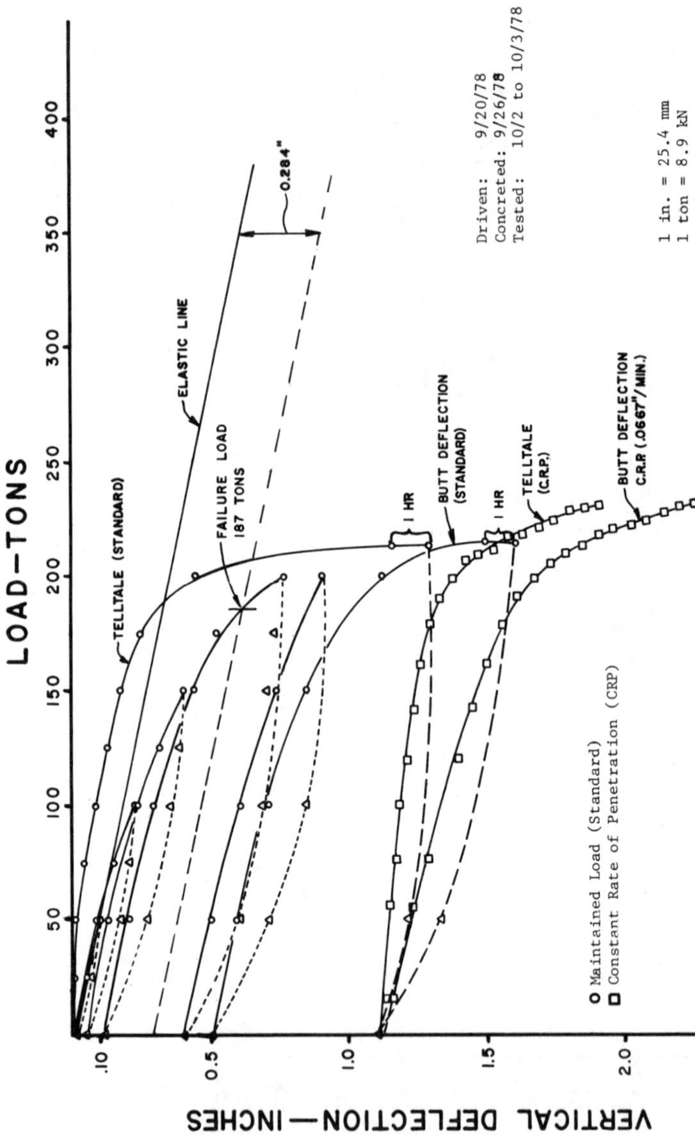

PIPE PILE

FIG. 16. RESULTS OF COMPRESSION TEST ON PILE C-4 SITE A

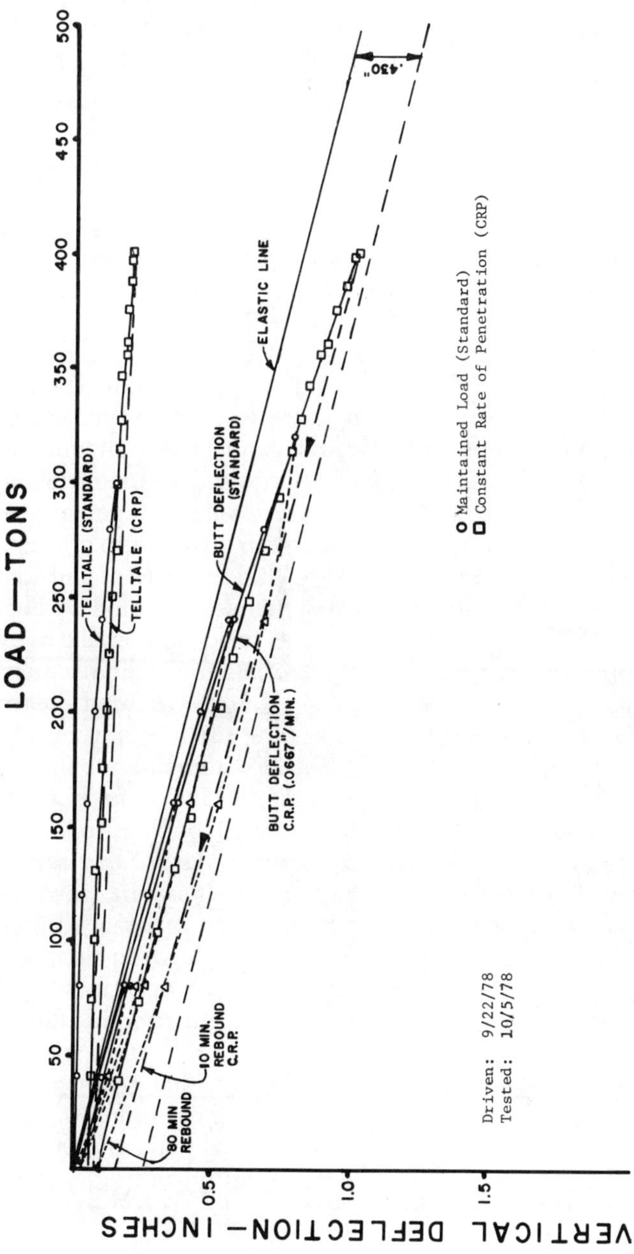

H - PILE

FIG. 17. RESULTS OF COMPRESSION TEST ON PILE C-5 SITE A

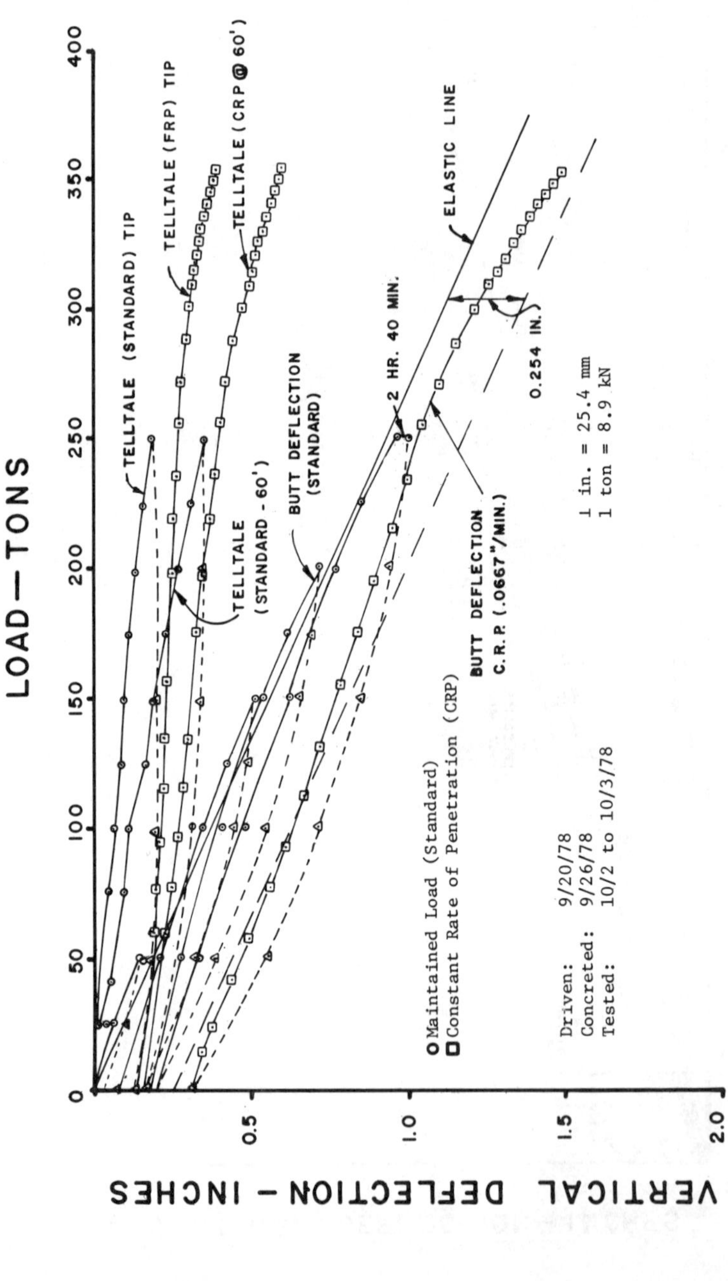

STEP TAPER PILE

FIG. 18. RESULTS OF COMPRESSION TEST ON PILE C-6 SITE A

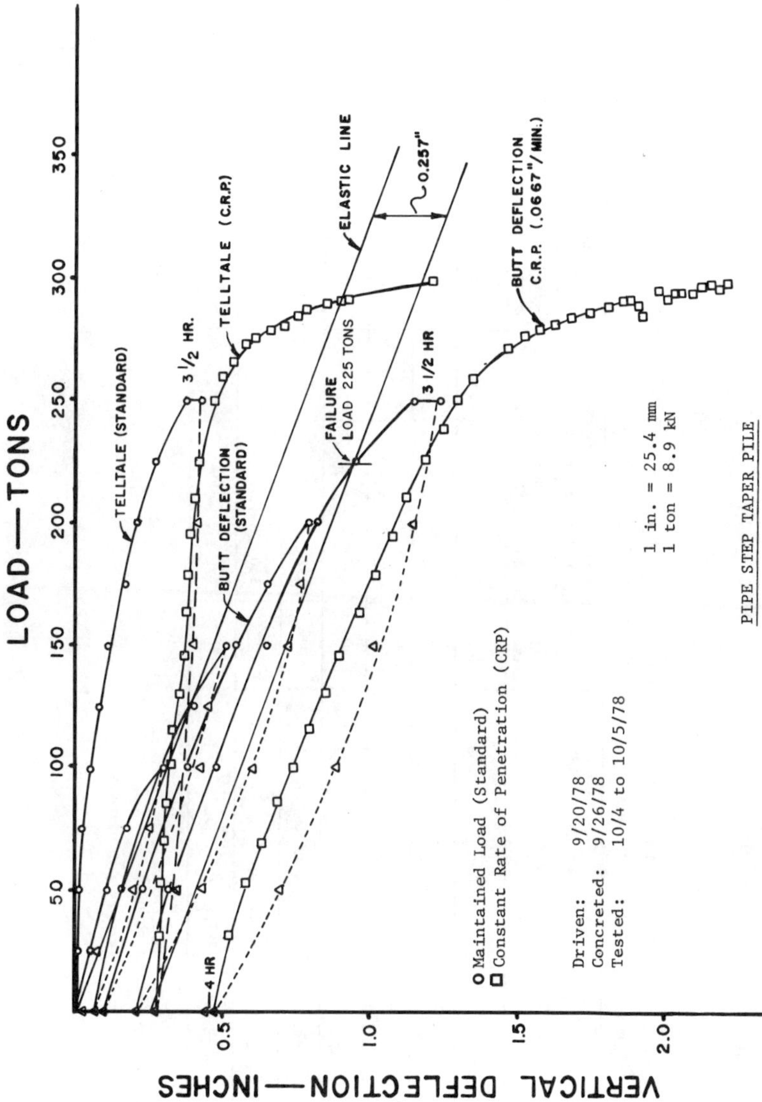

FIG. 19. RESULTS OF COMPRESSION TEST ON PILE C-7 SITE A

TABLE II

COMPARISON OF ACTUAL TEST FAILURE LOADS AND PILE CAPACITIES PREDICTED FROM DYNAMIC FORMULAE

Test Pile No.	Pile Type	Driving Hammer	Test Failure Load (Tons)	PREDICTED ULTIMATE PILE CAPACITY (TONS)				RATIO OF TEST CAPACITY TO PREDICTED CAPACITY			
				Pacific Coast	Janbu	Modified EN *	Wave EQ	Pacific Coast	Janbu	Modified EN *	Wave EQ
C-1	Pipe	80C	58	120	112	146	95	0.48	0.52	0.40	0.61
C-2	Pipe Step Taper	80C	210	347	524	448	337	0.61	0.40	0.47	0.62
C-3	Pipe Step Taper	80C	302	313	360	358	267	0.96	0.84	0.84	1.13
C-4	Pipe	80C	187	197	529	755	210	0.95	0.35	0.25	0.89
C-5	Steel H	3/0	> 400	290	1032	1490	435	**	**	**	**
C-6	Step-Taper	80C	> 350	313	360	360	355	**	**	**	**
C-7	Pipe Step Taper	80C	225	324	402	380	287	0.69	0.56	0.59	0.78

* ENGINEERING NEWS FORMULA

** ULTIMATE TEST CAPACITY NOT DETERMINED

1 ton = 8.9 kN

capacity of the test piles using the dynamic formulae varied over the ranges of 0.48 to 0.96 for the Pacific Coast formula, 0.35 to 0.84 for the Janbu formula, 0.25 to 0.84 for the Modified Engineering News formula and 0.61 to 1.13 for the Wave Equation.

Based on telltale readings and elastic properties of the piles, load distributions were computed. For the "short" Pipe pile (C-1), only approximately 10 percent of the total load of 100 tons (890 kN) was carried in side friction (See Table I for pile descriptions). For the "long" Pipe pile, approximately 25 percent of the 200-ton (1.8 MN) load was carried in the side friction. For the Pipe/Step-Taper piles (C-2, C-3, and C-7) approximately 30 percent of the 250-ton (2.2 MN) load was carried in side friction. Although the H-pile (C-5) and the Step-Taper Pile (C-6) were not loaded to failure, approximately 15 percent of the maximum load was carried in side friction for the H-pile and approximately 55 percent of the maximum load for the Step-Taper pile was carried in side friction.

TENSION TESTS

The tension load test results for Pile E-10 are presented graphically on Figure 20 as Vertical Deflection in Inches versus Load on Pile in Tons. A pile capacity of approximately 108 tons (960 kN) was obtained as the intersection of tangents to the CRU Curve. Unlike the compression test piles, contribution of the upper clay stratum was included, since only a 14-in. (35.6 cm) diameter earth auger was used.

LATERAL LOAD TESTS

The lateral load test results for Test Pile P-7 are presented graphically on Figure 21 as Total Load in Tons versus Horizontal Pile Deflection in Inches. The horizontal deflection at the ground line has been plotted versus lateral load. Based on the load-deflection curve, a horizontal ground-line deflection of 0.25 inches (0.6 cm) at a load of 10.2 tons (90 kN) was measured; and at a horizontal ground-line deflection of 0.5 inches (1.3 cm), a lateral load of 15.0 tons (130 kN) was measured.

Also shown on Figure 21 are the predicted deflections made using the COM 622 Computer Program. One of the major input data items affecting the predicted deflections is the P-Y curves. The ultimate lateral resistance has been assumed to be developed at a depth of 3D and 5D below the ground surface, where D is the pile diameter. The stiffness of the pile (EI) is calculated considering the additional value added by the six No. 9 bars, which is effective to 24 ft (7.3 m) below the top of the pile. The concrete compressive strength was considered as 6000 psi (4140 kPa). The predicted deflections correlate closely with actual deflections for the case assuming ultimate lateral resistance at 5D.

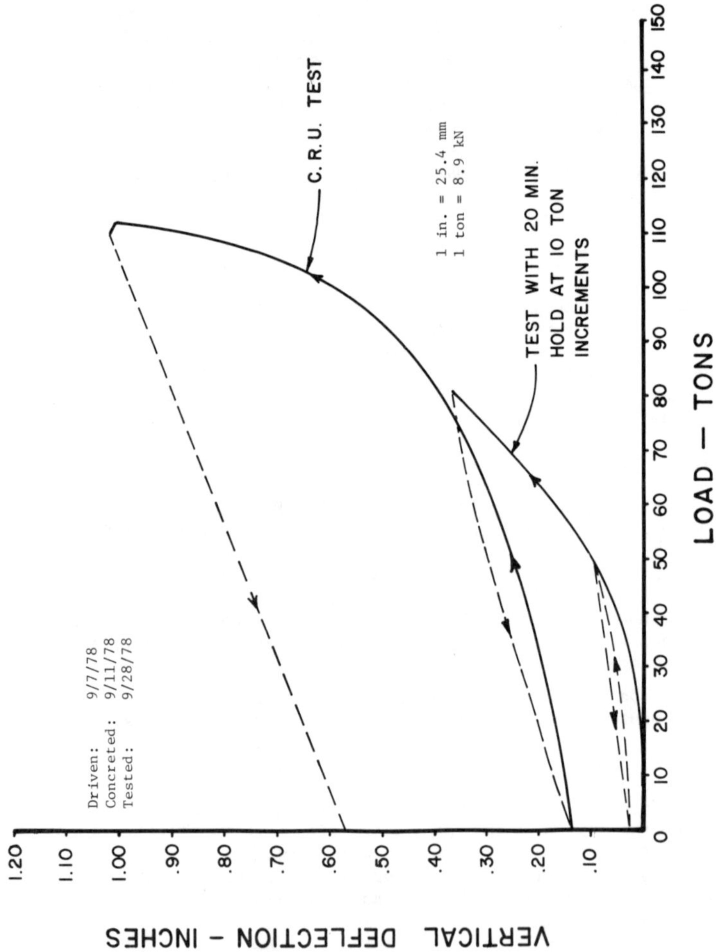

PIPE STEP TAPER PILE

FIG. 20. RESULTS OF UPLIFT TEST ON PILE E-10

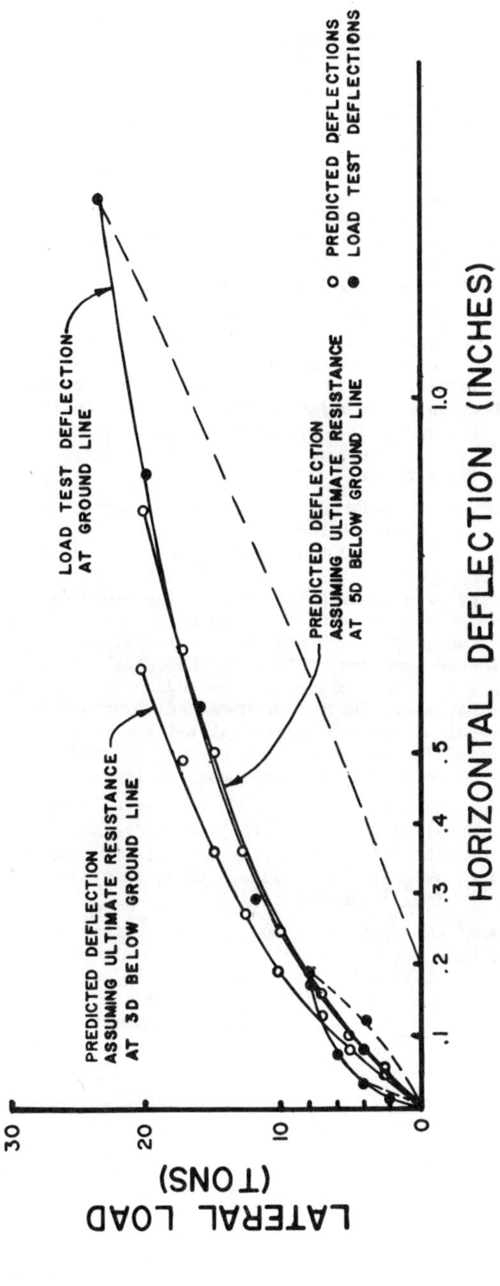

1 in. = 25.4 mm
1 ton = 8.9 kN

Driven: 9/25/78
Concreted: 9/28/78
Tested: 10/9/78

PIPE STEP TAPER PILE

FIG. 21. RESULTS OF LATERAL TEST ON PILE P-7

CONCLUSIONS

On the basis of the complete test pile program, the selected optimal pile type was the Raymond Step-Taper pile. The Raymond Step-Taper pile provided the most economical section of the pile types tested and also provided a factor of safety in excess of 3.0 for a 100-ton (890 kN) allowable pile. The composite pile (Pipe/Step-Taper) was selected as an alternate to the Step-Taper pile in areas where difficult driving and/or shell collapse would dictate the use of a more rigid pile section, such as in the dense gravel and cobble zone present between depths of approximately 65 and 70 ft (19.8 to 21.4 m). The composite pile provided a factor of safety in the order of 2.5 for a 100-ton (890 kN) allowable pile. Selection of both the Raymond Step-Taper and composite pile types would not require costly re-design and scheduling delays.

More than adequate pile capacity would be obtainable using H-piles, as evidenced by the inability to fail Test Pile C-5 at a load of 400 tons. Since the thickness, hardness and lateral extent of the limestone strata in the Cretaceous deposits is irregular, difficulties in both construction planning and actual construction would likely result from the use of H-piles. Additionally, costly re-design would be warranted for the H-piling alternate.

The steel pipe piling was considered to be the least desirable in view of both the economics and marginal load-carrying capacity (a factor of safety slightly less than 2.0 for a 100-ton (890 kN) allowable pile). On the basis of the test results for Pile C-1, a 100-ton (890 kN) capacity pile could potentially be obtained using a large diameter pipe or pre-cast concrete pile penetrating to a depth of 30 to 35 ft below existing grade. The use of "short" pile configurations, however, was considered less desirable than the deeper piles, due primarily to the presence of soft clay zones as indicated both by the test driving and soil borings. The presence of the compressible zones could potentially result in some differential settlement of the shallower foundation system. The deeper foundation system would also allow for greater resistance to potential earthquake loadings.

Prediction of pile driving behaviour soley on the basis of Standard Penetration Test N-values obtained in gravel can lead to erroneous assumptions. Prediction of pile driving resistance should be based on additional information, such as grain-size distribution, driving resistance of casing and drilling fluid losses.

Based on the results of the load tests and the predictions made by the Wave Equation, the Wave Equation results predicted ultimate pile capacities closer to the actual failure loads than the other methods used to predict ultimate capacity for site conditions. It must be emphasized that these estimates are only as good as the soil parameters used. Also, if load test results are to be correlated with Wave Equation analyses, the final driving resistance of the test pile should be accurately determined.

REFERENCES

1) Davisson, M. T., "Pile Load Capacity," Proceedings, Design, Construction and Performance of Deep Foundations, ASCE - University of California, Berkeley, August 1975.

RATIONAL APPROACH TO PILE FOUNDATIONS

MICHAEL M. BLENDY,[1] M.ASCE

The trend to higher capacity piles requires a rational and consistent approach to pile design, analysis and installation. The wave equation is a reliable method for selecting a suitable hammer, analyzing driving stresses and predicting ultimate pile capacity prior to field tests. Experience with pile foundations over the years has led us to conclude that the wave equation computer program analyses and load tests conducted to failure are the two most powerful tools the foundation engineer has available to him. The use of static capacity calculations or dynamic formulas can at best be approximations. Correlations between measured and predicted ultimate capacities for nearly 40 test piles are reasonably consistent. Pile driving problems using diesel hammers can be analyzed with a program that models both the thermodynamic and mechanical operations. Such programs are available and include WEAP (7) and DIESEL 1 (6).

PILING PRACTICE

Although many engineers say that experience and judgment alone are needed, there is a rational engineering approach to pile foundation analysis and design. Subsurface investigations are routinely conducted to define the foundation conditions. If foundation evaluations and analyses show that driven piles are required or feasible, then it is desirable to arrive at an economical pile foundation. Normal practice consists of selecting appropriate pile types and estimating static capacities for various lengths using methods that suit the specific site conditions (2, 9, 12, 13). The choice is narrowed until 1 to 3 pile types are considered competitive. This is based on local cost data to arrive at the lowest cost per ton of load supported. Although these soil and structural analyses provide the required cross section and an estimate of length for the design load, they don't tell us what hammer to use, how hard to drive the pile, and what capacity can be expected at a particular penetration depth and blow count.

Since 1970, the wave equation has been used routinely to answer these questions (10). Basically, the wave equation is used to describe how stress waves are transmitted in a long rod when a force is applied at one end of the rod. In application, the hammer-pile-soil system is idealized as a series of concentrated weights connected by weightless springs. Capabilities and details of wave equation analyses are well known (1, 3, 4, 8, 14). Performing a wave equation analysis for a particular hammer-pile-soil system provides significant information for a single blow of the hammer ram; prediction of the driving stresses in the pile or mandrel during impact, and an estimation of the resistance to penetration offered by the soil at the time of driving. Relaxation or freeze cannot be accounted for in the analysis.

---

1 Supervising Soils Engineer, Bechtel Incorporated, Gaithersburg, Maryland

After performing the wave equation analysis for each pile driven with the same hammer, tentative driving criteria are established. For large projects a preproduction pile load test program is conducted. This permits evaluation of installation methods and production rates, and allows determination of ultimate pile capacities prior to design or substantiation of the design capacity. This type of comparative test program often results in lower total foundation costs. On smaller projects, the preproduction pile load tests are sometimes conducted just prior to production driving using only one pile type.

Load tests are conducted to failure or to the limitation of the jack and reaction system. Normally, the jack and reaction have a capacity of 250 to 300% of the pile design load. Load tests involve three variables: time, load, and deformation. Time intervals are held constant. Load is applied in increments of 25% of design, and held for 1 hour. During the load increment, pile butt settlement is measured and recorded every 5 minutes. Two dial gages and a wire-scale arrangement are used. A single jack is preferred and a load test normally lasts 10 to 14 hours.

Interpretation of the load test data uses Davisson's method (5). This incorporates the physical properties of both the pile and the soil. Briefly, the entire pile length is assumed a free standing column. Scales for the load settlement curve are chosen so the elastic deflection of the pile column plots approximately 20 degrees from the load axis.

After the load settlement data is plotted, and the elastic compression line is constructed, an offset line is constructed parallel to the elastic compression line. The amount of offset is equal to 0.15 inches plus 0.10 B inches, where B is the nominal width of the pile tip in feet. The ultimate pile capacity is taken as the load where this offset line intersects the load-settlement curve. A typical example of this application is illustrated on Figure 1. Design capacity is taken at one-half the ultimate capacity.

Specifications for production piles include penetration and blow count criteria using the same hammer and accessories that were used to drive the test pile. Each production pile driving record is logged on a standard form by an experienced engineer or technician.

PREDICTED AND MEASURED CAPACITIES

A total of 39 load tests which have been performed at 9 sites are studied. A brief description of the general subsurface condition and pile types at each site is given as follows:

(Consistency of the soils is indicated by a range of Standard Penetration Resistance or N-Values in blows per foot.)

Site A is located in Missouri and consists of an upper 20 feet (6 m) of sand fill (N=20 to 40); 20 feet (6 m) of silt (N=5 to 15); and is underlain by more than 80 feet (24 m) of medium sand (N=30 to 70). Piles driven and tested included precast concrete, nominally 12 inches (300 mm) in diameter, and mandrel driven Raymond Step-taper piles with 8 inch (200 mm) tips and 17 inch (430 mm) butts.

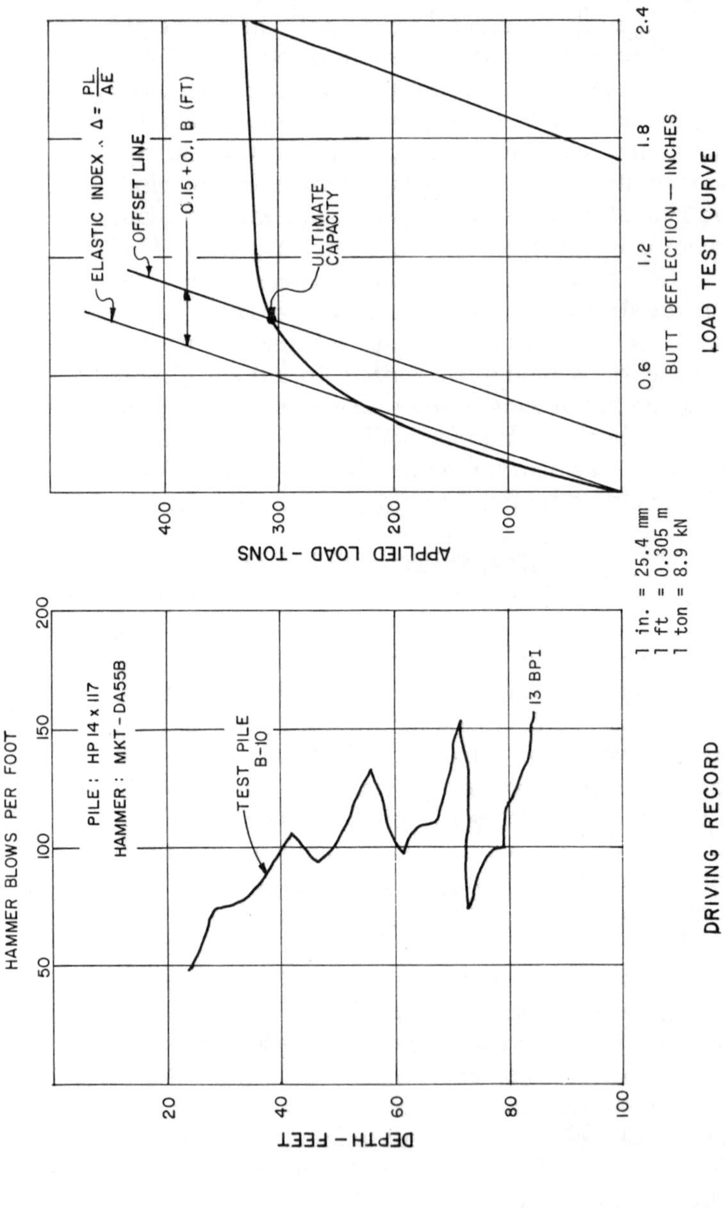

FIGURE-1 LOAD TEST INTERPRETATION PILE B-10

Site B is in Michigan where 16 inch (400 mm) O.D. x 0.312 inch (8 mm) wall closed-end pipe piles were used. The soil conditions consist of 20 feet (6 m) of silty clay (N-10 to 30) overlying dense glacial till (N=50 to 90).

Site C is loacted in Virginia where two pile types were tested. These were a closed end pipe pile 12 inch (300 mm) O.D. x 0.125 inch (3 mm) wall, and a concrete pile nominally 17 inches (430 mm) diameter. The soil profile consists of an upper 30 feet (9 m) of sandy clay fill (N=9 to 25); 30 feet (9 m) of sand and gravel (N=30 to 70); and then a deep deposit of clay (N=40 to 50).

Site D is in Mississippi where the subsurface conditions consist of an upper 30 feet (9 m) of silt fill (N=10 to 20); 60 feet (18 m) of silty clay (N=25 to 30); 10 feet (3 m) of sand and gravel (N=30 to 40); and then sandy silty clay (N=100+). Three piles were tested; HP 14 x 73, closed-end pipe 14 inch (350 mm) O.D. x 0.375 inch (10 mm) wall and a mandrel driven Raymond Step-taper pile with an 8 inch (200 mm) diameter tip and 17 inch (430 mm) diameter butt.

Site E, located in Connecticut, consists of a deep deposit of medium sand (N=10 to 40). Closed-end pipe piles 12 inch (300 mm) O.D. x 0.25 inch (6 mm) wall were used.

Site F is in Delaware where both 12 and 12.75 inch (300 and 320 mm) O.D. x 0.312 inch (8 mm) wall closed-end pipe piles were utilized. Soil conditions are: top 35 feet (11 m) of organic silt (N=1 to 7); 30 feet (9 m) of sand and gravel (N=15 to 30); and then more than 50 feet (15 m) of silty clay (N=20 to 50).

Site G is located in Arizona where HP 14 x 117 piles were used. The subsurface conditions consist of a top 50 feet (15 m) of silty sandy clay (N=15 to 40); underlain by silty clayey sand with gravel (N=30 to 50).

Site H is in New Jersey where soil conditions are: upper 35 feet (11 m) of hydraulic fill (N=3 to 6); 35 feet (11 m) of alternate layers of clay and sand (N=12 to 30); and then a deep deposit of silty sand (N=40 to 60). Closed-end pipe 12.75 inch (320 mm) O.D. x 0.375 inch (10 mm) wall was used

Site I is in Delaware where soil conditions consist of an upper 30 feet (9 m) of organic sandy clayey silt (N=2 to 4); 55 feet (17 m) of sandy silty clay (N=10 to 20); and then a thick deposit of silty sand (N=40 to 60). Closed-end pipe piles 12 inch (300 mm) O.D. x 0.25 inch (6 mm) wall; and Monotube piles with 8 inch (200 mm) tips and 12 inch (300 mm) butts were used.

Table 1 summarizes the test pile data and shows the comparison between measured and wave equation predicted ultimate capacities. Driving hammers are single or differential air/steam types with an assumed efficiency of 80 percent, except for 2 diesel hammers which were assumed to have an efficiency of 95 percent. The correlation between measured and predicted capacities is graphically shown on Figure 2. In general, the data on Figure 2 reveals that for these 39 test piles the measured ultimate capacity is greater than the wave equation predicted capacity. One can conclude from this data that if a pile is driven to a terminal resistance corresponding to a wave equation predicted capacity of about 2.3 times the design load, then the actual measured load test capacity should be about equal to or greater than twice the design load.

TABLE I --- SUMMARY OF TEST PILES

SITE	TEST PILE	PILE TYPE (1)	DRIVEN LENGTH FT.	PENET. DEPTH FT.	CROSS SECTION SQ. IN.	HAMMER	ENERGY FT-LB	CAPBLK MAT'L	DRIVE CAP KIP	CUSH. MAT'L	END BEAR. % (2)	SIDE FRICT. % (2)	FRICT. DIST. (2)	FINAL BLOWS	MEAS. CAP. TONS	W. E. CAP. TONS	REMARKS
A	SB-1	CONC.	80	78	180	VUL 80 C	24,450	MIC/AL	2.6	PLYWD.	0	100	UNIF.	38/FT.	215	180	
"	SB-2	"	70	68	"	"	"	"	"	"	"	"	"	39/	305	190	
"	SB-3	"	60	58	"	"	"	"	"	"	"	"	"	47/	260	220	
"	SB-4	"	40	35	"	"	"	"	"	"	"	"	"	9/	100	45	
"	SB-5	"	70	62	"	"	"	"	"	"	"	"	"	79/	350	275	JETTED 56 FT.
"	SB-7	"	40	29	"	"	"	"	"	"	"	"	"	20/	125	95	
"	R-1	STEEL M	84	78	74 AVG.	"	"	"	1.1	NONE	10	90	TRIAN.	44/	335	245	JETTED 20 FT.
"	R-2	"	76	68	69 "	"	"	"	"	"	"	"	"	26/	260	135	" 35 FT.
"	R-3	"	64	58	64 "	"	"	"	"	"	"	"	"	51/	280	270	
"	R-4	"	42	35	"	"	"	"	"	"	"	"	"	24/	130	135	
"	R-5	"	64	62	"	"	"	"	"	"	"	"	"	60/	360	305	JETTED 56 FT.
B	T-1	STEEL P	40	22	15.4	VUL 010	32,500	"	2.0	"	90	10	UNIF.	20/IN.	360	305	
"	M-11-4	"	25	23	"	"	"	"	"	"	"	"	"	10/	280	290	
"	F-16	"	25	24	"	"	"	"	"	"	"	"	"	22/	380	345	
"	44-A	"	30	24	"	"	"	"	"	"	"	"	"	10/	290	280	
C	PP-1	"	45	40	9.2	KOBE K-13	24,400	FOSTLN	1.0	"	"	"	TRIAN.	15/	165	170	
"	P-5	CONC.	35	33	230	VUL 014	42,000	MIC/AL	3.0	PLYWD.	"	"	"	33/	380	330	
"	P-95	"	36	34	"	"	"	"	"	"	"	"	"	50/	380	380	
"	P-111	"	35	33	"	"	"	"	"	"	"	"	"	50/	340	380	
D	H-2	STEEL H	110	100	21.5	RAY 4/0	48,750	"	2.2	NONE	80	20	UNIF.	30/	360	330	PREDRILLED 58 FT
"	H-4	"	80	75	"	"	"	"	"	"	"	"	"	30/	360	360	" 52 "
"	P-2	STEEL P	110	100	16.1	"	"	"	"	"	"	"	"	31/	350	290	" 50 "
"	P-4	"	80	70	"	"	"	"	"	"	"	"	"	27/	325	330	" 48 "
"	R-2A	STEEL M	110	100	64 AVG.	RAY 2/0	32,500	"	"	"	"	"	"	9/	390	420	" 48 "
"	R-2B	"	110	100	"	"	"	"	"	"	"	"	"	6/	350	340	" 48 "

TABLE I Continued

TABLE I --- SUMMARY OF TEST PILES (Continued)

SITE	TEST PILE	PILE TYPE (1)	DRIVEN LENGTH FT.	PENET. DEPTH FT.	CROSS SECTION SQ. IN.	HAMMER	ENERGY FT-LB	CAPBLK MAT'L	DRIVE CAP KIP	CUSH. MAT'L	END BEAR. % (2)	SIDE FRICT. % (2)	FRICT. DIST. (2)	FINAL BLOWS	MEAS. CAP. TONS	W. E. CAP. TONS	REMARKS
E	53	STEEL P	90	83	9.2	VUL 08	26,000	"	2.0	"	20	80	"	11/FT.	120	70	
"	98	"	90	82	"	"	"	"	"	"	"	"	"	9/	100	70	
F	9	"	60	56	12.2	VUL 010	32,500	OAK.	2.2	"	50	50	TRIAN.	70 /	120	130	
"	38	"	60	58	"	"	"	"	"	"	20	80	UNIF.	53 /	180	190	
"	212	"	60	54	11.5	"	"	"	"	"	"	"	"	28 /	140	150	
"	364	"	85	80	12.2	"	"	"	"	"	"	"	"	15/	110	95	
G	B-10	STEEL H	87	85	34.4	MKT DA55B	38,000	URETHN	1.4	"	70	30	"	13/IN.	300	290	
H	2 D	STEEL P	100	71	14.5	VUL 08	26,000	MIC/AL	1.8	NONE	40	60	UNIF.	34/FT.	155	120	
"	1 E	"		78	"	"	"	"	"	"	70	30	"	80/	215	155	
"	153 E	"	92	72	"	"	"	"	"	"	70	30	"	11/	235	155	
I	AP-135	"	70	68	5.1	LB-520	26,100	CONBEST	1.5	"	30	70	"	12/	95	60	
"	AP-250	"	86	84	9.2	"	"	"	"	"	30	70	"	72/	200	140	
"	AM-5	"	64	62	7.4AVG	"	"	"	"	"	50	50	"	87/	210	140	
"	EP-250	"	60	58	9.2	"	"	"	"	"	30	70	"	14/	75	60	
"	EM-7	"	60	58	6.3AVG	"	"	"	"	"	50	50	"	22/	100	80	

(1) PILE LEGEND: CONC. is "precast" concrete; STEEL M is steel mandrel; STEEL P is steel pipe; STEEL H is steel H

(2) Assumed soil resistance and distribution, Soil properties for wave equation analyses are: Quake = 0.1 in.; Damping = 0.05 sec/ft.; Point Damping = 0.15 sec/ft.

1 ft = 0.305 m 1 in. = 25.4 mm 1 in^2 = 645 mm^2 1 kip = 454 kg 1 ton = 8.9 kN

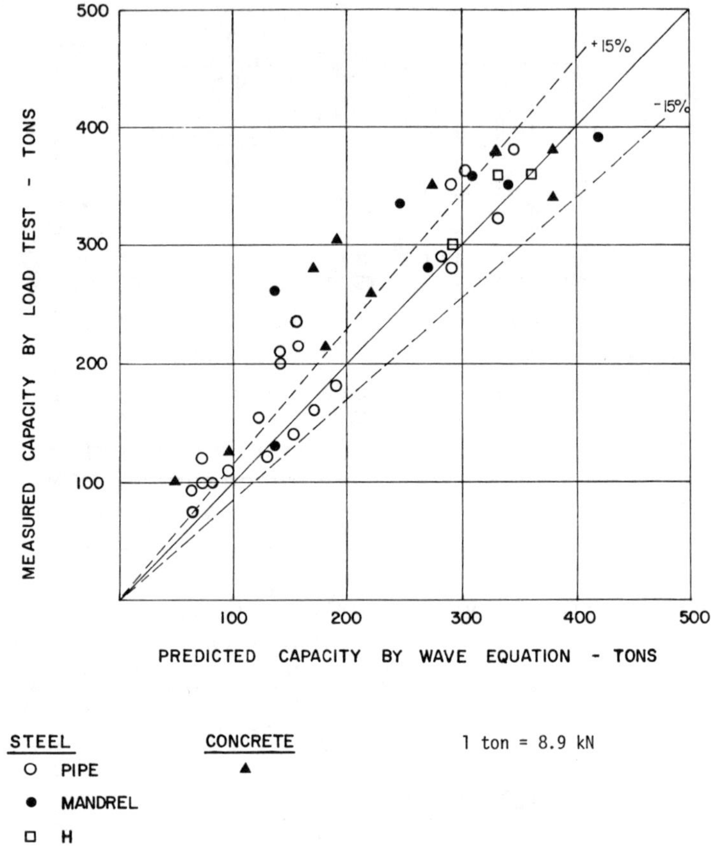

FIGURE-2 WAVE EQUATION AND LOAD TEST CORRELATION

DIESEL HAMMERS

Until 1976, we could not accurately analyze piles driven with diesel hammers using Lowery's version of the wave equation program (11). This program was originally developed at Texas A & M for the Texas Transportation Institute (TTI). Since 1976, Gobles' WEAP program (7) has been used. The WEAP analysis includes both the thermodynamic as well as mechanical operations. Wave equation analyses for the two test piles driven with diesel hammers using the TTI and WEAP programs are shown on Figure 3 for comparison. Two aspects of the results are apparent. First, the TTI version computes a higher ultimate capacity; and second, the TTI version computes significantly higher driving stresses. This high computed stress might rule out a hammer that probably would drive the pile satisfactorily. The test piles shown on Figure 3 were driven hard and no butt damage was evident. Experience has shown that pile damage generally does not occur until driving stresses of 45 to 50 ksi (310 to 345 MPa) are computed.

SUMMARY AND CONCLUSIONS

1. Wave equation analyses performed in conjunction with load tests conducted to failure are the two most powerful tools available to the foundation engineer.

2. Correlations between measured and predicted ultimate capacities for nearly 40 test piles indicate that if a pile is driven to a wave equation resistance equal to about 2.3 times the design load, then one would expect the ultimate measured capacity (Davisson Method) should be equal to or greater than twice the design load.

3. Diesel hammers appear to be more accurately modeled using a wave equation computer program that incorporates both the thermodynamic and mechanical operation.

ACKNOWLEDGMENT

The test pile data are from Bechtel projects with which the author has been involved since 1970. The opinions and conclusions expressed in this paper are those of the author.

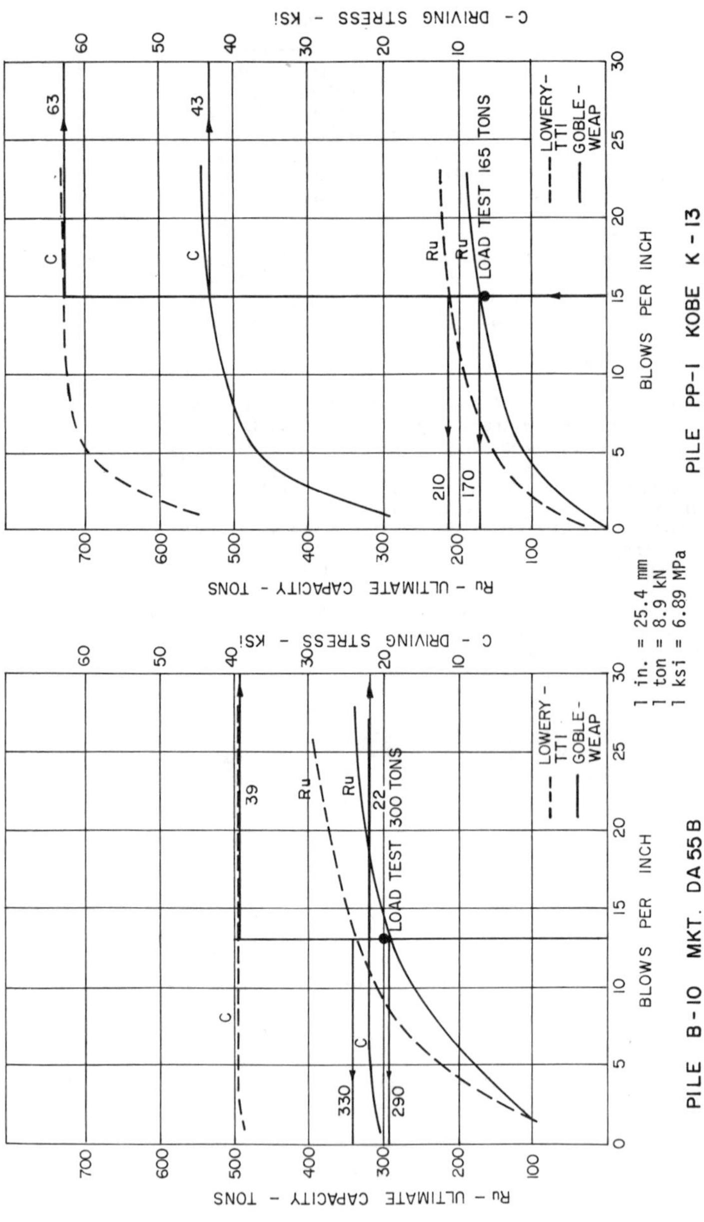

FIGURE-3 WAVE EQUATION ANALYSIS OF DIESEL HAMMERS

REFERENCES

1. Bender, C. H., Lyons, C. G., and Lowery, L. L., "Applications of Wave Equation Analysis to Offshore Pile Foundations," Offshore Technology Conference, 1969.

2. Blessey, W. E., "Allowable Pile Capacity Mississippi Deltaic Plain," Conference on Pile Foundations, Lehigh University, April 1970.

3. Coyle, H. M., Bartoskewitz, R. E., and Lowery, L. L., "Prediction of Static Bearing Capacity from Wave Equation Analysis," Offshore Technology Conference 1970.

4. Davisson, M.T., "Design Pile Capacity," Conference on Pile Foundations, Lehigh University, April 1970.

5. Davisson, M. T., "High Capacity Piles, Innovations in Foundation Construction," Illinois Section ASCE, May 1973.

6. Davisson, M. T., and Rempe, D. M., "Wave Theory Simplified," Piletalk Seminar, New Jersey, 1974.

7. Goble, G. G., and Rausche, F., "Wave Equation Analysis of Pile Driving - (WEAP), Users Manual," July 1976.

8. Hirsch, T. J., Lowery, L. L., Coyle, H. M., and Samson, C. H., "Pile Driving Analysis by One Dimensional Wave Theory: State of the Art," Highway Research Record No. 333 Pile Foundations, 1970.

9. Housel, W. S., "Pile Load Capacity: Estimates and Test Results," ASCE, July 1966.

10. Lowery, L. L., "Dynamic Behavior of Piling," Texas A & M University, May 1967.

11. Lowery, L. L., "Pile Driving Analysis - Users Manual," October 1970.

12. Mansur, C. I., and Hunter, A. H., "Pile Tests - Arkansas River Project," ASCE, September 1970.

13. Meyerhoff, G. G., "Ultimate Bearing Capacity of Driven or Displacement Piles," ASCE, January 1956.

14. Smith, E. A. L., "Pile Driving, Analysis by the Wave Equation," ASCE, August 1960.

BEARING CAPACITY OF STEEL CORE PILES

Håkan Bredenberg[1] and Bengt B. Broms, F.ASCE[2]

INTRODUCTION

The total length of all piles installed each year in Sweden is about 2 to 3 Mm (6.6 to 9.8 million feet). Mostly concrete piles are used as can be seen from the following table.

TABLE 1

PILE TYPES USED IN SWEDEN

	1974	1976
Concrete piles	76%	75%
Timber piles	21%	21%
Steel piles	3%	4%

More than 90% of the concrete piles are point bearing driven precast piles for apartments and industrial buildings. Timber piles are mainly used as embankment piles for roads while steel piles, primarily small diameter pipe piles and rails, are to a large extent used to support light structures, mainly single family housing.

Steel core piles which have a high bearing capacity are primarily used for heavy structures with high concentrated loads.

CONSTRUCTION OF STEEL CORE PILES

The load on a steel core pile is carried by a core consisting of round, smooth steel bars ø 75-250 mm (3-10 in) which are welded together. The steel core is driven through a casing and grouted. This method has become economical because of the reduction of the drilling costs that has occurred during the last ten years.

---

[1]Civil Engineer and Research Associate, Department of Soil and Rock Mechanics, Royal Institute of Technology, 100 44 Stockholm 70, Sweden.

[2]Ph.D., Professor, Department of Soil and Rock Mechanics, Royal Institute of Technology, 100 44 Stockholm 70, Sweden.

Steel core piles are generally designed in accordance with the requirements in the Swedish Building Code (SNB 1975, 23:6) for piles of Class A with respect to steel quality, straightness of the piles, buckling strength, point bearing capacity etc. The maximum allowable steel stress is 94 MPa (13 623 psi) according to this code.

A steel core pile is manufactured as follows

1. First a bore hole with 150 to 300 mm (6 to 12 in.) diameter is drilled to bedrock using a casing.
2. The steel core is then placed in the casing and driven to refusal.
3. Cement mortar is injected between the steel core and the casing.

Up to 25 m (83 ft.) long bore holes with 150 to 300 mm (6-12 in) diameter can usually be drilled without difficulties even in stony soils or through boulders and rock fill. The holes should extent at least 300 to 500 mm (12 to 20 in) into intact rock. The bore hole is stabilized by a casing which is left in the ground. The wall thickness is 3 to 8 mm (0.12 - 0.3 in).

The straightness of the borehole is checked before placing of the steel core by lowering a light to the bottom of the casing. A more accurate method is to use an inclinometer. With rotary drilling the deviations are usually small and negligible.

The steel core is normally driven to refusal with a free falling drop hammer. The area of the pile and of the hammer is usually the same. Guides are used to center the steel core in the casing during the driving. The casing should be thoroughly cleaned before the injection of cement mortar or concrete.

The pipe or hose for the injection should extend to the bottom of the casing. The injection pipe is gradually withdrawn while the pumping of the mortar is continued so that the mortar overflows the casing. The water/cement ratio of the mortar is normally about 0.4. Additives are used to reduce the shrinkage (about 1%).

USE OF STEEL CORE PILES

This type of pile has several advantages both economically and technically especially:

1.	At high concentrated loads	A large number of piles will be required within a small area and precast concrete piles are difficult to drive if the spacing is small. Damaged piles are difficult to replace.
2.	Close to adjacent structures which can be damaged by vibrations and heave from pile driving	This is often the case in large cities.

3.	When a certain penetration depth had to be reached	This case is common close to existing adjacent buildings.
4.	As foundation of structures or machines which are affected by small settlements	Examples are paper making machines and turbines.
5.	As foundations for structures or machines which generate vibrations or shock loads (impact)	Examples are forges and saw mills. The bearing capacity is affected by vibrations and shock loads.

Steel core piles are frequently used where the anticipated life of the structure will be large (300 years or more) and where the settlements and the effect of the installation of the piles on the surroundings as to be small.

This article describes the manufacture and the installation of steel core piles and how the bearing capacity of such piles can be calculated.

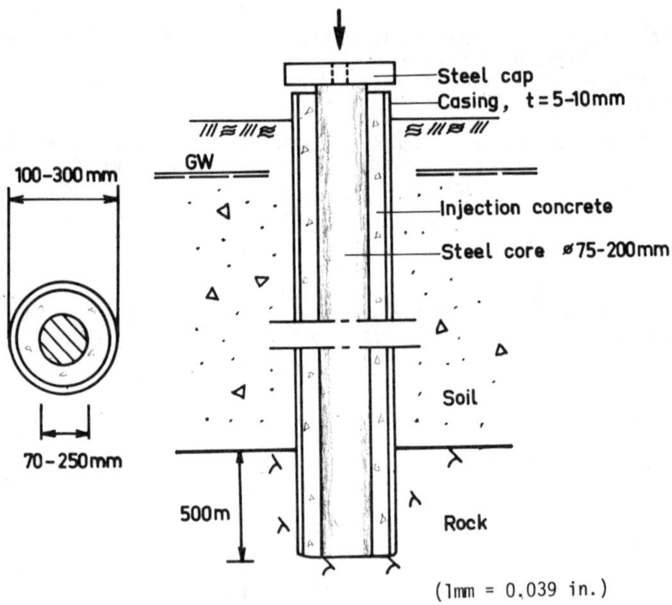

FIGURE 1 CROSS-SECTION OF A STEEL CORE PILE

ALLOWABLE LOAD

The allowable load on steel core piles is normally 0.5 MN to 3 MN (112 360 to 674 157 lbf). There are examples where the load has been up to 5 MN. The casing and the concrete or mortar are normally not considered in the calculations of the allowable load since they are only regarded as corrosion protection.

Since the allowable load is high it is important to check that the underlying rock can carry the load in the piles. The point resistance can be determined by measuring the set (mm/blow) when the pile is driven to refusal. The factor of safety should normally be at least 2.0 in granite or gneiss with respect to the dynamic point resistance. If, for example, the allowable point bearing capacity is 80 MPa then the pile should be driven until the point resistance during driving is at least 160 MPa.

The compressive strength (σ_{cyl}) as determined by unconfined compression tests on cylindrical samples with 50 mm height and diameter for Swedish granite is normally about 230 MPa (33 333 psi) (Rehnman and Broms, 1972). The ultimate strength of the rock (σ_{ult}) which decreases with increasing diameter of the rock point (d), can be estimated from the following relationship (Coates and Gynge, 1966)

$$\sigma_{ult} \approx 7\ \sigma_{cyl} / d^{0.2} \tag{1}$$

where d is diameter of the loaded area in cm. However, cracks and other defects in the rock at the pile point can appreciably influence the ultimate strength. The point resistance of every pile should therefore be checked.

DETERMINATION OF POINT RESISTANCE

The relationship between the point resistance F_B during driving and the permanent set s (mm/blow) can be determined as shown in Fig. 2a.

A rectangular stress wave is generated when the hammer strikes the pile if the cross-sectional area and the material in the hammer and in the pile are the same which is usually the case due to practical reasons. The properties of the resulting stress wave are as follows.

Length of stress wave: $2\ L$

Duration of stress wave: $t_o = \frac{2\ L}{c}$

Intensity of stress wave: $F_o = \frac{v}{2}\ Z$

where L is the length of the hammer, c is the velocity of the stress wave (in steel 5100 m/sec), v is the velocity of the hammer when it strikes the piles, Z is the dynamic impedance (AE/c) where A is the area and E is the modulus of elasticity (for steel E = 210 GPa).

Note: 1 MN = 112 tons 1 MPa = 145 psi 1 m = 3.28 ft. 1 GPa = 145×10^3 psi

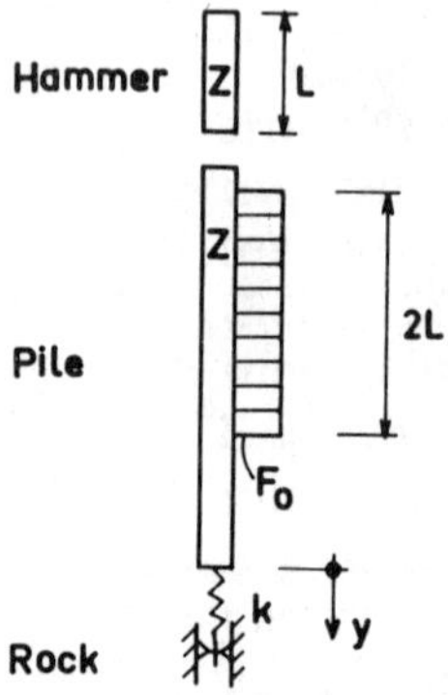

FIGURE 2 a MECHANICAL MODEL

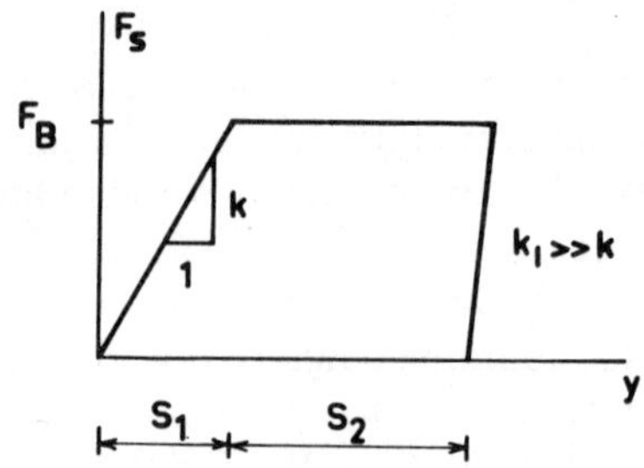

FIGURE 2 b ASSUMED FORCE-DISPLACEMENT RELATIONSHIP FOR PILE POINT

The behaviour of the pile point is assumed to be elastic-plastic as shown in Fig. 2 b with a spring constant k (MN/m) in the elastic range. The initial wave is reflected when it reaches the pile point. The intensity of the reflected wave is F_r. The point resistance (F_s) can be calculated from the following equilibrium equation (Bredenberg, 1977)

$$F_s(t) = F_r(t) + F_o(t) \qquad (2)$$

From the relationship between particle velocity in the pile and the force the following linear differential equation can be derived from Eq.(2) (Bredenberg, 1978):

$$\dot{y} + \frac{k}{Z} y = \frac{2 F_o(t)}{Z} \qquad (3)$$

where y is the displacement of pile point and $y = \frac{dy}{dt}$ is the velocity of pile point.

The solution of Eq. (3) for the initial condition y (o) = 0 gives

$$F_s(t) = 2 F_o \left(1 - e^{-\frac{k}{Z}t}\right) \qquad (4)$$

where $0 \leq t \leq t_o$.

Eq.(4) is illustrated in Fig. 3 a. For k = 0 and k = ∞ (infinitely soft and infinitely hard rock, respectively) the point force will be zero and +2 F_0 (compression), respectively.

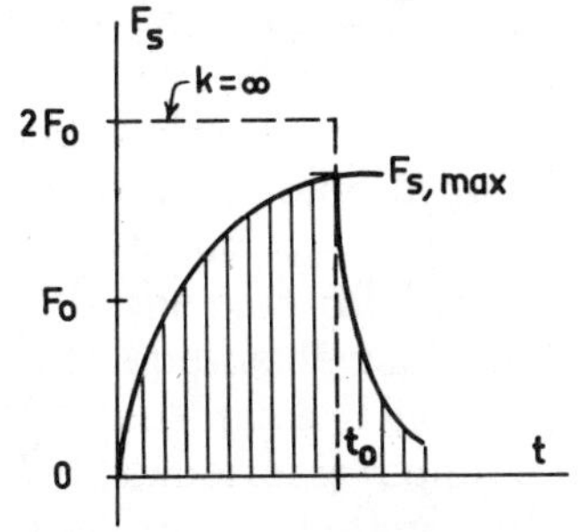

FIGURE 3 a POINT FORCE, $F_s(t)$, $F_s < F_B$

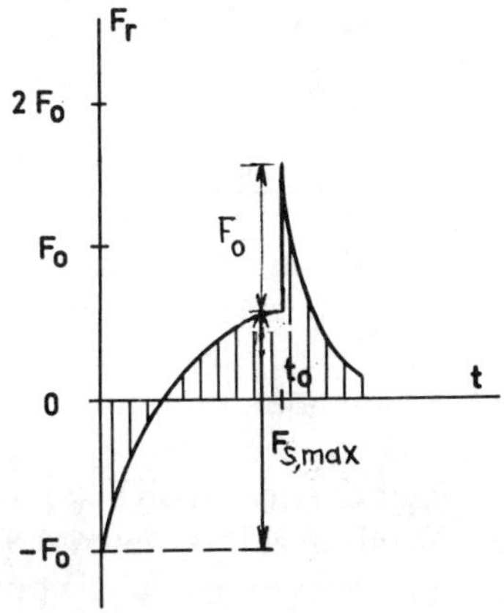

FIGURE 3 b REFLECTED FORCE, F_r (t)

At unloading ($t > t_0$) the following relationship is obtained

$$F_s(t) = F_{s\,max} e^{-\frac{k_1}{Z} \cdot t} \tag{5}$$

where

$$F_{s\,max} = 2 F_0 \left(1 - e^{-\frac{k}{Z} \cdot t_0}\right) \; ; \; F_{s\,max} < F_B \tag{6}$$

The reflection force can be determined from Eqs (4) and (2), for example

$$F_r(t) = F_0 \left(1 - 2 e^{-\frac{k}{Z} t}\right) ; 0 < t < t_0 \tag{7}$$

It should be noted that a tension is always obtained at t = 0 as can be seen from Fig. 3 b. The maximum tension force is $-F_0$ for an infinitely steep wave front.

PERMANENT PENETRATION

The pile point starts to accelerate downwards when the initial wave arrives to the pile point and the rock deforms elastically. If the intensity and the duration of the initial wave are sufficient the rock below the pile point will yield and deform plastically. The time t_1 required to reach yielding (F_B) can be computed from Eq.(4)

$$t_1 = -\frac{Z}{k}\ln\alpha \qquad (8\text{ a})$$

The corresponding deformation of the point is

$$s_1 = \frac{2F_o}{k}\left(1 - e^{-\frac{k}{Z}t_1}\right) \qquad (8\text{ b})$$

or

$$s_1 = \frac{2F_o}{k}(1-\alpha) \qquad (8\text{ c})$$

where $\alpha = 1 - \frac{F_B}{2F_0}$

The work $F_B s_2$ at the permanent set s_2 is equal to the difference in energy between the initial stress wave at the pile point (W_o) and the energy of the reflected wave (W_r) when the rock deforms plastically.

Then

$$F_B s_2 = \int_o^T (W_o - W_r)\,dt \qquad (9)$$

By inserting the relationships

$$W_o = \frac{1}{Z}\int F_o^2\,dt \qquad (10\text{ a})$$

and

$$W_r = \frac{1}{Z}\int F_r^2\,dt \qquad (10\text{ b})$$

into Eq.(9) then

$$F_B s_2 = \frac{1}{Z}\int_o^T (2F_o^2 - F_r^2)\,dt \qquad (11)$$

where T is the time for the plastic deformation of the rock below the pile point. If Eqs. (11) and (2) are combined then

$$s_2 = \frac{1}{Z}\int_o^T (2F_o - F_B)\,dt \qquad (12)$$

After integration

$$s_2 = \frac{T}{Z}(2F_o - F_B) \qquad (13)$$

since F_o and F_B are constant.

The significance of Eqs (12) and 13) is illustrated in Fig. 4 a.

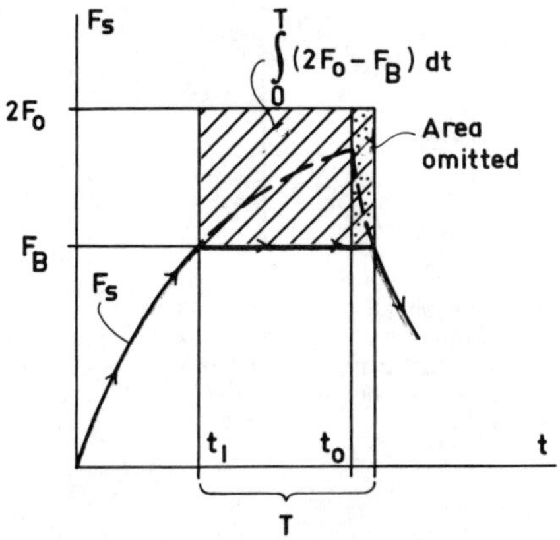

FIGURE 4 a POINT FORCE AT PERMANENT DISPLACEMENT

If the time interval $(T - t_o + t_1)$ at unloading of the point, is omitted since k_1 in Fig. 2 b is assumed to be very large and the value on t_1 from Eq. (8) is used, the following expression is obtained

$$T \approx t_o - t_1 = \frac{2L}{c} - \frac{Z}{k} \ln\alpha \qquad (14)$$

The relationship between point bearing capacity and the part of the permanent set denoted s_2 can then be calculated from Eqs (13) and (14):

$$s_2 = 2F_o\alpha \left(\frac{t_o}{Z} + \frac{\ln\alpha}{k}\right) \qquad (15\text{ a})$$

The total permanent set is then

$$s = s_1 + s_2 \qquad (15\text{ b})$$

Eq. (8 c) and (15 a) yields after some algebra

$$s = \frac{2F_o\alpha}{k}\left(\frac{1}{\alpha} - 1 + \frac{t_o k}{Z} + \ln\alpha\right) \qquad (15\text{ c})$$

The term $\frac{\ln\alpha}{k}$ can be neglected in the case k is very large. A linear relationship between F_B and s is then obtained from Eq.(15) since

$$\alpha = 1 - \frac{F_B}{2 F_0}$$

then from Eq. (15 a)

$$F_B \approx 2 F_o - \frac{Z c}{2 L} \cdot s_2 \qquad (16)$$

The shape of the reflected wave which is associated with a permanent set is shown in Fig. 4 b. In the figure is indicated how it is possible to determine the yield force F_B by measuring the reflection wave. The yield force F_B corresponds to the amplitude of the reflected wave for $0 < t < t_o$ as indicated in Fig. 4 b.

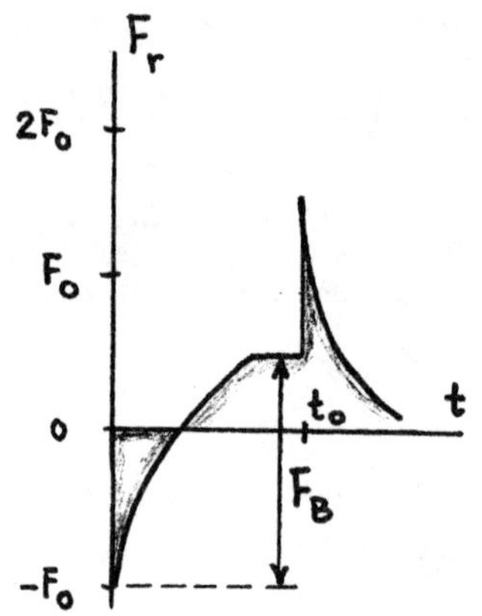

FIGURE 4 b

DAMPING OF THE STRESS WAVE IN THE PILE

It has been assumed in the derivation that the damping in the pile and the splices is small and negligible. If the contacts in the splices are perfect without any gaps the stress wave will not be affected as it travels down the pile if the damping is small.

The shape of an actual stress wave is not perfectly square. Time will always be required for the stress wave to reach its maximum and there will always be some oscillations behind the wave front. The oscillations are normally small in comparison with the intensity F_o of the stress wave.

The wave front is distorted as it travels down the pile. The rise time is normally short in comparison with the total length and the duration of the stress wave (2 L/c). The assumption that the shape of the stress wave is square is normally sufficient accurate for practical purpose. When a cushion is used which is generally the case for pneumatic hammers the stress wave will approximately be triangular or sinusoidal. In principle there are no difficulties to use the proposed method for other shapes of the initial stress wave.

CALCULATION OF POINT RESISTANCE IN ROCK

Only an indication of the point resistance is generally required to calculate the bearing capacity in rock since the point resistance is not affected much by the spring constant k. The modulus of elasticity (E_{rock}) of granite and gneiss normally varies between 10 GPa and 50 GPa. The spring constant k can be estimated from theory of elasticity. It is approximately equal to

$$k \approx E_{rock}\, d \qquad (17)$$

where d is the diameter of the steel core.

STRESS-WAVE MEASUREMENTS

In Fig. 5 is shown the initial and measured reflected stress waves for a ø 95 mm steel core which has been driven to refusal. The form of the initial wave is not perfectly square, as predicted by the theory due to imperfections of the impact surfaces and distortion in the measuring system. The maximum amplitude (0.8 MN) is close to the predicted value.

The maximum measured tension force, -0.39 MN, is only half of that calculated from Eq (7) partly due to the sloping front of the initial wave. If the initial stress wave increases linearly

$$F_i(t) = q\,t \qquad (18)$$

then the reflected wave will be (Bredenberg and Broms, 1978)

$$F_r(t_1) = 2q\frac{Z}{k}\left(e^{-\frac{k}{Z}t_1} - 1\right) + q\,t_1 \qquad (19)$$

The reflected wave reaches a maximum at the time t_m

$$t_m = \frac{Z}{k}\ln 2 \approx 0.693\ Z/k \qquad (20)$$

The maximum force can be calculated from the expression

$$F_{r,min} = q\frac{Z}{k}(1 - \ln 2) \approx -0.307\ qZ/k \qquad (21)$$

as indicated in Fig. 6.

Note: 1 GPa = 145 x 10^3 psi   1 mm = 0.039 in.   1 MN = 112 tons

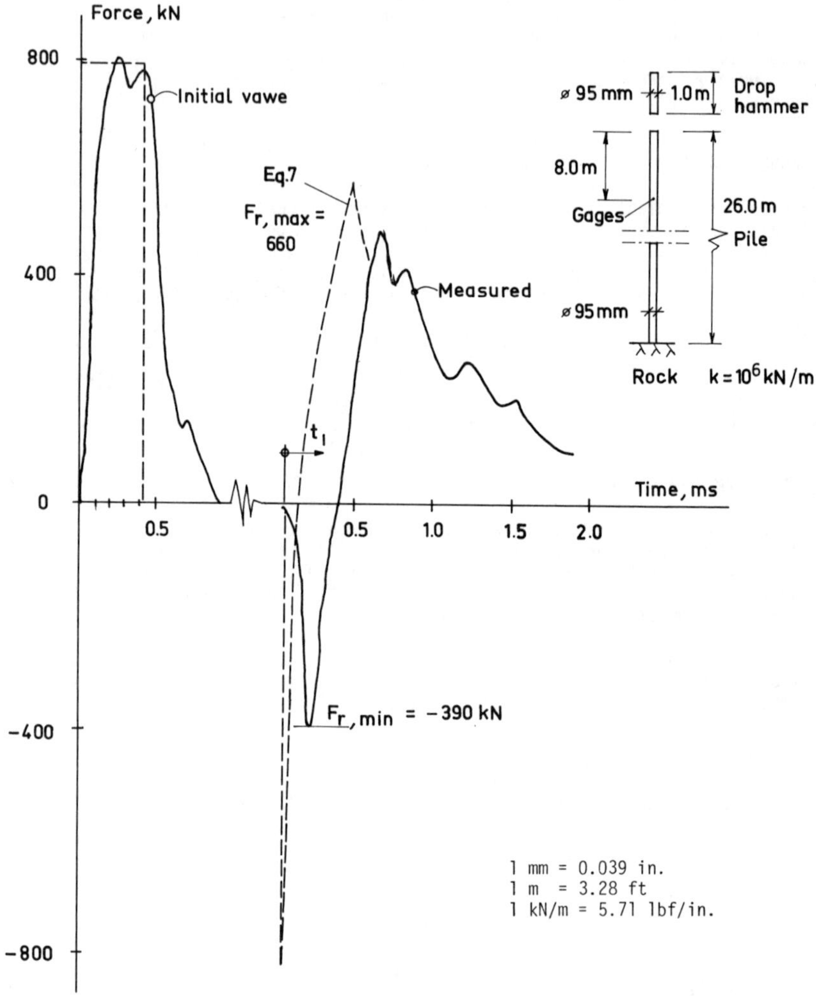

FIGURE 5 EXAMPLE OF MEASURED INITIAL AND REFLECTED WAVE FOR A STEEL PILE ON ROCK

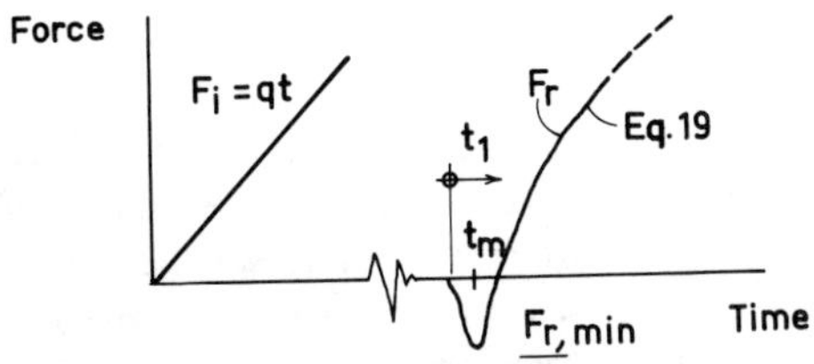

FIGURE 6 LINEARLY INCREASING INITIAL STRESS WAVE

The spring constant of the rock and the acoustic impedance of the pile ($Z = AE/c$) were estimated to 10^6 kN/m and to 270 kNs/m, respectively. The calculated maximum tension of the reflected wave occurred at $t_m = 0.18$ ms. The reflected stress wave was initially a tension wave even when the permanent set was less than 0.1 mm/blow. The agreement between measured and calculated values using the simple model in Fig. 6 was satisfactory for the assumed value on k (10^6 kN/m).

Note: 1 kN/m = 5.71 lbf/in. 1 mm = 0.039 in.

EXAMPLE

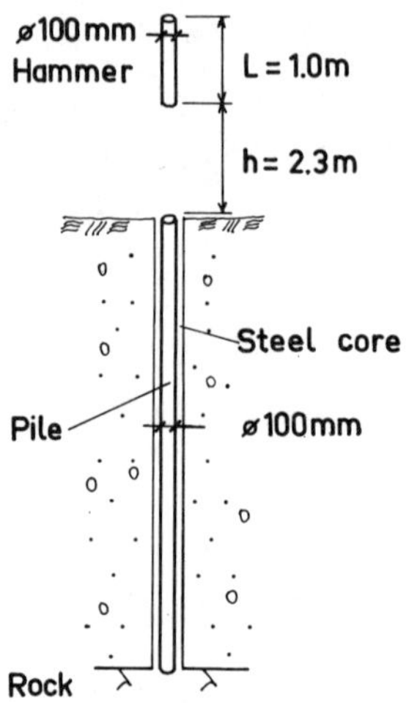

Calculate the maximum allowable permanent set s for a steel core pile with 100 mm diameter. The design load is 0.7 MN (70 Mp). A free falling 1.0 m long drop hammer with 100 mm diameter is used. The height of fall is 2.3 m. The modulus of elasticity of the rock E_{rock} is $50 \cdot 10^3$ MPa.

$$Z = \frac{AE}{c} = \frac{0.1^2 \cdot \pi}{4} \cdot \frac{2.1 \cdot 10^8}{5100} = 323 \text{ kN sec/m}$$

$$v = \sqrt{2\,gh} = \sqrt{2 \cdot 9.815 \cdot 2.3} = 6.7 \text{ m/s}$$

$$F_o = \frac{v}{2} \cdot Z = \frac{6.7}{2} \cdot 323 = 1080 \text{ kN}$$

$$k = \frac{E_{rock}\sqrt{A}}{1.35} \approx \frac{5 \cdot 10^7 \cdot 0.1}{1.35} = 3.7 \cdot 10^6 \text{ kN/m}$$

Note: 1 mm = 0.039 in. 1 MPa = 145 psi
1 m = 3.28 ft 1 kN = 225 lbf
1 MN = 112 tons 1 kN/m = 5.71 lbf/in.

For a factor of safety F of 2.0 then

$$F_B = 2 \cdot 0.70 = 1.40 \text{ MN}$$

$$\alpha = 1 - \frac{F_B}{2\,F_o} = 1 - \frac{1400}{2 \cdot 1080} = 0.35$$

$$\ln \alpha = -1.04$$

$$s_1 = \frac{2\,F_o}{k}(1-\alpha) = \frac{2 \cdot 1080}{3.7 \cdot 10^6}(1-0.35) \approx 3.8 \cdot 10^{-4} \text{ m}$$

$$s_2 = 2\,F_o\,\alpha\,(\frac{2L}{Zc} + \frac{\ln\alpha}{k} = 2 \cdot 1080 \cdot 0.35\,(\frac{2 \cdot 1}{323 \cdot 5100} - \frac{1.04}{3.7 \cdot 10^6}) = 7.05 \cdot 10^{-4}$$

$$s = s_1 + s_2 = (3.8 + 7.1) \cdot 10^{-4} \text{ m/blow} \approx 1.1 \text{ mm/blow}$$

Note: 1 MN = 112 tons   1 m = 3.28 ft   1 mm = 0.039 in.

SUMMARY AND CONCLUSIONS

1. Steel core piles drilled to rock are commonly used in Sweden. The settlements and the vibrations during the installation of the piles are small. The costs depend mainly on the depth to rock. The piles are protected against corrosion by filling the space between the steel core and the casing with mortar.

2. The design load for steel core piles is normally high (1 MN to 5 MN). The bearing capacity of every pile should therefore be checked by measuring the permanent set during the driving.

3. Calculations of the bearing capacity must include the effects of the elastic and the plastic deformations of the rock. If the rock at the pile point is assumed to be idealplastic then the calculated bearing capacity will be too high.

4. For some special cases (e.g. if the initial stress wave is rectangular) explicit formulas relating dynamic point force and permanent set can be derived. It is also possible to evaluate the point resistance by measuring the reflected wave during the final test driving.

5. A general relationship between permanent set and point force has been derived for a linear elastic-plastic pile point-model form force equilibrium and the principle of conservation of energy. Formulas are presented for calculation of the allowable set for steel core piles together with an example.

APPENDIX I - REFERENCES

1. Rehnman, S-E and Broms, B., 1972. "Bearing Capacity of End-Bearing Piles". Reprints and Preliminary Reports, Swedish Geotechnical Institute, No. 44, pp.

2. Coates, D.F. and Gyenge, M., 1966. "Plate-Load Testing on Rock for Deformation and Strength Properties". Testing Techniques for Rock Mechanics, Am. Soc. Test. Mat., STP, pp. 19-35.

3. Bredenberg, H., 1977. "Stötkrafter i pålar" (Impact forces in piles during driving). Byggmästaren, No. 5, Vol. 56, May 1977, pp. 32-37.

4. Bredenberg, H., and Broms, B., 1978. "Joints used in Sweden for Precast Concrete Piles". Conf. on design and construction of piles, Proc. Inst. of Civil Engineers, London. Submitted for publication.

APPENDIX II - NOTATION

A = area
c = velocity of stress wave
d = diameter
E = modulus of elasticity
F_B = point resistance (force) during driving
F_o = intensity (force) of initial stress wave
Fr = intensity (force) of reflected stress wave
F_s = point resistance (force)
h = height of fall of hammer
k = spring constant
L = length
s = permanent set
T = time for plastic deformation of rock
t = time
v = terminal velocity of hammer
W_o = energy of initial stress wave at pile point
W_r = energy of reflected stress wave
y = displacement
Z = AE/c = dynamic impedance
σ_{cyl} = compressive strength (of rock)
σ_{ult} = ultimate strength (of rock)

FIELD EVALUATION OF HAMMER EFFICIENCY AND PILE DRIVING CRITERIA

by

Dunstan D. S. Chen[(1)], James V. Toto,[(2)] M.A.S.C.E., I. H. Wong[(3)], M.A.S.C.E.

INTRODUCTION

In the evaluation and development of pile driving criteria, it is generally assumed that the efficiency of pile driving hammers range from 70 percent to 95 percent, and it is often assumed to be 80 percent. However, even an experienced engineer has difficulty judging the efficiency of a pile hammer. Hammer efficiencies are rarely calibrated for projects.

In the project described herein, the piling contractor provided three hammers of different makes, which offered an opportunity to enable the authors to compare the efficiencies of these hammers by evaluating the driving records of piles driven within a cluster with different hammers. The results of this comparison were supplemented by the Case method of dynamic pile analysis performed in the field. The results indicated that the efficiencies of the hammers ranged from 60 percent to as low as 25 percent, which were substantially lower than what would be assumed in general practice.

SUBSURFACE SOIL CONDITIONS

The site was originally located on a knoll overlooking the Ohio River (Figure 1). During the original site grading operation, material was excavated from the northern portion of the site, and used as fill at the southern (riverward) section of the site. In the filled area, ground elevation varies from 640 to 665. Boring information indicates that the fill is composed of a heterogeneous mix of sands, silts, clays, and gravels excavated from the upland portion of the site; the consistency and relative density of the fill soils varies from very soft to very stiff in some of the cohesive soils, and from loose to medium dense for the cohesionless soils. The thickness of the fill in the general area was found to vary from 8 feet (2.4 meters) to 37 feet (11.3 meters).

Beneath the fill in the northern portion of the site area, the predominant materials are sands and gravels. In the direction towards the river, the fills are underlain by a series of silty sands, silts, and clays, approximately 20 feet (6 meters) in thickness, of varying consistency. These deposits in turn are underlain by a medium dense to dense gravel and sand layer. Shaley limestone bedrock was at approximately elevation 560. A generalized subsurface soil condition is shown on Figure 2.

---

(1) Senior Engineer, and (2) Partner, Dames & Moore, Cranford, New Jersey;
(3) Consulting Engineer, Ebasco Services, Inc., New York, New York. Formerly with Dames & Moore Cranford, New Jersey.

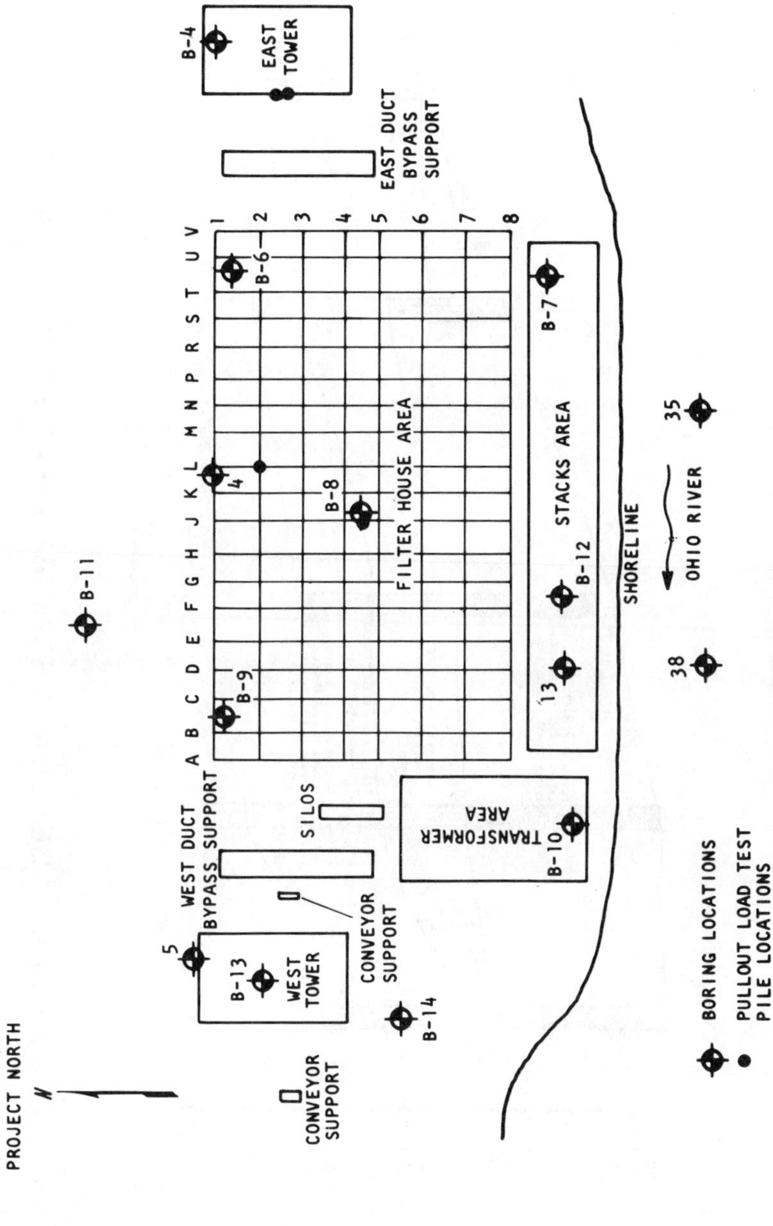

FIGURE 1, PLOT PLAN

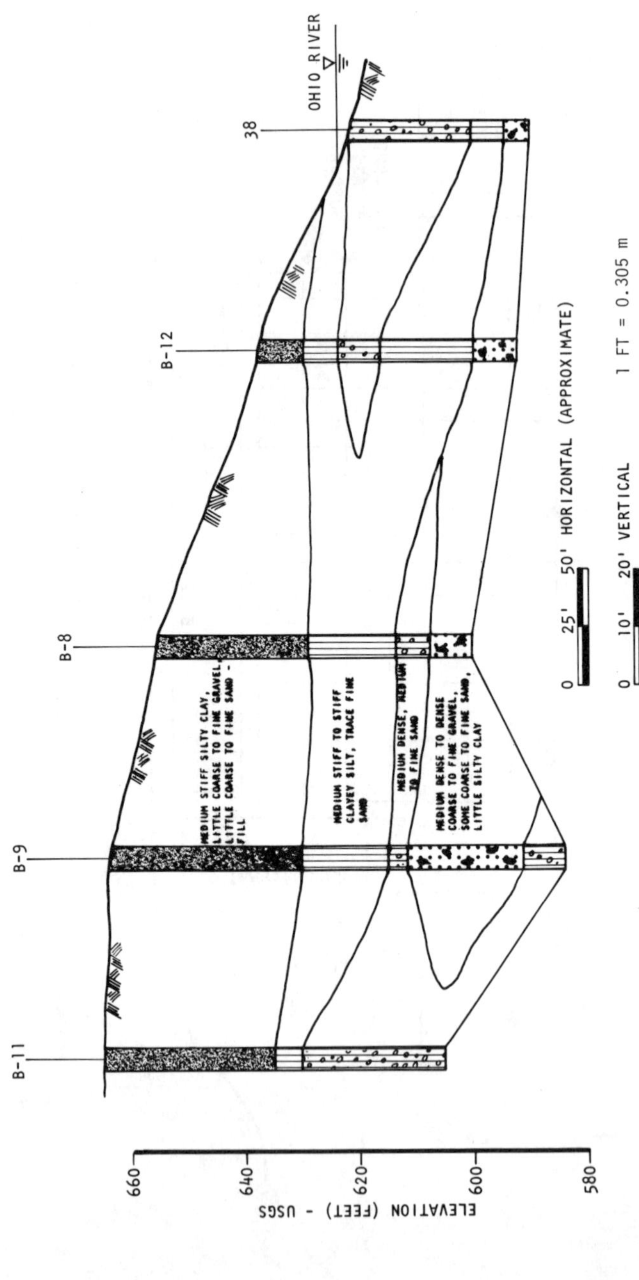

FIGURE 2, GENERALIZED SUBSURFACE SOIL CONDITIONS

The water table at this section of the Ohio River is approximately elevation +623 feet; however, the water table obtained in the observation wells varies in elevations, from 583 to 607, indicating that the ground-water table had been lowered by pumping from the two water supply wells within the plant.

Because of the heterogeneity of the fill material and the potential instability of the fill slopes in the plant, pile foundations were recommended for support of the major portion of the proposed facilities. The proposed facilities consisted of transmission towers, transformers, filter facilities and fan stacks. A total of approximately 1,000 piles was required for the entire project.

PILE AND HAMMER DATA

Ten-inch diameter (25.4 cm), one-quarter inch (6.4 mm) wall thickness, closed-end concrete filled high strength Grade 3 (Fy = 45 ksi) steel pipe piles were recommended to be used for supporting a downward load of 60 tons (534 KN), and an uplift load ranging from 15 tons (133 KN) to 25 tons (222 KN)per pile. It was recommended that piles be driven into the underlain natural dense sand and gravel layers using a single-acting steam or air hammer delivering not less than 19,500 foot-pounds (26.5 m-KN) of energy per blow.

In the production pile driving, both Grade 2 (Fy = 35 ksi) and Grade 3 steel pipe piles were used. In the early stage of construction the following hammers were provided by the contractor:

Make	Type	Rated Energy	Blows/ Minute	Boiler Pressure
Vulcan 06	Single Acting	19,500 ft-lb (26.5 m-KN)	60	100 psi (690 $KN/_m2$)
Vulcan 65C	Differential	19,200 ft-lb (26 m-KN)	117	150 psi (1035 $KN/_m2$)
Conmaco 65	Single Acting	19,500 ft-lb (26.5 m-KN)	60	100 psi (690 $KN/_m2$)

FIELD TEST PROGRAM

Initial analysis indicated that a pile length of approximately 60 feet (18.3 meters) was required to provide design uplift capacity. The piles were to be driven to a final resistance calculated according to Engineering News Formula to provide the required downward capacity. Test piles were driven at the site to estimate the average pile length required. Pullout load tests were performed on selected piles to confirm their pullout capacities. Compression load tests were considered. It was finally decided to utilize the Case method of dynamic pile analyses to estimate compressive pile capacity.

TEST DRIVING - Prior to the full production pile driving in the northern section of the site, a total of 11 piles along Column Line 3 in the Filter House Area (Figure 1) were chosen for test driving to estimate the required pile lengths in this area. These piles were driven using a Vulcan 06 hammer. Results indicated that piles in this area should be driven into the dense sand and gravel zone at approximately elevation 580 with a pile length of about 70 feet (21.3 meters).

PULLOUT LOAD TESTS - The results of analyses indicated that in most cases, the pile length is controlled by the design pullout capacity of the piles which ranged from 15 tons to 25 tons per pile. Load tests were performed on selected piles to confirm their pullout capacities. A total of 7 pullout load tests were performed on 4 piles at locations shown on Figure 1. The piles were load tested after a waiting period ranging from a few hours to as long as 46 days after they were driven. Results of these tests and other pertinent data are shown on Figure 3. It is seen from this figure that the pullout capacity of these piles increased as the elapsed time between driving and load testing increased, and eventually an average pullout resistance per unit length of pile on the order of 1.5 kips per foot (21.9 KN/m) to 1.6 kips per foot (23.4 $KN/_m$) was developed. This increase in skin friction of driven piles with elapsed time had been noted by other investigators (e.g., Terzaghi & Peck, 1967), and could be due to a combination of consolidation and thixotrophy. Figure 3 indicated that for the driven piles installed at the project site, close to 90 percent of the ultimate skin resistance was available about seven days after pile driving.

It is also observed from test results that the upward movement at failure load was small and ranged between 1/4 inch (6.4 mm) and 1/3 inch (8.5 mm). These relatively small movements required for full mobilization of skin friction are consistent with results reported by O'Neill and Reese (1972) and by Vesic (1967).

Based upon these load test data it was concluded that a minimum pile length ranging from 52 feet (15.9 meters) to 64 feet (19.5 meters) was required for a design pullout capacity ranging from 15-ton (133 KN) to 25-ton (222 KN) per pile, respectively, with an estimated safety factor of about 2 to 2.5.

HAMMER EFFICIENCY EVALUATION BY DIRECT COMPARISON OF DRIVING RECORDS

DEFICIENCY REVEALED BY DRIVING RECORDS - After the test driving of 11 piles using a Vulcan 06 hammer, the remaining piles along Column Line 3 were later driven by Vulcan 65C hammer and a Conmaco 65 hammer. Examination of the pile driving records revealed that substantially higher blows/foot were required to drive these piles to a comparable depth than the original 11 piles. The blows/foot were higher by a factor of 2 to 3 for the Vulcan 65C hammer and by a factors of 1.5 for the Conmaco 65 hammer (see Figure 4 and 5). These records led to the belief that the latter two hammers were deficient and prompted a comprehensive evaluation in order to establish reliable driving criteria for these two hammers without delaying the production pile driving. These objectives were achieved by the following evaluation program executed immediately following the revelation of probable hammer deficiences.

EVALUATION OF VULCAN 65C HAMMER - In order to check the driving characteristics of the Vulcan 65C hammer with respect to the Vulcan 06 hammer, the following steps were used to provide a direct comparison of the driving characteristics of these two hammers.

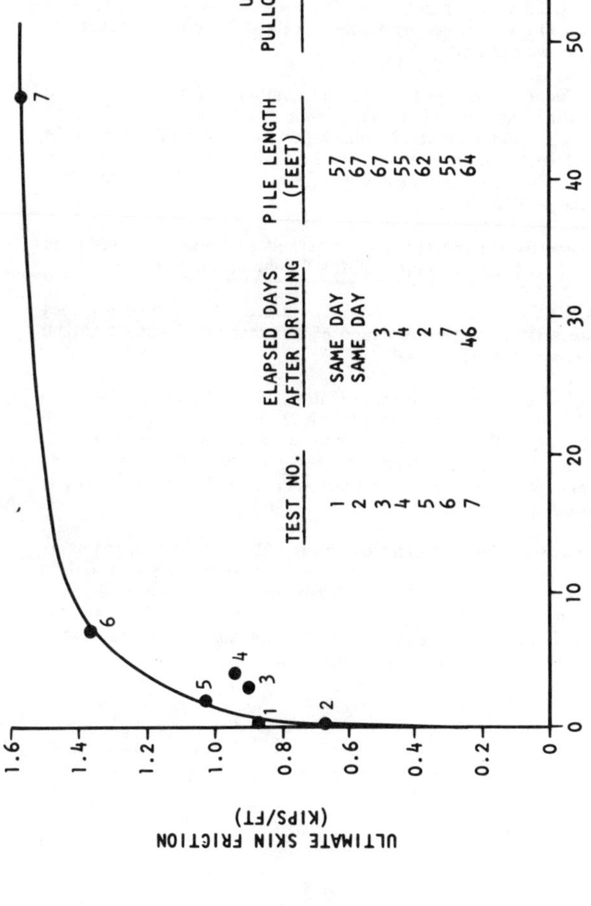

TEST NO.	ELAPSED DAYS AFTER DRIVING	PILE LENGTH (FEET)	ULTIMATE PULLOUT CAPACITY (TONS)	UPWARD MOVEMENT AT ULTIMATE LOAD (INCH)
1	SAME DAY	57	25	0.22
2	SAME DAY	67	22.5	0.28
3	3	67	30	0.28
4	4	55	26	0.22
5	2	62	32	0.25
6	7	55	37.5	0.33
7	46	64	50	0.27

1 FT = 0.305 m
1 IN. = 25.4 mm
1 TON = 8.9 kN
1 KIP/FT = 14.6 kN/m

FIGURE 3, RESULTS OF PULLOUT LOAD TESTS

(DIAMETER OF PILE =10 INCHES)

1. Drive Pile #343 at Location P-4, approximately 25 feet (7.6 meters) south of pile group P-3, utilizing the Vulcan 65C hammer from the ground surface to 35-foot depth. The recorded resistance for 32 feet (9.8 meters) to 35 feet (10.7 meters) were 56, 62 and 70 blows per foot. The hammer was then changed to a Vulcan 06 and pile driving continued. The recorded resistance to penetrate from depths of 36 feet (11 meters) to 38 feet (11.6 meters) were 35, 31 and 33 blows per foot. This pile was driven to a final depth of 55 feet (16.8 meters) with a final resistance of 61 blows per foot.

2. Retapping Pile #278 at P-3 location utilizing the Vulcan 65C hammer. This Pile #278 had been driven utilizing a Vulcan 06 hammer. A final resistance of 52 blows per foot had been achieved at a depth of 74 feet (22.6 meters). A resistance of 130 blows per 6 inches was recorded in driving this pile from depths of 74 feet (22.6 meters) to 74.5 feet (22.7 meters) with the Vulcan 65C hammer.

3. Drive Pile #274 deeper utilizing a Vulcan 06 hammer. This pile had been driven one day earlier utilizing the Vulcan 65C hammer and was stopped at 50 feet (15.2 meters) with a final resistance of 120 blows per foot. A resistance of 59 blows per foot driving from 50 feet (15.2 meters) to 51 feet (45.5 meters) was recorded with the Vulcan 06 hammer. This pile was driven to a final depth of 75 feet (22.9 meters) with a final resistance of 57 blows per foot.

4. Drive Pile #275 deeper utilizing a Vulcan 06 hammer. This pile also had been driven one day earlier utilizing the Vulcan 65C hammer and was stopped at 47 feet (14.3 meters) with a final resistance of 128 blows per foot. A resistance of 98 blows per foot driving from 47 feet (14.3 meters) to 48 feet (14.6 meters) was recorded with the Vulcan 06 hammer. This pile was driven to a final depth of 68 feet (20.7 meters) with a final resistance of 52 blows per foot.

All the driving records are plotted and presented on Figure 4. Based on these results, it was concluded that the efficiency of the Vulcan 65C hammer was low.

EVALUATION OF CONMACO 65 HAMMER - In order to provide a direct comparison of driving characteristics of the Conmaco 65 hammer with respect to the Vulcan 06 hammer, the following steps were used:

1. Piles #201 and #203 at the B-3 location (Piles #199 and #200 in the same pile cluster had been test driven previously to depths of 73 feet (22.3 meters and 76 feet (23.2 meters), respectively, with a final resistance of 54 and 61 blows per foot, respectively, utilizing a Vulcan 06 hammer) were both driven utilizing the Conmaco 65 hammer to a depth of 75 feet (22.9 meters); final resistance of 113 and 99 blows per foot, respectively, were recorded.

2. Piles #216 and #218 at the E-3 location (Piles #214 and #215 in the same pile cluster had been test driven previously to depths of 72 feet (22 meters) and 74 feet (22.6 meters), respectively, utilizing a Vulcan 06 hammer with a final resistance of 56 and 56 blows per foot, respectively) were driven utilzing the Conmaco 65 hammer to depths of 72 feet (22 meters) and 73.5 feet (22.4 meters), respectively, with a final resistance of 82 and 88 blows per foot, respectively.

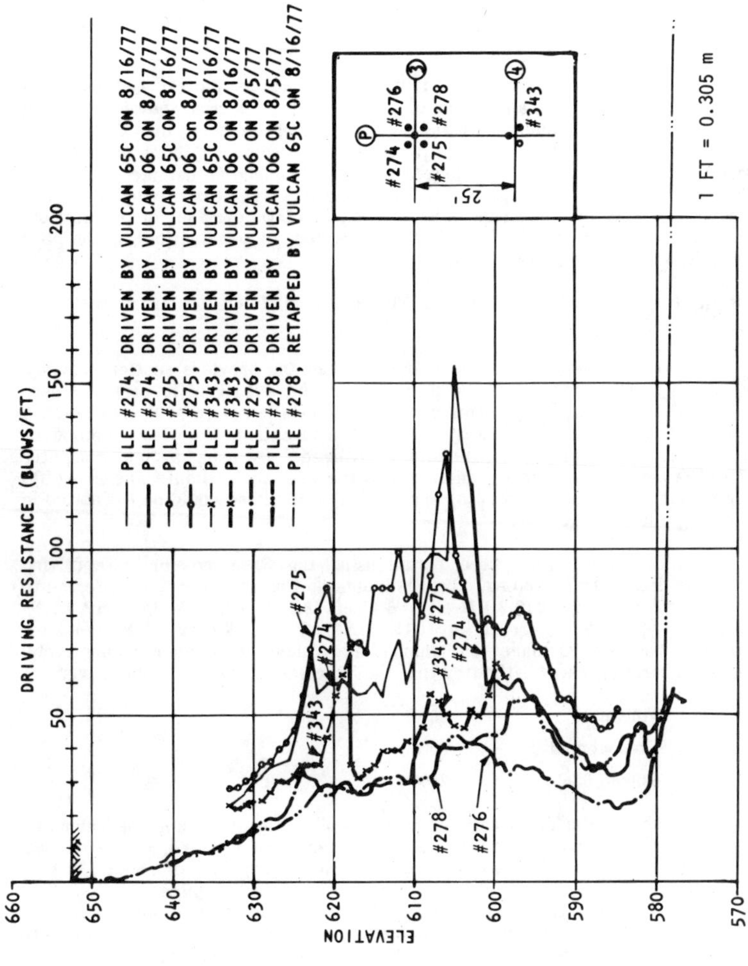

FIGURE 4. EVALUATION OF VULCAN 65C HAMMER

6

The driving records of these four-pile clusters are plotted and shown on Figure 5 for comparison. Examination of these driving records showed that the number of blows using a Conmaco 65 hammer was in general greater than that using a Vulcan 06 hammer by approximately 50 percent at the same depth of penetration. Based upon these observations, it was judged that the Conmaco 65 hammer was not as efficient as the Vulcan 06 hammer. It was also concluded that the driving resistance criteria developed for the Vulcan 06 hammer (i.e., 53 blows per foot of penetration) should be increased by 50 percent (i.e., 80 blows per foot of penetration), if the piles were driven using the Conmaco 65 hammer.

HAMMER EFFICIENCY EVALUATION BY THE CASE METHOD OF DYNAMIC PILE ANALYSIS

The driving test performed as described in the previous section provided only the relative efficiency of different hammers; the actual efficiency of these hammers and the downward capacities of the piles installed were still unknown. A downward pile load test was considered. As an alternative, the Case method of dynamic pile testing (Goble, G.G., 1970) was utilized. By using a field computer attached to the tester, the energy transferred to the pile, the ultimate pile capacity, and the maximum compressive force of impact at the pile head could be estimated in the field during pile driving.

A total of 10 piles were tested using the Case dynamic pile driving analyzer during driving or retapping. Results indicated that the energy transferred to the piles ranged from 5 kip-ft (6.8 KN-m) to 6 kip-ft (8.1 KN-m), 9.5 kip-ft (12.9 KN-m) to 15 kip-ft (20.4 KN-m), and 10 kip-ft (13.6 KN-m) to 18 kip-ft (24.4 KN-m) for the Vulcan 65C hammer, Conmaco 65 hammer, and Vulcan 06 hammer, respectively. Based upon test results, it was estimated that the efficiency of these hammers were:

Vulcan 06	-	60%
Conmaco 65	-	50%
Vulcan 65C	-	25%

The test results also indicated that the piles installed using the Vulcan 06 and Conmaco 65 hammers and were driven to resistances of approximately 53 blows per foot and 80 blows per foot, respectively, an ultimate capacity of about 120 tons (1068 KN)or higher were achieved at their final depth, while two piles installed using the Vulcan 65C hammer with final resistance of 123 blows per foot of penetration and 40 blows per 6 inches (15.2 cm) of penetration, respectively, indicated ultimate capacities achieved were about 120 tons (1068 KN) and 93 tons (823 KN), respectively.

CONCLUSIONS

Results of the pullout tests indicated that the skin friction of the piles increased with the elapsed time following pile driving. They also indicate that full skin friction was mobilized when the upward pile movement was about 1/4 inch (6.4 mm) to 1/3 inch (8.5 mm).

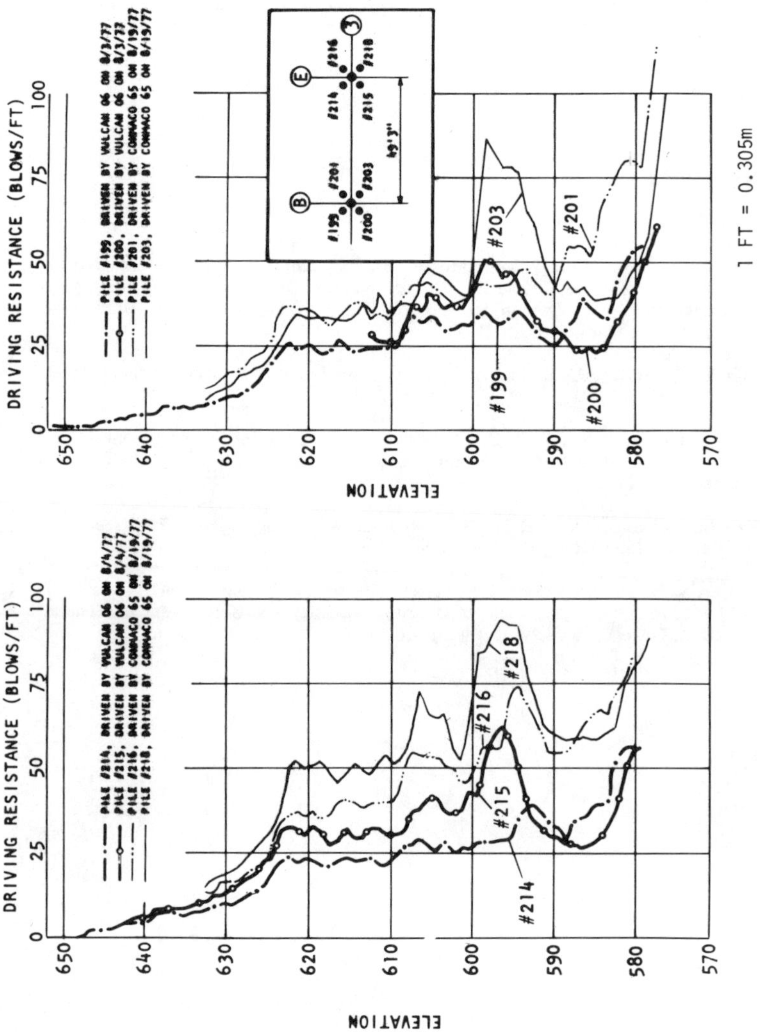

FIGURE 5. EVALUATION OF CONMACO 65 HAMMER

The efficiency of various pile driving hammers could be very low. For the three hammers evaluated in this study, the efficiency ranged from 25 percent to 60 percent, compared with the generally assumed value of about 80 percent.

The Case method of dynamic pile analysis indicated that the piles installed would develop the downward design capacities at an adequate factor of safety.

APPENDIX - REFERENCES

1. Goble, G.G., Rausche, F. and Moses, F., "Dynamic Studies on the Bearing Capacity of Piles, Phase III," Report No. 48, Division of Solid Mechanics, Structures and Mechanical Design, Case Western Reserve University, Cleveland, Ohio, August, 1970.

2. Goble & Associates, "A Report on Pile Dynamic Testing" Report for this project, 1977.

3. O'Neill, M.W. and Reese, L.C. "Behavior of Bored Piles in Beaumont Clay," Journal of the Soil Mechanics and Foundations Division ASCE, SM2, February 1972.

4. Terzaghi, K. and Peck, R.B. "Soil Mechanics in Engineering Practice," 2nd Edition, John Wiley, New York, N.Y. 1967.

5. Vesic, A.S. "Ultimate Loads and Settlements of Deep Foundations in Sand," in Proceedings of Symposium of Bearing Capacity and Settlement of Foundations, Duke University, Durham, N.C. 1967.

MEASURED RESPONSE OF

PILE SUPPORTED MACHINES

Gerald M. Diaz,[1] M. ASCE

INTRODUCTION

Large reciprocating compressors are common along pile lines and within industrial plants. These large machines can create significant levels of vibration and the foundation must be designed to minimize these vibrations.

Much work has been done to develop rational analysis techniques for the design of mat supported machine foundations. However, the analysis of pile supported machine foundations had to rely on approximate techniques which required simplifying assumptions for the effects of the soil-pile, pile group, and soil-cap behavior. Often, the simplifying assumptions result in a nonrational analysis. It is fortunate that in most cases the behavior of pile supported machine foundations has been adequate, given the techniques used.

Rational techniques have recently been proposed (1, 3, 4, 5, 6, 7, 8, 9) to estimate the pile-cap-soil impedence functions (stiffness and damping) for use in future analysis. However, these methods have only been checked by relatively small scale tests (4).

The purpose of this paper is to provide soil and vibration data from actual large operating pile supported machines to aid other researchers in checking their proposed design techniques. The occasion of the study was fortuitous in that three similar plants, each using at least two large four cylinder in-line vertical reciprocating compressors were to be constructed within a short time of each other in East Texas and Southern Louisiana. The soil investigations were planned to include measurements of dynamic properties of the soil (using field geophysical techniques) as well as conventional

[1]Associate-in-Charge, Southern California
Harding-Lawson Associates, Tustin, California

soil investigation data. The soil conditions at the sites were such that all sites required pile foundations. However, because of variation of the site conditions, a different pile type was used at each location.

Measuring the vibrations of the similar machines supported on different types of pile foundations provides a broad data base for checking new design techniques.

SITE CONDITIONS AND FOUNDATION TYPES

For ease of reference, the three sites are referred to in this paper by the name of their nearest town: Norco, and Plaquemine located in Southern Louisiana between Baton Rouge and New Orleans and Channel View located just east of Houston, Texas .

At the time of the study, all of the plants were in operation. All of the compressors were located within compressor buildings, which housed several other compressors of varying types.

The soil conditions at each site were investigated by borings advanced by conventional rotary wash drilling techniques. Soil samples were obtained using Sprague and Henwood split barrel samplers, and shelby tubes. Dynamic soil properties were determined from field measured shear and compressional wave velocities. Both down hole and surface seismic techniques were used. Generalized soil conditions at each site are indicated by soil profiles and data presented in Figures 1 through 3. A brief description of the conditions at each site and the foundation type used is as follows:

Norco

The Norco site is located just south of the Bonne Carre spillway, between the Mississippi River and Lake Ponchatrain. It is a marsh area which has recently been reclaimed by draining.

The natural soil conditions consist of an upper layer of medium stiff clay. This upper layer is underlain to a depth of about 50 feet (16.4 m) by soft and compressible clay which contains frequent lenses of silt. Stiff clays encountered below

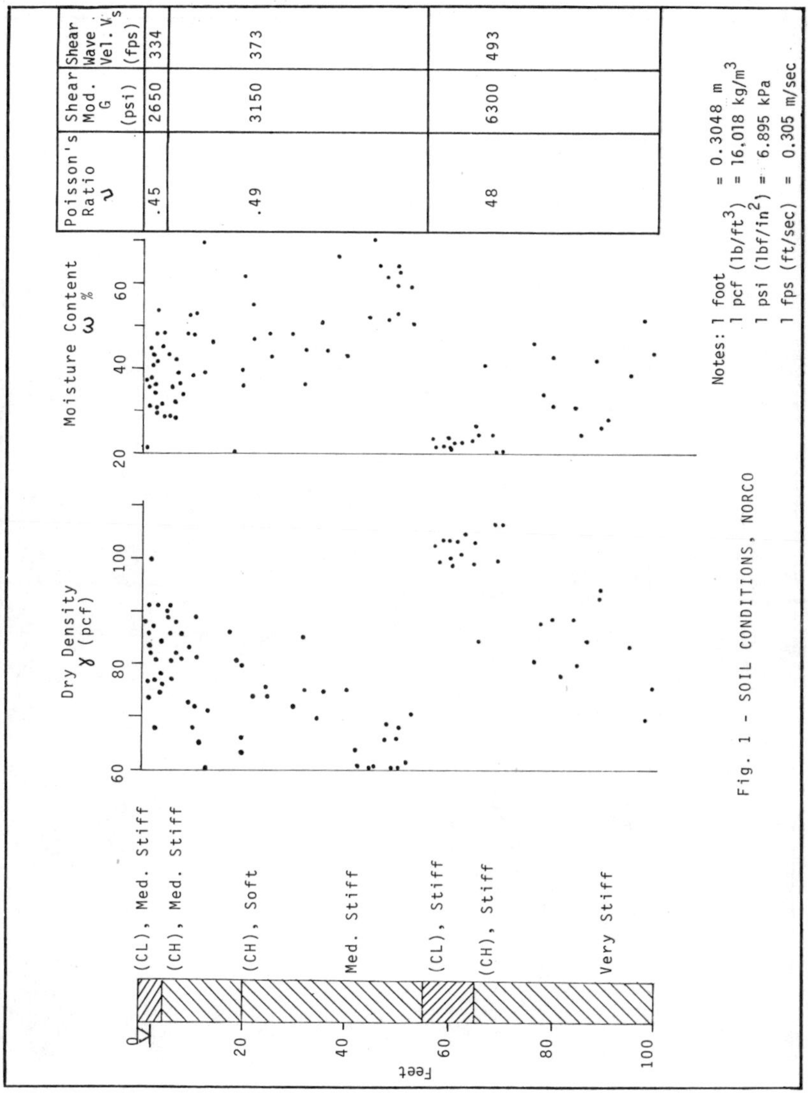

Fig. 1 - SOIL CONDITIONS, NORCO

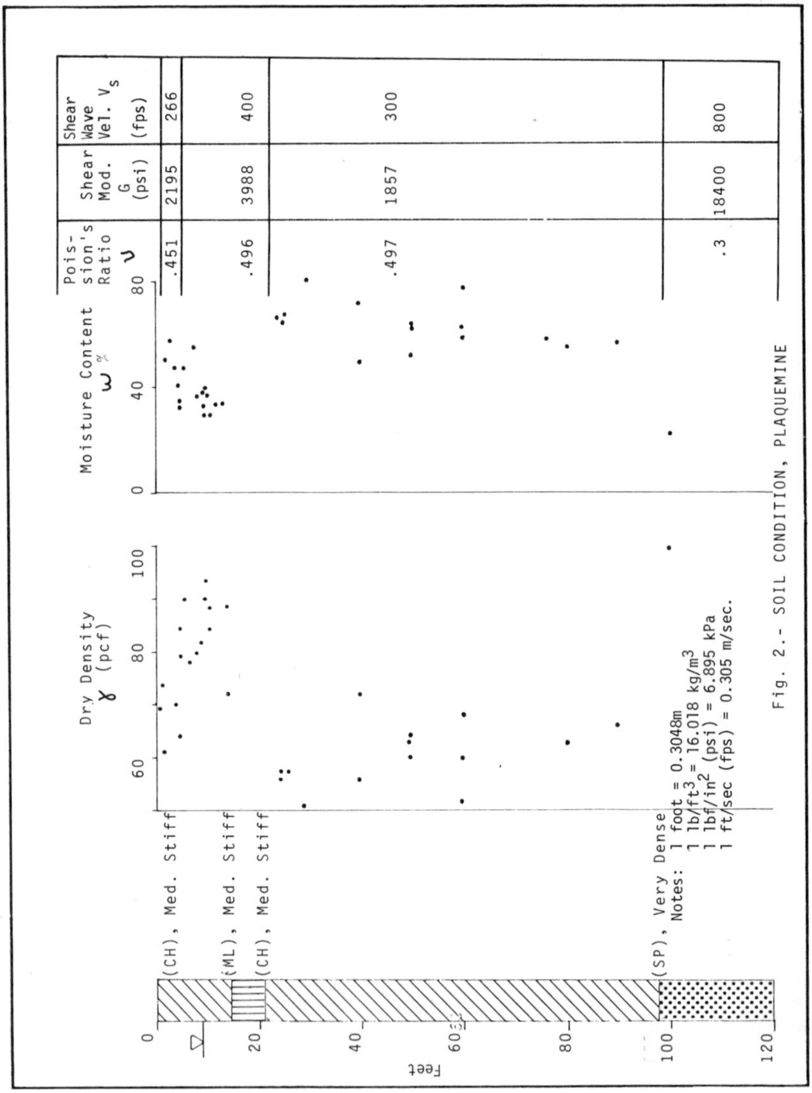

Fig. 2.- SOIL CONDITION, PLAQUEMINE

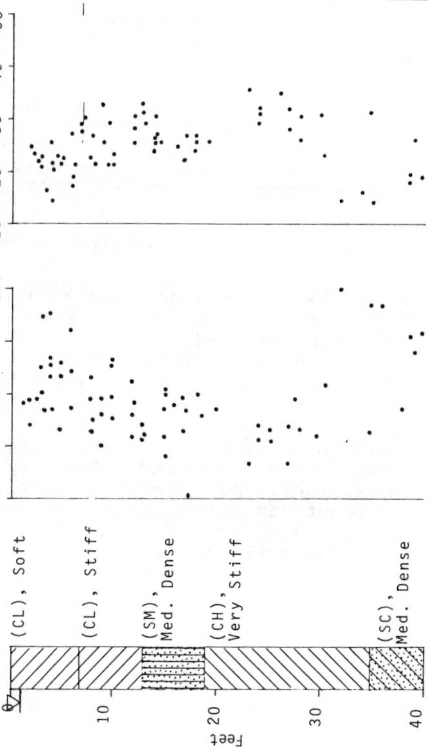

Poisson's Ratio ν	Shear Mod. G (psi)	Shear Wave Vel. V_s (fps)
.413	6530	500
.489	11100	650

Notes: 1 foot = 0.3048m
1 pcf (lb/ft^3) = 16.018 kg/m^3
1 psi (lbf/in^2) = 6.895 kPa
1 fps (ft/sec) = 0.305 m/sec

Fig. 3 - SOIL CONDITIONS, CHANNEL VIEW

55 feet (18 m). Groundwater levels were about one to four feet (.3 to 1.3 m) beneath the original ground surface.

The site was developed by adding approximately seven feet (2.3 m) of silty sand fill. The foundations were designed on the basis of friction piles gaining support in the stiff clays below 55 feet (18 m). Allowance was made in the design for down-drag loads caused by the new fill.

Plaquemine

The Plaquemine site is located near the west bank of the Mississippi River on old plantation land, developed by draining and clearing the higher ground adjacent to the river by constructing canals perpendicular to the river.

The soils generally consist of soft to medium stiff clay deposits to a depth of 95 to 100 feet (31 to 33 m). The clay contains numerous bits of organic matter. In the upper 10 to 15 feet (3.3 to 4.9 m) the clays are slightly overconsolidated, probably due to desiccation. The clay is underlain by very dense fine sand and silty sand which extended to the 127 foot (41.7 m) depth explored.

The water level was about three to four feet (1 to 1.3 m) below the natural ground surface. Driven piles gaining support by endbearing in the dense sands at a depth of approximately 100 feet (33 m) were used. The pile design considered the effects of down-drag. To reduce these effects, only minimum site grading was performed.

Channel View

The Channel View site is located just east of Houston, Texas, in the Texas Gulf Coastal Plain. The predominant underlying soils are stiff overconsolidated clays of the Beaumont formation. The ground surface is relatively flat. A small pond three feet (1 m) deep was located in the compressor building area.

The upper soils from the surface to depths of 13 to 17 feet (4.2 to 5.6 m) consist of silty clays. The clays are nonhomegenous, wet, compressible and relatively weak. Layers of gravelly clay were encountered in the upper six feet (2.0 m) and pockets of very soft silty clay were encountered throughout these upper soils. Competent soils consisting of medium dense silty sands, sandy clays and stiff clays were encountered below depths of 13 to 17 feet (4.2 to 5.6 m)

Groundwater levels were observed to be within one foot (.3 m) of existing ground surface.

The foundation system consists of drilled, cast-in-place concrete piles, gaining support by friction in the stiff clays below a depth of about 17 feet (5.6 m). The pond was filled and site grade was raised by filling two feet (.6 m) for drainage purposes. The piles design allowed for down-drag effects.

COMPRESSOR AND FOUNDATION CHARACTERISTICS

Norco, Plaquemine, Channel View

The Norco, Plaquemine and Channel View compressors are identical Sulzer, vertical, four cylinder, in-line reciprocating compressors (similar in configuration to most four cylinder automotive engines). Figure 4 is a photograph of one of the compressors. At the time of the studies, the compressor at Norco had been modified to operate on only three cylinders by removing the piston skirts from the first stage cylinder. The compressors are driven by large electric motors mounted on the foundation mats. The compressor characteristics are shown in Figure 5.

The compressors are supported by a cradle resting directly on the top of the pile cap. The pile caps and pile configurations are shown in Figure 6.

All of the pile caps are seven feet (2.3 m) thick and are embedded about six feet (2.0 m). The backfill is compacted sand.

FIELD VIBRATION MEASUREMENTS

Vibration Monitor

The vibration of each foundation was measured while the compressors were in normal operation. The vibrations were measured with a SINCO Model S-3 Vibration Monitor which uses pairs of three directional velocity transducers and a six channel recorder. The transducers have a natural frequency of 4.5 hertz. The recorder incorporates integrating and

Fig. 4 - <u>COMPRESSOR</u> - Norco, Plaquemine, Channel View

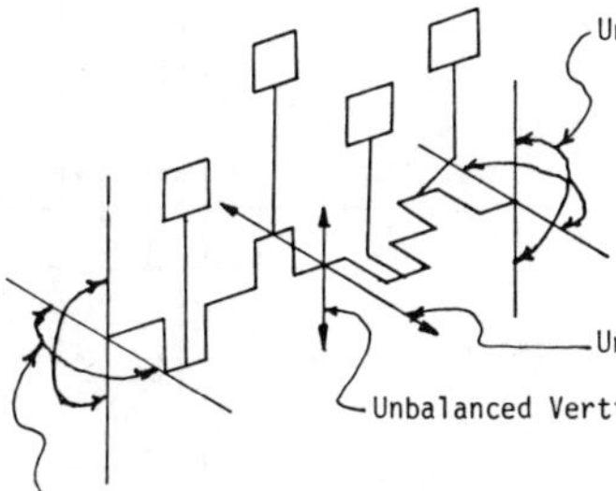

	Channel View & Plaquemine	Norco
No. of Operating Cylinders	4	3
Operating Speed (RPM)	360	360
Unbalanced Forces (lbs.)		
1st Order Vertical	4,980	6,440
2nd Order Vertical	2,415	1,635
1st Order Horizontal	0	0
Unbalanced Moments (ft.-lbs.)		
1st Order Vertical	259,000	252,300
2nd Order Vertical	94,000	91,000
1st Order Horizontal	95,500	95,700
Weights (lbs.)		
Compressor (w/pedestal)	133,000	132,700
Electric Motor (less pit)	30,500	30,500
Accessories	19,200	19,200
Foundation Material	1,106,000	1,106,000

Notes:
1 lb (lbf) = 4.448 N 1 lb = 0.4536 kg
1 ft-lb = 1.355 N·m

FIG. 5 - COMPRESSOR CHARACTERISTICS - Norco, Plaquemine, Channel View

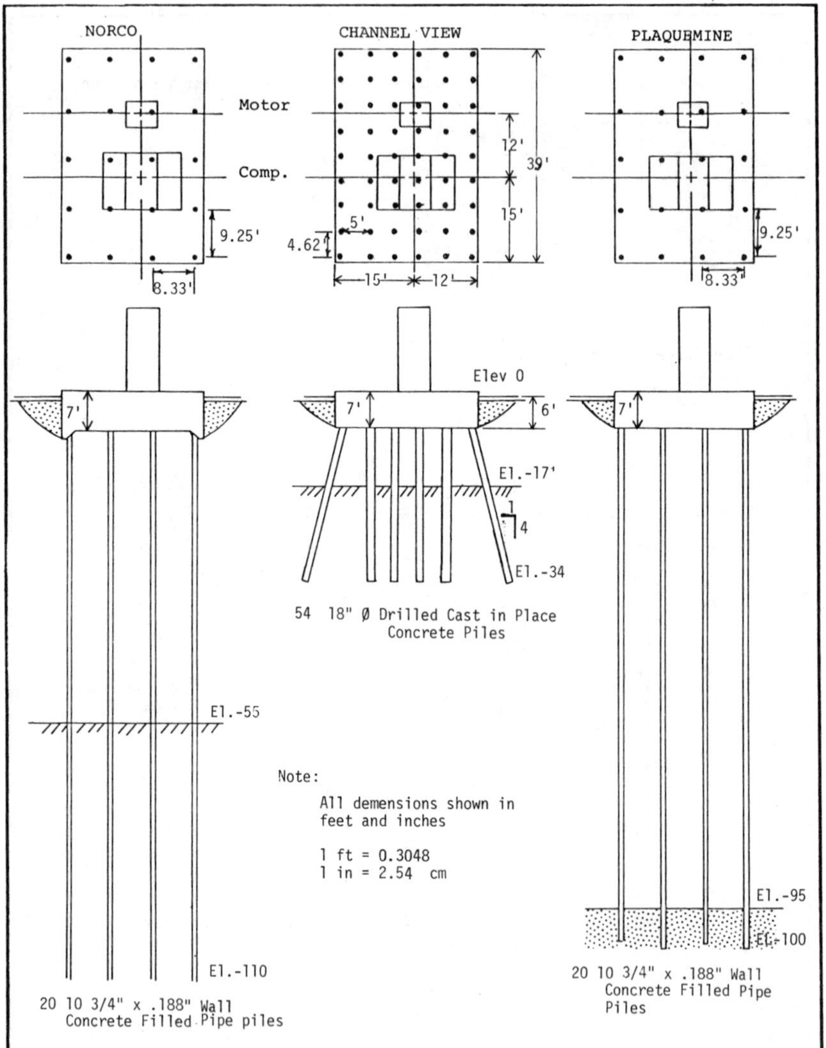

Fig. 6 - FOUNDATIONS: NORCO, PLAQUEMINE, CHANNEL VIEW

differentiating circuits which allow direct recording for either displacement, velocity or acceleration.

The six channel recorder had provisions for varying both chart speed and amplitude scale. After calibrations, each measurement was taken at a scale sensitivity of .01 inches per inch (0.01 mm/mm) to identify the overall pattern of movement. The scale was then switched to the maximum sensitivity which was .002 inches per inch (0.002 mm/mm) to enable more accurate determination of the amplitudes of vibration. Chart speeds of 5 inches per second and 10 inches (13.7 to 27.3 cm) per second were found to be satisfactory with five inches (13.7 cm) per second used for most of the readings.

Sensor Placement

The amplitudes of displacement were measured and recorded in three directions at two locations simultaneously. Placing pairs of transducers parallel to the long axis and parallel to the short axis enabled determination of the modes of vibration as well as the amplitude.

The sensors were used in pairs. The position of each reading was numbered and the sensor boxes labeled A and B to identify the position from which readings were taken. As an example, positions 1A and 1B would represent one pair of simultaneous readings, as would 2A and 2B, etc. The locations on the foundation at which readings were taken are shown on Figure 7.

Each three dimensional velocity transducer was housed in an aluminum box and constitutes a single sensor. The sensor boxes were drilled and tapped so that various brackets could be used to fasten the sensors to the foundations. However, it was found that satisfactory coupling of the sensors could be achieved by simply placing a 40 to 50 pound (18.1 to 22.7 kg) sand bag directly over the sensor. This method was used throughout including placing sensors on the foundation mat as well as on the cylinder heads and crank cases of the machines.

RESULTS

The guide to field records interpretation is shown on Figure 7. The results of the field measurements are shown in Figures 8 through 16.

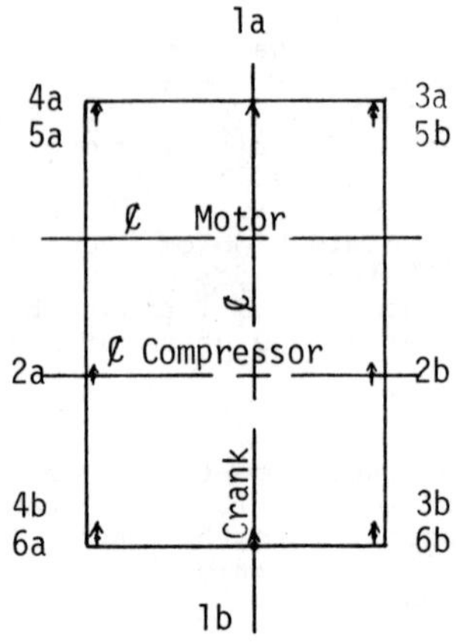

SENSOR LOCATIONS

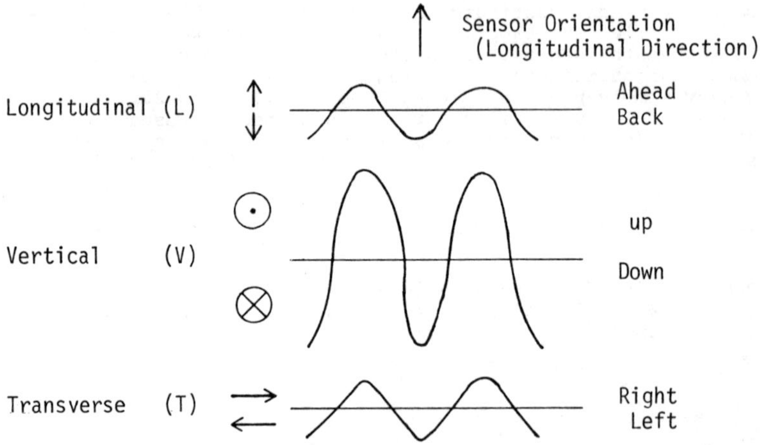

Fig. 7 - RECORD INTERPRETATION - Norco, Plaquemine, Channel View

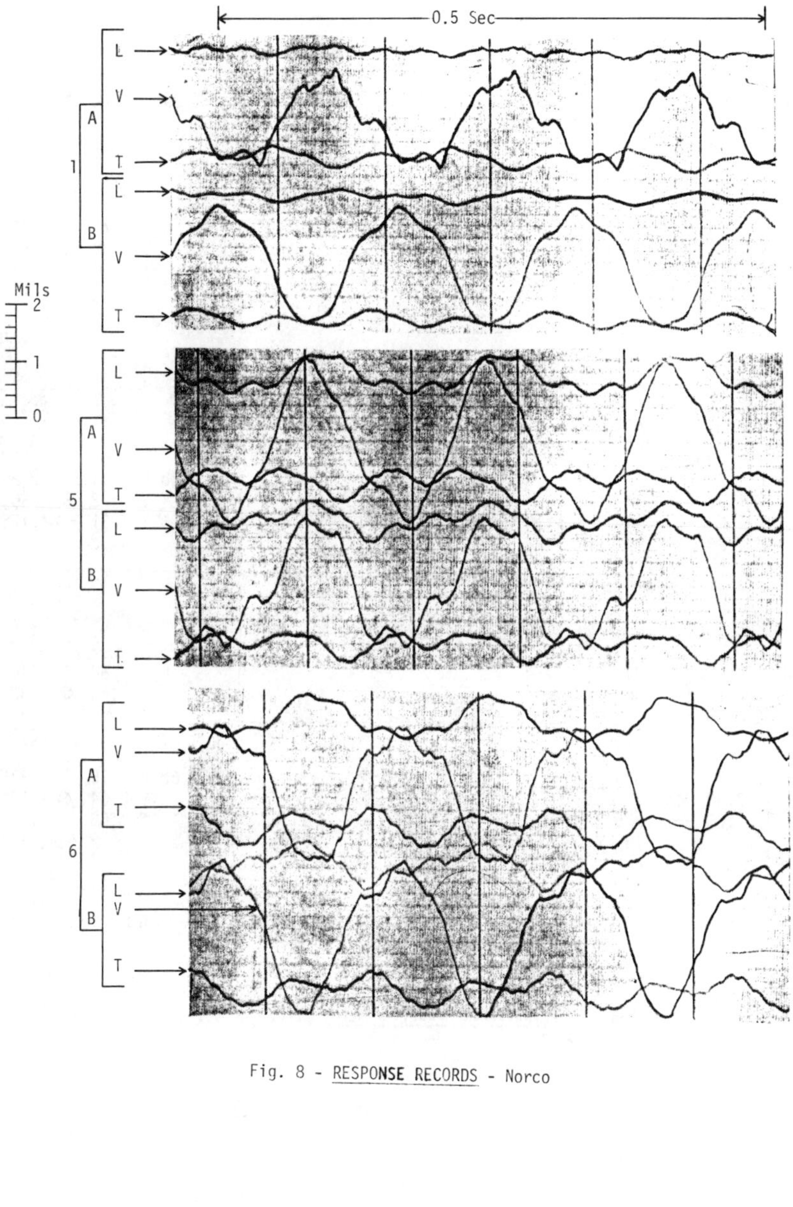

Fig. 8 - RESPONSE RECORDS - Norco

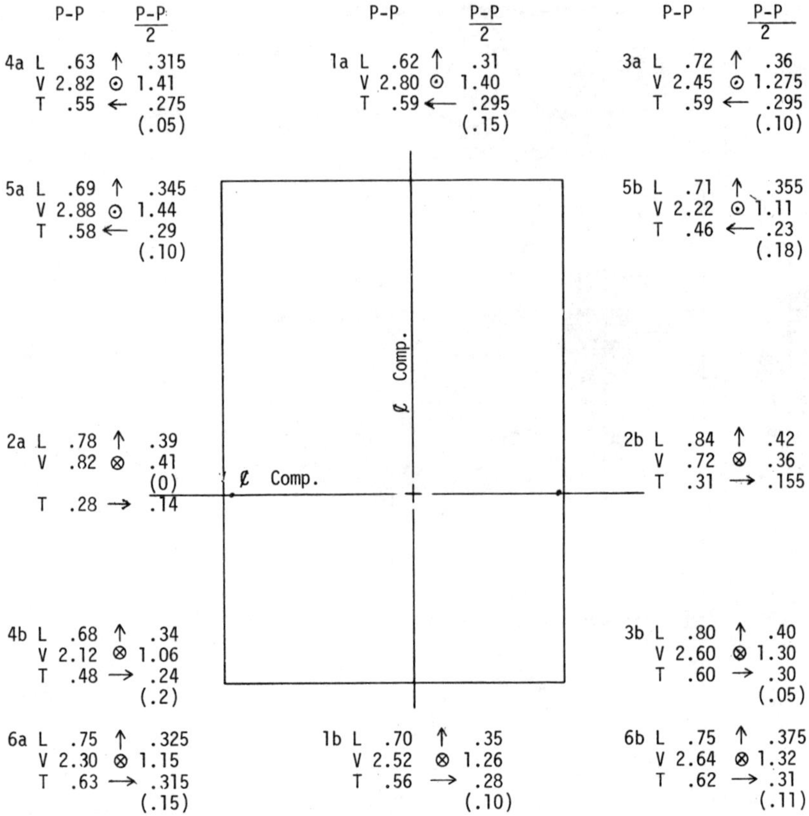

Notes:

1. P-P = Peak to Peak
2. Where there is a phase shift, the in-phase displacements are noted ( )
3. 1 mil = 0.0254 mm

Fig. 9 - MEASURED DISPLACEMENT (MILS) - Norco

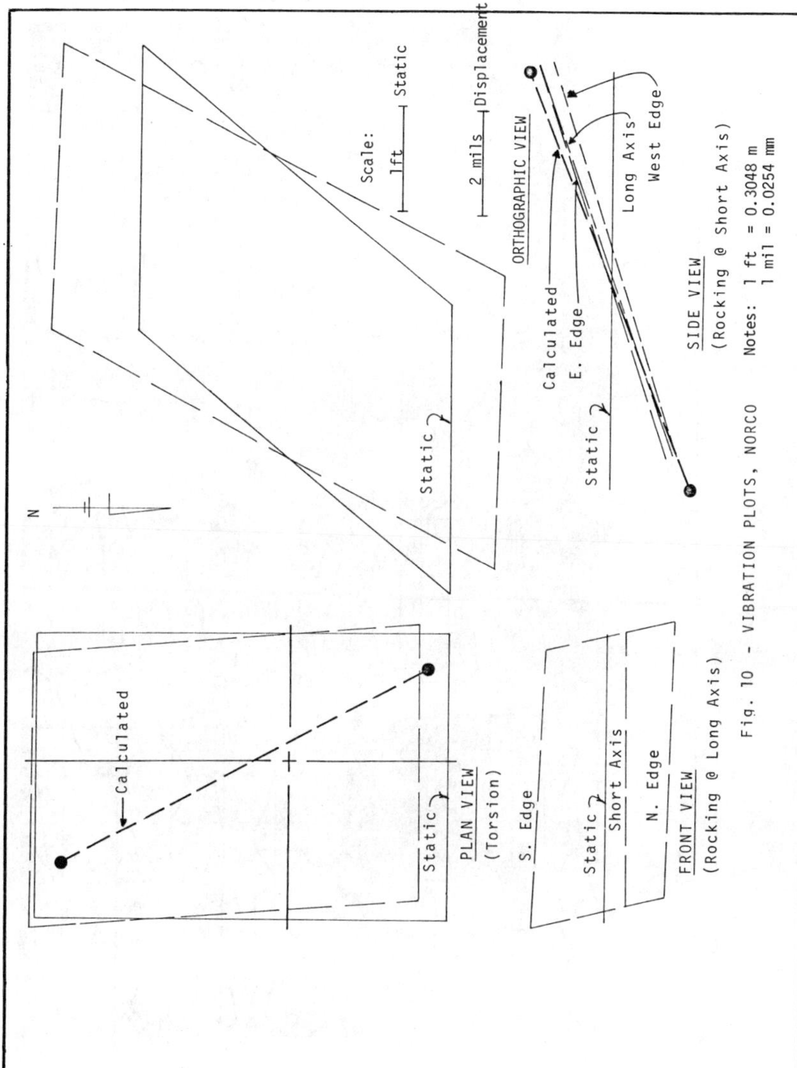

Fig. 10 - VIBRATION PLOTS, NORCO

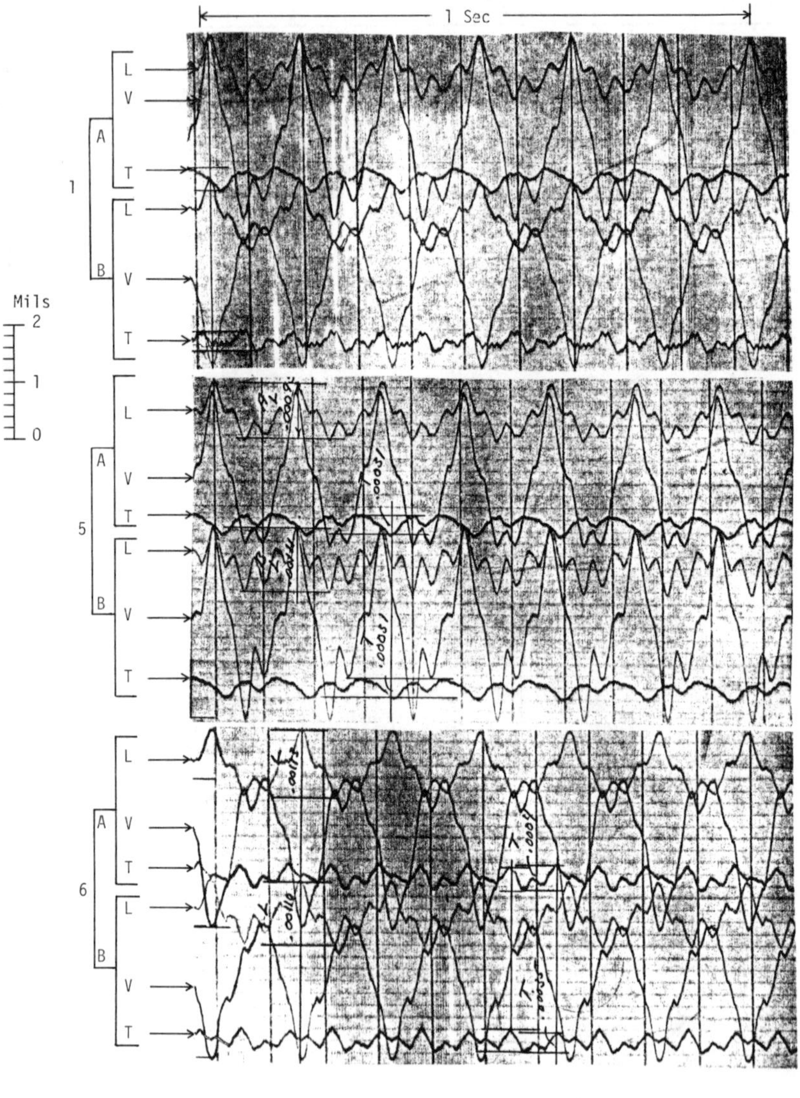

Fig. 11 - RESPONSE RECORDS - Plaquemine

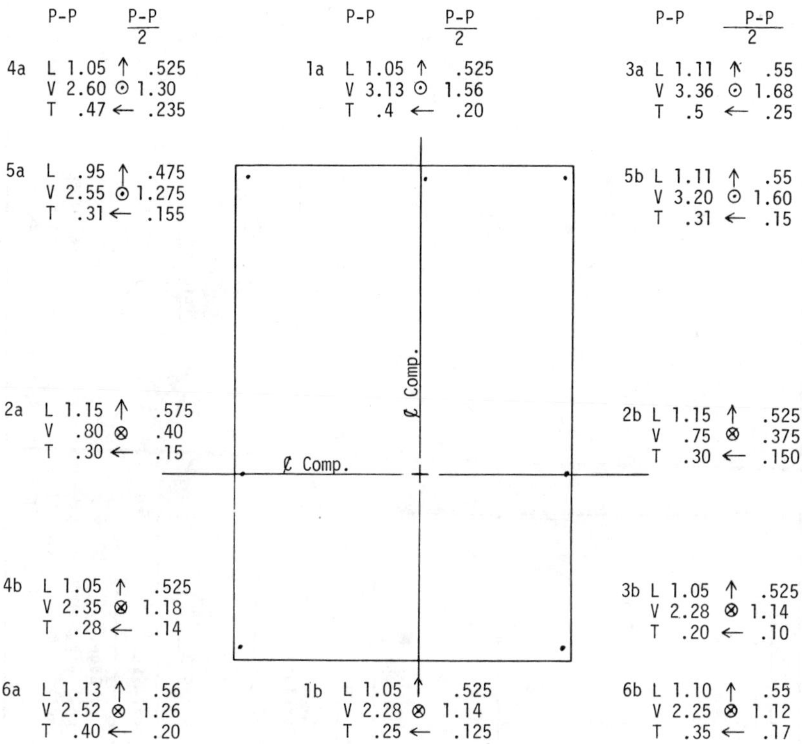

Note : 1 mil = 0.0254 mm

Fig. 12 - MEASURED DISPLACEMENT (MILS) - Plaquemine

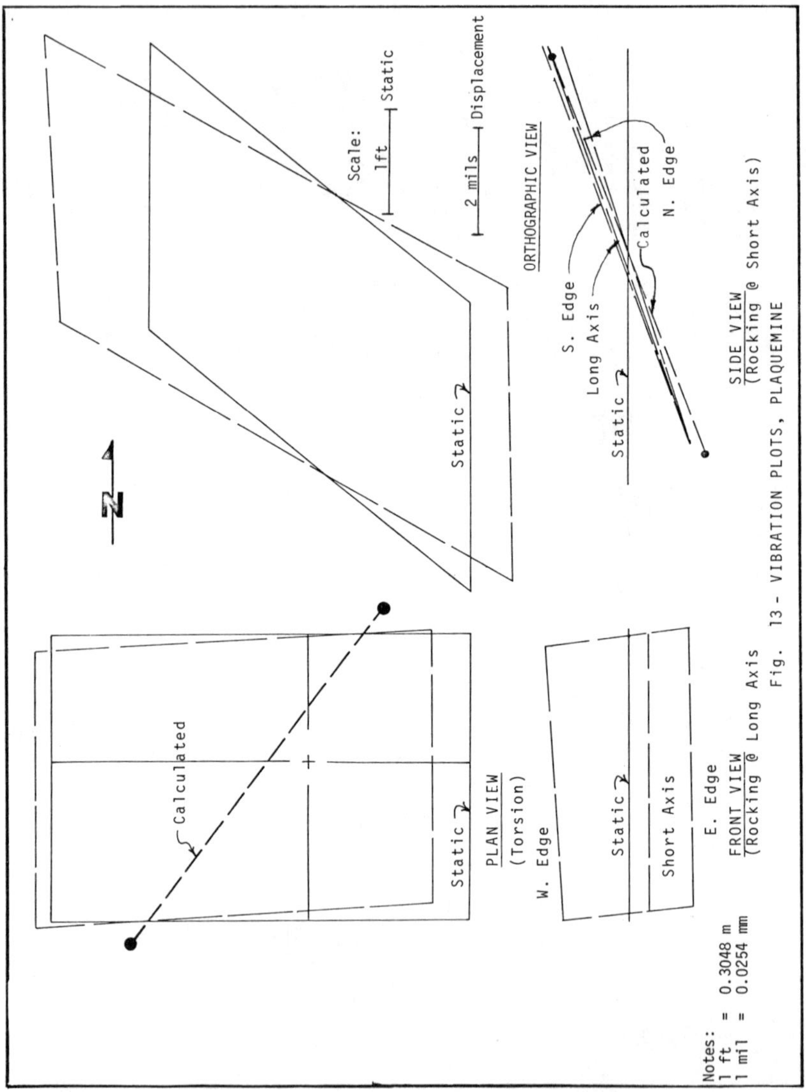

Fig. 13 - VIBRATION PLOTS, PLAQUEMINE

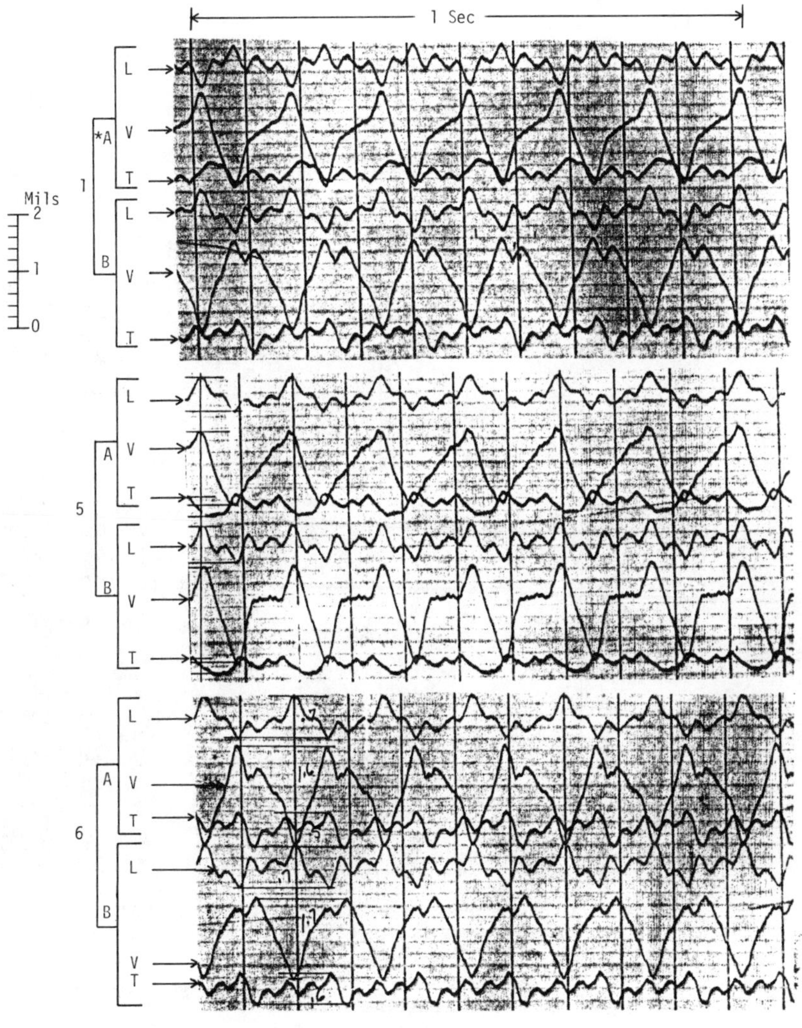

* Orientation Reversed

Fig. 14 - RESPONSE RECORDS - Channel View

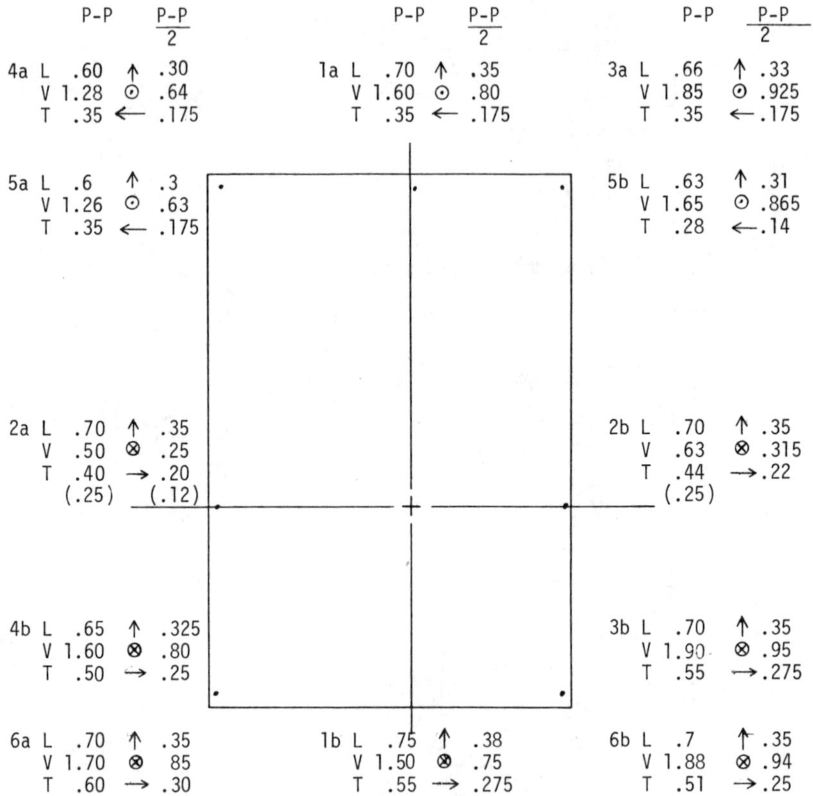

Notes: 1. Where there is a phase shift, the in-phase displacements are noted ( )
2. 1 mil = 0.0254 mm

Fig. 15 - MEASURED DISPLACEMENT (MILS) - Channel View

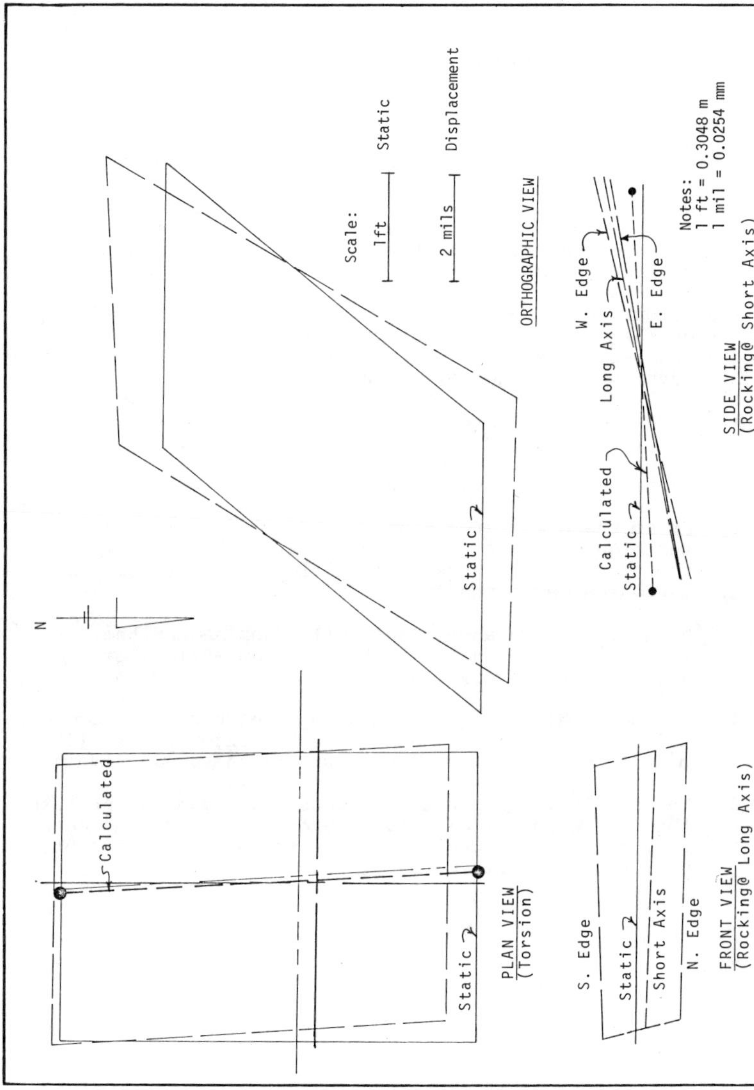

Fig. 16 - VIBRATION PLOTS, CHANNEL VIEW

For each compressor measured, the following are shown:

1. Copies of the field vibration records for typical readings parallel to the long and short axis for each compressor.

2. Measured amplitudes of vibration in three directions at each sensor location.

3. Scale plots of vibration amplitude in plan, front, side and orthographic views, to show graphically the modes of vibration.

4. Calculated response based on a proposed combination (3) of recently developed analysis techniques are shown on the scale plots to indicate how the data could be used.

The calculated response shown was based on a proposed combination (3) of recently developed analysis techniques. The method of analysis used was described in detail in the referenced paper and consisted of the following five steps:

1. Determine the pile-soil stiffness and damping of a single pile using the soil and pile properties determined in the field and laboratory, and procedures proposed by M. Novak, and Aboul-Ella (1, 11).

2. Using the calculated single pile stiffnesses and the techniques proposed by Reese et al (12), determine the overall stiffness of the pile group for the various modes of movement.

3. Using the calculated single pile damping and the formulae presented by Novak, and Aboul-Ella (11), determine the overall damping of the pile group.

4. Estimate the stiffness and damping effects of embeddment using the procedure proposed by Beredugo & Novak (2). add all these to the pile group stiffness and damping.

5. The resulting combined stiffnesses and damping are used together with the given unbalanced forces and moments to compute the dynamic response for various modes of vibration in terms of displacement.

The computed amplitudes of vibration at the machine operating speeds are then compared to the actual measured amplitudes.

CONCLUSIONS

Measured Response

The vibration measurements presented provide field observational data of actual machines supported on a variation of pile types. The foundation, soil and operational data can be used to compare to the results of various analytical methods of machine foundation design.

From inspection of these data, certain qualitative conclusions regarding the behavior of pile supported machines operating under steady state conditions can be made.

1. The consistency in the modes of vibration of these compressors indicate that the modes of vibration can be predicted.

2. The actual modes of vibration are consistent with the design unbalanced forces and moments.

3. Two compressors were monitored at each site. Each of the compressors had adjacent compressors of various types. The amplitudes indicate that the compressors are not significatly affected by the neighboring machines. However, this influence may account for variations in the vibration wave forms.

4. The Channel View foundation is considerably stiffer than those at Norco and Plaquemine and the amplitudes of vertical vibrations are less, as would be expected.

5. The Channel View foundation, unlike Norco and Plaquemine, includes batter piles. However, the amplitudes of torsional and translational vibration do not appear to be significantly reduced. This brings into question the effectiveness of using battered piles.

Calculated Response

The calculated response shows that reasonably accurate predictions of vibration amplitude can be achieved using easily measured soil properties together with the combination of analytical procedures described.

The relatively poor agreement between measured and calculated torsional response at the Norco and Plaquemine sites is attributed to the low torsional stiffness of the foundations and the fact that when the analysis was done the author ignored the effects of embedment for the torsional mode. For the flexible piles at these sites the effects of embedment would be relatively significant and probably controls the torsional response. By contrast, the Channel View foundation includes battered piles and has a much higher torsional stiffness such that the effects of embedment are relatively small and the torsional stiffness of the piles control. Very good agreement was obtained for the torsional mode for the Channel View site.

RECOMMENDED ADDITIONAL RESEARCH

The measured field responses were all under steady state operating conditions, and the natural frequencies of the pile supported compressors are on the order of three times the operating frequencies. At this low frequency ratio, the predicted and measured responses are both relatively small. This, in itself, would tend to make the comparison of the measured to calculated response appear to be close.

A better check on the analytical procedures used would be obtained if field observation data were available for machines which operate at speeds closer to the natural frequency for the foundation systems. Since machines are not usually designed to operate at close to the natural frequency, such data could probably only be obtained from shaking tests. Shaking tests should be performed on the total installation with the machines and all the appurtenances in place. Measurements for varying operating speeds would be useful, particularly if measurements could be on actual foundations that pass through resonance.

At the time the measurements were made, the slabs on grade surrounding the mats had settled relative to the mats. This indicates that the soil beneath the pile cap mats may have separated from the bottom of the mats, thus reducing or eliminating the effects of geometric damping due to the mat-soil contact. (This damping was totally neglected in the calculations.) The reduction of this damping may increase with time as the soil continues to settle and might have an effect on the long term performance of the compressors. Because of this, and because the machines themselves will tend to have a variation in the unbalanced forces due to normal wear, additional field measurements on the same machines should be made over a period of years.

ACKNOWLEDGMENTS

This study would not have been possible without the cooperation of Big Three Industries, the owner of the Norco, Plaquemine and Channel View plants. Financial support for the field costs and instrumentation were provided by Harding-Lawson Associates, Consulting Engineers, Geologists, and Geophysicists. Valuable assistance and guidance was provided by Dr. Michael W. O'Neill and Dr. Osman Ghazzaly, Associate Professors of Civil Engineering, University of Houston and Dr. Kenneth Stokoe, Associate Professor of Civil Engineering, University of Texas at Austin. Characteristics of the compressors were furnished by their manufacturers, Sulzer Bros., Inc.

APPENDIX I - REFERENCES

1. Aboul-Ella, F. and Novak, M. "Stiffness and Damping of Piles in Layered Media", Specialty Conference on Earthquake Engineering and Soil Dynamics, Pasadena, California, June 11-21, 1978

2. Beredugo, Y.O. and Novak, M., "Coupled Horizontal and Rocking Vibration of Embeded Footings", Canadian Geotechnical Journal, Vol. 9, No. 4, Nov. 1972

3. Diaz, Gerald M., and Azizi, Mir, "Calculated Response of Pile Supported Machines", International Symposium on Foundations for Equipment and Machinery, A.C.I., Annual Convention, Milwaukee, Wisconsin, U.S.A., March - 1979

4. Grigg, Ronald R. and Novak, Milos, "Dynamic Experiments with Small Pile Foundations", Canadian Geotechnical Journal, Vol. 13, No. 4, 372-385, 1976

5. Howell, John F. and Novak, Milos, "Torsional Vibration of Pile Foundations", A.S.C.E. Journal of Geotechnical Engineering Division, April - 1977

6. Nogami, T. and Novak, M., "Soil-Pile Interaction in Vertical Vibration", Earthquake Engineering and Structural Dynamics, Vol. 4, 277-293, 1976

7. Nogami, T. and Novak, M., "Resistance of Soil to a Horizontally Vibrating Pile", Earthquake Engineering and Structural Dynamics, Vol. 5, 249-261, 1977

8. Nogami, T. and Novak, M., "Soil-Pile Interaction in Horizontal Vibration", Earthquake Engineering and Structural Dynamics, Vol. 5, 263-281, 1977

9. Novak, Milos, "Dynamic Stiffness and Damping of Piles", Canadian Geotechnical Journal, 11, 574, 1974

10. Novak, Milos, "Vertical Vibration of Floating Piles", A.S.C.E., Journal of Engineering Mechanics Division, February - 1977

11. Novak, M. and Aboul-Ella, F., "PILAY, A Computer Program for Calculation of Stiffness and Damping of Pile Layered Media", The University of Western Ontario, London, Ontario, Report SACDA 77-33, December - 1977

12. Reese, L. C., O'Neill, M. W., and Smith, R. E., "Generalized Analysis of Pile Foundations", A.S.C.E., JSMFD, January - 1970

A STUDY OF AXIAL PILE LOAD TESTS

by

Shailendra N. Endley,[1] M.ASCE, Edward J. Ulrich Jr.,[2] M.ASCE
and Janice B. Gray,[3] A.M.ASCE

INTRODUCTION

Pile capacity is a function of pile type, shape, length, and soil characteristics. Installation technique also influences the pile capacity and the effect can become significant in certain soil conditions. The effect of these influencing factors is difficult to predict by analytical techniques due to the complex phenomenon of soil-pile interaction. Load tests are helpful in estimating pile capacity as well as in studying the effects of the influencing factors.

Results of seventeen axial pile load tests are presented and synthesized for five sites listed as A through E, in the Coastal Plain geologic province in Southeast Texas. Pile capacities predicted by accepted static methods are compared to observed failure loads and the influence of predrilling on pile capacity is addressed. The load test results include pile make-up, pile driving records, load deflection curves, and hammer properties. Generalized subsurface conditions are included for each site.

SOIL CONDITIONS

Soil conditions for each site consist of strong clays interspersed with sand layers and are generalized in Figs. 1 and 2. The deposits are typical of the Beaumont clay formation of Pleistocene geologic age. The undrained shear strength, c_u, of clays ranges from 0.5 tsf (47.9 kN/m^2) to 2.0 tsf (191.6 kN/m^2). Plastic limit and liquid limit range from 20 to 30 and 60 to 90, respectively. The low water contents and high shear strengths are characteristic of clay soils overconsolidated by desiccation. The groundwater table at the sites varies from 0 ft to 17 ft (5.12 m) below the ground surface and is indicated for each site on the appropriate illustration in Figs. 1 and 2. Sites A, B, and C are shown to have a major sand stratum underlying the clay soils. As indicated by the N-values in Fig. 1, the sands are in a medium-dense to dense condition. Before pile installation, sites A and B were excavated to a depth of about 12 ft (3.66 m) and Site C was excavated to a depth of about 20 ft (6.1 m).

PILE DESCRIPTION AND INSTALLATION

Pile make-up and total lengths of the 17 piles are shown in Fig. 3. Also shown on this illustration is a typical step-taper pile and the shell sizes. Pile designation and site are matched accordingly: Site A, Piles 1 through 8; Site B, Piles 9 through 11; Site C, Piles 12 and 13; Site D,

---

1. Geotechnical Engineer, McClelland Engineers, Inc., Houston, Texas
2. Project Manager, McClelland Engineers, Inc., Houston, Texas
3. Engineer, Houston Lighting and Power Company, Houston, Texas

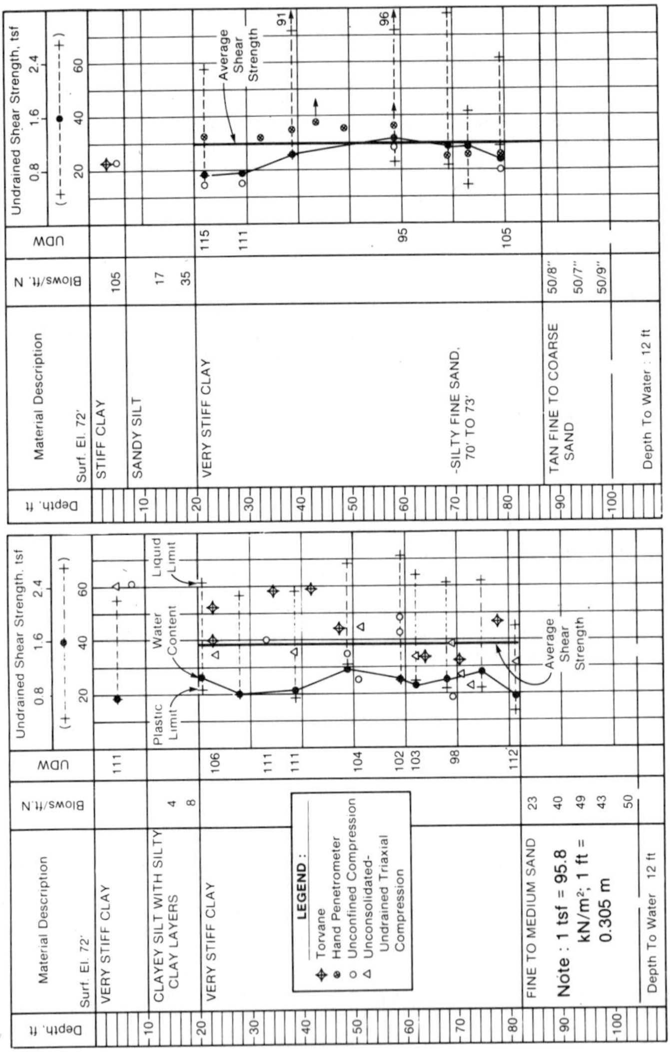

Fig. 1 — Generalized Subsurface Conditions, (Continued)

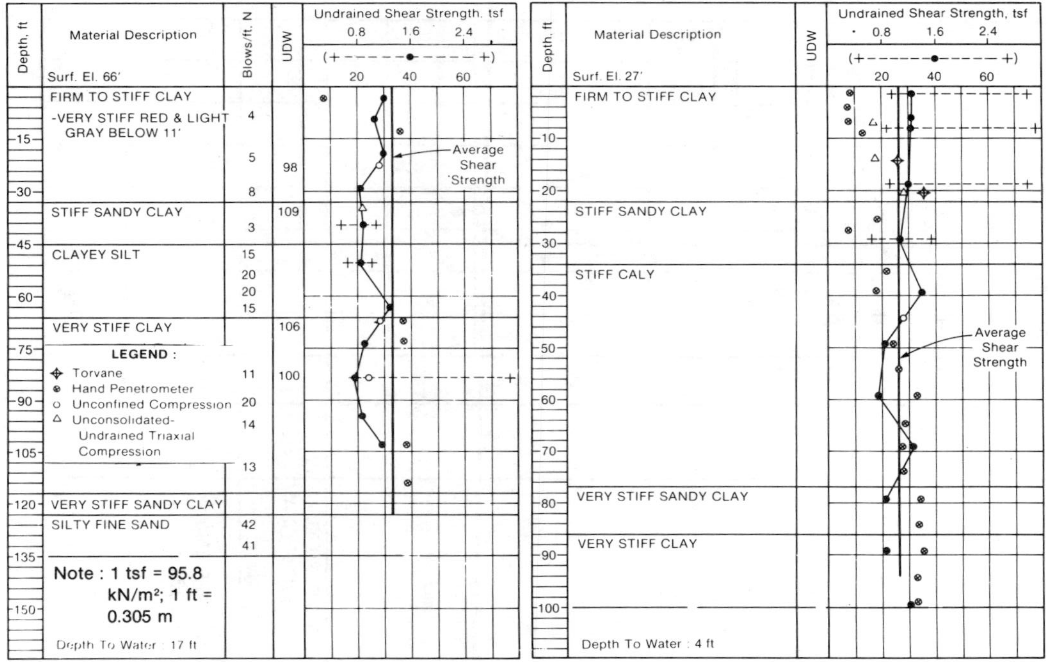

Site C Site D

Fig. 1 — Generalized Subsurface Conditions,

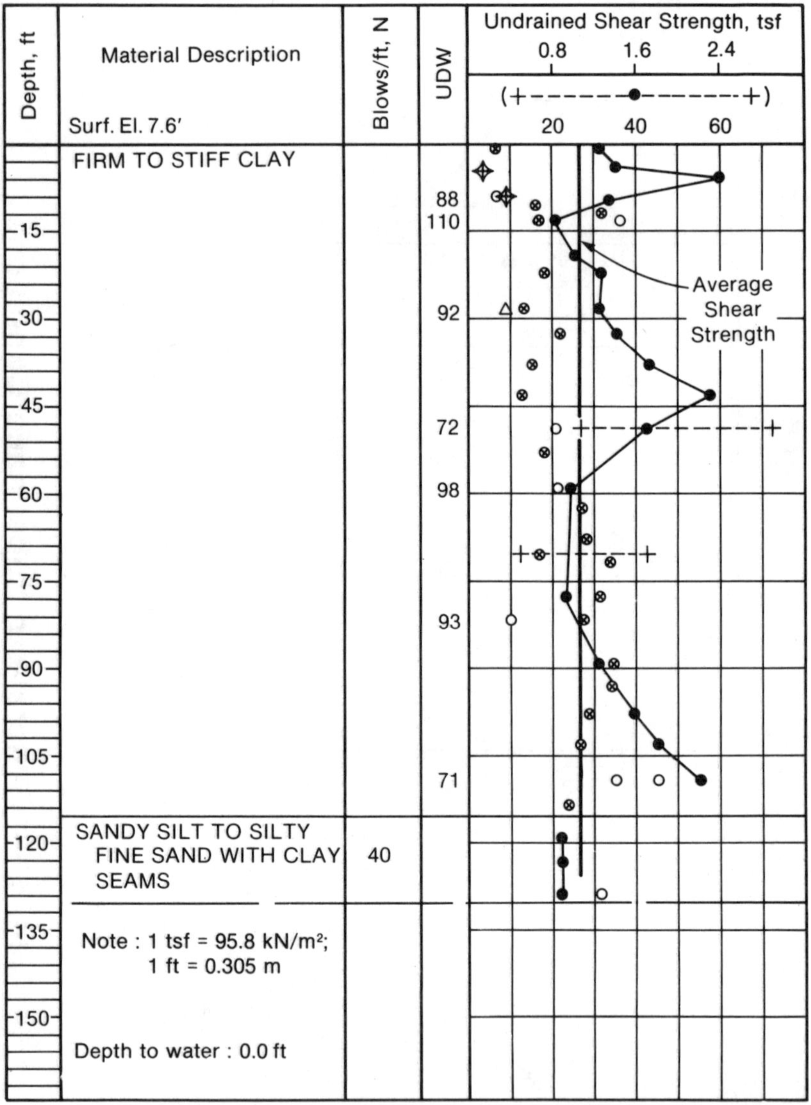

SITE E

Fig. 2 — Generalized Subsurface Conditions, Site E.

Pile No.	Length of Pile Sections, ft								Total Pile Length, ft	Site
	0	1	2	3	4	5	6	7		
1	12	12	12	12	12	12	1.0	-	73.0	A
2	12	12	12	12	12	12	1.0	-	73.0	
3	12	12	12	12	12	12	12	-	84.0	
4	12	12	12	12	12	12	12	8.0	92.0	
5	12	12	12	12	12	12	12	8.0	92.0	
6	12	12	12	12	12	12	12	8.0	92.0	
7	12	12	12	12	12	6	-	-	66.0	
8	12	12	12	12	12	12	2	-	74.0	
9	12	12	12	12	12	9	-	-	69.0	B
10	12	12	12	12	12	12	5	-	77.0	
11	12	12	12	12	12	12	6	-	78.0	
12	20(1)	12	12	12	12	12	12	3.0	95.0	C
13	28(1)	12	12	12	12	12	12	3.0	103.0	
14(2)	-	12	12	12	12	12	12	12	92.0	D
15(3)	16	16	16	16	10	-	-	-	106.0	E
16	10.75-in.-diam. closed end steel pipe pile								105.0	
17(3)	16	16	16	16	16	13	-	-	125.0	

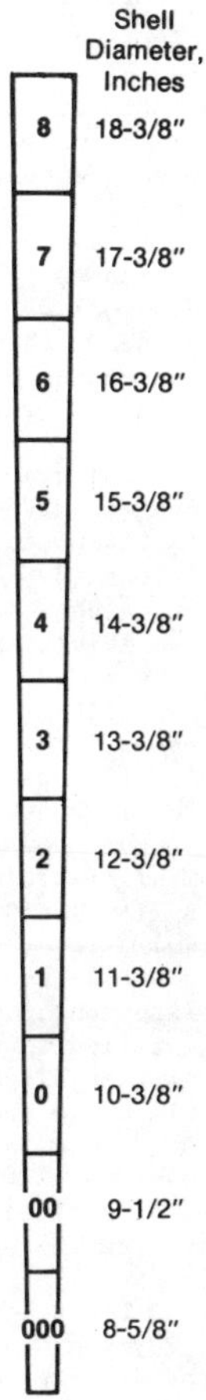

Notes :
(1) 10.75-in.-diameter closed end steel pipe pile section
(2) Also, 8 ft of No. 8 section
(3) Also, 16 ft of No. 000 and No. 00 sections

1 ft = 0.305 m
1 in. = 25.4 mm

Fig. 3 — Pile Make-up Description

Pile 14; and Site E, Piles 15 through 17. Pile 16 is a constant diameter, closed end pipe pile and the remaining ones are step tapered piles.

All the step taper piles were installed by predrilling a pilot hole. This technique helps in reducing the heave of the ground if there are many piles to be driven in an area by reducing the amount of displacement of soil due to driven piles. Predrilling of pilot holes also expedite the pile driving process and reduces the chances of damage of shell sections to some extent. About half the piles terminate at refusal in sand and the remaining piles terminate in clay. A 16-ft (4.88-m) shell section was used for Piles 15 and 17; 12-ft (3.66-m) sections were used for the remaining step-taper piles. Piles 12 and 13 had a 10.75-in. (273-mm) O.D. by 0.279-in. (7.09-mm) wall thickness pipe extending below the No. 1 shell. A 1-in. (25.4-mm) thick flat closure plate was flush welded at the tip of the pipe. The lengths of pipe sections were 20 ft (6.1 m) and 28 ft (8.54 m) for Piles 12 and 13, respectively.

The step taper piles were driven with an internal steel mandrel which engages drive-rings located at each change in shell diameter and extends to the tip of the pile. The mandrel was 96 ft (29.3 m) long, weighing about 10 tons (89 kN), with an aluminum-micarta driving block and shield of one-ton weight. Concrete was placed in the test piles immediately after driving. Pile driving resistances for each pile are shown in Table 1.

Piles 1 through 6 at Site A were installed using a Super Vulcan 80C hammer. The Vulcan 80C is a differential acting steam hammer with a rated energy of 24,450 ft-lbs (33,254 m-N). The hammer operated at about 96 to 100 blows per minute at a reported efficiency of 75 percent. The cushion assembly had a total thickness of 26 in. (0.660 m) and consisted of two steel plates at the top and bottom with 11 aluminum spacers measuring 14-1/4-in. (361.9 mm) by 1/2-in. (12.7 mm) and 11 micarta spacers measuring 14-1/2-in. (368.3 mm) by 1-in. (25.4 mm). The top steel plate was measured to be 14-3/8-in. (365.1 mm) by 3-1/2-in. (88.9 mm) and the bottom steel plate measured 14-3/8-in. (365.1 mm) by 6-in. (152.4 mm). The piles were installed in predrilled pilot holes made by a continuous, 10-in. (254-mm) diameter flight hollow auger emitting water at the tip. Pilot holes were drilled to about 50-ft (15.2-m) depth below the excavation grade. Piles 1 and 2 were installed so that the capacity derived from the upper clayey silt zone was eliminated. To achieve this, a 24-in (609.6-mm) diameter casing was installed in an oversized hole to about 10 ft (3 m) depth prior to driving the step-tapered pile.

Piles 7 through 13 were installed with predrilled pilot holes using a Raymond Model 80C differential acting steam hammer. This hammer has a rated energy of 24,450 ft-lbs (33,154 m-N) per blow and operated at a reported rate of about 100 blows per minute. Cushion assembly details are not available for these hammer pile systems. The pilot holes were drilled to 50 ft (15.2 m) with a wet rotary technique using a 10-in. (254-mm) diameter, fish-tail, bottom discharge bit. For Piles 12 and 13, the depth of the pilot hole was 76 ft (23.2 m) and the hole diameter is believed to be 13 in. (330 mm) from 0 to 35-ft (10.7-m) depth, 12 in. (305 mm) from 35 to 47-ft (10.7 to 14.3-m) depth, 11 in. (279 mm) from 47 to 57-ft (14.3 to 17.4-m) depth, and 9 in. (229 mm) from 57 to 76-ft (17.4 to 23.2-m) depth.

Table 1. - PILE DRIVING RESISTANCES

Depth, ft	Blows Per Foot																
50	13	7					5	5	6	7	6						
55	6	9	8				7	6	6	8	6					19	
60	6	11	6				6	6	7	8	9			12		21	
62	7	10	7		5	7	7	6	7	8	10			10		23	
64	8	10	12	8	5	7	7	7	7	7	9			12		25	
66	10	10	8	6	9	7	6	6	7	9	8			14		26	
69	6	10	7	10	7	6		8	10	11	11			15		32	
70	10	10	6	10	9	8		12		10	11			19		90	
73	10	10	6	8	9	10		15		11	11			19		148	
74			9	8	9	8		15		24	30			21	5	73	
76			11	10	20	12				33	40	5	5	18	15	60	4
77										33							
78			11	15	26	18					31	7	7	17	10	55	5
80			20	25	25	17						8	8	19	11	44	6
82			31	22	21	21						10	7	20	10	45	8
84			34	21	21	25						10	9	25	10	46	9
86				22	17	23						10	7	25	7	48	8
88				25	22	29						11	9	25	9	48	7
90				16	22	32						12	11	29	5	47	6
92				17	25	29						12	14	18	6	43	8
94													14		6	43	7
95												15					
96													15		7	46	8
98													21		8	44	7
100													25		8	47	8
102													90		8	47	11
103													*133		9	52	11
105																52	
106															*9		13
108																	15
110																	17
112																	11
114																	13
116																	13
118																	15
120																	16
122																	17
124																	23
125																	28
Pile No.	1	2	3	4	5	6	7	8	9	10	11	12	13	14	15	16	17
Pile Length, ft	73	73	84	92	92	92	66	74	69	77	78	95	103	92	106	105	125

* 10" Penetration

Note : 1 ft = 0.305 m

Piles 14 through 17 were installed by using a Raymond Model 65C hammer having a rated energy of 19,500 ft-lbs (26,442 m-N) per blow and operating at about 100 blows per minute. Cushion assembly details are not available for these piles. Hammer efficiency of about 65 percent was reported for these tests. Pilot holes were drilled for all piles except Pile 16, the closed-end steel pipe pile. The pilot holes were also drilled using wet rotary techniques with a bottom discharge, fish tail bit. The hole depth and bit diameter were 60 ft (18.3 m) by 12 in. (305 mm), 73 ft (22.2 m) by 8 in. (203 mm), 76 ft (23.2 m) by 6 in. (152 mm) for Piles 14, 15, and 17, respectively.

TEST LOADING PROCEDURE

The load tests were performed in general accordance with the specifications of ASTM D 1143-69. The load-movement observations are presented graphically in Fig.4; the pile load test procedures are summarized in the following paragraphs. All the piles were subjected to compressive loads except piles 3 and 4 which were subject to tensile loads.

Piles 1 through 6. The load in tons (1 ton = 8.9 kN) was applied in cycles as:

Cycle 1:	0, 24, 48, 24, 0
Cycle 2:	0, 24, 48, 72, 95, 72, 48, 0
Cycle 3:	0, 48, 95, 119, 142, 119, 95, 72, 48, 0
Cycle 4:	0, 48, 95, 142, 166, 190, 214, 142, 72, 42, 0

Each load was maintained until movement stabilized at 0.01 ft (3.05 mm) per hour.

Pile 7. This load test was performed about 56 days after pile installation. The loads were applied at 15-ton (133.5 kN) load increments to 120 tons (1068 kN) and maintained for at least 30 minutes. The load was released after every other increment and the pile was allowed to rebound for 15 minutes. After reaching 120 tons (1068 kN), the load was gradually reduced until the downward movement of the pile butt slowed significantly at 110 tons (979 kN).

Pile 8. This load test was also performed about 56 days after the installation. Loads were generally applied in 25-ton (222-kN) load increments until reaching 187 tons (1664 kN). Each load was maintained for at least 30 minutes. The load was released after every other increment and the pile rebounded for 15 minutes.

Pile 9. The load test was performed about 10 days after pile installation. Loads were applied in 15-ton (133 kN) increments until a load of 120 tons (1068 kN) was obtained. Each load was maintained for at least 20 minutes. The load was released from the pile every other increment and the pile was allowed to rebound for 15 minutes. After reaching 120 tons (1068 kN), the load was gradually reduced until the downward movement of the pile head ceased and an equilibrium was reached at 94 tons (837 kN). The load was then removed and a rebound reading taken after 12 hours.

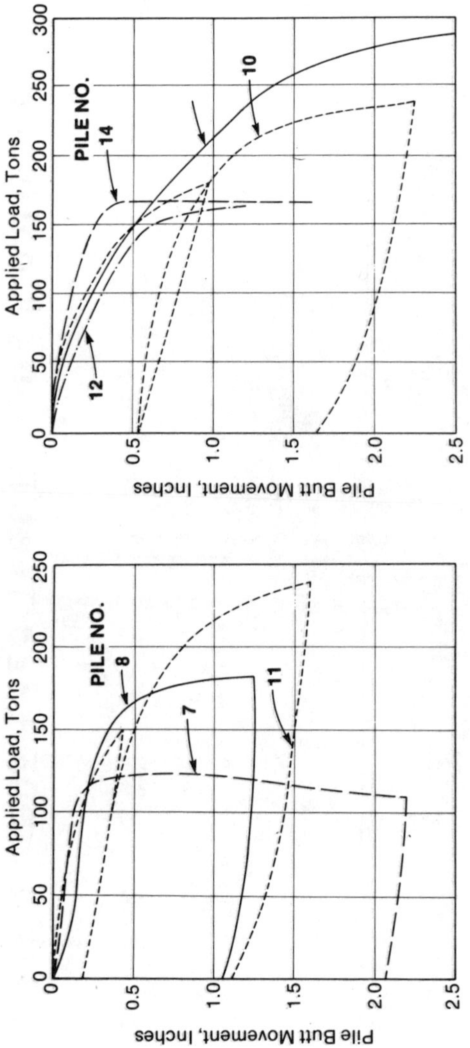

1 ton = 8.9 kN   1 in. = 25.4 mm

Fig. 4 — Pile Load Test Results (Continued)

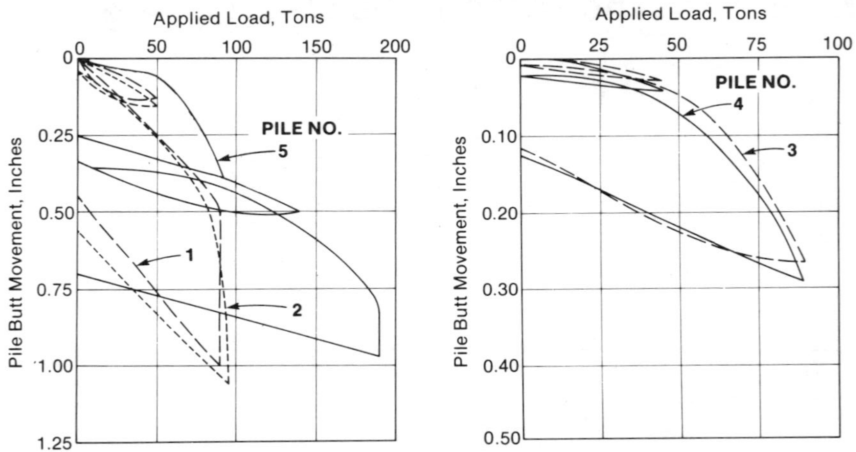

1 ton = 8.9 kN   1 in. = 25.4 mm

Fig. 4 — Pile Load Test Results (Continued)

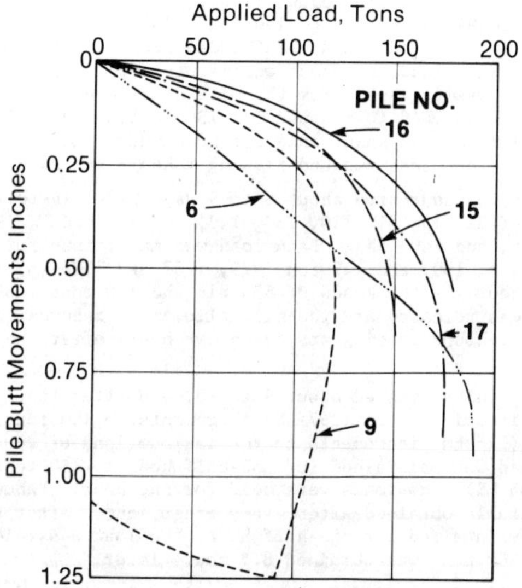

1 ton = 8.9 kN
1 in. = 25.4 mm

Fig. 4 — Pile Load Test Results

Pile 10. The pile was tested about 9 days after installation and loaded in approximately 25-ton (222-kN) increments to 185 tons (1646 kN), allowed to rebound for two days at zero load, and then reloaded incrementally to 240 tons (2136 kN). Each load increment was maintained for at least 20 minutes. The load was released from the pile after every other load increment and the pile was allowed to rebound for 15 minutes. The final load of 240 tons (2136 kN) was maintained for 45 minutes and then the load was released and a rebound reading taken after 20 minutes.

Pile 11. This pile was tested about 11 days after installation, loaded in 25-ton (222-kN) increments to 150 tons (1335 kN), allowed to rebound for one day at zero load, and then loaded incrementally to 240 tons (2136 kN). Each load increment was maintained for at least 20 minutes. The load was released from the pile after every other load increment and the pile allowed to rebound for 15 minutes. The final load of 240 tons (2136 kN) was maintained for 20 minutes and then the load was released and the final rebound reading taken after 20 hours.

Pile 12. The pile was tested about 8 to 9 days after installation and subjected to loads of 22, 55, 81, 110, 130, 149, and 168 tons (196, 489, 721, 979, 1157, 1326, and 1495 kN). Each load was maintained for at least one-half hour; the 110, 130., and 149-ton (979, 1157 and 1326-kN) loads were maintained for one hour. After loads of 55, 81, and 130 tons (489, 721 and 1157 kN), the load was released and the pile allowed to rebound for 15 minutes. The final rebound reading was taken two hours after the last load was applied.

Pile 13. The pile was tested about 9 to 10 days after the installation. Load was applied in 30-ton (267-kN) increments to 150 tons (1335 kn) and then in 20-ton (178-kN) increments to the maximum load of 290 tons (2581 kN). Each load was maintained for one-half hour to 150 tons (1335 kN). Above 150 tons (1335 kN), the loads were held for one hour. Rebound measurements were generally obtained after every other load. After the 290-ton (2581-kN) load was maintained for one-half hour, the load was released and a final rebound measurement was obtained 8.5 hours later.

Pile 14. The pile was tested 7 days after installation and load was applied in increments of 20 tons (178 kN). Loads were maintained for a period of at least one-half hour and after load applications of 40, 80, 120, and 160 tons (356, 712, 1068 and 1424 kN), the load was released and the pile was allowed to rebound for one-half hour before being reloaded to the next load increment. At the failure load of 165 tons (1468 kN) the load was released and the final rebound reading was taken after 15 minutes.

Piles 15 through 17. These piles were tested about 8 to 10 days after installation. Test piles were loaded to failure or a maximum of 200 tons (1780 kN) in 20-ton (178-kN) increments. In a few instances, these increments were reduced in the higher load ranges to provide additional information on the load-movement characteristics of the piles. Load applications in the lower load ranges were usually maintained for at least one-half hour; the higher test loads were generally maintained for either one or two hours, depending upon the rate of movement of the pile. In general, loads were released after every second load increment and the pile permitted to rebound for at least 15 minutes. The piles were allowed to rebound for about 2 to 10.5 hours after release of the final test load.

ANALYSIS OF DATA

The test data is analyzed by comparing the developed skin friction in clay and end bearing in sand with computed values. The analysis considers skin friction in clay and end bearing in sands. Piles 1 through 3, 7, 9, 12, and 14 through 17 were fully embedded in cohesive soils; these piles are analyzed for skin friction capacity and average adhesion value by subtracting the end bearing from the ultimate pile load. Clay end bearing was judged to be equal to 9 times the undrained shear strength, c_u, over the tip cross-sectioned area. The end bearing was considered to be zero for piles load tested in tension. The remaining piles, installed to refusal into the underlying sand strata, are analyzed for end bearing values in sand by considering the average unit skin friction to be equal to that determined for an adjacent pile not tipped in sand. Results of the analyses are given in Tables 2 and 3.

SKIN FRICTION

The methods used for the analysis of skin friction are the λ-Method(1), API RP 2A Method(2), and Effective Stress Method(3,4). The strength profile selected for various piles is shown on each illustration in Figs. 1 and 2.

λ-Method. The frictional capacity, Q_s, of a pile in clay at a given penetration is considered to be a function of both effective vertical stress and undrained shear strength, expressed as:

$$Q_s = \lambda\ (\bar{\sigma}_m + 2c_m)\ A_s \quad \ldots\ldots\ldots\ldots \quad (1)$$

where λ = dimensionless frictional coefficient (function of pile penetration)

$\bar{\sigma}_m$ = mean effective vertical stress between ground surface and the pile tip

c_m = mean undrained shear strength along the pile length

A_s = embedded surface area of the pile

API RP 2A Method. For overconsolidated clay soils, this method is based on the theory that unit skin friction, f, is a function of undrained shear strength, c_u, and can be expressed as follows:

$$f = \alpha\ c_u \quad \ldots\ldots\ldots\ldots \quad (2)$$

where the α factor varies with undrained shear strength, decreasing with an increase in shear strength. The undrained shear strength values provide for an α of 0.5 for the cases considered.

Effective Stress Method. In this method, the unit skin friction is defined as:

$$f = \beta\ \bar{p}_o \le c_u \quad \ldots\ldots\ldots\ldots \quad (3)$$

where $\bar{p}_o$ is the average effective overburden pressure along the shaft and

Table 2. - ANALYSIS OF FRICTIONAL CAPACITY IN CLAY

Pile No.	Site No.	Pile Length, ft	Pilot Hole Length, ft	Pilot Hole Dia., in.	Observed Total Capacity Q tons	Frictional Capacity Q_s, tons	Calculated Q_s			Observed Average Unit Friction f, tsf	Average c_u, tsf	$\alpha = f/c_u$
							λ Method	API Method	Effective Stress Method			
1	A	73	50	10	90	85	144	155	123	0.37	1.50	0.25
2	A	73	50	10	80	75	139	150	123	0.32	1.50	0.21
3	A	84	50	10	90**	90	-	-	-	0.34	1.50	0.23
7	A	66	50	10	110	106	152	162	118	0.47	1.50	0.31
9	B	69	50	10	115	109	139	140	125	0.47	1.20	0.39
12	C	95	76	*	155	150	200	226	260	0.45	1.35	0.33
14	D	92	60	12	155	150	183	186	244	0.40	1.04	0.38
15	E	106	73	8	140	138	148	156	230	0.44	1.06	0.41
16	E	105	-	-	175	168	153	159	230	0.57	1.06	0.54
17	E	125	76	8	170	168	190	197	337	0.43	1.06	0.41

* 9″ at tip; 11″, 12″ and 13″ reamer at 19′, 29′ and 41′ from tip

** Tension Test

Note : 1 ft = 0.305 m; 1 in. = 0.0254 m; 1 ton = 8.9 kN; 1 tsf = 95.8 kN/m²

Table 3. - ANALYSIS OF END BEARING IN SAND

Pile No.	Site No.	Pile Length, ft	Pilot Hole Length, ft	Pilot Hole Dia., in.	Observed Total Capacity Q, tons	End Bearing Q_p, tons	Calculated End Bearing, Q_p, tons		N_q from Load Test Q_p	N_q from API
							Limiting q_l	No Limit on q_p		
4	A	92	50	10	90**	-	-	-	-	-
5	A	92	50	10	189	99	58	67	59	40
6	A	92	50	10	192	88	58	72	57	40
8	A	74	50	10	170	53	55	55	40	40
10	B	77	50	10	218	98	55	55	72	40
11	B	78	50	10	216	95	58	60	64	40
13	C	103	76	*	270	110	31	41	55	20

* 9″ at tip; 11″, 12″ and 13″ reamer at 19′, 29′ and 41′ from tip
** Tension Test

Note : 1 ft = 0.305 m; 1 in. = 0.0254 m; 1 ton = 8.9 kN

β, the skin friction factor, is defined by the expression:

$$\beta = K_s \tan \phi \quad (4)$$

where K_s is the average coefficient of earth pressure on the pile shaft and ϕ is the angle of internal friction for clay. For driven piles in stiff overconsolidated clay, K_s is roughly 1.5 times greater than K_o where K_o is the coefficient of earth pressure at rest. K_o may be approximated by:

$$K_o = (1 - \sin \phi) \sqrt{R_o} \quad (5)$$

where R_o is the overconsolidation ratio of clay. The value of β can scatter in a wide range depending upon the clay shear strength, overconsolidation ratio and pile type. A lower bound value of 0.5 for β is used in this study.

Meyerhof has also suggested a semi-empirical relationship for estimating skin friction in clays.

For driven piles:

$$f = 1.5\, c_u \tan \phi \quad (6)$$

For bored piles:

$$f = c_u \tan \phi \quad (7)$$

By utilizing a ϕ value of 20 degrees for the stiff to very stiff clays the expressions reduce to

For driven piles:

$$f = 0.55\, c_u \quad (8)$$

For bored piles:

$$f = 0.36\, c_u \quad (9)$$

The results are summarized in Table 2.

END BEARING IN SAND

The end bearing developed in Piles 3 through 6, 8, 10, 11, and 13 during load tests was calculated by subtracting the skin friction, Q_s, from the total pile capacity. For Piles 5 and 6, the end bearing was calculated by subtracting the load developed by Pile 4 which was driven adjacent to them and load tested in tension. The end bearing developed by Piles 10 and 11 was evaluated by subtracting the skin friction developed in Pile 9, installed adjacent to them. Similarly, Piles 12 and 13 were paired to evaluate the end bearing in Pile 13.

The evaluated end bearing from pile load tests were then compared by estimating the end bearing with the relationship:

$$Q_p = q_p A_p = \bar{p}_o N_q A_p \quad (10)$$

where q_p = unit end bearing

N_q = bearing capacity factor

$\bar{p}_c$ = effective overburden pressure

A_p = pile end area

The bearing capacity factor, N_q, is mainly a function of angle of internal friction and increases with an increasing ϕ. The unit end bearing, q_p, increases with depth up to a limiting value, q_ℓ, and remains constant in a homogenous sand due to arching and other factors. The maximum value of q_ℓ is also mainly a function of ϕ. Other factors which might influence the value of N_q and q are compressibility of soil, method of pile installation, and pile shape and taper. Wide ranges of values of N_q and q_ℓ are suggested by various authors(4). In this study, API values of N_q and accepted limiting values of q_ℓ are used for estimating Q_p. Values of N_q are also calculated from the load test values of end bearing. Results of the end bearing analyses are present in Table 3.

DISCUSSION OF RESULTS

Table 2 summarizes the analysis of piles end bearing in clay by comparing the predicted skin friction determined by some accepted methods and the skin friction computed by load testing. The predicted values of Q_s are found to be higher than the Q_s values calculated from the load tests in all cases considered. Values of α (ratio of unit skin friction to undrained shear strength) computed from the load tests range from 0.25 to 0.41 with most of the values below 0.4. Comparing these α values with the suggested values by Meyerhof (equations 8 and 9), it can be seen that the observed α values fall in the category of bored piles. These results indicate that the presence of pilot hole for the step-taper piles might be causing the reduction in the pile capacity. These numbers suggest the step-taper piles with pilot holes are behaving more like bored (or low displacement piles) rather than driven piles. It will be interesting to note that α for Pile 16 is 0.54 and for Pile 15 it is 0.41, where Pile 16 is a pipe pile driven without pilot hole and Pile 15 is a step-taper pile driven with a pilot hole and both piles were driven at the same site. The α values determined from this study are also presented on Fig. 5 along with data compiled by Tomlinson(5).

Results of the analysis of piles end bearing in sand are given on Table 3. Computed end bearing using the API criteria underestimated actual end bearing for all piles except Pile 8. Bearing capacity factors derived from the load test results suggest the API factors are conservative. Fig. 6 shows the comparison of bearing capacity factors developed from the load test data, the API values and the results compiled by Meyerhof(4).

CONCLUSIONS

The results of 17 pile load tests in overconsolidated clays and sands at sites along the Texas Gulf Coast, are reviewed and synthesized. Pile skin friction determined by the λ-Method, API-Method and Effective Stress Method overpredicted the actual skin friction capacity. The predrilled pilot holes may contribute to reducing the available skin friction. The

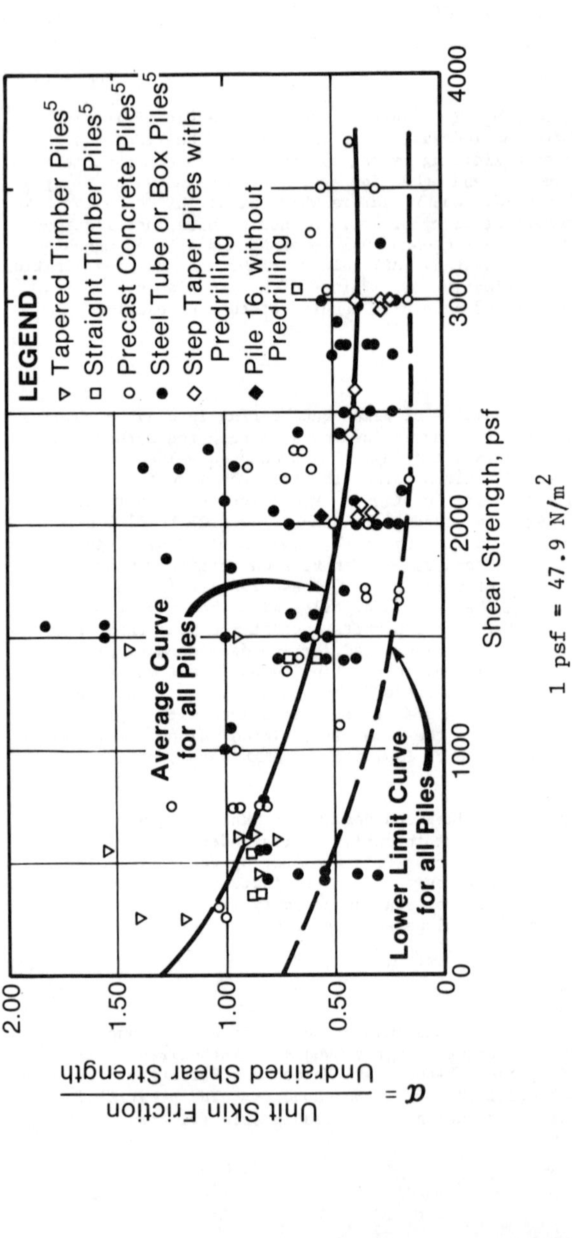

Fig. 5 — Relationship Between Shaft Skin Friction of Piles Driven into Clay Soils and the Shear Strength of the Clays

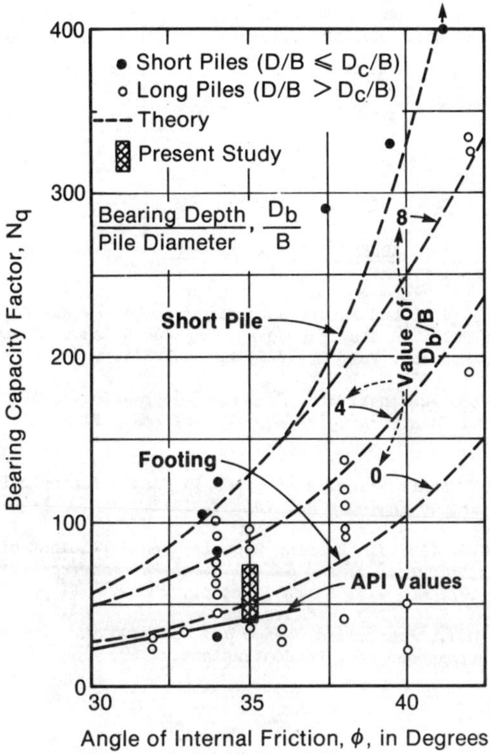

Fig. 6 — Bearing Capacity Factors for Driven Piles in Sand

computed α values indicate that the step-taper piles installed with a pilot hole tend to behave like a low displacement pile. It may be worthwhile to study the effect of diameter and length of pilot hole on the mobilized skin friction. An optimum size of pilot hole could probably be obtained by load tests so that the advantage of pilot holes is still there without compromising the pile capacity. The piles terminating in sand show that API criteria provide conservative estimates of pile end bearing.

APPENDIX I - REFERENCES

1. Vijayvergiya, V.N. and Focht, John A., Jr., "A New Way to Predict the Capacity of Piles in Clay," *Proceedings, 4th Offshore Technology Conference,* Houston, Vol. 2, 1972, pp. 865-874.

2. American Petroleum Institute, *Recommended Practice for Planning, Designing and Constructing Offshore Platforms,* API RP 2A, 10th edition, 1979.

3. Burland, J., "Shaft Friction of Piles in Clay - A Simple Fundamental Approach," *Ground Engineering,* Vol. 6, 1972.

4. Meyerhof, G.G. (1976), "Bearing Capacity and Settlement of Pile Foundations," *Journal of the Geotechnical Engineering Division,* ASCE, Vol. 102, No. GT3, March 1976, pp. 197-228.

5. Tomlinson, M.J., *Foundation Design and Construction,* 2nd edition, Sir Isaac Pitman and Son, London England, 1969.

APPENDIX II - NOTATION

The following symbols are used in this paper:

α = ratio of unit skin friction to undrained shear strength

β = skin friction factor

λ = dimensionless frictional coefficient

$\bar{\sigma}_m$ = mean effective vertical stress

ϕ = angle of internal friction

A_p = pile end area

A_s = surface area of pile

B = pile diameter

c_m = mean undrained shear strength

c_u = undrained shear strength

D_b = bearing depth

D_c = depth at which limiting value of end bearing has reached

f = unit skin friction

K_o = coefficient of earth pressure at rest

K_s = coefficient of earth pressure

N_q = bearing capacity factor

$\bar{p}_o$ = effective vertical stress

q_ℓ = limiting value of unit end bearing

q_p = unit end bearing

Q_p = end bearing of pile

Q_s = frictional capacity of pile

R_o = overconsolidation ratio of clay

tsf = tons per sq ft.

UDW = unit dry weight

INDICATOR PILE AND LOAD TEST PROGRAM FOR A SEWAGE TREATMENT PLANT

By Oliver H. Gilbert, Jr.[1], M. ASCE
Mahmut Otus[2], M. ASCE
and Joseph Corollo[3]

INTRODUCTION

San Francisco has undertaken a $1.9 billion wastewater management program to minimize ocean and bay overflows resulting from the existing combined sanitary sewage and storm water collection system.

Part of this program involves the upgrading of the existing Southeast Water Pollution Control Plant from primary to secondary treatment. The plant is located near San Francisco Bay, adjacent to the shoreline of Islais Creek, on an infilled marsh area (Figure 1).

The $115,000,000 expansion includes Primary Sedimentation Tanks, Aeration Tanks, and Secondary Sedimentation Tanks, with appurtenant structures such as pipe tunnels, headworks, sludge control room, effluent channels, and connecting pipe conduits. Also included in the expansion are a Pretreatment Building, an Oxygen Production Building, a Chlorination Building, and several Laboratory, Shop and Control buildings. The locations of the major facilities are shown in Figure 2.

SUBSURFACE CONDITIONS

The 38 borings drilled for the geotechnical investigation revealed the following generalized stratigraphy beneath the 40-acre (16 hectare) site:

1. Fill (predominantly granular)
2. Bay Mud (soft to medium-stiff organic silty clay)
3. Sands (alluvial outwash deposits) interbedded with
4. Old Bay Mud (stiff, overconsolidated marine deposit)
5. Colluvium and slopewash (predominantly stiff clays with rock fragments)
6. Franciscan Bedrock.

[1]Senior Project Engineer, Woodward-Clyde Consultants, San Francisco, California
[2]Senior Associate, Woodward-Clyde Consultants, San Francisco, California
[3]Section Engineer, Structural Design Section, Bureau of Engineering, D.P.W., City and County of San Francisco

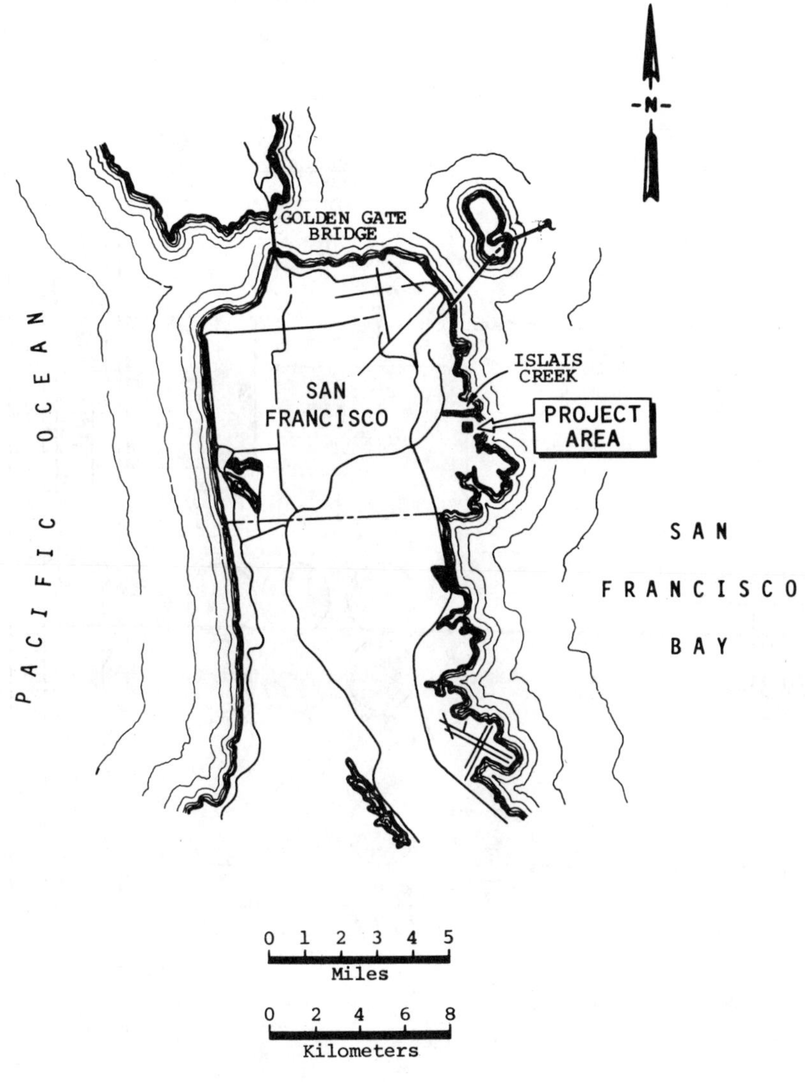

Figure 1 - LOCATION PLAN

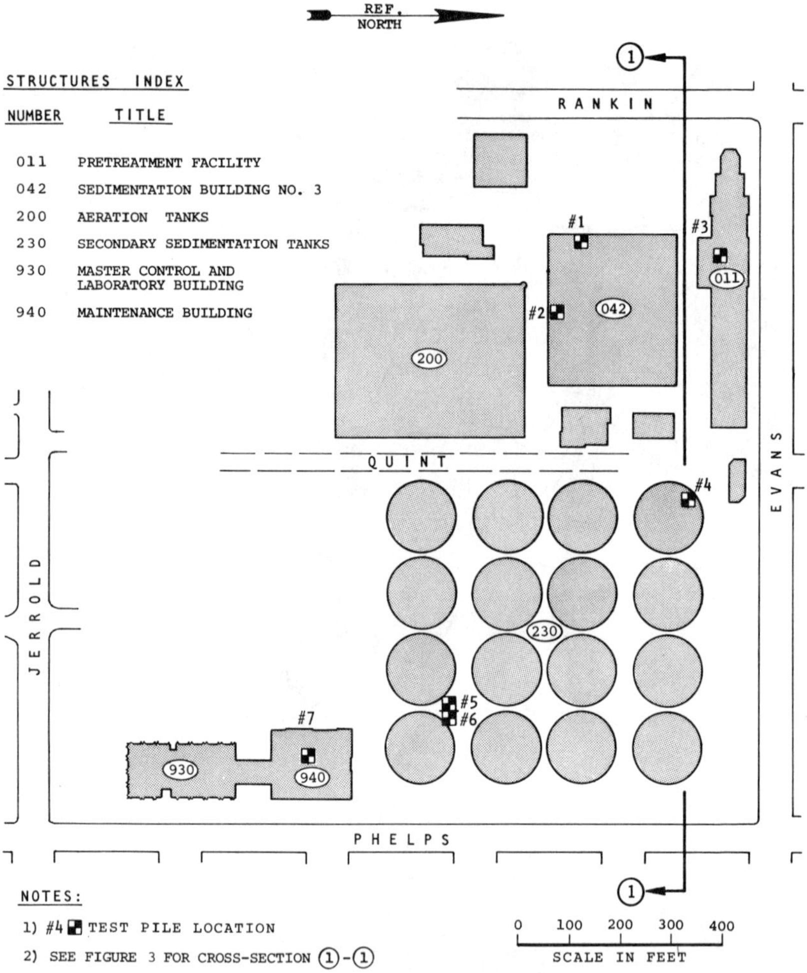

Figure 2 - SITE PLAN (1 ft = 0.305 m)

The fill contains random, but significant concentrations of obstructions--abandoned foundations, timber piles and utilities, and buried scrap metal. Groundwater is typically encountered between depths of 4 ft to 10 ft (1.2 m to 3 m).

The site stratigraphy was developed during the Quaternary Period in response to worldwide sea level fluctuations during glacial advance and retreat.

The thickness of the different layers varies significantly across the site. The fill and Bay Mud overlie medium dense to dense sands at depths ranging from 20 ft to 60 ft (6 m to 18 m). The sands are underlain by--and interbedded with--stiff to very stiff, overconsolidated silty clays, known locally as old Bay Mud.

The borings drilled at the site indicated the presence of two distinct sand layers. For brevity, these will be referred to as the "upper sands" and the "lower sands."

The "upper sands" directly underlie the Bay Mud and have a base elevation ranging generally from 940 to 930*. The thickness of the layer encountered in the test borings varies from a maximum of 43 ft (13 m) to a minimum of 5 ft (1.5 m) near the Evans-Rankin street corner.

The "lower sands" have a base elevation ranging from 897 to 884. The thickness of the layer encountered in the test borings varies from a maximum of 42 ft (12.8 m) to a minimum of 20 ft (6 m).

The old Bay Mud beneath the sand layers overlies colluvium which in turn overlies bedrock of the Franciscan Assemblage, at depths ranging from 100 ft to 230 ft (30.5 m to 70 m).

A generalized subsurface profile is shown in Figure 3.

FOUNDATION SUPPORT

It was decided to support the new structures on end-bearing piles driven into either the "upper sands" or the "lower sands." Twelve-inch (30.5 cm) square precast prestressed concrete piles with a gross design load of 100 tons (890 kN) each were selected as the optimum size for foundation support. The gross loading includes significant downdrag loads as well as the structural loads.

West of Quint Street the "upper sands" thin out as indicated in Figure 3. There was some uncertainty about whether the "upper sands" would prove inadequate for the pile design loads, requiring the piles to extend downwards to the "lower sands." Further, approximately 10,000 piles were required for the support of the new structures.

*Project Datum is Mean Sea Level +1008.5 ft (307.5 m)

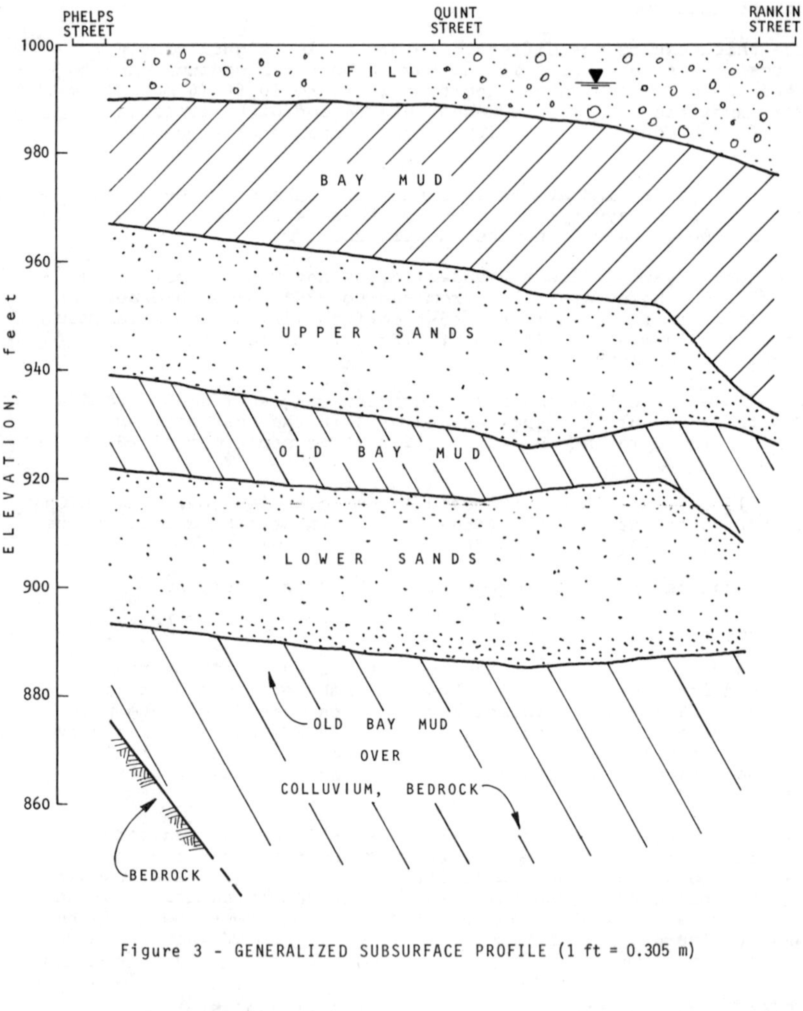

Figure 3 - GENERALIZED SUBSURFACE PROFILE (1 ft = 0.305 m)

Based on the above considerations, it was decided that it would be cost-effective to refine the design pile tip elevations by implementing a comprehensive indicator pile/load test program prior to general bidding.

INDICATOR PILE/LOAD TEST PROGRAM

A total of 109 indicator piles and 7 load test piles were driven across the site. All of the piles were driven with a Delmag D-30 diesel hammer having a maximum rated energy of 54,250 ft-lbs (73 kJ).

The locations of the load test piles are shown in Figure 2. Sixty-five of the indicator piles were driven west of Quint Street, where the "upper sands" were thinning out, and the balance were driven east of Quint Street. The indicator piles were uniformly distributed in the areas of the major structures.

All indicator piles were placed in 16 in. (40.6 cm) diameter holes predrilled through the fill and then were driven to a nominal refusal of 100 blows per ft (30 cm) of penetration. Load test piles were driven in predrilled and cased holes extending through the fill, terminating in the underlying Bay Mud. The driving criteria for the load test piles were individually modified based on the results of driving records and dynamic measurements on indicator piles driven in the area surrounding the load test piles. Several of the test piles were intentionally stopped at only moderate driving resistances (e.g., Test Piles 2, 3, and 4) in order to broaden the range of load test capacities.

Two of the piles (Test Pile 1 and Indicator Pile 107) were jetted through the upper sand layer and then driven into the lower sand layer.

Dynamic Monitoring

All of the 7 load test piles and 53 of the indicator piles were dynamically monitored utilizing an electronic pile driving analyzer system. The analyzer system was initially developed at Case Western Reserve University by Dr. George Goble (1, 2). The method involves measurement of force and acceleration at the pile top during driving. This is done by attaching piezoelectric transducers to the pile. Specifically, two strain gages and two accelerometers are mounted approximately 3 ft (1 m) below the pile top so as to allow sufficient clearance for the hammer and driving cap. Installation at this point eliminates the hammer and cushioning system as a variable. At each hammer impact, signals from the transducers are fed to a small field computer.

The system consists of the pile driving analyzer (computer), digital recorder, tape recorder, oscilloscope, strain transducers, accelerometers and all necessary connections and accessories for monitoring the piles. The equipment is illustrated schematically in Figure 4.

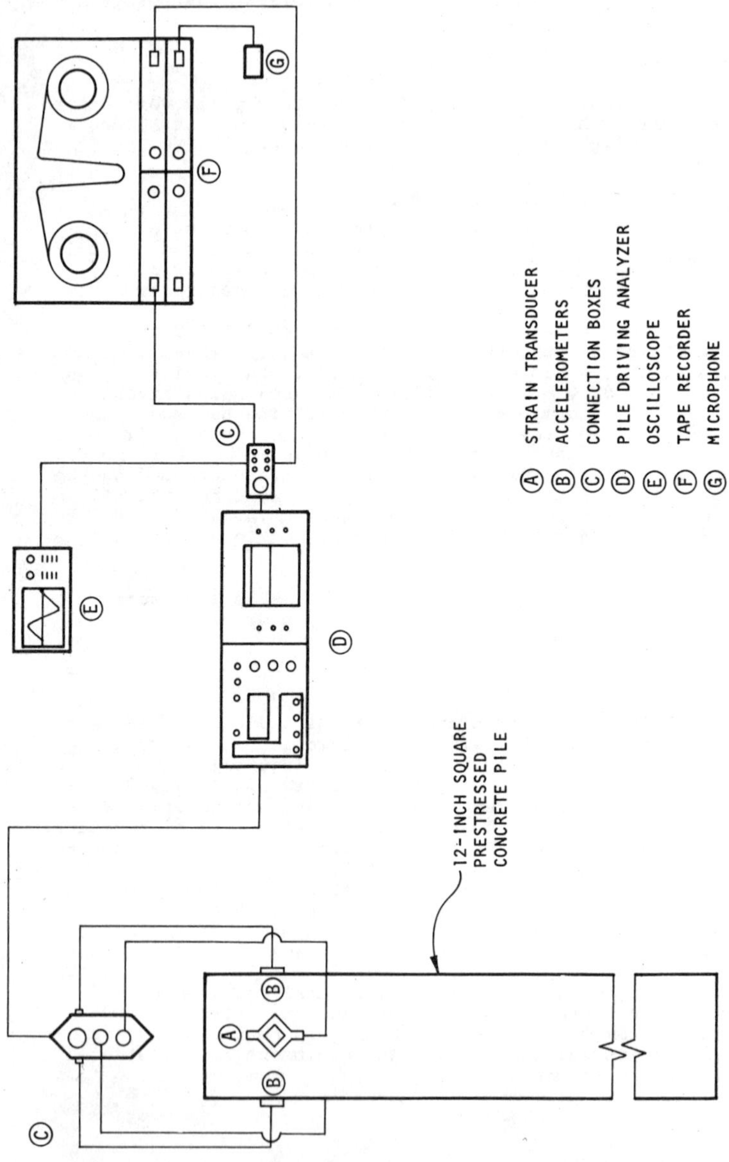

Figure 4 - DYNAMIC PILE ANALYZER SYSTEM

The analyzer is designed to provide the following data in the field within milliseconds after hammer impact:

1. Predicted ultimate bearing capacity.
2. Transferred energy to the pile.
3. Maximum compressive force in pile.
4. Maximum tension force in pile.
5. Maximum impact velocity.
6. Maximum acceleration.

Any one of the six items can be visually shown on a digital display on the front of the instrument. In addition, any three of the items can be printed on a paper tape to provide a permanent record. Acceleration and force values are also taken from the computer and recorded on magnetic tape for later processing. This latter aspect is quite important if malfunctions occur in the field since data can be taped even if the computer if malfunctioning.

Force-time and velocity-time traces are displayed on the oscilloscope; typical traces are shown in Figure 5.

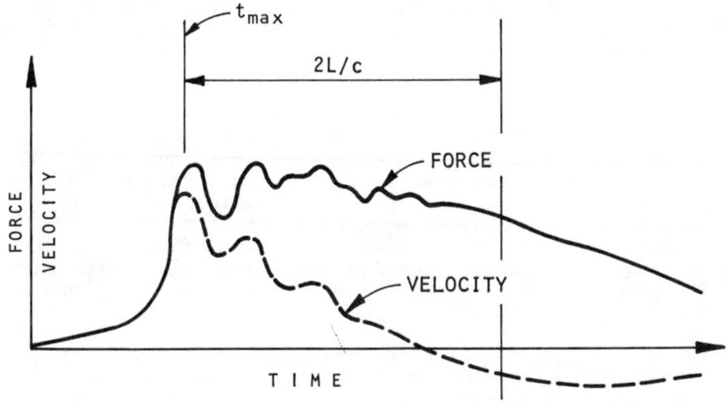

Figure 5 - Typical Force-Time and Velocity-Time Traces

The field computer estimates the pile capacity based on the following relationship:

$$R(t) = \frac{F(t_{max}) + F(t_{max}+2L/c)}{2} + \frac{Mc\{V(t_{max}) - V(t_{max}+2L/c)\}}{L\quad 2} - D \qquad (1)$$

Where: R(t) = time varying soil resistance
F(t) = time varying hammer force
L = pile length
c = wave propagation speed
M = pile mass
V(t) = time varying pile top velocity
D = dynamic damping of the system, a function of both pile tip velocity and a soil damping factor, J.

The analysis assumes that the pile is a rigid body struck by a time varying hammer force, F(t) which is resisted by the interia of the pile and soil resistance forces, R(t), according to Newton's second law.

The development of the procedure has shown that the resistance can be more accurately predicted by taking the force and velocity measurements over an average time period. Consequently, the force is averaged over a time beginning at t_{max} (generally taken as the force at the time of maximum velocity) and at a time 2L/c later (2L/c is the time the stress wave travels from the pile top to the tip and then returns to the top).

Input data that must be provided in the field thus includes the length of the pile (L), the weight (mass) of the pile (M) and the wave velocity (c).

Hammer energy transferred to the pile can be calculated from the expression:

$$E(t) = \int_0^t F(t)V(t)dt \qquad (2)$$

where energy (E), force and velocity are all functions of time. This is the energy actually available to the pile and eliminates losses that have occurred within the hammer and cushioning system.

At the completion of the field program, the analyzer equipment was brought back to the office in order to correlate the data with the load test results. The taped data was then played back through the analyzer's computer using varying J values (the soil-dependent damping factor) until the predicted capacity values were in reasonably good agreement with the load test results. Repeating this process for all seven of the load tests yielded a range of J values from 0.16 to 0.23. All of the monitored indicator piles were then re-analyzed using J factors of 0.16 and 0.23.

The effect of varying the J-factor between 0.16 and 0.23 on the predicted capacity values decreased as harder driving (higher blowcounts) was encountered. Typically the difference was between 15 and 20 percent when the driving resistance was over 30 blows per foot (30 cm). This difference decreased to 10 to 15 percent when the driving resistance was over 60 blows per foot (30 cm). The capacity curves presented in Figures 6 through 8 represent the average of the two values for J = 0.16 and J = 0.23.

Load Tests

The vertical compression pile load tests were conducted in accordance with ASTM Designation D1143-74, "Standard Method of Testing Piles Under Axial Compressive Load." The standard loading and measuring procedures of the method were used.

A calibrated pressure gage and 500 ton (4450 kN) hydraulic jack system were used to apply the compressive load. Vertical movements of the pile top were measured by two dial gages fixed to a pair of reference beams adjacent to two sides of the pile. Test Piles 1, 2, 3, 5, and 6 also contained two internal strain rods (telltales). Movement of each telltale was measured by a

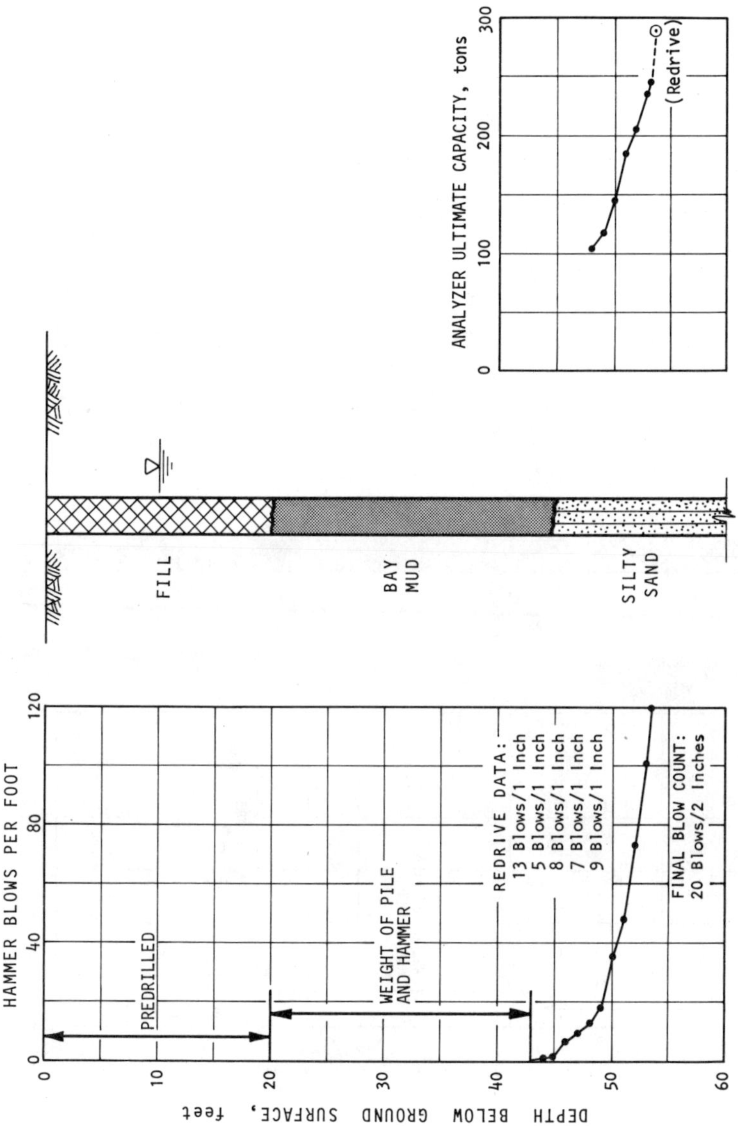

Figure 6 - ANALYZER RESULTS, INDICATOR PILE #29 (1 ft = 0.305 m; 1 ton = 8.9 kN)

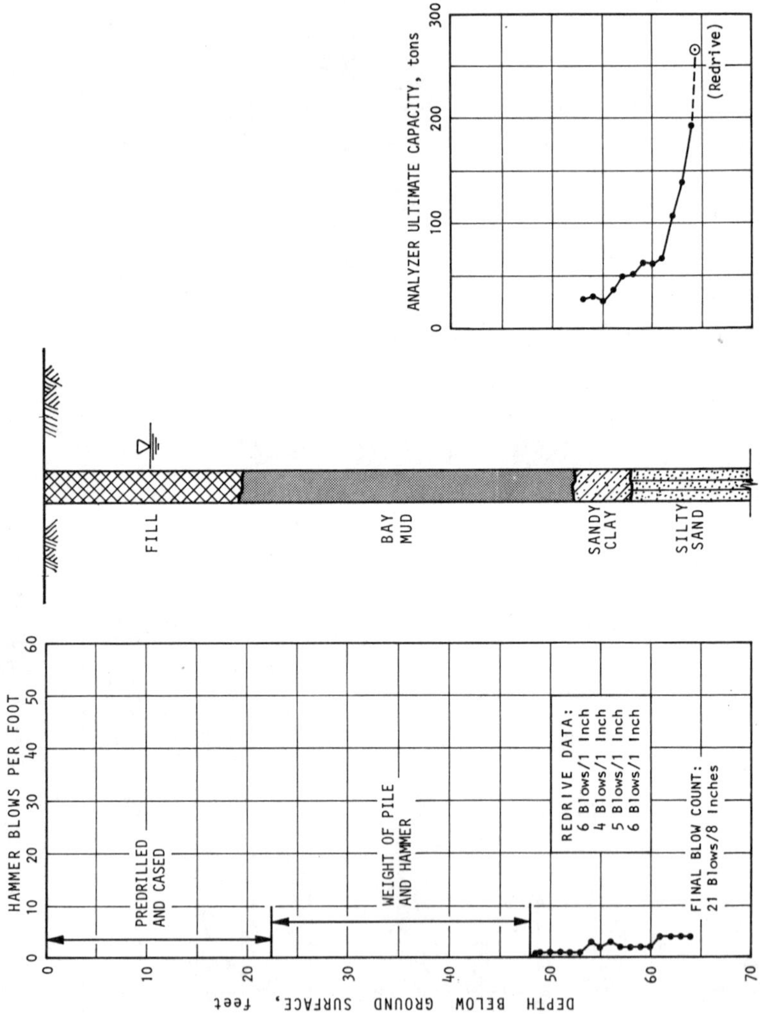

Figure 7 - ANALYZER RESULTS, LOAD PILE #2 (1 ft = 0.305 m; 1 ton = 8.9 kN)

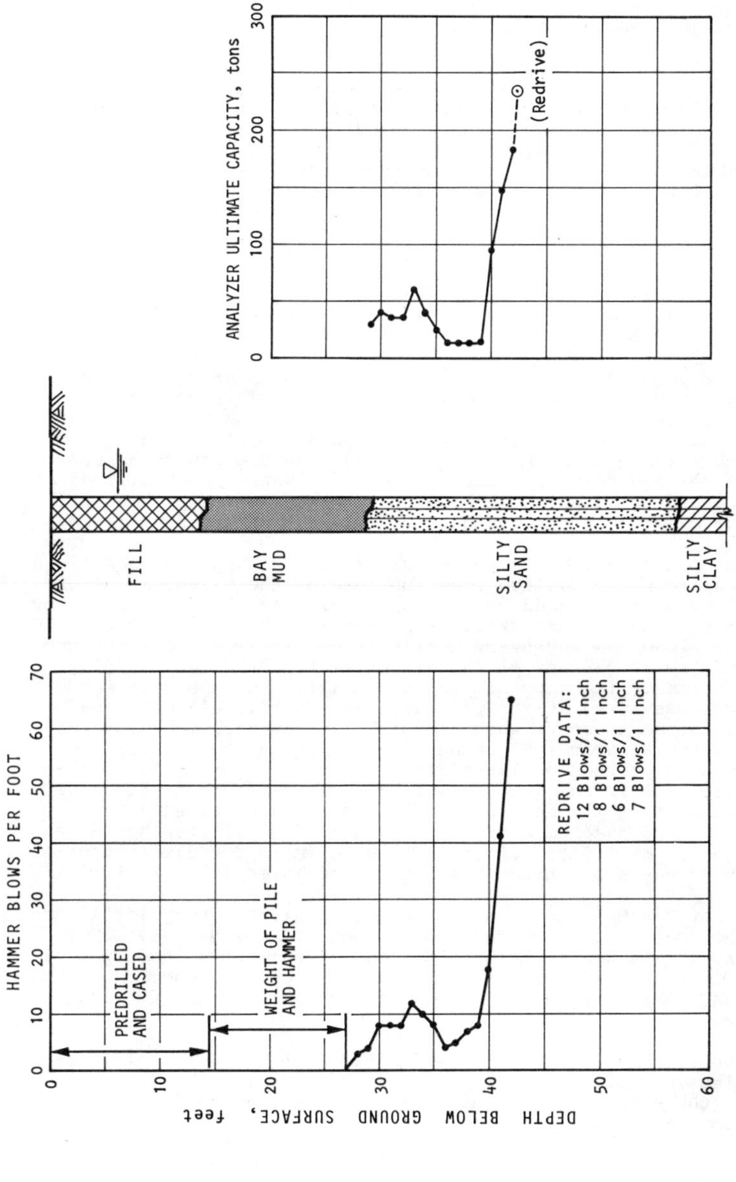

Figure 8 - ANALYZER RESULTS, LOAD PILE #6 (1 ft = 0.305 m; 1 ton = 8.9 kN)

dial gage mounted to the compression members on top of the pile. Test Pile 2 also had a pair of internal electric strain gages embedded just above the pile tip.

In general, the test piles were loaded in 25 ton (222 kN) increments to 200 tons (1780 kN) (200 percent of design load), which was sustained for 24 hours. Then the piles were unloaded in 50 ton (445 kN) decrements and allowed to rebound under zero load for 12 hours. The piles were then reloaded in increments until rapid settlement (plunging) occurred. Test Pile 1 failed structurally prior to attaining ultimate capacity.

DISCUSSION

Results of Pile Load Tests

Typical results of the load tests are presented graphically in Figures 9 and 10.

The piles sustained the design load of 100 tons (890 kN) with vertical movement of the pile tops of 0.3 in. (7.6 mm) or less. Pile behavior at maximum applied compressive load varied. Most piles exhibited rapid settlement at a definite load, and loads above this value could not be sustained with the equipment used for the tests. For a few test piles there was a level of load which caused total vertical pile top movement greater than one inch (25.4 mm), but the movement was gradual rather than rapid.

Because of the variable behavior of the piles, and the lack of an easily discernible failure load for some of the piles, a uniform criteria was established to define the ultimate load sustained by each pile. The failure criteria selected is based on the premise that the tip of an end bearing pile in cohesionless soil must be displaced by 10 percent of its effective tip diameter to fully mobilize its ultimate point resistance (5). Because pile movements were measured at the pile top and not tip, ultimate load was considered to occur when the pile top downward movement was equal to the elastic shortening of the pile due to the applied load plus 10 percent of the effective tip dimension. Since the test piles are 12 in. (30.5 cm) square, 10 percent of effective tip dimension is 1.2 in. (30.5 mm). Using this criteria, test piles involved reached ultimate load when the pile top had moved on the order of 1.5 to 1.8 in. (38 to 46 mm). The ultimate pile capacities based on this criteria are tabulated in Table 1.

Effects of Bay Mud - It would have been desirable to predrill and case the load test piles down to the base of the Bay Mud layer. However, it was feared that even slightly eccentric loading could result in structural pile failure on such long unsupported column sections. The structural failure of Test Pile 1 supports that concern.

Consequently, the casing was terminated just below the fill, and the piles were driven through the Bay Mud layer. The load test results therefore include the resistance of the Bay Mud around the pile shaft.

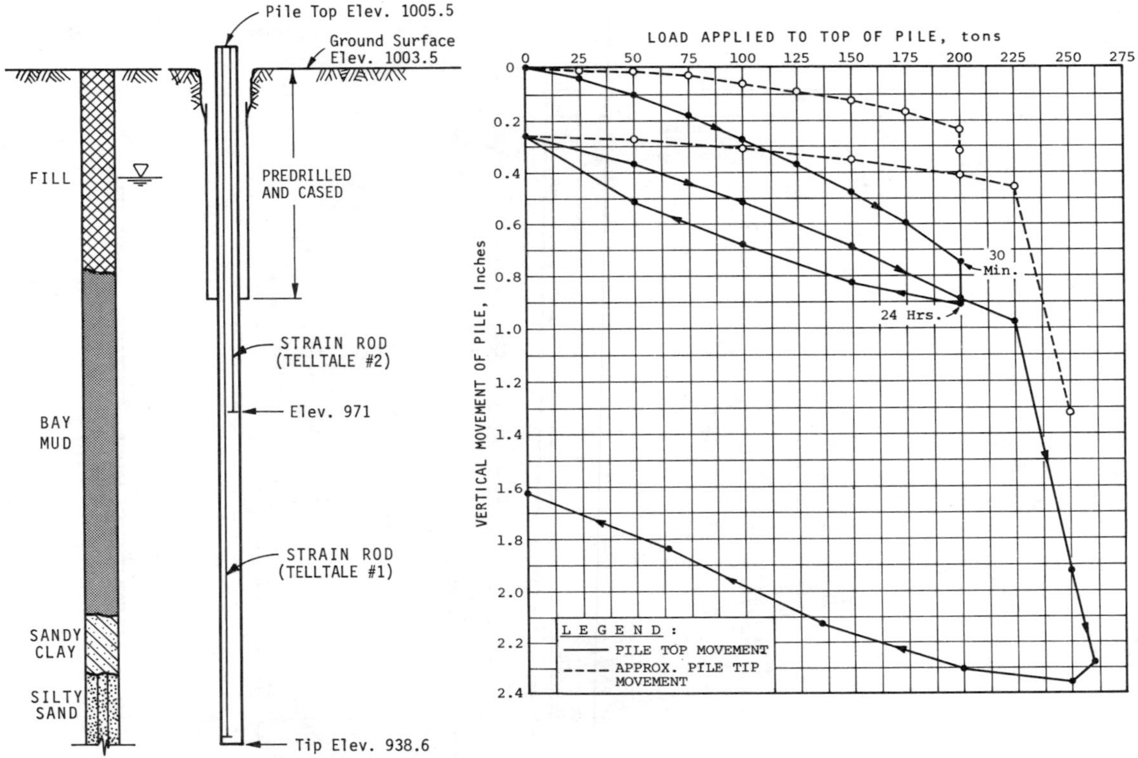

Figure 9 - LOAD TEST PILE #2 (1 in. = 25.4 mm; 1 ton = 8.9 kN)

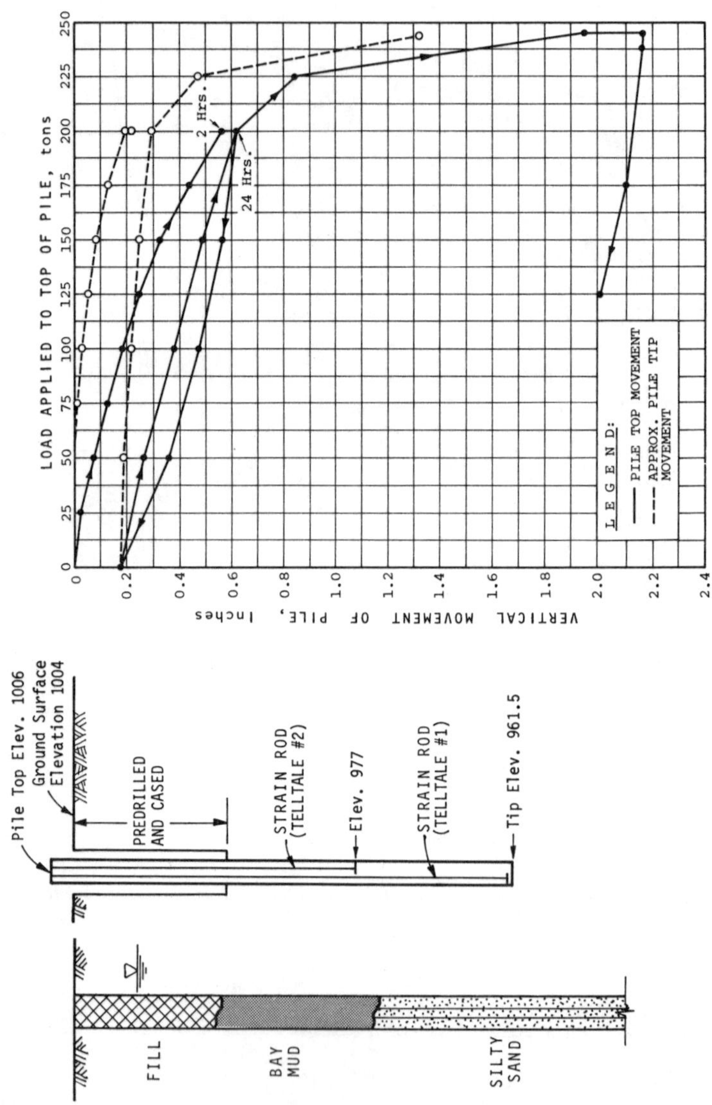

Figure 10 - LOAD TEST PILE #6 (1 in. = 25.4 mm; 1 ton = 8.9 kN)

In order to evaluate the bearing capacity of the sand layer(s), it is necessary to estimate the amount of shaft resistance that results from the shear resistance of the Bay Mud.

During pile driving, the Bay Mud is completely remolded, and offers negligible shaft resistance. Subsequent to driving, the mud recovers a certain amount of shear strength from thixotropic effects, and also regains shear strength from the effects of reconsolidation of the soil surrounding the pile shaft. The time required for a strength regain of 70 to 90 percent may require a period ranging from 10 to 30 days (4).

Evaluation of the Bay Mud shaft resistance was based on analysis of several independent sets of data: telltale strain data, internal tip strain gage data, and prior data regarding the shear strength of the Bay Mud. The values tabulated in Table 1 are the best estimates of Bay Mud shaft resistance based on the various data sources and interpretations.

Dynamically Monitored Piles

As discussed previously, the results of the analyzer monitoring were "calibrated" against the pile load test results by varying the single major independent parameter of the system--the damping factor J--to improve the correspondence between predicted and actual ultimate capacities (see Table 1).

Driving Behavior

A review of the data suggests a significant difference in driving (and capacity) behavior for that portion of the site west of Quint Street compared to the areas east of Quint.

West of Quint, the piles "took-up" rather quickly in the bearing sand layer, without a large amount of "set-up" or "freeze" on re-driving. Indicator Pile 29 (Figure 6) is a representative record for the driving behavior west of Quint. As discussed previously, Load Test Pile 2 (west of Quint) was intentionally stopped at a relatively low driving resistance, so its driving behavior (Figure 7) is not representative of the general trend observed.

East of Quint the piles penetrated the bearing layer to depths of 15 to 20 ft (4.5 to 6 m) with only moderately hard driving. However, redriving showed significant "set-up" both in blow-count resistance and analyzer capacities*, which were verified by the load tests. Load Test Pile 6 (Figure 8) is representative of this driving behavior. This kind of driving behavior is not uncommon for medium-dense to dense sands, as compared to very dense sands.

Predictive Criteria

Based on the above analyses, the following predictive criteria were established for the dynamically monitored piles:

*Redrives were within 24 hours or less, so the influence of the Bay Mud would be negligible.

Table 1

Test Pile No.	Length, ft	Tip Elevation ft	Load Test Ultimate Capacity tons	Estimated Bay Mud Shaft Resistance, tons	Estimated Bearing Layer Capacity tons	Predicted Analyzer Capacity, Tons	
						Initial	Redrive
1*	89	917.3	300+	60	240+	205	250
2	67	938.6	250	50	200	195	265
3	57	946.9	200	30	170	160	170
4	52	950.0	220	40	180	85	-
5	49	957.4	350	15	335	220	-
6	45	961.5	245	15	230	185	235
7	45	961.7	365	10	355	145	-

*Test pile 1 failed structurally approximately 1/2 minute after applying 300-ton load, terminating load test

Note: 1 ft = 0.305 m; 1 ton = 8.9 kN

West of Quint: 200 tons (1780 kN) analyzer capacity
East of Quint: 150 tons (1335 kN) analyzer capacity
+50 tons (445kN) set-up

The results of the dynamically monitored piles were extrapolated to adjacent indicator piles that were not monitored. Based on these data, recommended contours were developed for design pile tip elevations. Figure 11 shows these contours in the area west of Quint where the variation in tip elevations was most significant. With the exception of the cross-hatched areas shown on Figure 11, all of the piles were to bear within the "upper sands." Piles in the cross-hatched areas were to be jetted through the "upper sands" and bear in the "lower sands."

Order Lengths and Pay Items

The Bureau of Engineering established pile tip elevations and pile order lengths on a structure-by-structure basis, based on the recommended pile tip elevation contours. The structural tolerance for pile butt elevations was generally less than 3 in. (7.6 cm), plus or minus. Since it was anticipated that pile buildups would be more costly than pile cutoffs, the order lengths were somewhat biased towards cutoffs, as opposed to buildup.

The pile subcontract contained the following pay items: furnish piles (per l.f.)*; drive piles (ea.)*; pile cutoff (ea.); and pile extension (buildup) (ea.). These fairly explicit pay items were intended to minimize uncertainties or "second guessing" among prospective bidders.

The refined tip elevations developed from the program reduced the total lineal feet of piling from an initial estimate of 500,000 l.f. to 390,000 l.f. (152,500 to 118,900 l.m.), for an estimated construction savings of $750,000. Further, the results of the program were available to all prospective bidders, and probably helped to reduce the uncertainties normally associated in bidding a pile job. The difference between the Engineer's estimate of $6,000,000 for the foundation piling and the $4,350,000 bid of the successful bidder probably reflects the reduction of uncertainties, at least in part.

Driving Criteria

The following driving criteria were developed exclusively for the Delmag D-30 hammer:

1. Drive piles to designated tip elevation and stop driving if driving resistance is 40 blows per ft (30.5 cm) or more.

2. Drive piles below designated tip elevation until minimum resistance of 40 blows per ft (30.5 cm) is attained.

---

*Also includes replacement piles for piles encountering obstructions.

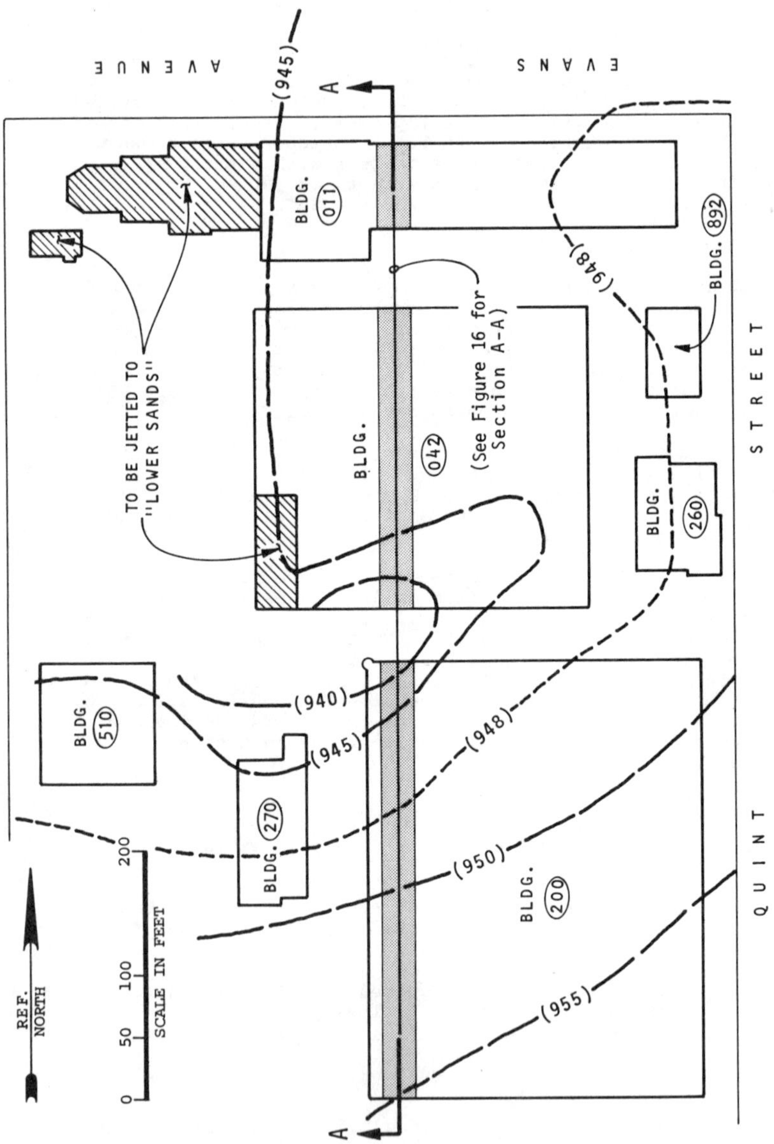

Figure 11 - RECOMMENDED PILE TIP ELEVATIONS (1 ft = 0.305 m)

3. Terminate driving above designated tip elevation when "refusal" driving of 110 blows per ft (30.5 cm) (or more) is achieved.

The criteria for other hammers was to be established by the driving of a minimum of ten additional indicator piles for each alternative hammer.

WAVE EQUATION COMPARISON

The scope of the program did not include wave equation analyses. However, in preparation for this paper, a limited number of wave equation analyses were run for Test Pile 2.

The analyses utilized the WEAP wave equation program developed by Messrs. George G. Goble and Frank Rausche (3) for the Federal Highway Administration. The program incorporates a fairly realistic thermodynamic model for diesel hammers. Figure 12 presents a comparison of the force-time trace for Test Pile 2 to the force-time trace predicted by the WEAP program. It can be seen that the agreement is quite good.

The following soil parameters were used in the analyses:

Quake	0.1 in. (2.5 mm)
Shaft Damping, J_s	0.10
Point Damping, J_p	0.20 and 0.25

Figure 13 presents a comparison of wave equation predictions with the estimated driving resistance (load test data corrected for estimated Bay Mud resistance and "freeze"). While the analyses were strictly applicable only to Test Pile 2, the comparison appears satisfactory.

In Figure 14, the wave equation predictions are compared with analyzer predictions for dynamically monitored piles surrounding Test Pile 2. While there is some experimental scatter, the two predictive methods appear to be relatively consistent.

DELIVERED ENERGY

In a limited number of cases, the ram stroke of the hammer was observed for dynamically monitored piles. The energy transferred to the piles (as computed by the analyzer) is plotted versus ram stroke in Figure 15. The lack of correlation between stroke and delivered energy for diesel hammers has been noted previously (2), but seems worth repeating.

PRODUCTION DRIVING

Production pile driving commenced in September, 1978. At the time of writing, approximately 7000 of the 10,000 foundation piles have been driven. Of the piles driven, 1.5 percent are replacement piles, and 2.5 percent required buildup. Approximately 40 percent of the piles required cutoff, for reasons discussed previously.

The resident engineering staff and pile driving inspectors have been provided by DeLeuw-Greeley-Hyman, the Construction Managers for the project for the City and County of San Francisco.

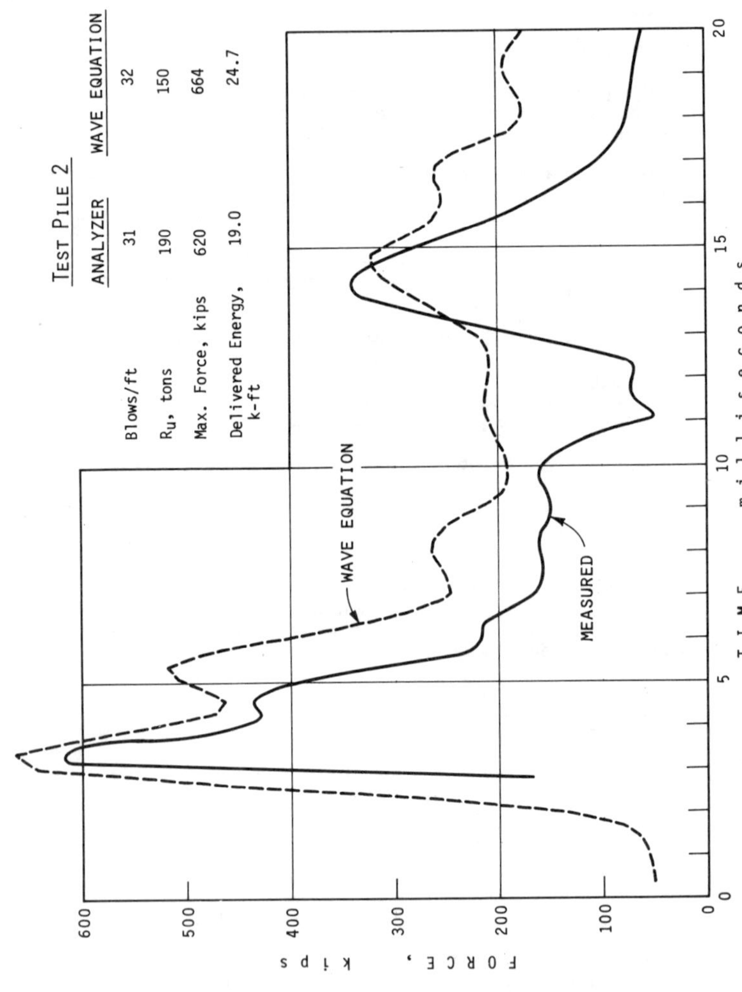

TEST PILE 2

	ANALYZER	WAVE EQUATION
Blows/ft	31	32
R_u, tons	190	150
Max. Force, kips	620	664
Delivered Energy, k-ft	19.0	24.7

Figure 12 - MEASURED FORCE - TIME vs. WAVE EQUATION
(1 ft = 0.305 m; 1 ton = 8.9 kN; 1 kip = 4.45 kN; 1 k-ft = 1.35 kJ)

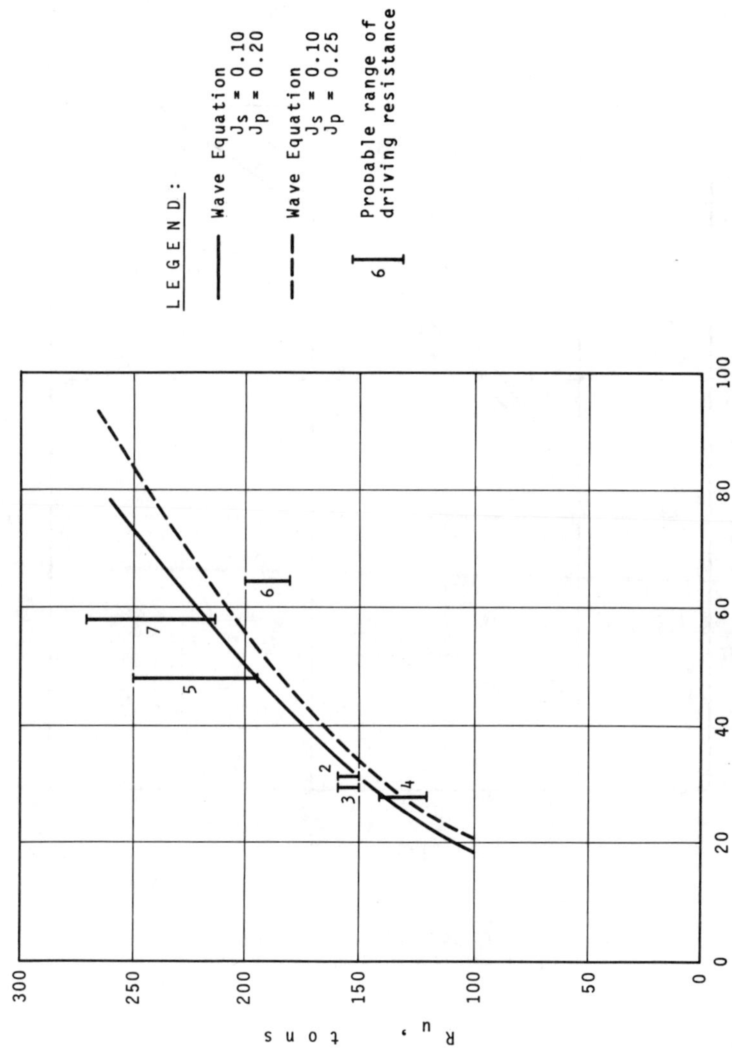

Figure 13 - WAVE EQUATION vs. ESTIMATED DRIVING RESISTANCE
(1 ft = 0.305 m; 1 ton = 8.9 kN)

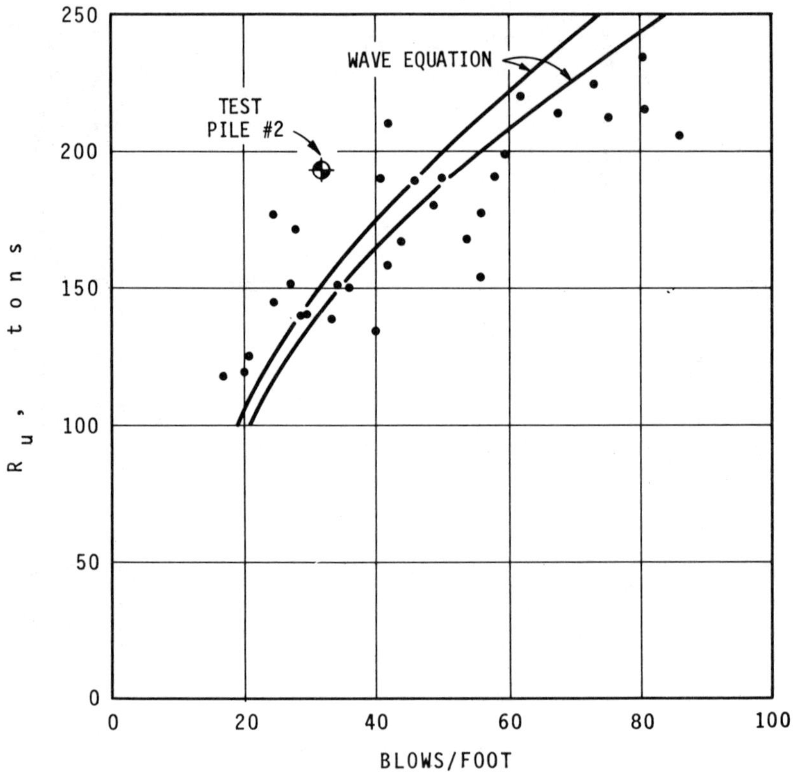

Figure 14 - WAVE EQUATION vs. ANALYZER PREDICTIONS
(1 ft = 0.305 m; 1 ton = 8.9 kN)

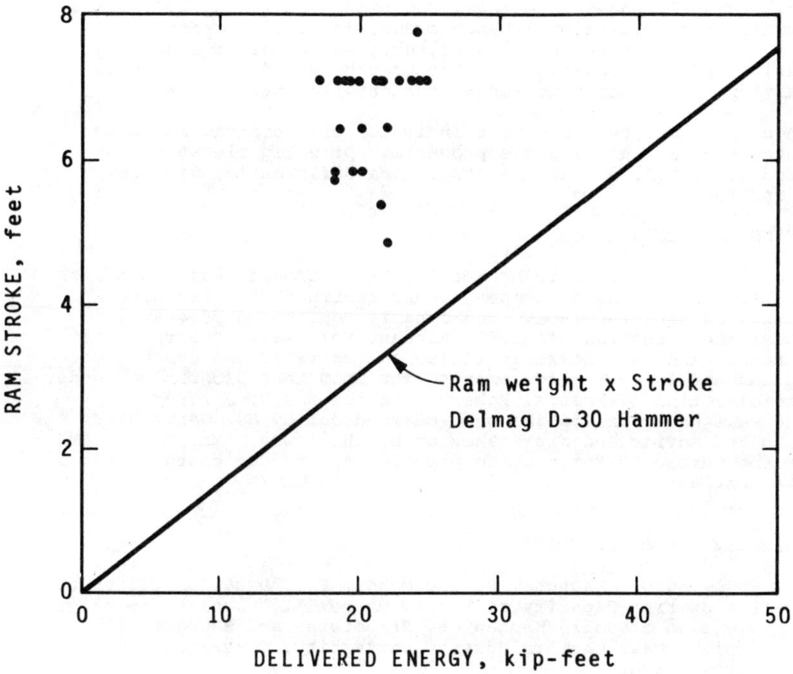

Figure 15 - RAM STROKE vs. DELIVERED ENERGY
(1 ft = 0.305 m; 1 k-ft = 1.35 kJ)

Space does not allow for a detailed analysis of production pile driving results. However, Figure 16 compares production pile tip elevations versus recommended tip elevations for a 30 ft (9 m) wide (3-pile bent) strip cutting north-south through the major structures west of Quint Street.

The Delmag D-30 hammer was not used in production driving. Rather, four different hammers were used at different times: 1) Delmag D30-02; 2) Kobe K-25; 3) BSP B-25; and 4) Linkbelt 660.

Figure 16 indicates that the production piles generally follow the "predicted" tip elevations, especially in the Elevation 935-940 "valley" between Buildings 200 and 042. The higher production tip elevations in Building 011, the northerly portion of Building 042, and the southerly portion of Building 200, are a result of a relaxation (change order) in driving criteria. If the minimum required driving criteria was attained above the designated tip elevation, the contractor could elect to stop driving if he agreed to absorb the cutoff cost.

In conclusion, the load test/indicator pile program has been successful in refining the production pile tip elevations. Production pile driving is proceeding efficiently, with few surprises.

ACKNOWLEDGMENTS

The indicator pile driving and load test program was accomplished by Peter Kiewit Sons' Company under contract with the City and County of San Francisco. The dynamic monitoring program was under the direction of Dr. D. Michael Holloway. Messrs. Scott Huntsman and Gary Locke installed and operated the dynamic measuring and analyzing equipment. The load test program was under the direction of Messrs. Robert Johnston and R. Jeffrey Dunn. The wave equation studies were carried out by Ms. Carol Ries with the advice and assistance of Dr. Holloway. Mr. Simon Tam of the Bureau of Engineering provided day-to-day coordination of the contract.

APPENDIX I - REFERENCES

1. Goble, G.G., Rausche, F. and Moses, F., "Dynamic Studies on the Bearing Capacity of Piles, Phase III," *Report No. 48*, Division of Solid Mechanics, Structures and Mechanical Design, Case Western Reserve University, Cleveland, Ohio, August 1970.

2. Goble, G.G., Kovacs, W.D. and Rausche, F., "Field Demonstration: Response of Instrumented Piles to Driving and Load Testing," *Proceedings of the Specialty Conference on Performance of Earth and Earth-Supported Structures*, American Society of Civil Engineers, Vol. 3, June, 1972, pp. 3-38.

3. Goble, G.G. and Rausche, F. "Wave Equation Analysis of Pile Driving," *FHWA Report No. IP-76-14*, U.S. Department of Transportation, Federal Highway Administration, Washington, D.C., July, 1976.

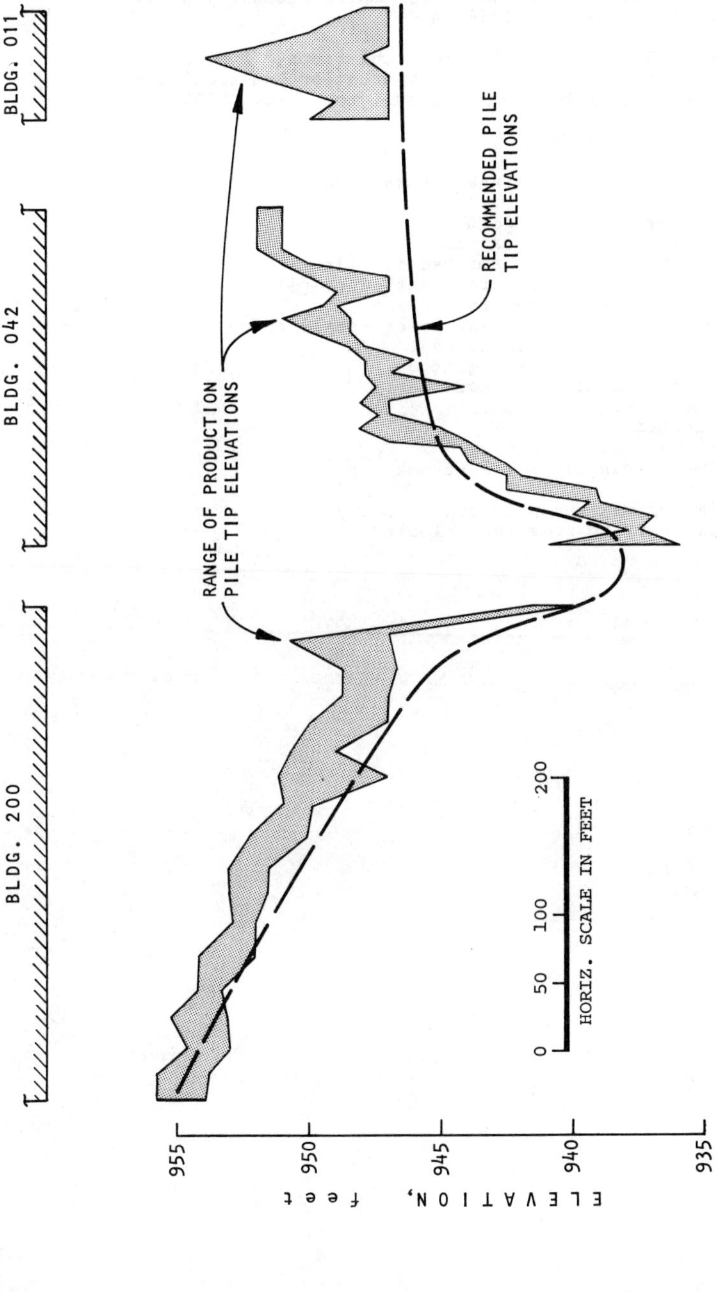

Figure 16 - RECOMMENDED vs. PRODUCTION PILE TIP ELEVATIONS (1 ft = 0.305 m)

4. Seed, H.B. and Reese, L.C., "The Action of Soft Clay Along Friction Piles," *Transactions*, American Society of Civil Engineers, 1957, pp. 731-754.

5. Vesic, A.S., "Design of Pile Foundations," *Synthesis of Highway Practice No. 42*, Transportation Research Board, National Research Council, Washington, D.C., 1977.

APPENDIX II - NOTATION

The following symbols are used in this paper:

c = wave propagation speed
D = system (analyzer) damping
E = maximum energy transmitted to pile
$E(t)$ = time varying energy transmitted to pile
$F(t)$ = time varying hammer force
J = soil damping factor (analyzer--shaft and point not differentiated)
J_p = point damping (wave equation)
J_s = shaft damping (wave equation)
L = pile length
M = pile mass
R_u = ultimate pile capacity
$R(t)$ = time varying soil resistance
t = time
t_{max} = time of maximum pile top velocity
$V(t)$ = time varying pile top velocity

INSTRUMENTED DRILLED SHAFTS IN CLAY-SHALE

by Patrick M. Goeke,[1] A.M. ASCE and Paul A. Hustad,[2] Aff. ASCE

INTRODUCTION

During the past 15 years extensive research has occurred in the area of the soil-structure interaction of drilled shaft foundations. This research was concentrated on drilled shafts in clay (5,8) and hard rock sockets (10,13). However, less is known about the behavior of drilled shafts in soft rocks such as shale or clay-shale.

When designing drilled shafts in clay-shale, it is common practice in many design firms to increase the diameter of the shaft base with a bell and to neglect skin resistance along the sides of the shaft. However, load tests on drilled shafts in clay have indicated that the side shaft resistance often exceeds the tip capacity, and thus, neglecting the side resistance in many cases is too conservative.

In 1976, Burns & McDonnell Engineering Company was retained to design a 400 megawatt power station in southeastern Oklahoma. During the design phase of the project, two, full-scale, instrumented drilled shafts were axially load tested to investigate the side and end bearing resistance of shafts supported in the clay-shale found beneath the proposed powerhouse area. The tests were conducted on 30 inch (76 cm) diameter shafts which penetrated 11 and 18.5 ft. (3.3 m and 5.6 m) into the clay-shale.

The principle aims of this paper are to present the results and conclusions from the field tests, to correlate these tests with other tests in soft rocks, and to review an existing design procedure for clay-shales.

SOIL CONDITIONS

The subsurface conditions in the powerhouse area were of major concern as construction was concentrated in this area and individual column loads as large as 4600 kips (20,470 kN) required support from the foundation system. Figure 1 presents an idealization of the subsurface conditions in the immediate powerhouse area. A thin mantel of highly plastic clay overlies the Caddo and Kiamichi Formations of Cretaceous Age. The Caddo Formation, which represents the uppermost bedrock material capable of supporting the heavy powerhouse loading, is composed primarily of a brown to gray clay-shale that contains occasional thin limestone seams. The underlying

[1] Geotechnical Engineer, McClelland Engineers, Inc., St. Louis, Mo., formerly Geotechnical Engineer, Burns & McDonnell Engr. Co., Kansas City, Mo.

[2] Manager, Geotechnical Dept. Burns & McDonnell Engineering Company, Kansas City, Mo.

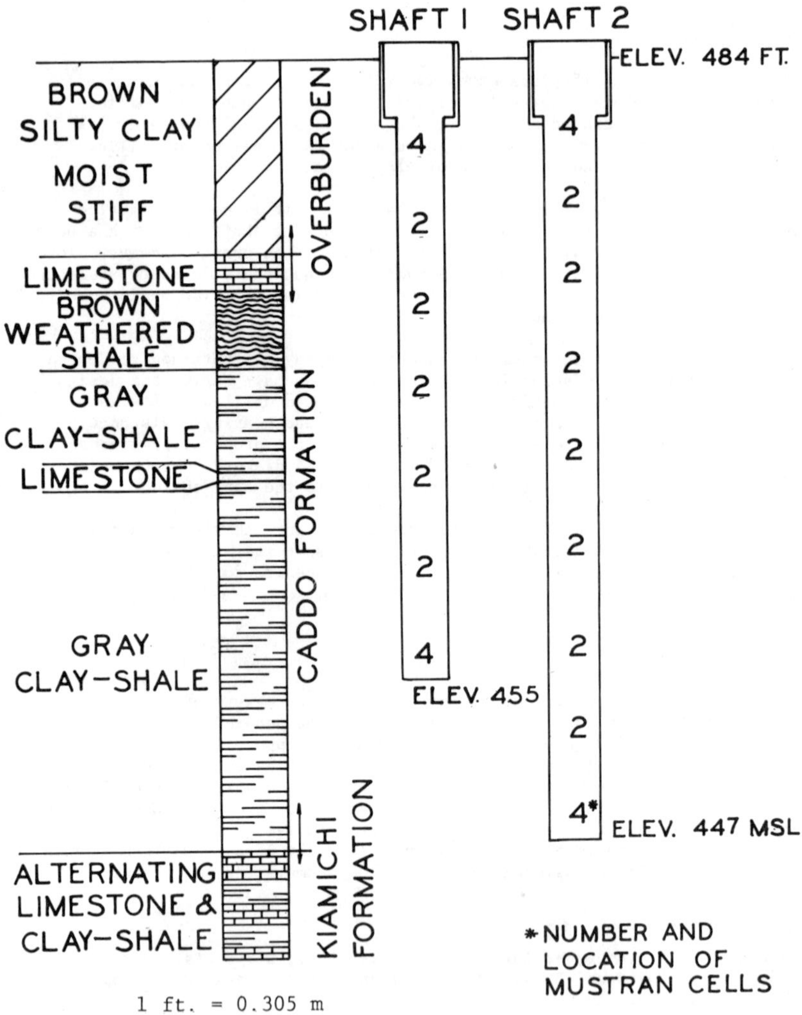

FIGURE 1. - SUBSURFACE PROFILE AND SHAFT EMBEDMENTS

Kiamichi Formation consists of thin, alternating layers of limestone and clay-shale.

Numerous laboratory unconfined compression and quick triaxial compression tests were performed on undisturbed Pitcher tube and NX-size core samples that were recovered from the clay-shale of each formation. All triaxial tests were run at a confining pressure equal to the estimated, effective overburden pressure. The triaxial and unconfined compression test results are in close agreement with one another. These tests reflect a wide variation indicating shear strengths ranging from 1.5 to 6.5 tsf (144 kPa to 623 kPa) as shown in Fig. 2. Although visual inspection of the bedrock samples during the subsurface investigation indicated increasing competence with depth, the test results indicate only a slight increase in shear strength with depth.

The erratic laboratory test results can be partially attributed to swelling of the shale in the core barrel during coring and to partial drying of the samples which resulted in micro-fissures forming in the samples. As a result of the inherent disturbance in sampling soft rocks, it is the writers' opinion that the shear strengths determined from the laboratory tests were considerably lower than the in-situ shear strength. The geotechnical properties are tabulated in Table 1, with the average and range of laboratory test results given for each stratum.

TEST SHAFT CONSTRUCTION DETAILS

Three anchor shafts and two test shafts were dry drilled using a Watson 3000 drill rig equipped with a 30 in. (76 cm) diameter H&T rock auger. Each anchor shaft was drilled a minimum of 7 ft. (2.1 m) into the underlying Kiamichi Formation to provide the necessary reaction capacity. Shafts 1 and 2 were drilled to depths of 29 ft. and 36.5 ft. (8.8 m and 11.1 m), respectively, as shown in Fig. 1. Shaft No. 1 derived total support from the Caddo Formation, and Shaft No. 2 developed side resistance from the Caddo Formation and end bearing resistance from the upper portion of the Kiamichi Formation. The two bearing elevations were selected for the test shafts to evaluate the bearing capacity of the proposed foundation bearing elevations and to determine the variation in side resistance capacity at varied penetration into the clay-shale.

Preconstruction excavation in the powerhouse area required the removal of approximately 11 ft. (3.4 m) of overburden before the plant foundations were installed. To simulate the proposed foundation system, each test shaft was isolated from the existing ground surface to a depth of approximately 11 ft. (3.4 m) to eliminate side resistance along the shaft above the planned design cutoff elevation. The isolation system consisted of styrofoam and a cardboard form.

To reduce variables between the test shafts and the plant foundations, the construction methods used to install the test shafts were the same as specified for the powerhouse foundation. The shafts were dry augered in an open hole and the concrete was placed with a tremie pipe on the same day the

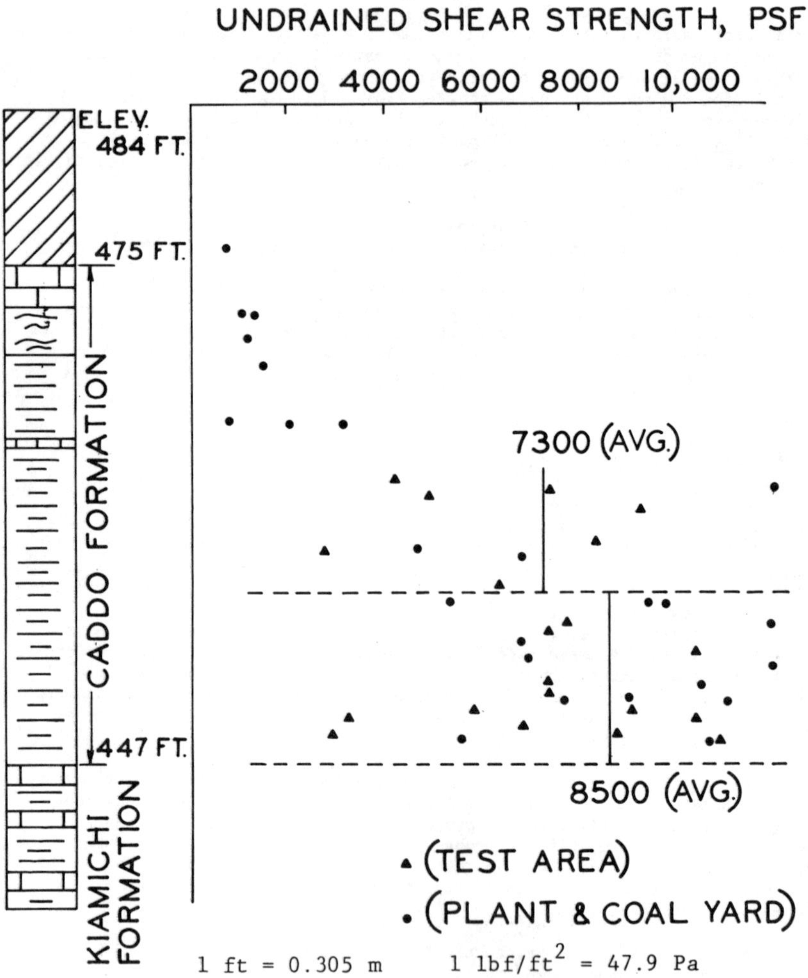

FIGURE 2. - SHEAR STRENGTH PROFILE

Table 1 - GEOTECHNICAL PROPERTIES

Soil Description	Depth, In Feet		Moisture %	Dry Density, In Pounds Per Cubic Foot	Atterberg Limits LL	Atterberg Limits PL	Compressive Strength, In Tons per Square Foot
Brown Silty Clay	0-8.7	Ave.	27	94	67	22	1.7
		Range	12-34	83-107	64-73	17-24	0.7-2.5
Gray Limestone	8.7-10.5	Ave.	NA	NT	NA	NA	NT
		Range					
Silty Clay to Weathered Clay-Shale	10.5-18.7	Ave.	23	98	58	19	1.2
		Range	17-32	96-102	55-62	18-21	0.6-2.6
Caddo Formation Gray Clay-Shale	18.7-37.3	Ave.	20	109	58	21	4.0
		Range	11-27	99-118	53-62	18-25	1.0-13.3
Kiamichi Formation Clay-Shale	below 37.3	Ave.	16	112	NT	NT	8.8
		Range	13-20	110-118			4.5-10.5
Limestone	below 37.3	Ave.	NA	159	NA	NA	8700*
		Range		151-166			1900-13,800*

Note: 1 ft. = 0.305 m, 1 lb/ft^3 = 16 Kg/m^3, 1 ton/ft^2 = 95.8 kPa, 1 psi = 6.9 kPa, NT=Not Tested, NA=Not Applicable, *values in psi

shaft was drilled. Each test shaft was dewatered and hand cleaned before the concrete was poured. Seepage was minimal, occurring at a rate of approximately 4 in. (10 cm) per hour. The permanent water table was located at great depth and the seepage resulted from isolated zones of perched water within the Caddo Formation.

INSTRUMENTATION

Three dial indicators, capable of measuring vertical movement to 0.001 in. (0.025 mm), were mounted on a stationary wooden reference frame at 120° intervals around the top of the shaft. The axial load applied to the shaft was measured by a calibrated Bourdon pressure gage and a BLH Model GT-CG electrical pressure transducer. The pressure transducer measured the hydraulic pressure in two, 600 ton (5340 kN) jacks to a sensitivity of 10 psi (69 kPa), which corresponds to a load of approximately 1 ton (8.9 kN).

Mustran cells (2) were embedded in the shaft at various depths (Fig. 1) to determine the distribution of axial load in the shaft. Mustran cells, which contain electrical resistance gages, measure the strain in the shaft under the applied load. The strain data is used to compute the axial load at the various levels in the shaft. The top level of Mustran cells, which was located in the portion of the shaft isolated from the soil, were used to obtain a calibration curve for the cells located at other levels in the shaft. The calibration curve correlates the strain output of the top level of cells with the known load applied to the shaft by the hydraulic jacks. The strain output from the Mustran cells and the electrical pressure transducer was recorded using a Vishay-Ellis data recording system. This system scanned and recorded up to 30 channels at a rate of approximately 1 channel per second.

TEST PROCEDURE AND RESULTS

A "quick" load test procedure was used to apply the axial load to the shaft. This procedure adds set increments of load to the shaft in prescribed increments of time. For Shaft 1, loads were applied to 550 tons (4895 kN) in 25 ton (222 kN) increments, at which point the increment was decreased to 10 tons (89 kN). After each increment of load was added to the shaft, two sets of settlement and strain readings were taken. The first set of readings was taken 30 seconds after the load was applied to the shaft, and the final set of readings was taken 2 minutes later. On the average, load increments were applied at approximately 4 minute intervals.

Shaft 1 was loaded to an ultimate capacity of 560 tons (4984 kN). The ultimate capacity of the shaft is taken as the load at plunging failure, which is defined as the point at which additional load cannot be applied to the shaft without experiencing continuous movement. At this load, continuous pumping of the hydraulic jacks is required to maintain the load.

The shape of the load-settlement curve for Shaft 1, presented in Fig. 3, indicates the clay-shale behaved as a stiff, plastic material. At a settlement of less than 0.1 in. (2.54 mm) 50 percent of the failure load is resisted, and at 0.5 in. (12.7 mm) of settlement, 95 percent of the failure load is resisted.

Load distribution curves near failure are obtained by plotting the load in the shaft determined from the Mustran cells as shown in Fig. 4. The nearly vertical lines from 0 to 19 ft. (0.0 m to 5.8 m) indicate that little load is transferred in the upper 19 ft. (5.8 m) of the shaft. This is to be expected, as the shaft is essentially isolated to a depth of 11 ft. (3.4 m) and the clay-shale between 11 ft. and 19 ft. (3.4 m and 5.8 m) is soft, highly weathered. A thin limestone seam at a depth of 19 ft. (5.8 m) marks the beginning of the more competent clay-shale, and this is confirmed by the increase in the amount of load transferred below that depth.

By integrating and differentiating the load distribution curves, curves can be developed as a function of depth showing load transfer in side resistance versus side movement of the shaft. The load transfer versus movement curves, shown in Fig. 5, indicate the load transfer generally increases with increasing depth. The peak load transfer value for Shaft 1 is 5.2 tsf (498 kPa), with 0.25 in. (6.4 mm) of movement required to mobilize the full side resistance. The average load transfer, determined by taking the ratio of the load transferred between depths of 19 ft. and 29 ft. (5.8 m and 8.8 m) and the surface area of the shaft through that length, is 4.2 tsf (402 kPa).

The α factor is defined as the ratio of the maximum unit load transfer in side resistance to the shear strength of the soil. The computed value for α is influenced by the quality of the samples, the type of test and averaging method used to determine the undrained shear strength, and those factors which affect the measured load transfer. Laboratory unconfined and quick triaxial compression tests indicate that the average undrained shear strength is 3.95 tsf (378 kPa). For Shaft 1, an α factor value of 1.06 is obtained when the average laboratory shear strength is used. Previous investigators have reported values generally ranging between 0.25 and 0.70 for drilled shafts in stiff clays; however, in some cases values greater than 1.0 have been reported. It is reasonable to assume that the value of α should not exceed 1.0, as inherent disturbance of the soil along the sides of the drilled shaft will reduce the soil shear strength.

To overcome problems associated with determining the clay-shale shear strength in the laboratory, Skempton's (9) bearing capacity equation is applied to the Shaft 1 results to calculate the average in-situ clay-shale shear strength. For N_c equal to 9 and the maximum measured bearing stress of 49 tsf (4694 kPa), the undrained shear strength is calculated to be 5.4 tsf (517 kPa). This shear strength is higher than the average laboratory value, however, it does fall within the range of test values determined in the laboratory. Using the calculated, in-situ shear strength determined above, the average α value for Shaft 1 is 0.78.

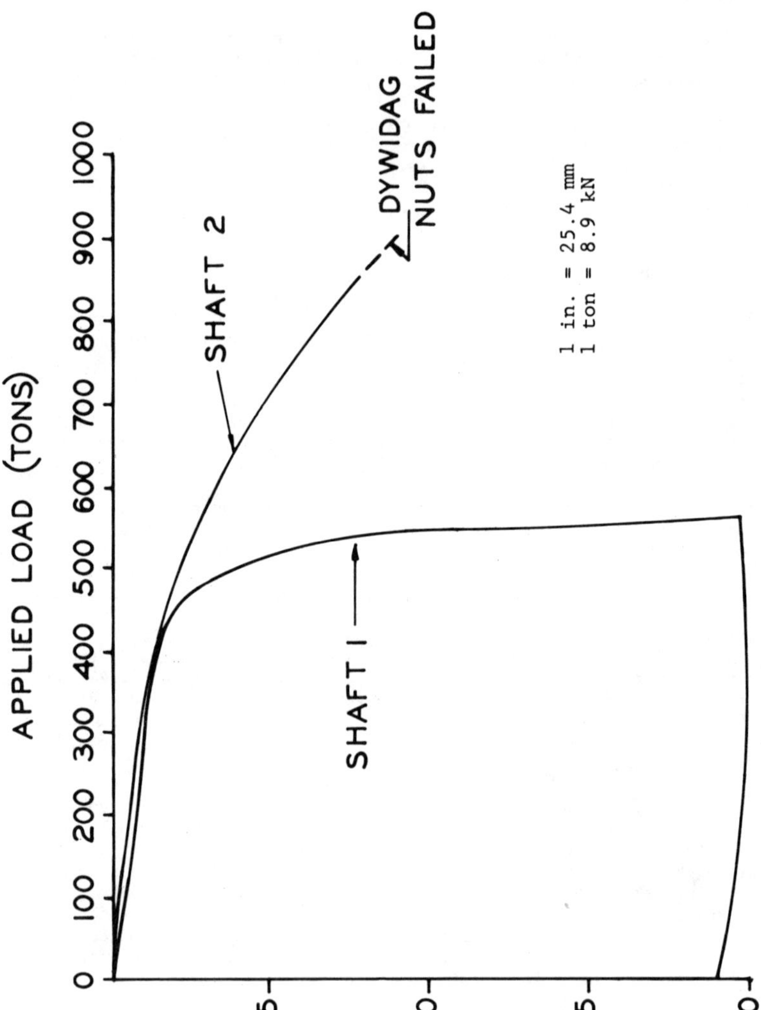

FIGURE 3. - LOAD-SETTLEMENT CURVES

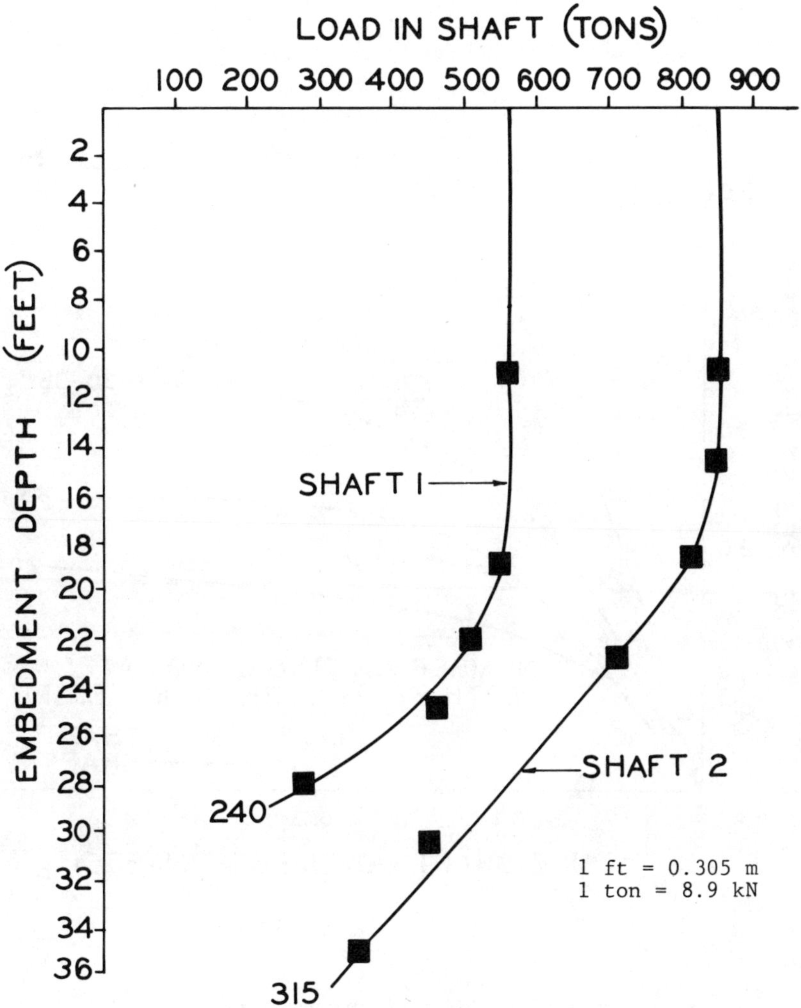

FIGURE 4. - AXIAL LOAD DISTRIBUTION CURVES

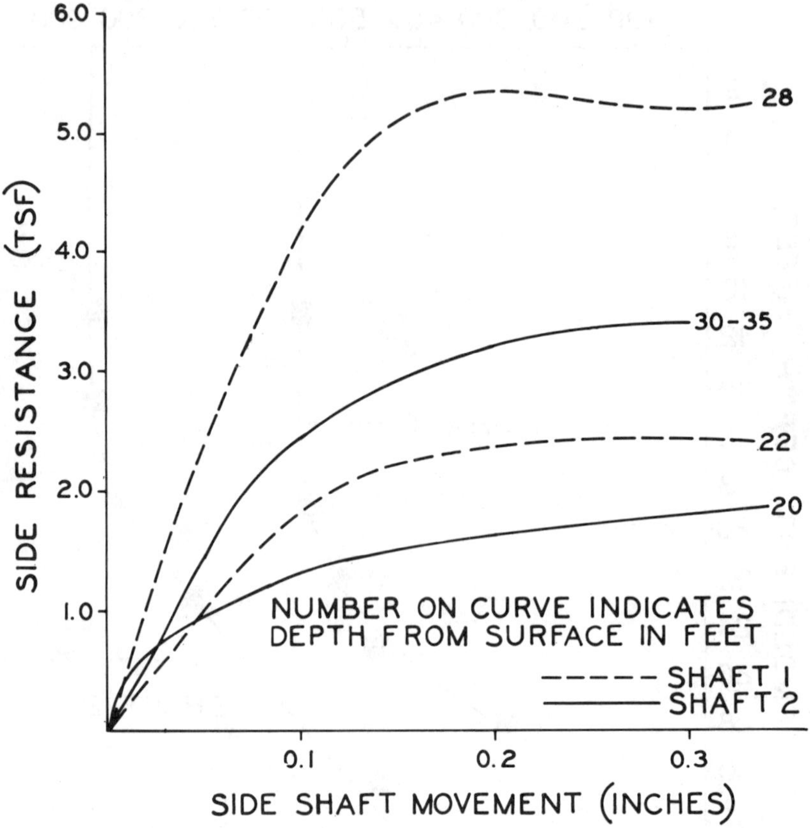

1 in. = 25.4 mm   1 ton/ft^2 = 95.8 kPa

FIGURE 5. - SIDE SHAFT RESISTANCE

Shaft 2 was loaded to 900 tons (8010 kN), however, due to failure of the reaction system the shaft was not loaded to failure. The load-settlement curve, shown in Fig. 3, indicates that at an applied load of 850 tons (7565 kN), 0.79 in. (20 mm) of settlement occurs. Although the shaft was not loaded to failure, the side resistance capacity of the shaft is fully mobilized (see Fig. 5). Because the remaining capacity would be developed in end bearing, the load-settlement curve for the shaft base can be extrapolated to failure.

The ultimate axial capacity of Shaft 2 is estimated by extrapolation to be 1125 tons (10,013 kN). The load distribution curve for the 850 ton (7565 kN) load is shown in Fig. 4. As for Shaft 1, the upper 19 ft. (5.8 m) of the shaft transfers minimal load. At the applied load, the load distribution curve indicates 63 percent of the load is carried through side resistance, and the remaining 37 percent by tip resistance.

The load transfer versus movement curves, plotted in Fig. 5., indicate that a peak load transfer of 3.9 tsf (374 kPa) is developed after 0.25 in. (6.4 mm) of shaft movement. The general trend is for the load transfer through side resistance to increase with depth to approximately 25 ft. (7.6 m) and thereafter remain constant.

Using the average load transfer for Shaft 2, the in-situ shear strength calculated from Shaft 1, and Skempton's (9) equation, the α factor for Shaft 2 is 0.68.

SUMMARY OF RESULTS

Of the two shafts described, one was loaded to plunging failure. Shaft 1 was loaded to an ultimate capacity of 560 tons (4984 kN) and Shaft 2 was loaded to 900 tons (8010 kN). The ultimate capacity of Shaft 2 is estimated to be 1125 tons (10,013 kN). The load-settlement curves show that 50 percent of the ultimate shaft capacity is resisted at settlements of 0.15 in. (3.8 mm) and 0.3 in. (7.6 mm) for Shafts 1 and 2, respectively.

The measured tip capacities of Shafts 1 and 2 are 49 tsf and 64 tsf (4694 kPa and 6131 kPa), respectively, with the ultimate tip capacity of Shaft 2 estimated to be 117 tsf (11,209 kPa). The tip movements required to develop 50 percent of the tip capacity are 0.1 in. and 0.5 in. (2.5 mm and 12.7 mm) for Shafts 1 and 2, respectively.

The load transfer curves indicate peak load transfers of 5.2 tsf and 3.9 tsf (498 kPa and 374 kPa) for Shafts 1 and 2, respectively. Movements of less than 0.25 in. (6.4 mm) are required to mobilize the peak load transfer, which generally increases with depth. The test results indicate that the magnitude of the load transfer does not decrease significantly with large shaft movements. The average load transfer is 4.2 tsf and 3.7 tsf (402 kPa and 354 kPa) for Shafts 1 and 2, respectively. Using the calculated in-situ shear strength of 5.4 tsf (517 kPa), determined from Skempton's (9) equation, the average shear strength reduction factor (α) is 0.73.

COMPARISON OF PREVIOUS LOAD TEST DATA

Seven load tests of drilled shafts in clay-shale are known to have been reported in the literature. The results of these tests are summarized in Table 2. Not all of the investigators computed load transfer and α values; and in some cases it was necessary to compute the values shown in Table 2. The α values reported range from 0.4 to 1.5. For comparison purposes it was necessary, in some cases, to determine the in-situ shear strength from Skempton's (9) bearing capacity equation. A review of the literature indicates that difficulty in determining the clay-shale shear strength is a major reason for the wide range in α. However, the average values of load transfer reported indicate that the side resistance of a drilled shaft in clay-shale is significant, and when neglected may result in an overconservative design.

EXISTING DESIGN PROCEDURE

Aurora and Reese (1) proposed a design procedure for drilled shafts in clay-shale based on the results of four load tests performed in 1976. The design procedure involves the following equation:

$$Q_{ult} = Q_b + Q_s \quad \ldots\ldots\ldots\ldots\ldots \quad (1)$$

where,

Q_{ult} = ultimate axial capacity

Q_b = ultimate base resistance

Q_s = ultimate side shaft resistance

The proposed design procedure was limited to shafts with penetration lengths of five feet or less in clay-shale and it was recommended that an α value of 0.75 be utilized for calculating ultimate side shaft resistance. Also, Skempton's (9) equation was recommended for calculating the ultimate base resistance. The suggested N_c value was 8.0.

Table 3 provides a limited comparison of full-scale field load tests performed on drilled shafts in clay-shales and the computed ultimate load capacities determined using the design procedure recommended by Aurora and Reese (1). In all cases the computed capacity was less than the experimental capacity. The ratio of the experimental to computed capacity represents a measure of the accuracy which ranges between 1.06 and 1.32.

The comparisons illustrated in Table 3 are limited because several of the tests listed in Table 2 were not loaded to failure, although, shaft movements are judged to be sufficient to fully mobilize maximum side shaft resistance. In other cases the data are not comprehensive enough to justify comparison by this method.

Table 2 - LOAD TEST RESULTS

Investigator	Shear Strength, In Tons Per Square Foot	Unit Tip Resistance, In Tons Per Square Foot	Maximum Unit Transfer, In Tons Per Square Foot	α
Aurora and Reese (1)	7.4#	64	7.2	1.0
Vijayvergiya (12)	6.8*	61	4.8	0.7
Matich and Kozicki (7)	2.5#	NA	2.6	1.0
Thornburn (10)	5.6*	50	2.5	0.4
Henley (6)	8.8#	147	4.6	0.5
Corp of Engineers (3)	3.8#	NA	2.8	0.7
Van Doren et al (11)	1.8*	16	2.7	1.5
Writers	5.4*	49	4.2	0.8

Note: 1 ton/ft^2 = 95.8 kPa, *determined from Skempton's equation
#laboratory test value

Table 3 - COMPARISON OF EXPERIMENTAL VERSUS COMPUTED CAPACITIES

Investigator (1)	Shaft No. (2)	Construction (3)	Experimental * Axial Capacity, In Tons (4)	Computed * Axial Capacity, In Tons (5)	Ratio 4/5 (6)
Aurora and Reese (1)	MT1	Casing	517	396	1.31
Aurora and Reese (1)	MT2	Slurry	521	404	1.29
Aurora and Reese (1)	MT3	Dry	570	485	1.17
Aurora and Reese (1)	DT1	Casing	290	237	1.22
Vijayvergiya et al (12)	1	Dry	1000	757	1.32
Writers	1	Dry	560	529	1.06

Note: 1 ton = 8.9 kN, *ultimate values

The ratio of experimental to computed capacity for the writers' case is 1.05. This comparison is based on the use of the in-situ, undrained shear strength, which is calculated using Skempton's (9) bearing capacity equation. As previously shown, this value is 5.4 tsf (517 kPa). If the average laboratory value of 3.95 tsf (378 kPa) is used, the computed capacity is 386 tons (3435 kN) and the ratio is 1.45. It is evident that for clay-shales the computed capacity is directly related to the shear strength of the shaft embedment material. However, in harder shales and other rocks the maximum side shaft resistance is controlled by the bond strength between the concrete and rock (4). Under such conditions, the allowable side shaft resistance (adhesion) is expressed as a percent of the shaft concrete strength.

CONCLUSIONS

Clay-shales are classified as soft rock and are geographically present in considerable thickness throughout much of the United States. The common design practice of installing drilled shafts in clay-shale usually involves enlarging the shaft base to provide a larger bearing surface and thereby, increase the allowable design capacity of the shaft. In many cases, the side shaft resistance is neglected or given a minimal value and the shaft is considered to be primarily end bearing. While this design practice has proven, in most cases, to be satisfactory from the standpoint of foundation performance, it may be uneconomical when compared to straight shafts where appropriate allowance is given to side shaft resistance development.

This investigation indicates that for drilled shafts, installed by the dry method, the maximum side shaft resistance is approximately 70 percent of the in-situ, clay-shale shear strength. Furthermore, this side shaft resistance is fully developed at movements that are much smaller than the movement required to fully mobilize the end bearing resistance. Thus, drilled shaft designs based solely on end bearing and that penetrate clay-shale for at least ten feet (3.05 m), in reality support all; or a major portion of the design load through the development of side shaft resistance.

Conventional laboratory tests performed on samples recovered with Pitcher samplers or through diamond bit coring may underestimate the average shear strength of clay-shales by as much as 30 percent when the bedding planes and laminations are horizontal. The disparity between laboratory and in-situ shear strengths strongly suggest that full-scale field load tests are justified for most major projects. The α values recorded in the literature for both stiff clays and clay-shales range from as low as 0.25 to more than 1.0. Although many factors are recognized to affect the value of α, the major influencing factor appears to be the method utilized to approximate the in-situ shear strength of the material in which the shaft embedded.

The compressibility characteristics of stiff clays and clay-shales are equally difficult to measure in the laboratory and additional field monitoring of drilled shafts under long term, sustained loads is necessary before factors of safety can be rationally developed to reflect settlement criteria.

ACKNOWLEDGEMENT

The Association of Drilled Shaft Contractors, Inc., provided the hydraulic jacks and necessary truss reaction system for the field load tests. Appreciation is also expressed to Jim Anagnos, Jene Hayes, and Lymon C. Reese for their advice and assistance in implementing the load test program. Their tireless efforts were invaluable to the project construction schedule.

APPENDIX I. - REFERENCES

1. Aurora, Ravi P. and Reese, Lymon C., "Behavior of Axially Loaded Drilled Shafts in Clay-Shales," Research Report 176-4, Center for Highway Research, The University of Texas at Austin, 1976.

2. Barker, Walter R. and Reese, Lymon C., "Instrumentation For Measurements of Axial Load In Drilled Shafts," Research Report 89-6, Center for Highway Research, The University of Texas at Austin, 1969.

3. Corps of Engineers, "Investigations For Building Foundations in Expansive Clays," Vol. 1, U.S. Army Engineer District, Fort Worth, Texas, Apr., 1968.

4. D'Appolonia, E., D' Appolonia, D. J., and Ellison, R. D., "Drilled Piers," Foundation Engineering Handbook, H. F. Winterkorn and H. Fang, ed., Van Nostrand Reinhold Co., New York, N.Y., 1975, pp. 610-611.

5. Engeling, D. E. and Reese, Lymon C., "Behavior of Three Instrumented Drilled Shafts Under Short Term Axial Loading," Research Report 176-3, Center for Highway Research, The University of Texas at Austin, 1974.

6. Henley, A. D., "Investigation of Side Shear Stress Transfer for Drilled In Foundation Shafts in Weak Rock," Proceedings of the Fifth Annual Symposium on Engineering Geology and Soils Engineering, held at Pocatello, Idaho, Apr., 1967.

7. Matich, M. A. and Kozicki, P., "Some Load Tests on Drilled Cast-In-Place Concrete Caissons," Canadian Geotechnical Journal, Vol. IV, No. 4, Nov., 1967, pp. 367-375.

8. O'Neill, Michael W. and Reese, Lymon C., "Behavior of Axially Loaded Drilled Shafts in Beaumont Clay," Research Report 89-8, Center for Highway Research, The University of Texas at Austin, 1970.

9. Skempton, A. W., "The Bearing Capacity of Clays," Proceedings of the Building Research Congress, 1951, Division I., Building Research Congress, London, 1951.

10. Thorburn, S., "Large Diameter Piles Founded on Bedrock," Large Bored Piles, Institution of Civil Engineers, 1966, pp. 121-129.

11. Van Doren, L. M., Hazard, S. G., Stallings, J. R., and Schnacke, D. P., "Drilled Shaft Foundation Test Program," Project No. 35W-87-I-35W-1(58)48, State Highway Commission of Kansas, Van Doren-Hazard-Stallings-Schnacke Engineers-Architects, Topeka, Kansas, Jul., 1967.

12. Vijayvergiya, V. N., Hudson, W. R., and Reese, Lymon C., "Load Distribution of a Drilled Shaft in Clay-Shale," Research Report 89-5, Center for Highway Research, The University of Texas at Austin, 1969.

13. Webb, D. L., "Piling in Weak Rock-Session 2," Piles in Weak Rock, The Institution of Civil Engineers, 1977, pp. 209-210.

APPENDIX II. - NOTATION

The following symbols are used in this paper:

N_c = bearing capacity factor;
Q_b = ultimate base resistance;
Q_s = ultimate side shaft resistance;
Q_{ult} = ultimate axial capacity; and
α = ratio of maximum unit side shaft resistance to clay-shale shear strength.

TIMBER PILE STUDY, EXISTING STRUCTURE

By Frank D. Grefsheim[1], Member, ASCE

INTRODUCTION

Timber pile foundations of the Old Post Office Building in Washington, D.C. are being evaluated prior to proposed renovations which are anticipated to cost about $20,000,000. This historic large wall bearing stone-faced 10-story structure has a tower extending about 156 feet above the roof. The existing untreated pile foundation was driven to below the groundwater level at the time of construction in 1890. There has been extensive development of the immediately surrounding areas since that time and the groundwater level has been depressed because of subdrainage related to developments with deep basement levels, and the recent Washington METRO area subway construction. The hydrostatic water table is now about 12 to 20 feet below the butt end of piles based on recent readings taken around the Old Post Office.

These conditions are known to have resulted in deterioration of pile material and it was therefore decided to make a study of the existing pile foundations. Available construction plans were reviewed and analyzed, subsurface data was developed, piles were examined in the field, and core samples of the piling were obtained for further microscopic and biologic examinations in the laboratory. This data was analyzed to determine the need for any modification of the foundations before beginning renovations.

The available construction plans indicate the pile cap design and pile tip elevations. Estimated existing loads for individual piles were calculated to be about 20 to 30 tons. For selected piles and transient wind load conditions maximum loads of about 35 to 38 tons per pile are estimated. Subsoil data available from original test pit records indicate the pile tips are in a competent bearing stratum consisting of dense sand and gravel. Test borings that were made verified this and provided data for use in developing estimated pile capacities. This was accomplished using static pile formulas and primarily N-value correlations.

Variations in the groundwater level were developed based on available data including readings indicated on the original drawings, later readings taken for the adjacent subway construction, and observations of well points installed as part of the recent subsurface investigation. Preliminary evaluations indicated adequate pile capacities but at least some of the piles could have deteriorated because of the lowered groundwater.

Initial examinations of test pits indicated sound piling except at one test pit location. After assimilation of this testing, which included a subsoil investigation, it was decided to make a comprehensive test pit investigation to evaluate pile soundness and determine the lateral extent of the badly decayed conditions at one test pit.

A total of fourteen test pits have been made, one in 1972, five in 1977, and 8 more in 1979. There was little or no deterioration in the pits examined in 1977 but the piles of the 1972 pit, which was re-excavated in 1979, showed significant deterioration. At the re-excavated pit, deterioration was noted extending to within a few inches of the pile center. Basically sound conditions were noted in the last eight

Note: 1 ft = 0.305 m   1 ton = 8.9 kN

[1]Associate, Schnabel Engineering Associates, Bethesda, Maryland

test pits of 1979 except in the area immediately around the old 1972 test pit. Data and analysis herein include results of microscopic and biologic tests on the core samples, and estimated pile capacities for the basically sound pilings generally noted.

Based on the test pit and laboratory examinations, an estimate was made of the reliable sound piling. Estimated allowable pile loads were determined including considerations of fiber stress limitations based on a reduced effective sound pile diameter. Preliminary recommendations for underpinning were developed for noted decay areas and the columns with relatively high pile loads.

Existing Foundation System - The Old Post Office Building was constructed over the fill and soft alluvial deposits of the downtown Washington, D.C. area, and the foundation system consists of driven untreated timber piles. Pile caps consist of 12" x 12" timbers. The cap timbers are supported directly on the pile butts, and there are four successively smaller courses of cap stones extending above the concrete and timber to form the column pad.

Figure 1 is a cross section through a typical pile cap and column pad, as shown by the available foundation plans and later verified by test pit observations. The pile butts are indicated at El -3.61 and the top of column pad is at El +4.39. Exterior grades vary from El +8.44 to +12.28 as shown by the available site excavation plan, and the pile butts are therefore at least about 12 feet below the exterior grade. This soil cover is significant because of protection from oxygenation that would contribute to potential deterioration of the pile material. The question of deterioration and capacities of both the piling and the pile caps was evaluated based on various subsurface and laboratory investigations including test borings, water observation wells, test pits, and examinations of the pile material using scanning electron micrographs and agar plate readings. Details of these investigations are described herein.

Subsoil and Groundwater Conditions - Five test borings and four water observation wells were made in 1977 indicating the following general subsurface profile:

Stratum A:	At the ground surface to a maximum depth of 20 feet, El -8	Brown and gray sand, clay, cinders, and brick, FILL; generally loose (N = 2 to 16)
Stratum B:	At the ground surface or below Stratum A to depths of 26 to 35 feet, El -16 to -22	Brown and gray SILTY CLAY (CL) or SANDY CLAY (CL) with layers of silty sand at some locations; generally medium to stiff consistency (N = 5 to 25)
Stratum C:	Below Stratum B to depths of 45 to 56.5 feet, El -35 to -43	Brown and gray fine to coarse SANDY GRAVEL (GP or GP-GM) with fine sandy clay layers at some locations; compact to very compact (N = 26 to 100/3")
Stratum D:	Below Stratum C to depths of 55 to 56.5 feet, the maximum depth of the borings	Brown and gray DISINTEGRATED ROCK (SC); very compact (N = 67 to 100/3")

Note: 1 in. = 25.4 mm 1 ft = 0.305 m

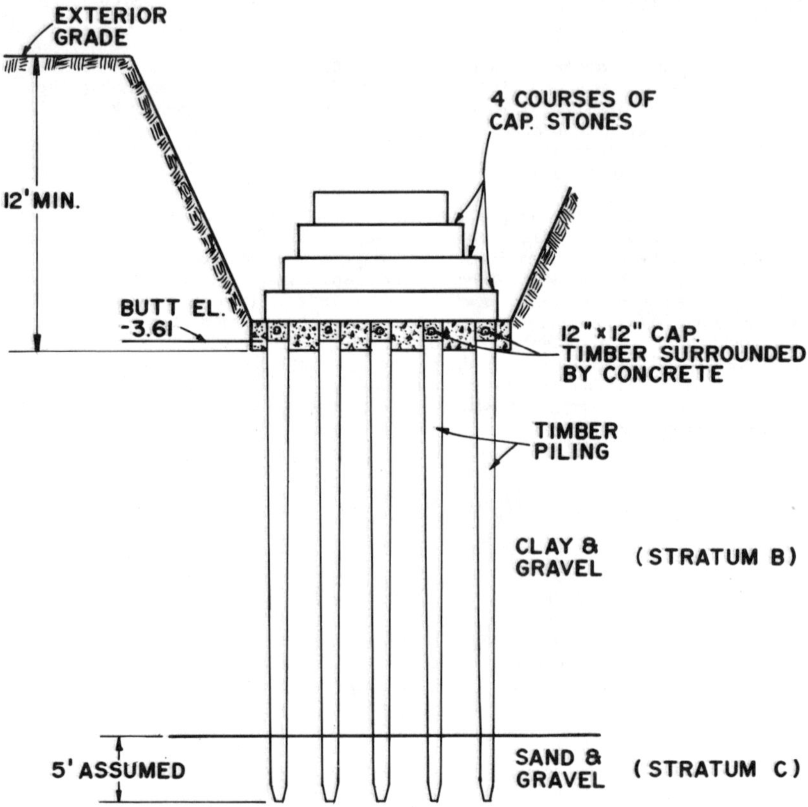

1 in. = 25.4 mm   1 ft = 0.305 m

FIGURE 1

The fill of Stratum A is from development of the existing building. The natural soils of Strata B and C include undifferentiated terrace deposits with a basal layer of coarse grained Cretaceous age sediments. The fill material is generally loose. The clayey soils in the upper levels of Stratum B are generally overconsolidated while at lower elevations within this stratum there are soft zones of slightly overconsolidated or nearly normally consolidated material. Stratum C grades into the coarse grained Cretaceous sediments. The underlying Stratum D is disintegrated rock resulting from in-place weathering of the parent bedrock.

Soil laboratory testing was conducted to aid in evaluating shear strength and compressibility for static analysis of pile bearing and considerations of potential pile cap bearing. Testing and evaluations were primarily for the natural soils. The coarse granular soils were evaluated based on standard penetration tests or N-values which are included in the listing of general soil descriptions.

Stratum B - Predominantly Clayey Soil

From cohesive samples of Stratum B we have estimated preconsolidation stresses varying from less than 0.5 tsf to just over 1.0 tsf in excess of the existing overburden stresses. A summary of index properties and soil compressibility parameters is as follows:

Soil Classifications	CL, ML and SM
Liquid Limits of Plastic Samples	21 to 26
Liquidity Indices	Negative to 0.6
Estimated Compression Index, C_c	0.1
Estimated Recompression Index, C_r	0.02

From consolidated-undrained triaxial compression tests, we have developed the following variations of shear strength values:

Internal Friction, ϕ	19 to 26 degrees
Cohesion, c	700 to 1100 psf

Stratum C - Sandy Gravel

Soil classifications are GP or GP-GM, and N values are consistently very high. For consideration of end bearing of piling, we have used the following variation of N-value within Stratum C:

Depth	N-Value
Upper 5 feet	25 blows per foot
Below 5 feet	50 blows per foot

The underlying disintegrated rock, Stratum D, and the bedrock are not a limiting factor in the evaluation of foundation bearing characteristics.

Groundwater - Groundwater conditions at the site are indicated by the following based on various available data:

Note: 1 ton/ft^2 = 96 kN/m^2 1 lbf/ft^2 = 48 N/m^2 1 ft = 0.305 m

Date	Groundwater Elevation	Remarks
1890	-1.405 or -2.25	From original Post Office Design Plans
1968	-22.0 to -24.1	Metro test borings (before de-watering)
March, 1974	-24.0 to -36.9	1st readings after Metro de-watering was started
July, 1974	-26.5 to -34.3	Last readings during the de-watering
January, 1978	-23.7 to -16.1	Water observation well levels recorded for the subject investigation

Additional readings of groundwater available from Metro records and also the recent test pits indicate high perched water levels up to about El -4.0 to 0.0. The generally lower levels indicated by the water observation wells are believed caused by penetrating the basically impervious Stratum B layer. The relatively high level near El 0.0, as given on the original 1890 design drawings is probably influenced by perched water.

The perched water table has been noted to vary significantly in the test pits which were extended to about El -6 to -10. Some pits required continuous pumping while at others, little or no water was encountered.

The available data indicates a current static water table below about El -16 and perched water varying up to about El 0. The timber piles can therefore be considered to be only partially submerged. Further test pit investigations were therefore made to examine the possible fungus or dry rot damage that could result under the water conditions described above.

Test Pits, Sampling and Evaluation - In the initial evaluations of 1977 and early in 1979 timber piling was exposed at 5 test pits essentially along the outside wall lines, and also at a 6th previous test pit location near the middle. The outer pits were completed in 1977 and the pit near the middle was re-excavated in 1979, removing loose backfill that had been placed after its original excavation in 1972. Additional test pit excavations were made later in 1979 at eight locations. Figure 2 shows all the test pit locations as summarized below:

Test Pit Designation	Excavation Year
A thru E	1977
F	1972, 1979
G thru N	1979

Photographs of the test pits and piling are given on Figure 3. The exposed timber could be visually examined, and cores of the timber were obtained using an increment borer. Core samples 3.2 mm diameter were recovered along the pile diameter extending through the pile between opposite outer faces.

Pile core samples were examined to note existing wood conditions, and to determine presence of viable wood destroying organisms. Microscopic examinations were made by scanning electron micrographs taken from prepared samples at various points along the cores. Bacteria found were isolated and observed to note any active

Note: 1 ft = 0.305 m 1 mm = 0.04 in.

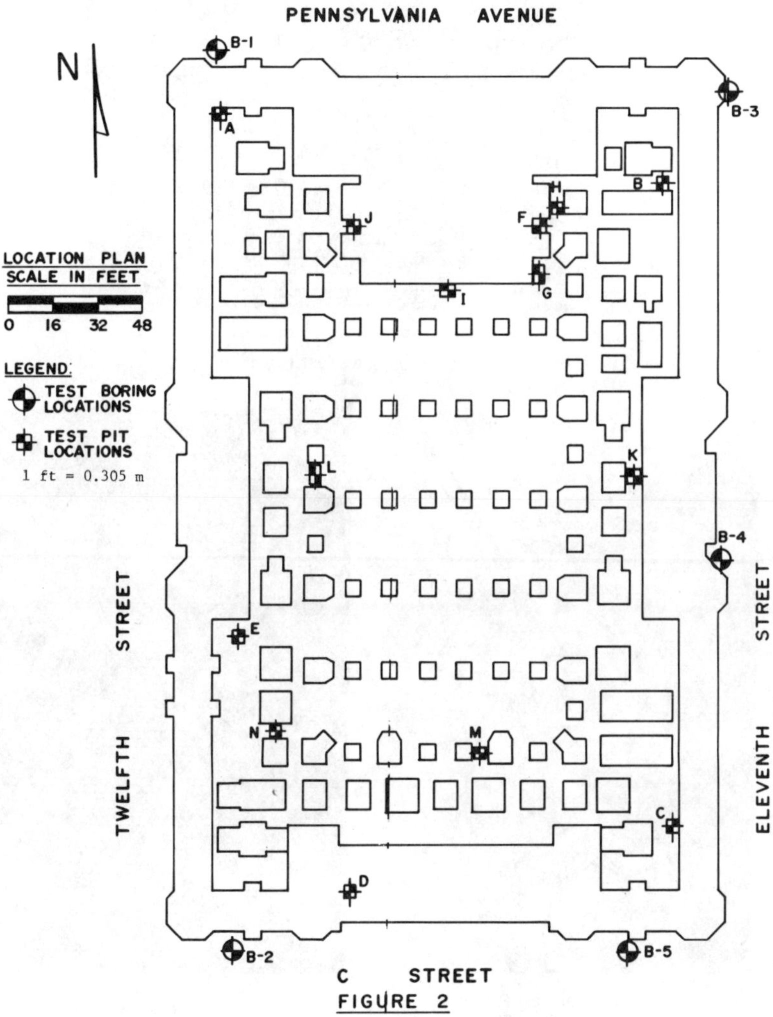
PENNSYLVANIA AVENUE
N
LOCATION PLAN
SCALE IN FEET
0 16 32 48
LEGEND:
TEST BORING LOCATIONS
TEST PIT LOCATIONS
1 ft = 0.305 m
B-1
B-2
B-3
B-4
B-5
A
B
C
D
E
F
G
H
I
J
K
L
M
N
TWELFTH STREET
ELEVENTH STREET
C STREET

FIGURE 2

Digging at Test Pit H

Typical Test Pit with
Gravel Base Showing
Concrete Pile Cap Encasement

Typical Sound
Timber Pile

FIGURE 3

wood decaying fungus. This testing and laboratory evaluations were conducted on a total of 39 core samples in the testing of 1977 and 1979.

The piles exposed by the test pits were noted to be southern yellow pine, one of the following species: loblolly, longleaf, pond, shortleaf or slash. Except for evidence of creosote in one sample, the piles are not treated and piles installed at about the turn of the century were typically untreated.

Deterioration of the pile material was noted to be very slight at the five 1977 test pits, but substantial deterioration was noted at the 1972 pit when it was re-excavated in 1979. Cores from the 1977 pits indicated slight deterioration in the outer 1 cm zone and, from bacteria found, it was noted that there were no active wood destorying fungi. However, for the re-excavated 1972 pit, substantially more deterioration was found. A core sample from one of the piles in this pit indicated decay extending to within a few inches of the pile center. Active wood destroying fungi were also found in the laboratory investigations.

Because of the deterioration noted at one of six test pit locations, a more comprehensive test pit investigation was undertaken consisting of the eight 1979 test pits designated as Test Pits G thru N. These pits were located to determine approximate horizontal limits of the decayed condition at Test Pit F, and to provide better coverage of the remaining foundations. Twenty-three pile core samples were examined and tested for this phase of the investigation.

With the exception of Test Pit F and the adjacent Test Pits G and H, the core samples tested were noted as varying from entirely sound, or 100 percent, to about 60 to 70 percent sound. Generally, 70 to 80 percent sound conditions were noted in the samples recovered.

The observation of culture samples indicated mold and bacteria for many of the samples but active wood decaying fungi were identified only in Test Pit F and the nearby Test Pit H. The presence of bacteria and mold does not indicate active decay, and it is possible that these contaminants resulted from the sampling operations.

Analysis of Suitability of Foundations - The existing building has been examined to note any settlement damage, and existing pile loads were estimated based on available building plans and examination of building materials. These findings were used along with further test boring and test pit investigations as reviewed in the foregoing to determine adequacy or possible repair requirements for the existing foundations. Pile capacities were estimated to evaluate the question of feasibility of supporting the estimated proposed pile loads as given above. An area of significant pile material decay was identified. Underpinning requirements are given for selected areas, because of extensive decay and also the relatively high estimated pile loads of nearly 40 tons. Results of this evaluation and remedial measures are described in the following.

a. Existing Building Condition and Estimated Pile Loads

Many portions of the building were unoccupied for a time and deterioration was evident ranging from minimal to significant. This included minor cracking apparently from moisture penetration and general lack of maintenance. Numerous cracks, roof leaks, broken and missing slates and missing or open windows were noted, all of which contribute to water damage. The cracking and other deterioration noted do not appear to be structurally significant nor indications of settlement.

Existing pile loads are estimated to range from about 20 to 25 tons

---

Note: 1 cm = 0.39 in. 1 ton = 8.9 kN

except for columns surrounding the court area for which existing pile loads are estimated at 37.5 to 38.5 tons. Also, wind loading results in a maximum transient pile load of 35 tons for piles on the perimeter of the tower. The proposed remodeling is estimated to increase the pile load by a maximum of 1 ton per pile. Final loads are generally less than 30 tons per pile and a maximum of 38.5 tons for selected locations.

b. Pile Capacities and Safety Factors

Allowable pile loading depends on soil considerations and allowable wood fiber stresses.

1. Soil Considerations

The existing piles are considered to be a combination of friction and end-bearing in the soils of Strata B and C. Considering the estimated weakest conditions of the variable clayey soils of Stratum B, we would estimate the following variations of load distributed to the pile tips.

Total Pile Load	Pile Tip Load
40 tons	29 tons
30 tons	19 tons
20 tons	9 tons

The above listing is considered to indicate the maximum probable pile tip loading. The full tip load would be mobilized along with or after full utilization of skin friction bearing along the pile length.

Our analysis is based primarily on static pile capacity relations but we have also examined available records of test piles installed for the original construction. This data indicates fairly steady increases in driving resistance as the test piles were installed and there are no indications of an abrupt decrease in pile penetration. We therefore conclude that a significant portion of the load, at least as indicated by the above distribution of pile loads, is carried by skin friction.

Also, the records of test pile installations indicate the pile tips were extended at least into the relatively dense soils of Stratum C. Tip elevations of -21.0 to -30.8 are indicated for seven test piles for which driving records are available.

We believe the pilings were driven adequately into the Stratum C materials. For calculation of skin friction and end bearing of the piles, we have used the relations and analysis as described in the following:

End Bearing, Stratum C, Dense Sand & Gravel

$$q_p = \frac{0.6ND_b}{B}$$

Note: 1 ton = 8.9 kN 1 ft = 0.305 m

where:

q_p = unit ultimate point resistance in tons per square foot

N = standard penetration resistance (50)

D_b = depth of embedment (3 feet)

B = average pile diameter for embedded portion (8 inches, effective sound pile)

For the values included with the above listing, q_p = 135 tsf and the ultimate end bearing value would be 47 tons for a minimum tip diameter of 8 inches.

Skin Friction, Stratum B Medium to Stiff Clayey Soil (N greater than 9)

$$f_s = 0.4+0.015(N-10)$$

where:

f_s = unit skin friction resistance in tons per square foot

N = standard penetration resistance (10)

For the above relation and using an average pile diameter of 11.5 inches for the upper 5 feet near the butt end within the Stratum B clayey soil, the total skin friction resistance would be 6 tons.

Skin Friction, Stratum B Soft Clayey Soil (N less than 9)

$$f_s = 0.045N$$

Using N = 2, an average pile diameter of 11 inches, and a pile length of 12 feet within this portion of Stratum B, the total skin friction resistance would be 3 tons.

Skin Friction, Stratum C, Dense Sand & Gravel

$$f_s = 0.02N$$

The skin friction of Stratum C would be fully developed for an embedment of 15B and for lesser embedments a reduction is assumed based on a linear variation. For 3 feet of embedment in soil with N = 50 and using a tip diameter of 8 inches, the resulting unit skin friction would be 0.3 tons per square foot. This results in a skin friction resistance of 2.4 tons for a 3 foot length of embedded section with an average outside diameter of 10.2 inches.

Summation, Load Distribution Along Pile, and Group Action

The total ultimate bearing resistance from this analysis would be 58.7 tons with 11.5 tons total skin friction. Based on this analysis, we estimate an allowable loading of 30 tons and a factor of safety of 2.0. For estimating the load distribution along the pile, the

Note: 1 ton/ft^2 = 96 kN/m^2 1 ton = 8.9 kN 1 in. = 25.4 mm 1 ft = 0.305 m ultimate skin friction of about 11 tons has been considered to be developed before load is transmitted to the pile tip.

Since the piles for this project are driven into a dense sand and gravel, we have considered essentially 100 percent efficiency. Reduction of the skin friction portions of the total pile bearing would be minor compared to the available end bearing. Allowable loading for pile groups is estimated as 30 tons times the number of piles in the group.

Soil bearing considerations indicate an allowable load of 30 tons with a factor of safety of 2.0. The allowable loading described in the foregoing applies for structurally sound piles. Limitation of pile material stresses and estimated decayed condition could further limit allowable pile loads, particularly for this existing foundation system.

2. Pile Material Considerations

For our analysis of the pile material, we have examined deterioration to estimate the effective sound pile cross section. Allowable wood fiber stresses have been considered with respect to the estimated maximum load transferred to the limiting cross section at the pile tips and assuming full load applied at the larger butt end.

Except for the area of Test Pit F, which is considered separately, the piles are generally sound. Varying amounts of decay were noted, but we estimate at least 70 percent sound pile material at the butt end. The value 70 percent has been applied to pile diameter, and effective sound pile cross sectional area is based on the reduced diameter. The potential of pile decay would decrease with depth because conditions of moisture, temperature and oxygen would be less favorable for decaying fungi. We estimate at least 80 percent sound piling at and below the firm bearing of Stratum C to which the pile tips have penetrated.

For the above estimated minimum pile sections, and considering the load reduction along the pile from the maximum near the butt, we estimate maximum pile fiber stresses as follows for various estimated pile loads:

Pile Load	Maximum Pile Fiber Stress
40 tons	1444 psi
30 tons	1082 psi
20 tons	722 psi

The allowable pile fiber stress for the wood species considered herein would be 1285 psi according to the current design standard as given by ASTM D-2899. There is an implied factor of safety of about 1.2 in development for this allowable pile fiber stress. However, some reduction applies for the subject piling because of continuous long term loading more than 10 years duration. For consideration of this factor, and also to include consideration of increases for transient wind loads, we have used recent research and state-of-the art comments from the literature.

Note: 1 ton = 8.9 kN 1 $lbf/in.^2$ = 6.9 kN/m^2

Development of Allowable Fiber Stresses

Recent research[1] on long term behavior of wood materials in piles, which we understand has been tentatively accepted by ASTM for use in a revision of ASTM D-2899, utilizes the following relation:

$$S_a = \frac{L_d}{f.s.} WPS'_c$$

where:

S_a = allowable working stress on axially loaded timber piles
L_d = load duration factor = 0.6
f.s. = factor of safety = 1.2
W = strength-reduction for treatment = 1.0 in this case
P = strength-reduction for imperfections = 0.85
S'_c = basic reference strength

Values used are as listed, including the treatment factor of 1.0 based on untreated piles. The basic reference strength, S'_c , has been taken as the lowest of the southern pine species: loblolly, longleaf, shortleaf, slash, and pond, or 2504 psi. This listing of wood species has been established based on the microscopic examination of the core samples.

Using the above relation, we have determined an allowable working stress of 1064 psi. Some increase may be applied because of higher butt strength. However, since there is some question as to critical area of load application along the pile length, we have conservatively ignored this increase.

Increase for Transient Wind Loading

Wood is able to absorb considerable overloads for a short time and some increased allowable stress would therefore apply for consideration of the temporary wind loading for the periphery piles of the tower area. This question is treated only qualitatively by ASTM D-2899. For our analysis, we have used a 33-1/3% increase of allowable fiber stress based on a recent state-of-the-art presentation[2].

To calculate a factor of safety, we have used a reference working stress, $S = L_d \, WPS'_c$ which is S_a x (factor of safety, 1.2). This reference working stress of 1277 psi has been increased by 33-1/3% to 1703 psi for consideration of the transient wind load. The unit stress of 1263 psi for 35 tons load on an 8.4 inch diameter section results in a calculated safety factor of 1.35 or 1.4 as used below.

Note: 1 lbf/in^2 = 6.9 kN/m^2 1 ton = 8.9 kN

3. Recommendations for Pile Capacity

Considering both the soil limitations and the pile fiber stresses as reviewed herein, we estimate safety factors as follows for long term loads:

Pile Load	Estimated Factor of Safety
30 tons	1.2
25 tons	1.4
20 tons	1.8

Maximum transient pile loads of 35 tons, which are estimated for wind load conditions for piles around the tower, are estimated to have a factor of safety of about 1.4.

A maximum allowable pile load of 30 tons has been recommended for sustained loading, and an allowable load of 35 tons applies for the transient wind load condition. Remedial measures are therefore necessary as discussed below for estimated sustained loads of more than 30 tons per pile, and also for the area of Test Pit F where excessive pile deterioration has been found.

Preliminary Recommended Remedial Measures - Underpinning is recommended for columns with high estimated pile loads, and for the area of badly decayed existing piles. Additional soil treatment may also be necessary in the badly decayed area because of existing fungi. Some consideration has been given to alternate field settlement observations to note performance of existing unimproved foundations. This is discussed herein, but not recommended as a substitute for underpinning. Areas recommended for underpinning are noted below.

a. Areas Requiring Underpinning

Estimated new loads of 38.5 tons per pile apply to the small columns around the court area. This condition applies for 16 of the columns around the court area as shown on Figure 4.

The area of decayed piling around Test Pit F is estimated to have affected the southeast corner of the tower area and two adjacent columns. The locations recommended for underpinning are shown on Figure 4.

b. Future Performance Observations, Final Underpinning Areas, and Additional Remedial Measures

We have not determined conclusively an explanation for the anomalous condition at Test Pit F and the existing test pit data is not sufficient to delineate damaged areas of the existing foundations. Recommended limits of the tower underpinning as given on Figure 4 are based on the field investigations reviewed herein but final adjustment of the limits of underpinning in this area should be made as this part of the underpinning work is conducted. This will involve tunneling under the tower pile cap which should best be accomplished as part of that work.

Note: 1 ton = 8.9 kN

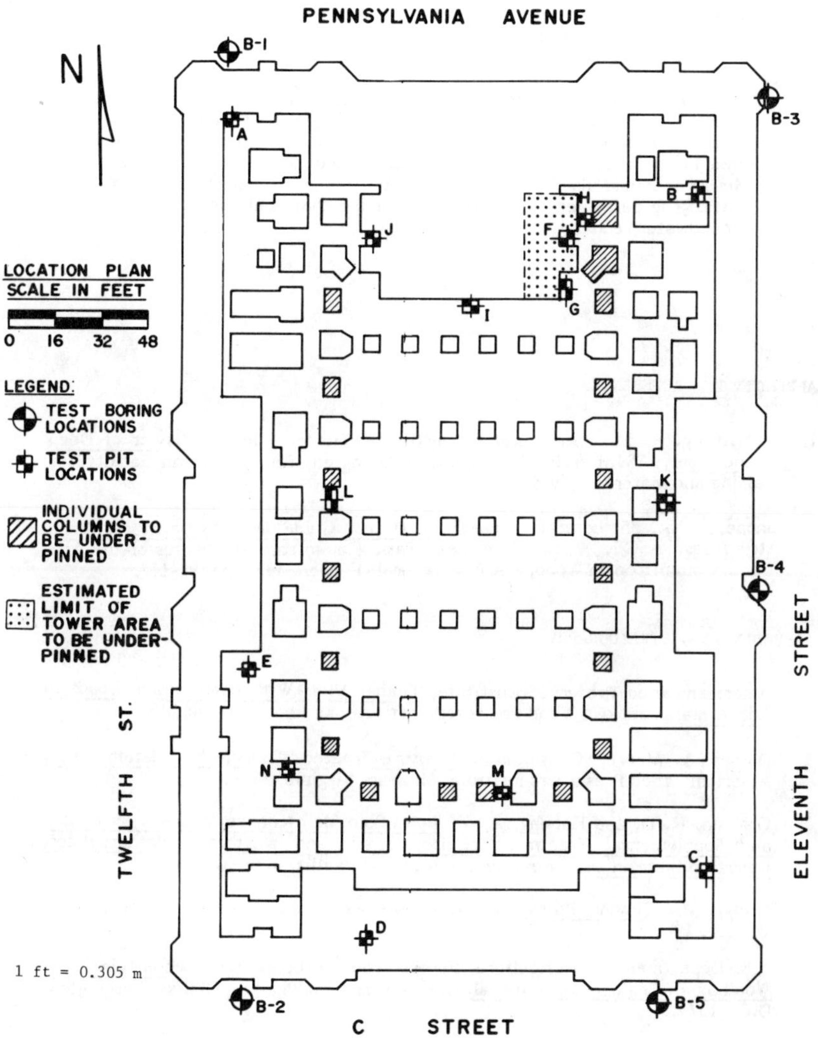
PENNSYLVANIA AVENUE
N
B-1
B-3
A
B
H
F
J
G
I
LOCATION PLAN
SCALE IN FEET
0 16 32 48
LEGEND:
TEST BORING LOCATIONS
TEST PIT LOCATIONS
INDIVIDUAL COLUMNS TO BE UNDER-PINNED
ESTIMATED LIMIT OF TOWER AREA TO BE UNDER-PINNED
K
L
B-4
E
N
M
C
D
TWELFTH ST.
ELEVENTH STREET
1 ft = 0.305 m
B-2
B-5
C STREET

FIGURE 4

Although no existing wood destroying organisms have been identified, conditions for their development and activity may be present and we have recommended periodic field observations and various additional remedial measures. Settlement points have been recommended for selected column locations and near the finished tower area. Settlement points should be read on a regular basis to note any possible movement of the foundations. During the proposed renovation, bi-monthly readings should be adequate and a time interval of about a year may then be selected for future readings.

The underpinning recommended is necessary to correct the inadequate foundation condition established by the investigation and analysis reviewed herein. Because some decay has been noted, soil treatment should also be used in the area of Test Pit F to handle the active fungi of this area. Measures should also be taken to maintain the perched water above the pile butts. The permanent test pits are available that could be utilized for the groundwater recharge stations.

APPENDIX I - REFERENCES

1. Armstrong, R. M., "Structural Properties of Timber Piles", Behavior of Deep Foundations, ASTM STP 670, Raymond Lundgren, Ed., American Society for Testing and Materials, 1979.

2. Shope, J. G., "Structural Lumber", A Design Guide and Commentary, Wood Structures, ASCE, Wood Structures, Task Committee on Status-of-the-Art: Wood, Committee on Wood, ASCE Structural Division, 1979, pp. 98-117.

APPENDIX II - BIBLIOGRAPHY

1. American Wood Preservers Institute, Timber Piles With High Design Loads In California, American Wood Preservers Institute, McLean, Virginia, 1969.

2. Dames & Moore, Consultants, Pressure-Treated Timber Foundation Piles, American Wood Preservers Institute, McLean, Virginia, 1976.

3. Graham, R. D. and Helsing, G. G., Wood Pole Maintenance Manual: Inspection and Supplemental Treatment of Douglas-Fir and Western Redcedar Poles, Research Bulletin 24, Forest Research Lab, Corvallis, in press.

4. Pribyl, W., Timber Piling Handbook, Northwest Piling Company, Scappoose, Oregon, 1969.

5. U.S. Department of Agriculture, Forest Products Laboratory, Wood Handbook: Wood as an Engineering Material, U.S. Government Printing Office, Washington, D.C., 1974.

APPENDIX III - NOTATION

B = average diameter of embedded portion

D_b = depth of embedment

fs = unit skin friction resistance

L_d = load duration reduction factor

N = standard penetration resistance

P = strength reduction factor for imperfections

q_p = ultimate unit point resistance

S = Sa x f.s. = reference working stress

S_a = allowable axial working stress

S'_c = basic reference strength

W = strength reduction factor for treatment

SHAFT RESISTANCE OF ROCK-SOCKETED DRILLED PIERS

by Robert G. Horvath[1], M.A.S.C.E. and T. Cameron Kenney[2], M.A.S.C.E.

INTRODUCTION

Large diameter drilled pier foundations, which provide an effective and economical means of transmitting large concentrated structural loads through overburden soils to underlying rock, are being used in many areas of the world. Usually these piers are founded on competent rock with their load capacity governed by the structural strength of the concrete pier rather than the strength of the rock. Recently, drilled piers socketed into weak rock have been used more frequently.

The significant increase of load capacity that can be achieved by socketing the pier into weak rock has been recognized for many years (5). However, the load transfer behaviour or distribution of load support of drilled piers has not been adequately understood until recently.

Rock-socketed piers derive their load capacity from two components (Fig. 1):

1) shearing resistance at the concrete-rock interface around the vertical cylindrical surface of the socket (hereafter called shaft resistance), and

2) end-bearing resistance at the base of the pier.

Factors which control the load-carrying capacity of a socketed pier include the quality and strength of the rock and the efficiency of the load-transfer mechanisms operating between the concrete pier and its rock socket.

Analytical studies (4,15, 22) and load test measurements (3, 9, 12, 23) indicate that shaft resistance accounts for a large percentage of the total load capacity of a drilled pier foundation.

Shaft resistance is therefore an important consideration for determining the load capacity and load transfer behaviour of rock-socketed piers. Shaft resistance as used herein refers to the total load resistance along the pier-rock interface parallel to the axis of the pier, and includes all components of the shear resistance at the interface (i.e., friction and adhesion).

Load test data on shaft resistance of drilled pier foundations and model piers were compiled from published literature and unpublished reports. Drilled-shaft contractors, geotechnical engineering consultants and research organizations were contacted for data. This information will be presented with the hope that it will lead to a better understanding of load transfer

---

[1] Graduate Student and Research Assistant, Dept. of Civil Eng., Univ. of Toronto, Toronto, Ontario, CANADA

[2] Professor, Dept. of Civil Eng., Univ. of Toronto, Toronto, Ontario, CANADA

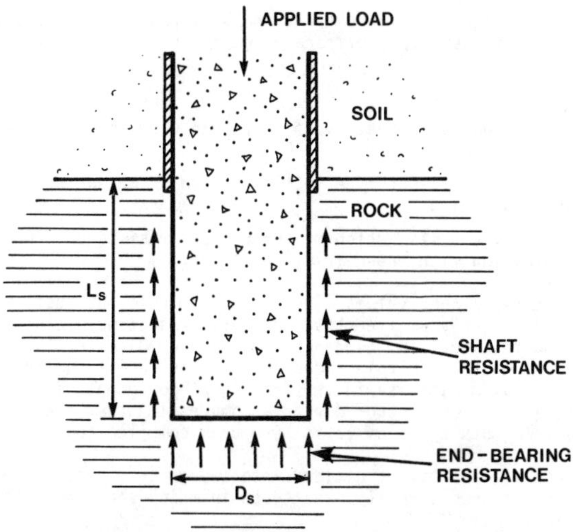

(a) APPLIED AND RESISTING FORCES

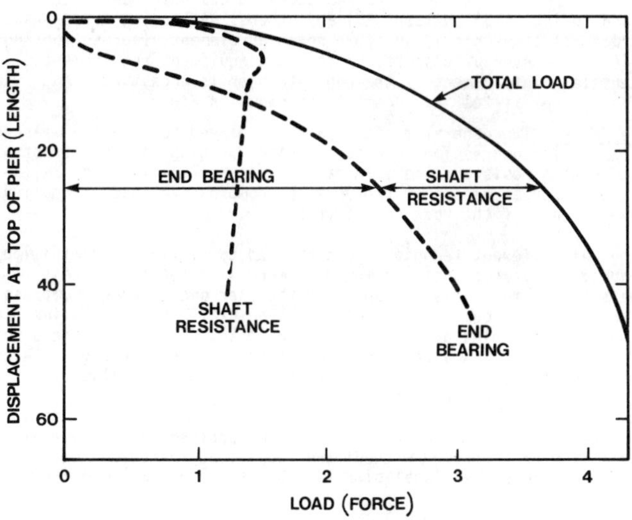

(b) LOAD - DISPLACEMENT BEHAVIOUR

FIGURE 1 – ROCK-SOCKETED DRILLED PIER

behaviour in socketed pier foundations and will prove useful for estimating shaft resistance values for preliminary foundation design.

TESTING METHODS

In general, two basic test types have been used to determine the shaft resistance of socketed piers:

1) tension or pull-out type tests, limited to measuring shaft resistance, and
2) compressive tests, which measured shaft resistance alone or in combination with end-bearing resistance.

The shaft resistance values determined from pull-out tests would be expected to be slightly smaller than the shaft resistance values determined by compression testing. The tendency for the pier diameter to decrease during a pull-out test would result in a decrease of the radial confining stress and thus a corresponding decrease in the frictional component of the shaft resistance. The opposite effect would occur during a compression test. However, in this report data from both types of tests have been compiled together.

Load test methods may be divided into two groups; tests in which shaft resistance is measured directly, and tests in which shaft resistance is determined indirectly.

Shaft Resistance Measured Directly

Load tests which measure shaft resistance directly include all pull-out tests and compression tests which are designed to eliminate the end-bearing component of load support.

Pull-out Tests. Pull-out tests on rock-socketed piers have been described by Matich and Kozicki (13), Wilson (25) and others. A loading frame (usually a heavy structural steel beam), suitable reaction elements (spread footing or reaction piers) and hydraulic jacks are the main equipment requirements. Adequate tensile reinforcement of the concrete pier is necessary. An illustration of typical test arrangements is shown in Figure 2.

Compressive Tests. Two general methods can be employed to directly measure shaft resistance in compression load tests; piers or plugs are constructed so that end-bearing resistance is eliminated or is negligible (Fig. 3a), and plugs are constructed so that the vertical compressive load is applied upward from the base of the rock socket (Fig. 3b).

A number of different techniques can be used to eliminate or minimize the end-bearing resistance during downward compression loading of test piers. The simplest method is to construct the pier over a weak, compressible material, such as styrofoam (19). Another method utilizes an isolation plug with appropriate shear pins at the base of the socket (14). Creating a void at the base of the pier was attempted by constructing the pier over a layer of styrofoam and subsequently dissolving this layer (20), and also by placing the pier concrete over a plug of dry ice (23).

Load tests in which the compressive load was applied to the base of the plug or pier were performed without the usual load frame and reaction elements required for conventional pier load tests. The loading device,

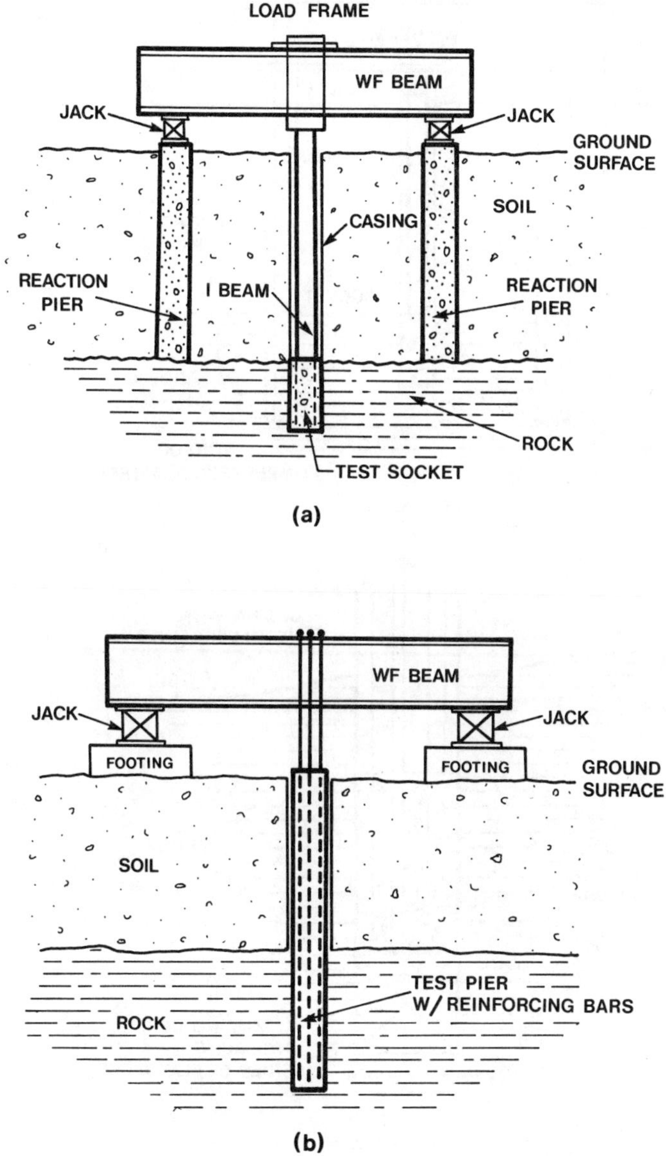

FIGURE 2 – TYPICAL PULL-OUT (TENSION) LOAD TEST ARRANGEMENTS.

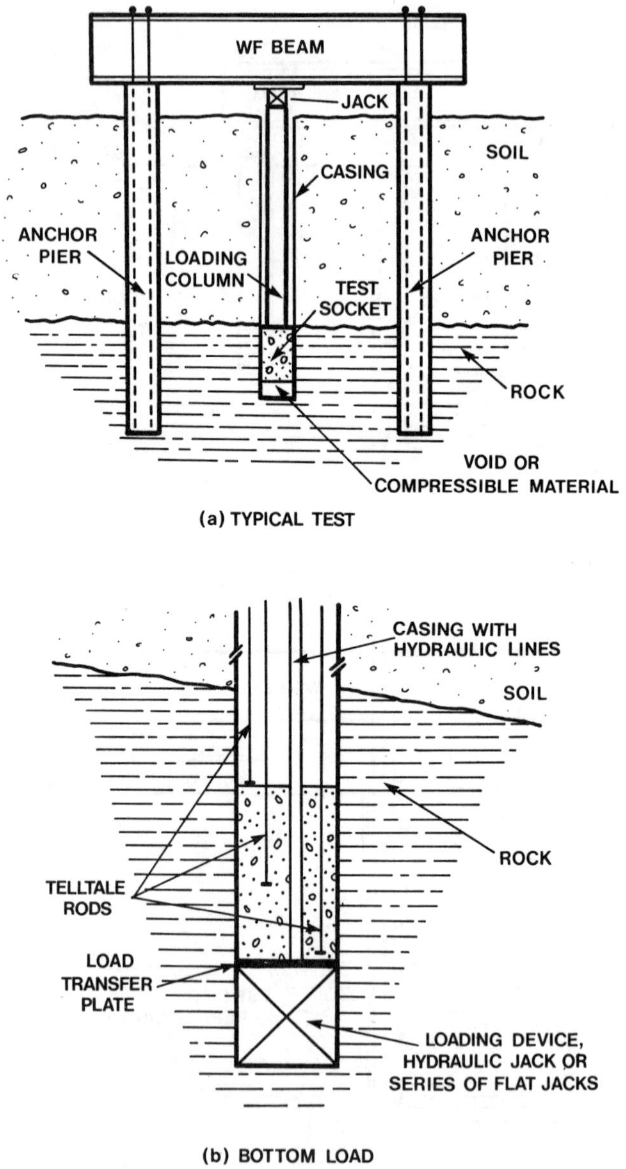

FIGURE 3 – ARRANGEMENTS OF COMPRESSION LOAD TESTS FOR DIRECT MEASUREMENT OF SHAFT RESISTANCE.

which can be a hydraulic jack (7) or a series of flat jacks (pers. comm., I.B. Donald, 1978), is placed at the bottom of the rock socket and the concrete pier or plug is constructed above the device. In this test there is no vertical resistance at the top of the plug and the shaft resistance is determined directly from the applied load. The advantages of this test method are simplicity and low cost.

The average value of shaft resistance for tests which measure shaft resistance directly is determined by means of the following equation:

$$s_r = \frac{Q_T}{\pi DL} \quad \text{....................(1)}$$

where s_r is average shaft resistance,
Q_T is applied load (tension or compression)
D is diameter of pier socket
L is length of pier socket

Shaft Resistance Determined Indirectly

Shaft resistance can also be determined by indirect methods from the results of conventional pier load tests in which the applied load is resisted by a combination of shaft and end-bearing resistances. Shaft resistance is determined either by deducting the measured end-bearing load or by determining the load distribution within the pier.

End-bearing resistance can be measured using suitable load cells at the base of the pier. The end-bearing resistance is then subtracted from the applied load to obtain the total shaft resistance. Shaft resistance can be determined by the equation:

$$s_r = \frac{Q_T - Q_B}{\pi DL} \quad \text{....................(2)}$$

where Q_B is load supported in end-bearing (measured)

The second indirect method of determining shaft resistance requires a fully-instrumented test pier so that load distribution along the length of the pier can be obtained. Based on measured strains, the distribution of shaft resistance along the pier can be calculated for each condition of loading. From this information average shaft resistance can be determined.

SUMMARY OF SHAFT RESISTANCE DATA COMPILED

The compiled shaft resistance data for rock-socketed piers and anchors includes information from more than 40 sources. Load test data from over 50 sites located mainly in Australia, Canada, England and the United States, and laboratory model test results have been reviewed. Table 1 summarizes the data distribution with respect to the types of tests reported. The results generally pertain to rock sockets with essentially straight-sided cylindrical walls parallel to the axis of the applied load.

The dimensions of the rock sockets varied considerably. The data have been grouped according to size as follows:

1) large scale piers - diameter $\geq$ 16 in (410 mm)
2) small scale piers - diameter < 16 in (410 mm)
3) rock anchors - length/diameter > 10 (generally)

Table 2 summarizes the information regarding test socket dimensions.

TABLE 1. Distribution of Load Test Methods

Test Description	Direct Measurement of Shaft Resistance: Pull-out Test	Direct Measurement of Shaft Resistance: Compressive Test	Indirect Determination of Shaft Resistance	Unclassified
Large scale piers*	8	20	13	8
Rock anchors*	21	-	-	-
Small scale piers*	1	26	-	-
Small scale piers **	4	46	-	-

* Field tests ** Laboratory tests

TABLE 2. Size Range of Rock Sockets Tested

Test Description	Diameter (D) in (mm)	Length (L) ft (m)	L/D
Large scale piers*	16 to 48 (410 to 1220)	1.8 to 40 (0.6 to 12)	1 to 29
Rock anchors*	1.5 to 7 (38 to 178)	2 to 14 (0.6 to 4.3)	8 to 140
Small scale piers*	3.3 to 12 (85 to 305)	0.4 to 5.2 (0.13 to 1.60)	1 to 7
Small scale piers**	0.5 to 10 (13 to 254)	0.1 to 0.8 (.03 to .25)	1 to 3

* Field tests ** Laboratory tests

Shale, including clay shale, siltstone, and mudstone, was the most common rock type encountered in the literature review. The information available concerning rock type, compressive strength and shaft resistance is summarized in Table 3.

TABLE 3. Rock Types and Strength Properties Reported

Rock Type	No. of Tests	Unconfined Compressive Strength psi (MPa)	Mobilized Shaft Resistance psi (MPa)
Shale or Mudstone	50	50 to 16,000 (0.35 to 110)	17 to 440 + (0.12 to 3 +)
Sandstone	8	1,000 to 3,500 + (7 to 24 +)	25 to 940 (0.52 to 6.5)
Limestone or Chalk	17	150 to 1,000 + (1 to 7 +)	17 to 418 + (0.12 to 2.8 +)
Igneous	4	50 to 1,500 + (0.35 to 10.5 +)	18 to 920 (0.12 to 6.3)
Metamorphic	8		68 to 273 + (0.47 to 1.9 +)

CORRELATION OF SHAFT RESISTANCE DATA

Correlation of the shaft resistance data is based on the authors' interpretation of the information collected. Complete information regarding engineering properties of the rock and concrete, construction details, etc., was often not available, and it was frequently necessary to use estimated values for rock and/or concrete strength. Deere's (2) engineering classification system for rock, as modified by Peck (16), was used for estimating the engineering properties of the rock (Fig. 4).

Shaft Resistance versus Unconfined Compressive Strength

The correlation of shaft resistance to strength of material was the primary relationship considered for this paper for the following reasons:

1) The maximum potential shaft resistance at the pier-rock interface is controlled by the shear strength of the weaker of the two materials.
2) The controlling material strength is dependent on material type, degree of weathering, jointing, etc.. Thus, these factors are dependent parameters included in the correlation.
3) Information concerning material strength is frequently available or can be easily obtained.
4) Other investigators have used material strength for correlation with shaft resistance.

The unconfined compressive strength of the weaker material (f'_w), either concrete or rock, has been used to represent the controlling material strength.

The relationship between shaft resistance (s_r) and the controlling unconfined compressive strength (f'_w) is shown in Figure 5 for large diameter

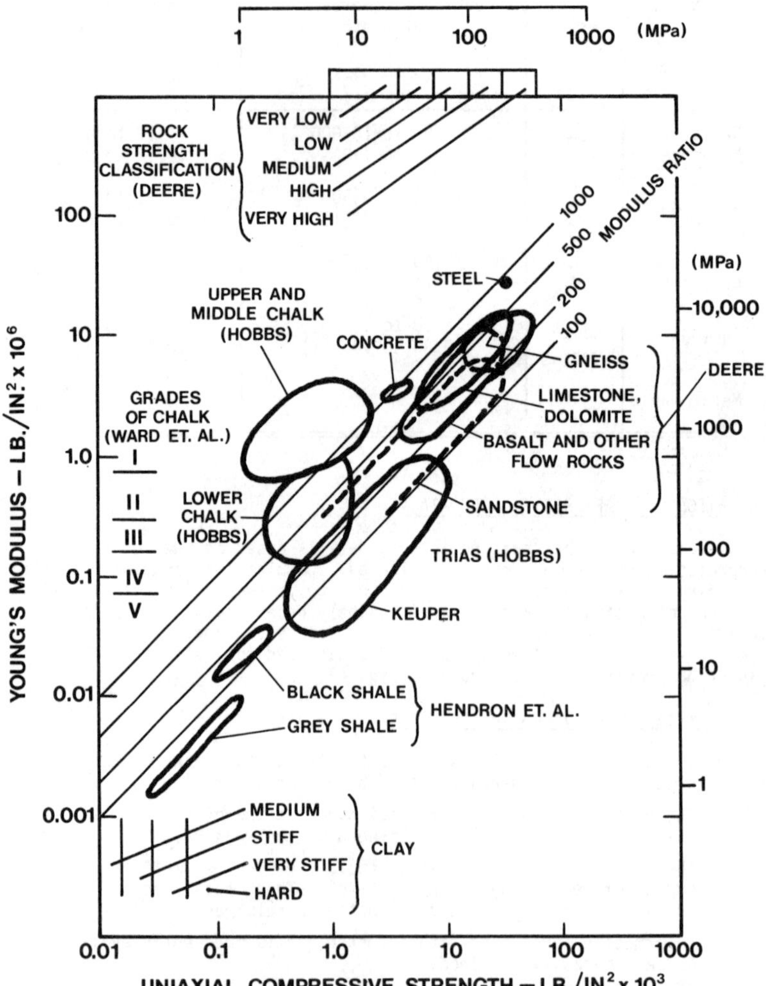

FIGURE 4 – ENGINEERING CLASSIFICATION FOR INTACT ROCKS (AFTER DEERE, 1968)

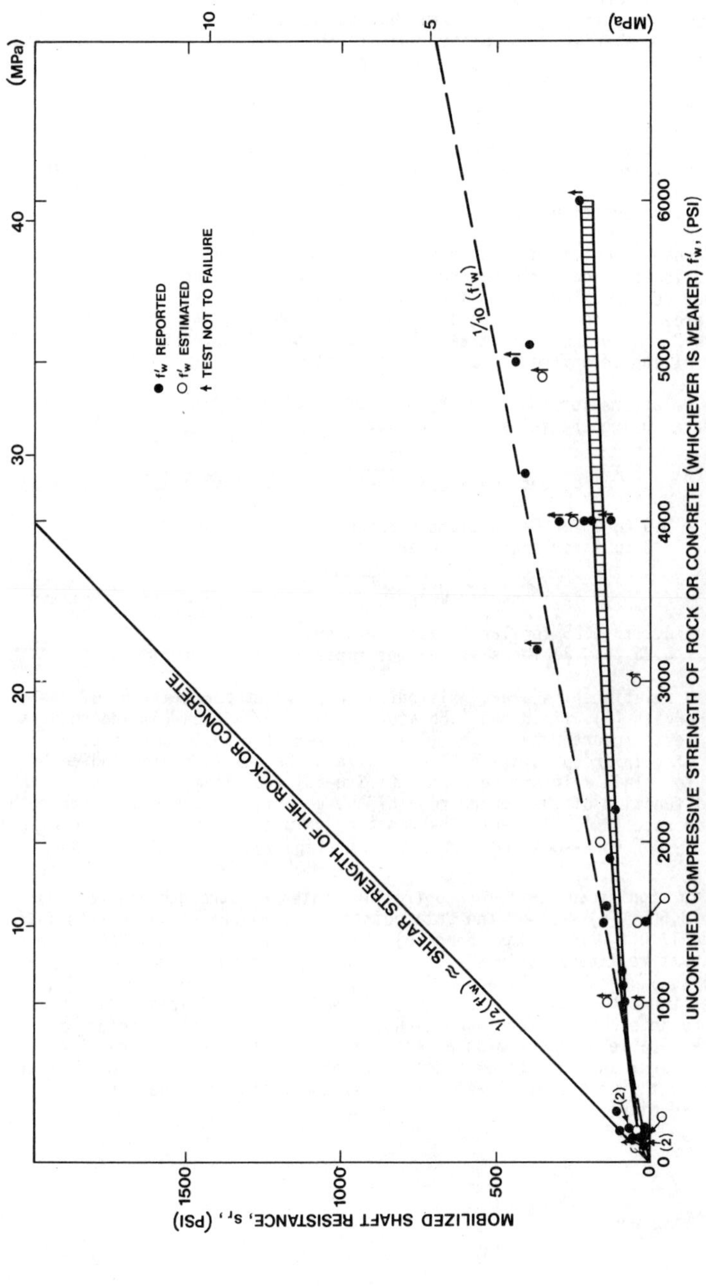

FIG. 5. — FIELD LOAD TEST DATA, LARGE DIAMETER PIERS, D ≥ 16 IN. (410 MM)

piers and in Figure 6 for small diameter piers and anchors. The data points are included. Solid points indicate that the material strength properties were reported, whereas open points indicate that material strength values were estimated.

In instances where a range of compressive strength values (f'_w) was reported, data points were plotted as follows; a) for single load tests, the average value of f'_w was used with s_r, and b) for a series of load tests in the same rock, s_r values were distributed proportionately within the reported range of f'_w values using a linear relationship (e.g., s_r minimum and f'_w minimum constitute one point).

The shaded region on the plots indicates the correlation range which was established by determining the curves of best fit (linear regression of a power curve) using the maximum and minimum strength values. These regions are also shown on a log-log plot after Rosenberg and Journeaux (18) in Figure 7 to further illustrate the relationship between shaft resistance and controlling unconfined compressive strength.

The relationships indicated by the curves of best fit can be closely approximated using the following equations:

$$s_r \text{ (psi)} = a\sqrt{f'_w \text{ (psi)}} \qquad \text{(3a)}$$

where a = 2.5 to 3 for large diameter piers
= 3 to 4 for small diameter piers

$$s_r \text{ (MPa)} = b\sqrt{f'_w \text{ (MPa)}} \qquad \text{(3b)}$$

where b = 0.2 to 0.25 for large diameter piers
= 0.25 to 0.33 for small diameter piers

These relationships are consistent with previous correlations of field data by Horvath (8), field and laboratory results of Gibson (6) and results of model tests reported by Kenney (11). Further, the analogous question in the area of reinforced concrete also parallels the relationships indicated. That is, for plain reinforcing bars (ACI 318-63) the ultimate bond strength (u_u) is a function of the square root of the unconfined compressive strength of the concrete (f'_c). In fact, the range of values of u_u for plain reinforcing bars and s_r for rock-socketed piers overlap (Fig. 7).

It is significant that for controlling material strength greater than 500 psi (3,500 kPa) most of the data points for straight-sided sockets fall below the 1/10 f'_w line (Figs. 5 and 6); that is, $s_r < 1/10\ f'_w$. Comparison of the shaft resistance values s_r to the shear strength of the weaker material (s_w), which can be approximated by $s_w = 1/2\ f'_w$ (Figs. 5 and 6), indicates the potential for substantially increasing mobilized shaft resistance in socketed pier foundations. Also, deformed reinforcing bars which are capable of developing at least twice the bond resistance allowed for plain bars, and results of model tests on shaped piers (Fig. 6) further demonstrate the potential for significantly increasing the shaft resistance of socketed piers.

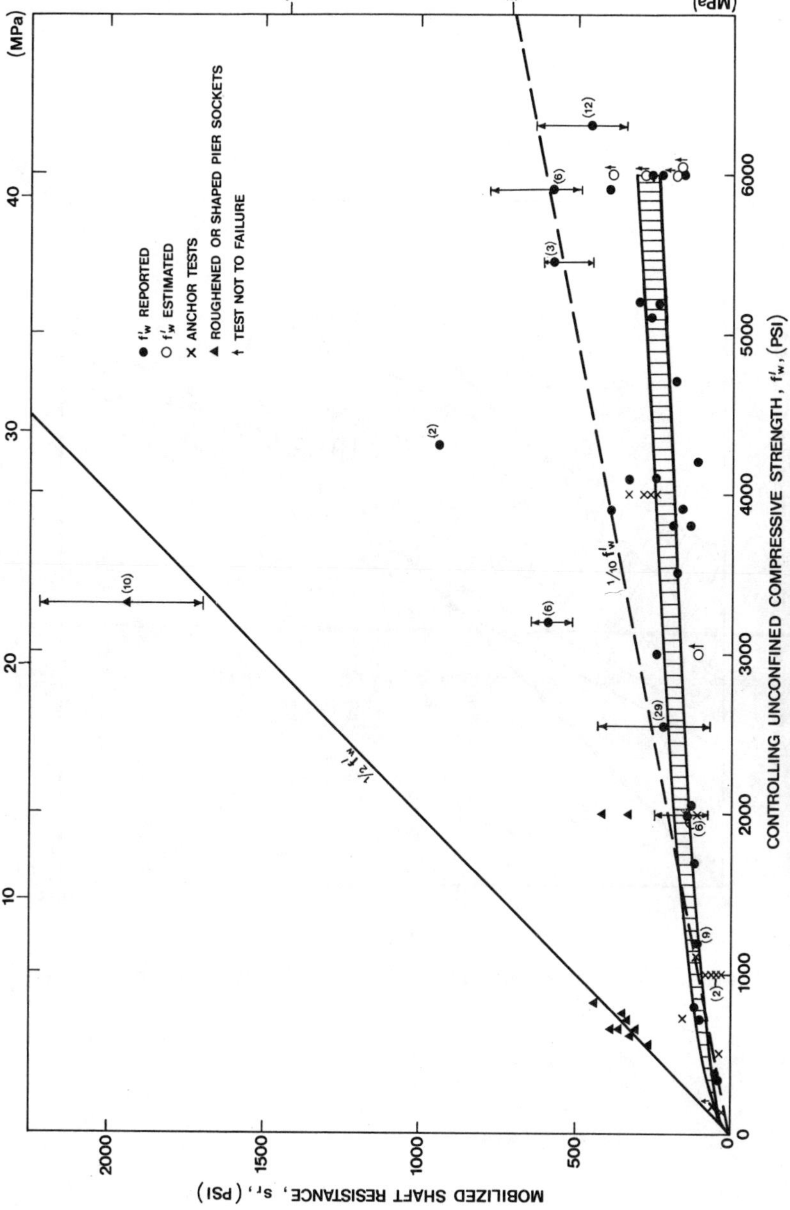

FIG. 6. — LOAD TEST DATA, SMALL DIAMETER PIERS & ANCHORS, D<16 IN.(410 MM)

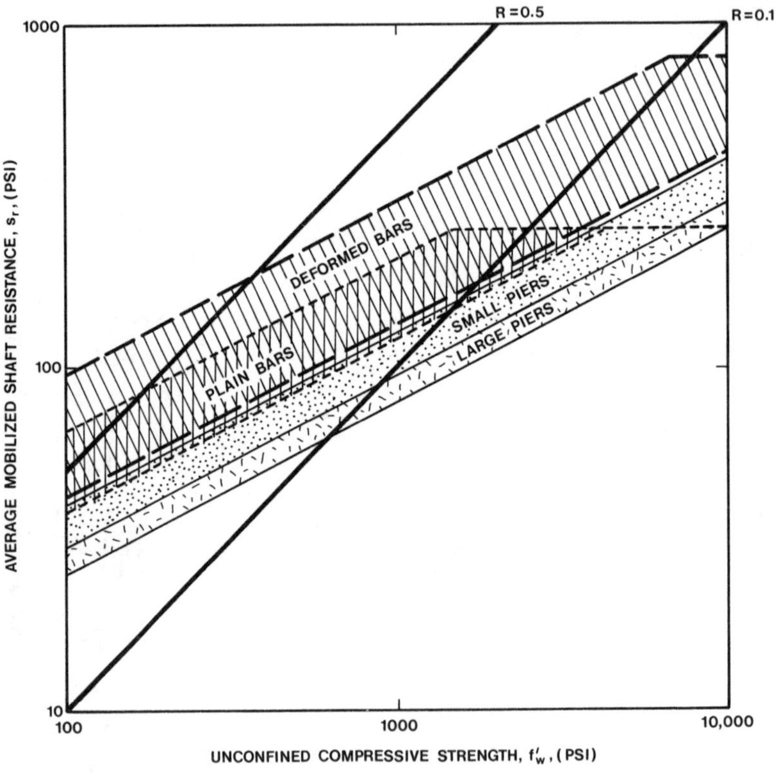

FIGURE 7 — SHAFT RESISTANCE VS. UNCONFINED COMPRESSIVE STRENGTH

Another interesting relationship concerning shaft resistance and controlling compressive strength was observed. The strength ratio (R_s), where

$$R_s = s_r/f'_w \qquad \text{....................(4)}$$

apparently decreases with increasing controlling compressive strength (f'_w). This relationship is apparent in Figures 5 and 6. However, the curve shown in Figure 8 illustrates this point more clearly. The data points fall approximately into three groups as shown in Table 4.

TABLE 4. Strength Ratio (R_s) Relationship for Straight-sided Piers

UNCONFINED COMPRESSIVE STRENGTH f'_w	STRENGTH RATIO R_s
Less than 500 psi (3.5 MPa)	Greater than 0.2
500 to 2500 psi (3.5 to 17 MPa)	0.2 to 0.1
Greater than 2500 psi (17 MPa)	0.1 to 0.025

Shaft Resistance versus Diameter of Pier Socket

An apparent reduction in the magnitude of mobilized shaft resistance with increasing socket diameter has been indicated by some laboratory test data (17). This mode of behaviour is also suggested by the case of reinforced concrete where the American Concrete Institute recommends bond values inversely proportional to bar diameter (ACI 318-63).

The data compiled herein were reviewed to examine the effect of socket diameter on mobilized shaft resistance in socketed piers. The average value of strength ratio (R_s) for each socket diameter tested was plotted to determine possible trends, rather than a specific relationship. The data indicate that strength ratio (R_s) apparently decreases with increasing diameter (Figure 9). However, within the size range for large diameter socketed piers ($D \geq 16$ in) the effect of socket diameter appears to be negligible.

Shaft Resistance versus Roughness of Socket Wall

Roughness of the socket wall must influence the magnitude of mobilized shaft resistance in rock-socketed piers. Lack of published information concerning socket roughness, however, precludes thorough evaluation of this factor.

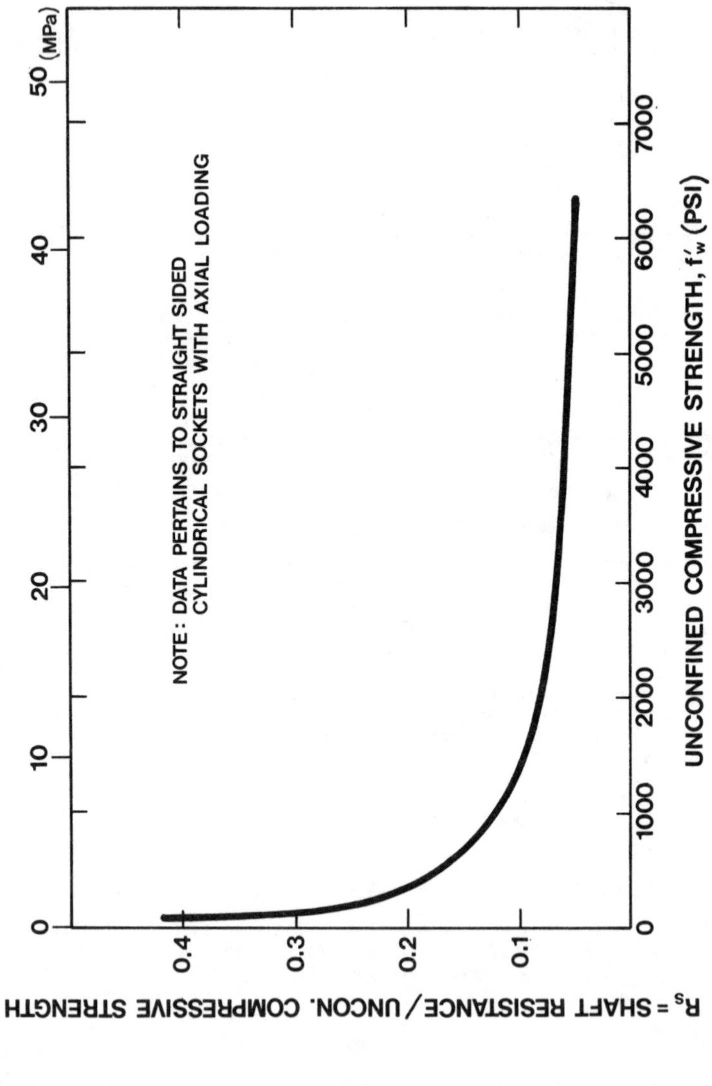

FIGURE 8 —STRENGTH RATIO VS. UNCONFINED COMPRESSIVE STRENGTH FOR STRAIGHT SIDED SOCKETS.

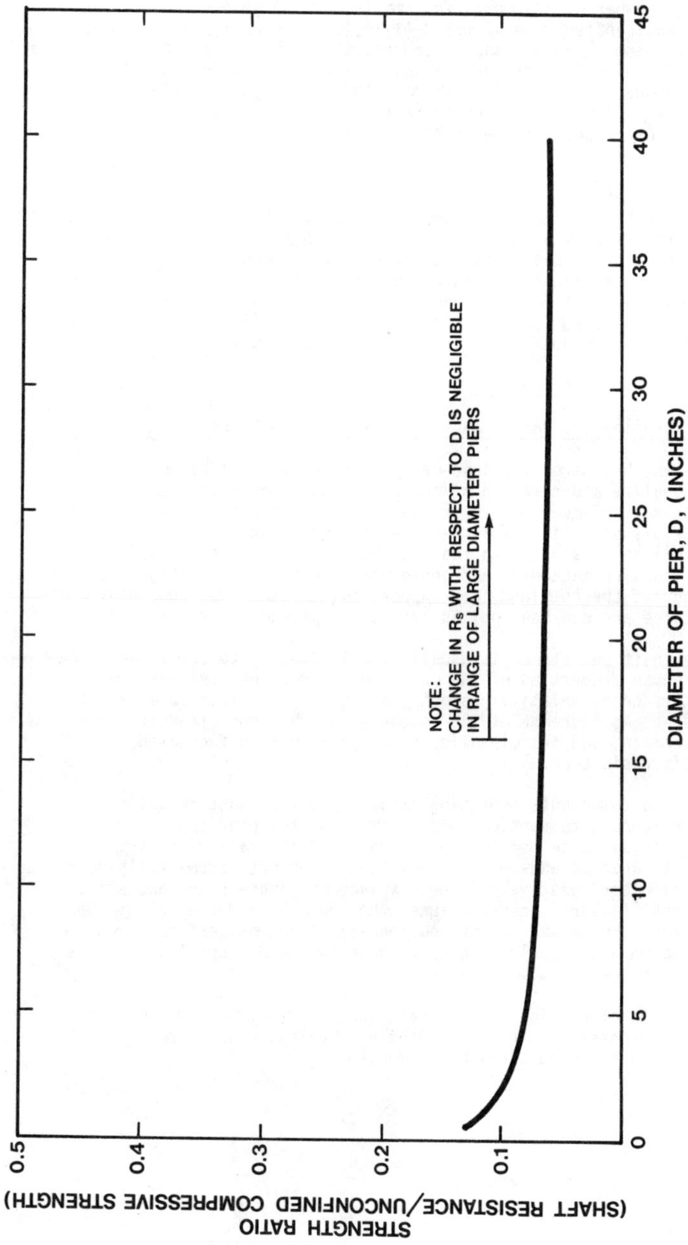

FIGURE 9-AVERAGE STRENGTH RATIO, R_S, FOR PIERS OF VARIOUS DIAMETERS.

A limited number of test data for small-diameter grooved model piers reflects the anticipated general trend (1,10,21). Three general categories were studied: smooth, rough and shaped sockets. The results are summarized in Figure 10 in which the strength ratio (R_s) is plotted against a qualitative measure of roughness. The field test data for large diameter piers ($D \geq 16$ in), for values of $f'_w > 1000$ psi, suggests that R_s for most rock-socketed piers would fall somewhere between the smooth and rough categories.

Limitations

The relationships discussed above are based on the authors' interpretation of test data reported by others. The lack of complete information and the wide scatter of plotted data limits the use of the proposed relationships. The shaft resistance values and tentative correlations should be used with caution and careful engineering judgement. It is suggested that the information only be used for preliminary design evaluation. Actual design values of shaft resistance should be verified by suitable field testing.

DISCUSSION

Methods of Improving the Performance of Rock-Socketed Piers

Factors which control the load-carrying capacity of a socketed pier include the quality and strength of the rock and the efficiency of the load-transfer mechanisms operating between the concrete pier and its rock socket. The full potential of rock-socketed piers (with respect to load-support capability) has not been realized because of intrinsic inefficiency of the load-transfer mechanism and conservative engineering design practice. The efficiency of the combined load-support system (that is, combination of shaft resistance and end-bearing) is less than optimum for two reasons:

1) The shaft resistance is usually small compared to the shear strength properties of the concrete and rock because a) the interface is relatively smooth, and/or b) the interface is not clean. An increase of the roughness of the concrete-rock interface, and its cleanliness, would result in increased shaft resistance.

2) The two components providing axial support are not compatible with respect to vertical displacement of the pier (Fig. 1b). Shaft resistance becomes fully mobilized at small pier displacements, whereas end-bearing resistance does not become fully mobilized until after relatively large displacements take place and slippage ('bond' failure) occurs between the concrete and rock along the side of the socket. Improved load-displacement performance of socketed piers would result if these two load support components were to develop simultaneously.

On the basis of previous analytical, laboratory model, and field test studies, two concepts for the improvement of the load-displacement behaviour of rock-socketed piers have evolved.

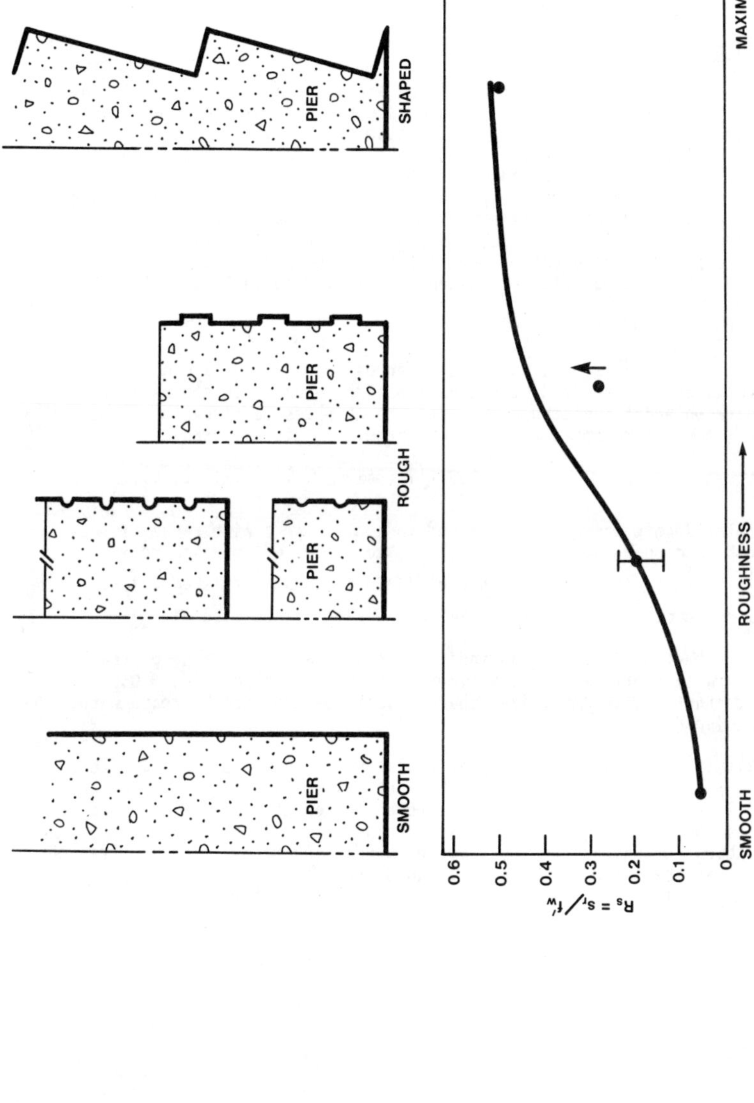

FIGURE 10 — SHAFT RESISTANCE VS. ROUGHNESS OF SOCKET WALL.

Shaped Socket Wall

Increasing the magnitude of the mobilized shaft resistance along the socket wall would be accomplished by roughening or shaping the socket wall to take full advantage of the shear strengths of the pier and rock materials, while placing less reliance on the displacement resistance of the interface. Laboratory model tests on shaped piers (1, 10, 21) indicate that mobilized shaft resistance can be increased by a factor of at least 2. Figures 6 and 10 illustrate this point. The anticipated improvement of the load-displacement behaviour of a shaped socket is shown in Figure 11a.

Preload Base of Socket

An increase of the magnitude of the applied load transferred to the base of the socket at small displacements could be achieved by applying an expansion load between the pier and base of the socket prior to (or during) application of the foundation load. Laboratory tests on model piers (11) showed that preloading the base of the socket can stiffen the load-displacement behaviour of piers socketed into weak rock. Also, the portion of applied load transferred to the socket base through end-bearing resistance was significantly increased at small relative displacements. The anticipated improvement of the load-displacement behaviour for a socketed pier that has been preloaded is illustrated in Figure 11b.

CONCLUSIONS

1) On the basis of a study of numerous load-test data concerning rock-socketed piers constructed by conventional procedures, an empirical relationship between shaft resistance, s_r, and the compressive strength of the weaker interface material (rock or concrete), f'_w, is proposed:

$$s_r(\text{psi}) = 2.5\sqrt{f'_w(\text{psi})} \quad \ldots\ldots\ldots\ldots\ldots\ldots\ldots (5)$$

This relationship was generally found to apply over a wide range of rock strengths and rock types and corresponds to values of approximately $s_r = \frac{1}{20}$ to $\frac{1}{10}\, f'_w$. In the absence of field test data, this empirical relationship can be used for preliminary design of socketed piers.

2) The test data further indicated that shaft resistance generally becomes fully mobilized at relatively small pier displacments, < 0.25 in (6 mm). Further, the mobilized shaft resistance does not decrease with larger displacements.

3) The magnitude of the shaft resistance values reported were very small compared to the strength of the controlling material. The obvious way to increase mobilized shaft resistance in socketed piers is to groove or shape the socket walls, which is comparable to using deformed reinforcing bars in concrete to increase the bond. The material shear strength, rather than the interface adhesion, would then be mobilized to support the applied pier loads.

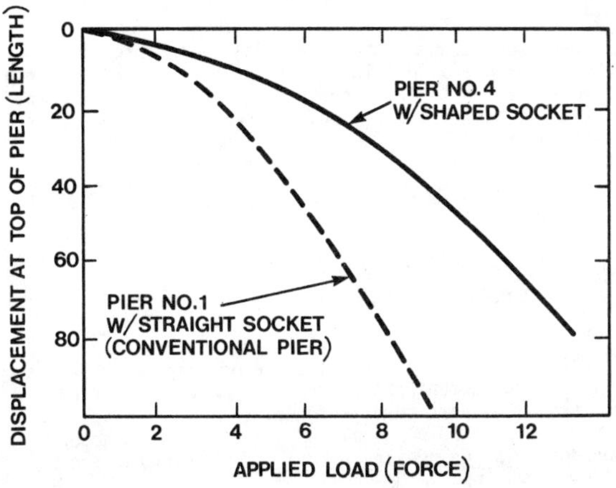

(a) PIER WITH SHAPED SOCKET

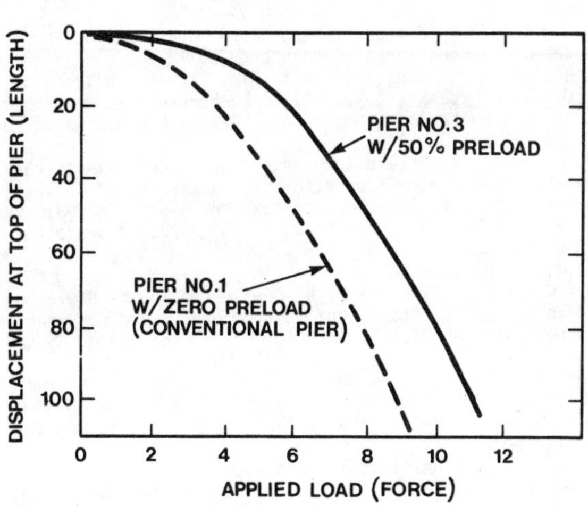

(b) PIER WITH PRELOAD

FIGURE 11 – ANTICIPATED LOAD-DISPLACEMENT BEHAVIOR OF MODIFIED PIERS VS. CONVENTIONAL PIERS.

ACKNOWLEDGEMENTS

This study was undertaken at the University of Toronto, Toronto, Canada. The many consulting engineers, drilled shaft contractors and research organizations who responded to the request for information are gratefully acknowledged. Without their cooperation, the compilation and correlation of data would have been impossible. The writers would also like to express their appreciation to Mr. Peter Kozicki of Western Caissons Limited and Mr. William A. Trow of the Trow Group for their review and comments on the paper.

APPENDIX I - REFERENCES

1. CRUICKSHANK, D.J., 1976. "The influence of shape and surface roughness on the load-displacement characteristics of rock-socketed piers". B.A.Sc. thesis, University of Toronto.
2. DEERE, D.V., 1968. "Geological considerations". Chapter 1 in Rock Mechanics in Engineering Practice by K.G. STAGG and O.C. ZIENKIEWICZ, Wiley, New York. pp. 1-20
3. DVORAK, A., 1966. "Tests of anisotropic shales for foundation of large bridges". Proc. 1st Int. Cong. Rock Mech., LISBON, Vol. 2, pp. 537-541.
4. ELLISON, R.D., D'APPOLONIA, E. and THIERS, G.R., 1971. "Load-deformation mechanism for bored piles". Journal of the Soil Mechanics and Foundations Division, ASCE. Vol. 97, No. SM4, Proc. Paper No. 8052, April 1971, pp. 661-678.
5. E.N.R., 1937. "Drilled piles seated in rock sockets". Engineering News Record, Sept. 9, 1937.
6. GIBSON, G.L., 1973. "Field and laboratory investigation for design of rock-socketed caissons for a multi-storey building founded in sound limestone". Master of Engineering thesis, Carleton University, Ottawa.
7. GIBSON, G.L. and DEVENNY, D.W., 1973. "Concrete to bedrock bond testing by jacking from bottom of a borehole". Canadian Geotechnical Journal. Vol. 10, No. 2, pp. 304-306.
8. HORVATH, R.G., 1978. "Field load test data on concrete-to-rock bond strength for drilled pier foundations". University of Toronto, Department of Civil Engineering Publication 78-07 (ISSN 0316-7968).
9. JACKSON, W.T., PEREZ, J.Y. and LACROIX, Y., 1973. "Foundation construction and performance for a 34-storey building in St. Louis". Geotechnique. Vol. 24, No. 1, pp. 63-90.
10. KENNEY, T.C., 1977. "Factors to be considered in the design of piers socketed into rock". Proc., conf. on the design and construction of deep foundations, Canadian Society for Civil Engineering, Sudbury, Ontario.
11. KENNEY, T.D., LEE, L.C., and GAGG, C., 1975. "Laboratory investigation of preloading as a method of improving the load-displacement performance of socketed piles". University of Toronto, Department of Civil Engineering Publication 75-09.
12. LADANYI, B., 1977. "Friction and end bearing tests on bedrock for high capacity socket design: discussion". Canadian Geotechnical Journal. Vol. 14, pp. 153-155.

13. MATICH, M.A.J. and KOZICKI, P., 1967. "Some load tests on drilled cast-in-place concrete caissons". Canadian Geotechnical Journal. Vol. 4, No. 4, Nov.
14. MOORE, W.W., 1964. "Foundation design". ASCE, Civil Engineering. Jan. 1964, pp. 33-35.
15. OSTERBERG, J.P. and GILL, S.A., 1973. "Load transfer mechanism for piers socketed in hard soils or rock". Proc., 9th Canadian rock mechanics symposium, Montreal.
16. PECK, R.B., 1976. "Rock foundations for structures". Proc. specialty conf. on rock engineering for foundations and slopes. Boulder, Colorado. ASCE, New York.
17. PELLS, P.J.N., 1978. "Investigation into the allowable loadings for bored piles founded on Hawkesbury sandstone". Progress Report 1 and 2, Investigation Report No. S226, School of Civil Engineering, University of Sydney.
18. ROSENBERG, P. and JOURNEAUX, N.L., 1976. "Friction and end bearing tests on bedrock for high capacity socket design". Canadian Geotechnical Journal. Vol. 13, pp. 324-333.
19. SEYCHUCK, J.L., 1970. "Load tests on bedrock". Canadian Geotechnical Journal. Vol. 7, No. 4, pp. 454-469.
20. SPANOVICH, M., undated. "Caisson-shale interaction: some load test results". Unpublished report - Engineering Mechanics, Inc., Consulting Geotechnical Engineers, Pittsburg.
21. SUGAMORI, H.T., 1977. "Laboratory study of load-displacement behaviours of shaped rock-socketed piers," Undergraduate project, Department of Civil Engineering, University of Toronto.
22. TOLLIN, B.I., 1976. "An analytical study of rock-socketed piers using the finite element method". M.Eng. thesis, University of Toronto.
23. WATSON III, F.X., 1971. "An investigation of the development of skin friction of drilled piers in decomposed rock". PhD. thesis, Georgia Institute of Technology, Atlanta, Georgia.
24. WHITAKER, T. and COOKE, R.W., 1966. "An investigation of the shaft and base resistances of large bored piles in London clay." Symposium, large bored piles, London, 1966.
25. WILSON, L.C., 1976. "Tests of bored and driven piles in cretaceous mudstone at Port Elizabeth, South Africa." Geotechnique, Vol. 26, No. 1.

APPENDIX II - NOTATION

LIST OF SYMBOLS

C_o uniaxial compressive strength of rock
D diameter of socket
f'_c compressive strength of concrete
f'_w unconfined compressive strength of weaker material, either concrete or rock
L length of socket
Q_T applied load
Q_B load supported in end-bearing
R_S strength ratio, ratio of the shaft resistance to the unconfined compressive strength of the weaker material
s_r average shaft resistance
s_w shear strength of the weaker material

APPENDIX III — Reported Values of Shaft Resistance for Rock-Socketed Piers and Anchors

Number Reference Location	*Rock Type*	*Shaft Resistance in kPa (psi)*	*Remarks*
PC3 (Seychuck, 1970) Canada	SHALE, horizontally laminated, thin upper weathered zone, sound, with occasional hairline fractures wholly or partially healed with calcite and thin mud seams. C_o = 35-60 MPa (5000-9000 psi)	3040* (440)	Pier compression test D_s = 483 mm (19 in) L_s = 0.9 m (3 ft); L/D = 2 Construction of socket by coring pilot hole & churn drilling. Styrofoam in bottom to eliminate end-bearing.
PC4 (Trow, 1972) Canada	SHALE, sound, horizontal bedding planes	2460* (356)	Pier compression test D_s = 838 mm (33 in) L_s = 0.7 m (2.25 ft); L/D = 0.8 constructed using tremie concrete, 7 in. slump f'_c = 40,700 kPa (5900 psi) Styrofoam in bottom to eliminate end-bearing. 1035 kPa (150 psi) used for design.
PC5 (MTC, 1968) Canada	SHALE, platey bedded, slightly calcareous, upper 0.2 m weathered C_o = 12.4-18 MPa (1800-2600 psi) $E_r \simeq$ 1040 MPa (150 ksi)	830 (120)	Pier compression test D_s = 635 mm (25 in) L_s = 0.91 m (3 ft); L/D = 1.4 f'_c = 30,700 kPa (4450 psi) Bond value determined from combined test and end-bearing only test.
PC6a (Spanovich, undated) U.S.A.	SHALE, soft, silty, with occasional indurated clay seams, hardness = 2-3	770 (peak) (111) and 900 (ult.) (130)	Pier compression tests D = 457 mm (18 in) L_s = 0.9 m (3 ft); L/D = 2 seepage water in shaft, auger clean only Void at bottom of socket f'_c = 20,700 kPa (3000 psi)
PC6b	(same)	656 (95)	D = 610 mm (24 in) L_s = 0.91 m (3 ft); L/D = 1.5
PC6c	(same)	1533 (222)	D = 762 mm (30 in) L_s = 0.91 m (3 ft); L/D = 1.2
PC6d	(same)	580 (84)	D = 610 mm (24 in) L_s = 1.52 m (5 ft); L/D = 2.5
PC6e	(same)	620 (90)	D = 762 mm (30 in) L_s = 1.52 m (5 ft); L/D = 2
PC10a (Vogan, 1977) Canada	SHALE, competent, bedded dipping 35° to 40°, numerous near-vertical hairline fractures cemented with calcite C_o = 62-110 MPa (9000-16000 psi) (core tests from adjacent site)	932* (135)	Pier compression test D = 610 mm (24 in) L = 2.9 m (9.5 ft); L/D = 4.8 constructed by churn drilling water flowing from fractures f'_c = 28.8 MPa (4000 psi)
PC10b	(same)	387* (56)	(same as above) D = 406 mm (16 in) L_s = 1.52 m (5 ft); L/D 3.8 Pressure Chamber failed during loading thus test terminated

Number Reference Location	*Rock Type*	*Shaft Resistance in kPa (psi)*	*Remarks*
PC14 (Ladanyi, 1977) Canada	SHALE, horizontally stratified, broken RQD = 0 to 39	345* (50)	Pier compression test D = 884 mm (34.8 in) L = 4.6 m (15 ft); L/D = 5.2 Designed for 1035 kPa (150 psi)
PC16 (Thornborn, 1966) U.K.	SHALE, fragmented	242 (35)	Pier compression test D = 915 mm (36 in) L_s = 3.35 m (11 ft); L/D = 3.7 excavated by bucket auger with rock teeth Polystyrene pad on base to eliminate end-bearing.
PC24 (Osterberg & Gill, 1973) U.S.A.	SHALE, black C_o = 9.7-12.5 MPa (1400-1800 psi)	1040 (150)	Pier compression test D = 1220 mm (48 in) L = 1.22 m (4 ft); L/D = 1
PC25 (Mason, 1960) U.S.A.	SHALE, weak C_o = 1.5 MPa (210 psi)	417 (60)	Pier compression test D = 610 mm (24 in) L = 1.83 m (6 ft); L/D = 3
PC28 (Pells et al, 1978) (Groundtest, 1976) Australia	SHALE, fresh, horizontal bedding RQD = 90	2900 (420)	Test method not stated D = 400 mm (16 in) L = 1.0 m (3.3 ft); L/D = 2.5 test 2.9 m (10 ft) below rock surface
PC31 (Pells et al, 1978) (Coffey & Partners, 1975) Australia	SHALE, fractured, spacing 20-100 mm (1-4 in) C_o = 21-51 MPa (3000-7300 psi)	2800 (400)	Test method not stated D = 450 & 900 mm (17.7 & 35.4 in) L – 0.7 & 1.3 m (2.3 & 4.3 ft) L/D = 1.6 & 1.4
PC32 (Dvorak, 1966) Czechoslavakia	SHALE (ordivician), folded, faulted, argillaceous C_o = 60 MPa (8640 psi) E_r = 170 MPa (25,000 psi)	750* (108)	Pier Compression Test D = 250 mm (9.8 in)
PT1 (Matich & Kozicki, 1967) Canada	SHALE, massive with bedding in places dipping to 70°, frequent weathered zones, hardness 2-3, disintegrates easily. C_o = 690 kPa (100 psi) γ = 23.6 kN/m^4 (150 pcf) $w_n \simeq$ 15%	310 (45)	Pier pull-out test D_s = 610 mm (24 in) L_s = 6 m (19.8 ft); L/D = 10 Rifled groove 150 mm wide by 75 mm deep by 450 mm pitch. Designed for bond resistance only.
PT3 (Rosenbergy & Journeaux, 1976) Canada	SHALE, joints spaced at 2 in. C_o = 20,700 kPa (3000 psi) core lengths 76-127 mm (3-5 in)	1690 (245)	Pier pull-out test D = 203 mm (8 in) L = 0.9 m (3 ft); L/D = 4.5 f'_c = 34,500 kPa (5000 psi)
PT6 (Freeman et al, 1972) Canada	SHALE (Dundas) C_o = 9.6-34.6 MPa (1380 to 5000 psi) Brazilian Tensile Test = 200 psi	2600* (375)	Pier pull-out test D = 1065 mm (42 in) L = 3.2 m (10.5 ft), L/D = 3 Design bond value = 1000 kPa (150 psi)

Number Reference Location	*Rock Type*	*Shaft Resistance in kPa (psi)*	*Remarks*
PT7 (Morton, 1969) Canada	SHALE (Collingwood), dark grey to black, brittle, sound, well bedded, lightly calcareous	1390 (200)	Pier pull-out test D_s = 760 mm (30 in) L_s = 1.0 m (3.25 ft); L/D = 1.3 f'_c = 27.7 MPa (4000 psi)
A2a (MTC, 1964) Canada	SHALE, layered, generally sound, unweathered	345 (peak) (50) to 242 (ult.) (35)	Shallow anchor pull-out tests D = 76 mm (3 in) L = 0.6 m (2 ft); L/D = 8 f'_c = 17,200 kPa (2500 psi) Expanding agent added to concrete
A2b	(same)	345 (50)	D = 76 mm (3 in) L = 0.9 m (3 ft); L/D = 12
A2c	(same)	443 (64)	D = 76 mm (3 in) L = 1.2 m (4 ft); L/D = 16
A15 (Saliman & Schaefer, 1968) U.S.A.	SHALE, very weathered	390 (56)	Anchor Pull-out test D = 60 mm (2.375 in) L = 1.5 m (5 ft); L/D = 25
A16a (Saliman & Schaefer, 1968) U.S.A.	SHALE, fractured, poor	240 (35)	Anchor Pull-out test D = 60 mm (2.375 in) L = 1.5 m (5 ft); L/D = 25
A16b	SHALE, fractured, average	350 (50)	Anchor Pull-out test D = 60 mm (2.375 in) L = 1.5 m (5 ft); L/D = 25
A16c	SHALE, fractured, good	750 (110)	Anchor Pull-out test D = 60 mm (2.375 in) L = 1.5 m (5 ft); L/D = 25
A17 (Abraham & Porzig, 1973) Germany	SHALE and SANDSTONE (Greywacke) C_0 = *49-78 MPa (7000-11000 psi)* E_r = 2000-4000 MPa (280-560 ksi) ν = 0.4	1630 (234) to 4380 (631)	Anchor Pull-out test D = 101-116 mm (4-4.6 in) L = 1-1.3 m (3.3-4.3 ft) L/D = 10
MC1 (9 tests) (Webb, 1976) South Africa	SHALE, weathered, dark grey, widely spaced joints, thinly laminated C_o = 8.5 MPa (1200 psi)	750 (108)	Model Compression Tests D = 50 mm (2 in) L = 0.075 m (0.25 ft) L/D = 1.5
PC8a (Aurora & Reese, 1976) U.S.A.	CLAY SHALE, shear strength = 690 kPa (100 psi)	412 (60)	Pier compression tests D_s = 740 mm (29.1 in) L_s = 1.35 m (4.4 ft); L/D = 1.8 constructed by casing method using auger.
PC8b	(same)	370 (54)	D_s = 740 mm (29.1 in) L_s = 1.35 m (4.4 ft); L/D = 1.8 constructed by slurry displacement method

Number Reference Location	*Rock Type*	*Shaft Resistance in kPa (psi)*	*Remarks*
PC8c	(same)	690 (100)	D_s = 750 mm (29.5 in) L_s = 1.52 m (5 ft); L/D = 2 Constructed by dry method
PC9 (Aurora & Reese, 1976) U.S.A.	CLAY SHALE, Shear strength = 310 kPa (50 psi) (estimated from static cone tests)	278 (40)	Pier compression test D_s = 890 mm (35 in) L_s = 1.37 m (4.5 ft); L/D = 1.5 constructed by casing method
PC29 (Pells et al, 1978) (Ervin, 1976) Australia	MUDSTONE, weathered, highly fractured, frequent clay seams C_o = 1.5-3 MPa (210-430 psi)	800 (115)	Pier compression test D = 1092 mm (43 in) L = 1.52 m (5.0 ft) L/D = 1.4
A3 (Wilson, 1976) South Africa	MUDSTONE, weak, clayey, no secondary cementation or induration C_o = 1090 kPa (160 psi) S_o' = 152 kPa (20 psi); ϕ' = 45° γ = 19.5 kN/m³ (124 pcf) w_L = 44%, PI = 26, W_n = 13% linear shrinkage = 9.6%	310 (45)	Anchor pull-out test D = 100 mm (3.9 in) L = 14 m (46 ft); L/D = 140 No failure established
PT4a (Wilson, 1976) South Africa	(same as A3)	120 (17)	Pier pull-out tests D = 900 mm (35.4 in) L_s = 1 m (3.3 ft); L/D = 1.1
PT4b	(same as A3)	184 (27)	D = 900 mm (35.4 in) L_s = 1 m (3.3 ft); L/D = 1.1
MT1 (Parkins & Donald, 1972) Australia	MUDSTONE (Melbourne a) C_o = 2.4 MPa (340 psi)	285 (41)	Model pull-out tests D = 13 mm (0.5 in) L – not given
	b) C_o = 2.5 MPa (360 psi)	306 (44)	(same)
	c) C_o = 4.9 MPa (710 psi)	660 (95)	(same)
	d) C_o = 5.6 MPa (800 psi)	800 (115)	(same)
PC30 (Pells et al, 1978) (Soil mech, 1972) Australia	SILTSTONE & CLAY SHALE, moderately fractured with highly fractured zones.	1200 (170)	Test method not stated D = 711 mm (28 in) L = 3.35 m (11 ft); L/D = 4.7
A5 (Moss, 1971) New Zealand	SANDSTONE & SILTSTONE, alternating zones C_o = 4.9 MPa (1000 psi), mean value	520* (75)	Anchor pull-out test D = 101 mm (4 in) L = 25.8 m (85 ft); L/D = 90 Anchor holes reamed and flushed clean, pressure grouted

Number Reference Location	*Rock Type*	*Shaft Resistance in kPa (psi)*	*Remarks*
PC1 (Moore, 1964) U.S.A.	SANDSTONE, with shale, highly fractured, unevenly weathered, extremely variable strength.	1015 (147)	Pier compression test D_s = 508 mm (20 in) L_s = 2.6 m (8.5 ft); L/D = 5 excavated with auger, cleared by flushing, tremie concrete with 6 in slump, f'_c = 34.5 MPa (5000 psi). Mechanical device to eliminate end bearing 485 kPa (70 psi) used for design
PC13A (Vogan, 1977) Canada	SANDSTONE, weakly cemented, with some shale layers C_o = 12-24 MPa (1700-3500 psi)	900 (130)	Plug compression tests D = 245 mm (9.63 in) L = 0.64 m (2.1 ft); L/D = 2.6 Rotary drilled, tricone bit hand cleaned, tremie concrete f'_c = 37,000 kPa (5400 psi)
PC13b	(same)	830 (120)	D = 245 mm (9.63 ft) L = 0.76 m (2.5 ft); L/D = 3.1
PC13c	(same)	1250 (181)	D = 245 mm (9.63 in) L = 0.95 m (3.1 ft); L/D = 3.9
PC27 (2 tests) (Pells et al, 1978) MacKenzie, 1969) Australia	SANDSTONE, intact	6500 (940)	Pier Compression test D = 267 mm (10.5 in) L = 0.58 & 0.07 m (1.9 & 3.2 ft) L/D = 2.2 & 3.6 f'_c = 30 MPa (4,300 psi)
A13 (Drossel, 1970) U.S.A.	SANDSTONE,	1730 (250)	Anchor Pull-out test D = 150 mm (6 in) L = 0.92 m (3 ft); L/D = 6 f'_c = 42 MPa (6000 psi) Bond = 300 KPa (45 psi) Used for Design.
PC18 (Buttling, 1976) U.K.	CHALK, weak, grades III to IV average C_0 = 1036 kPa (150 psi)	190 (28)	Pier compression test D_s = 1050 mm (42 in) L_s = 8 m (27 ft); L/D = 8 Constructed using bentonite slurry and tremie concrete
PC19 (Buttling, 1976) U.K.	CHALK	205* (30)	Pier compression test D_s = 1050 mm (42 in) L_s = 10 m (33 ft); L/D = 9 Constructed using bentonite slurry.
PC20 (Buttling 1976) U.K.	CHALK, hard to very hard	120* (17)	Pier compression test D_s = 900 mm (35.4 in) L_s = 10 m (33 ft); L/D = 11
PC21 (Lord, 1976) U.K.	CHALK	146 (21)	Pier compression test D = 406 mm (16 in) L = 6.1 m (20 ft); L/D = 15
PT5a (Palmer, 1976)	CHALK, Medium dense, fragmented, stiff clayey chalk matrix, numerous flints	242 (35)	Pier pull-out tests D = 762 mm (30 in) L_s = 1.5 m (5 ft); L/D = 2

Number Reference Location	*Rock Type*	*Shaft Resistance in kPa (psi)*	*Remarks*
PT5b	(same)	180 (26)	(same)
A1 (Littlejohn, 1970) U.K.	CHALK, stiff to hard, Grades I, II, III C_o = 995-4970 kPa (144-720 psi) N = 20-80	215-1070 (31-155)	Anchor pull-out test D = 76-102 mm (3-4 in) L = 3-9 m (10-30 ft); L/D > 30 recommends SF = 2
A12 (Associated Tunnelling 1973) U.K.	CHALK	450-1350 (65-180)	Anchor pull-out test D = 120-150 mm (4.8-5.5 in) L = 1.5-3 m (5-10 ft)
PC2 (Jackson et al, 1973) U.S.A.	LIMESTONE, cherty, horizontal bedded, weathered layers, clay seams. RQD = 36-59	1630* (236)	Pier compression test D_s = 760 mm (30 in) L_s = 1.5 m (5 ft); L/D = 2 percussion drill, tremie concrete f'_c = 41,400 kPa (6000 psi) E_c = 30,400 MPa (4400 ksi)
PC7 (D'Appolonia, 1967) U.S.A.	LIMESTONE, Stratified	2070* (300)	Pier compression test D = 610 mm (24 in) L_s = 3 m (10 ft); L/D = 5 14WF11 core 700 kPa (100 psi) recommended for design.
PC23 (Spencer, White & Prentice, undated) U.S.A.	LIMESTONE, (Niagara),	1740* (250)	Pier compression test D = 460 to 760 mm (18 to 30 in) L = 1.5 to 4.6 m (5 to 15 ft) L/D = 3.3 to 6
A8 (Losinger & Co. Ltd., 1966) Switzerland	LIMESTONE (Tertiary),	2830 (410)	Anchor pull-out test f'_c = 48 MPa (7000 psi)
A9 (Ruttner, 1966) Switzerland	LIMESTONE,	4550 (655) to 4800 (690)	Anchor pull-out test D = 101 mm (4 in) L = 1.5 to 5 m (5 to 16 ft) L/D = 15 to 50 f'_c = 15.5 to 23.5 MPa (2200 to 3400 psi)
A10 (Saliman & Schaefer, 1968) U.S.A.	LIMESTONE and SHALE (Soft Rock)	270 (40)	Anchor pull-out test D = 62-102 mm (2.5-4 in) L = 1.8 m (6 ft); L/D = 18+
A11 (Brown, 1970) Canada	LIMESTONE, dolomitic, layered	1800 (260)	Anchor pull-out test D = 38 mm (1.5 in) L = 0.2 m (0.67 ft) L/D = 5.3 f'_c = 16 MPa (2300 psi)

Number Reference Location	*Rock Type*	*Shaft Resistance in kPa (psi)*	*Remarks*
PC22 (Jordon & Dobie, 1976) U.K.	KEUPER MARL, high to moderately weathered, closely to widely fractured E_r = 500 MPa (72 ksi)	360 (52)	Pier compression test D = 600 mm (23.6 in) L = 3.3 m (11 ft); L/D = 6
PC26 (Davis, 1974) U.K.	KEUPER MARL, zone II C_o = 1.3 MPa (190 psi) Er 1320 MPa (190 ksi)	230 (33)	Pier compression test D = 760 mm (30 in) L = 4.0 m (13 ft); L/D = 5.3
PC15 (Rosenberg & Journeaux, 1976) Canada	Andesite, hard, highly fractured C_o = 10,400 kPa (1500 psi) BxL core recovery 33-75%	1100 (160)	Plug compression test D = 457 mm (18 in) L = 0.56 m (1.83 ft); L/D = 1.2 Styrofoam pad on base to eliminate end bearing
PC17 (Webb, 1976) South Africa	DIABASE, fissured, highly weathered C_o = 332-560 kPa (50-80 psi) E_r = 87 MPa (13 ksi) S_o = 125 MPa (80 psi), ϕ = 18° W_L = 51-80%, W_p = 29-43% W_n = 23-32%, Clay fraction = 12-28%	122 (18)	Pier compression test D = 615 mm (24.2 in) L = 12.2 m (40 ft); L/D = 20
A4 (Schmidt, 1956) U.S.A.	Unclassified	1990 (286)	Anchor pull-out test D = 178 mm (7 in) L = 3 in (10 ft); L/D = 17 expandable grout used Working load "bond" = 350 kPa (50 psi)
A6 (Parker, 1958) U.K.	BASALT, medium hard	6370 (920)	Anchor pull-out test D = 64 mm (2.5 in) L = 0.61 m (2 ft); L/D = 9.6 Pure cement grout bond design value = 1.9 MPa (280 psi)
PC11a (Vogan, 1977) Canada	SLATE, shaley, with closely spaced near vertical features uncemented, red	1885* (273)	Plug compression tests D = 229 mm (9 in) L = 0.64 m (2.1 ft); L/D = 2.8 Rotary drilled with tricone bit. Walls clean, surface fractured Tremie concrete, f'_c = 34,500 kPa (5000 psi)
PC11b	(same)	1167* (169)	D = 229 mm (9 in) L = 1.05 m (3.4 ft); L/D = 4.6
PC11c	(same)	1070* (155)	D = 229 mm (9 in) L = 1.13 m (3.7 ft); L/D = 4.9
PC12a	(same), green	470 (68)	D = 229 mm (9 in) L = 0.98 m (3.2 ft); L/D = 4.3
PC12b	(same), green	1070* (155)	D = 229 mm (9 in) L = 1.62 m (5.3 ft); L/D = 7.1

Number Reference Location	*Rock Type*	*Shaft Resistance in kPa (psi)*	*Remarks*
PT2 (ENR, 1937) U.S.A.	MICA SCHIST, soft, faulted, fractured	2660 (385)	Pier pull-out test D_s = 305 mm (12 in) L_s = 0.8 m (2.6 ft); L/D = 26 Excavated with cable tools and reaming bit, smooth walls
A7a (Littlejohn & Truman-Davies, 1974) U.K.	SLATE, (shillet), with thin quartzitic greywacke layers and thin igneous intrusions, strongly folded, core recovery = 80 to 100%	1800 (260)	Anchor pull-out test D = 140 mm (5.5 in) L = 3 m (10 ft); L/D = 21 f'_c = 28 MPa (4000 psi) neat cement grout, w/c = 0.45 design bond value = 600 kPa (85 psi)
A7b	same	2270* (330)	

Key to Tables

A – Anchor Pull-out Test (Tension), small diameter
PT – Pier or Plug Pull-out Test (Tension), large diameter
PC – Pier of Plug Compression Test, large diameter
M – Model or small diameter tests

* indicates test not to failure

REFERENCES FOR APPENDIX III

1. ABRAHAM, K. and PROZIG, R., 1973. "Die Felsanker des Pumps peicherworkes Waldeck II", Baumaschine and Bautechnik, 20(6), pp. 209-220.

2. ASSOCIATED TUNNELLING CO. LTD., 1973. "Report on Ground Anchor Tests - Thames Flood Prevention". Unpublished (14 pp) Lowton St. Mary's near Warrington, England.

3. AURORA, R.P. and REESE, L.C., 1976. "Behavior of axially loaded drilled shafts in clay-shales". Research report 176-4, Center for Highway Research, The University of Texas at Austin.

4. BROWN, D.G., 1970. "Uplift Capacity of grouted rock anchors". Ontario Hydro Research Quarterly 22(4), pp. 18-24.

5. BUTTLING, S., 1976. "The influence of shaft and end loads in piles in chalk using strain gauge instrumentation". Geotechnique. Vol.26, No.1.

6. D'APPOLONIA, E., 1967. Unpublished report, E. D'Applonia Consulting Engineers, Inc., Pittsburg, Pennsylvania.

7. DAVIS, A.G., 1974. "Contribution to discussion IV, Rocks". Proc. Cambridge conf. settlement of structures, 757-59. London: British Geotechnical Society.

8. DROSSEL, M.R., 1970. "Corrosion proofed rock ties save building from earth pressure". Construction Methods, May, pp. 78-81.

9. DVORAK, A., 1966. "Tests of anisotropic shales for foundation of large bridges". Proc. 1st Int. Cong. Rock Mech., LISBON Vol.2, pp. 537-541.

10. E.N.R., 1937. "Drilled piles seated in rock sockets". Engineering News Record, September 9, 1937.

11. FREEMAN, C.F., KLAJNERMAN, D. and PRASAD, G.D., 1972. "Design of deep socketed caissons into shale bedrock". Canadian Geotechnical Journal. Vol.9, No.1, pp. 105-114.

12. GIBSON, G.L., 1973. "Field and laboratory investigation for design of rock socketed caissons for a multi-storey building founded in sound limestone". Master of Engineering thesis, Carleton University, Ottawa.

13. JACKSON, W.T., PEREZ, J.Y. and LACROIX, Y., 1973. "Foundation construction and performance for a 34-storey building in St. Louis". Geotechnique. Vol.24, No.1, pp. 63-90.

14. JORDEN, E.E. and DOBIE, M., 1976. "Discussion: Session 2 - Rocks other than chalk". Symposium, piles in weak rock, The Institution of Civil Engineers, London. pp. 215-217.

15. LADANYI, B., 1977. "Friction and end bearing tests on bedrock for high capacity socket design: discussion". Canadian Geotechnical Journal. Vol.14, pp. 153-155.

16. LITTLEJOHN, G.S., 1970. "Soil Anchors". Proc. Conf. on ground engineering, Institution of Civil Engineering, London. pp. 33-44.

17. LITTLEJOHN, G.S. and TRUMAN-DAVIES, C., 1974. "Ground anchors at Devonport Nuclear Complex". Ground Engineering, 7(6), pp. 19-24.

18. LORD, J.A., 1976. "Discussion: Session 1 - Chalk". Symposium, piles in weak rock. The Institution of Civil Engineers, London. pp. 180-185.

19. LOSINGER AND CO., 1966. "Prestressed VSL rock and alluvium anchors". Technical Brochure, Berne, Switzerland, 15 pp.

20. MASON, R.C., 1960. "Transmission of high loads to primary foundations by large diameter shafts". Paper to ASCE convention, New Orleans, March 1960.

21. MATICH, M.A.J. and KOZICKI, P., 1967. "Some load tests on drilled cast-in-place concrete caissons". Canadian Geotechnical Journal. Vol.4, No.4, November

22. MOORE, W.W., 1964. "Foundation design". ASCE, Civil Engineering, January 1964, pp. 33-35.

23. MORTON, J.D., 1969. "An assessment of deep pile foundations bearing on shale bedrock with special reference to the Hamilton-Toronto region of Southern Ontario". M.Eng. thesis, University of Toronto.

24. MOSS, J.D., 1971. "A high capacity load test for deep bored piles". Proc. 1st Australia-New Zealand conf. geomechanics, Melbourne. Vol.1, pp. 261-267.

25. M.T.C., 1964. "Anchor pulling tests, Highway #402, District #4, Hamilton". Ministry of Transportation and Communications - Ontario, Foundation investigation and design report. W.J. 64-F-7 -- W.P. 231-58-1.

26. M.T.C., 1968. "Pile load tests on concrete caissons at Highway #401 and Airport Road. District No.6 Toronto". Ministry of Transportation and Communications - Ontario, Foundation investigation and design report. W.J. 68-F-58 -- W.P. 396-65.

27. OSTERBERG, J.O. and GILL, S.A., 1973. "Load transfer mechanism for piers socketed in hard soils or rock". Proc., 9th Canadian rock mechanics symposium, Montreal.

28. PALMER, D.J., 1976. "Discussion: Session 1 - Chalk". Symposium, piles in weak rock, The Institution of Civil Engineers, London, pp. 189-192.

29. PARKER, P.I., 1958. "The raising of Dams with particular reference to the use of stressed cables". Proceedings, 6th congress on large dams, New York.

30. PARKIN, A.K. and DONALD, I.B., 1975. "Investigations for rock socketed piles in Melborne Mudstone". Second Australia-New Zealand conference of eomechanics, Brisbane, pp. 195-200.

31. PELLS, P.J.N., DOUGLAS, D.J., RODWAY, B., THORNE, C., and McMAHON, B.K., 1978. "Design loadings for foundations on shale and sandstone in the Sydney region". Research report No. R315, The University of Sydney, School of Civil Engineering, Sydney, March

32. ROSENBERG, P. and JOURNEZUX, N.L., 1976. "Friction and end bearing tests on bedrock for high capacity socket design". Canadian Geotechnical Journal. Vol.13, pp. 324-333.

33. RUTTNER, A., 1966. "Amuendung von Vorgespannten Felsenkern (System BBRV) bei der Erhahung der Spullersee - Telsperren". Schweizerische Bauzeitung 84(4), pp. 83-8.

34. SALIMAN, R. and SCHAEFER, R., 1968. "Anchored footings for transmission towers". ASCE annual meeting and regional meeting on structural engineering, Pittsburgh, Penn., September 30-October 4, Preprint 753, 28 pp.

35. SCHMIDT, A.E., 1956. "Rock anchors hold TV tower on Mt. Wilson", Civil Engineering - ASCE Vol.26, January.

36. SEYCHUCK, J.L., 1970. "Load tests on bedrock". Canadian Geotechnical Journal. Vol.7, No.4, pp. 454-469.

37. SPANOVICH, M., undated. "Caisson-shale interaction: some load test results". Unpublished report - Engineering Mechanics, Inc., Consulting Geotechnical Engineers, Pittsburg.

38. SPENCER, C.B., 1941. "Drilled-in caissons with heavy H-cores carry 16-storey navy warehouse". Engineering News Record. September 25, 1941. pp. 3-6.

39. SPENCER, WHITE & PRENTICE, INC., undated. "A system of foundation columns anchored in rock sockets".

40. THORBURN, S., 1966. "Large diameter piles founded on bedrock". Symposium, large bored piles, London.

41. TROW, W., 1972. Unpublished report, William Trow Associates, Limited, Toronto.

42. VOGAN, R.W., 1977. "Friction and end bearing tests on bedrock for high capacity socket design: discussion". Canadian Geotechnical Journal. Vol.14, pp. 156-158.

43. WEBB, D.L., 1976. "The behaviour of bored piles in weathered diabase". Geotechnique. Vol.26, No.1.

44. WILSON, L.C., 1976. "Tests of bored and driven piles in cretaceous mudstone at Port Elizabeth, South Africa". Geotechnique, Vol.26, No.1.

CASE HISTORY
PILE FOUNDATION IN FLOOD PLAIN SOILS

BY
K.A. KESSLER[1]/M. ASCE

INTRODUCTION

Unit No. 4 of the Clay Boswell Steam Electric Station is a 500 MW expansion to an existing coal-fired generating plant. The plant is owned by the Minnesota Power and Light Company and is located on the east bank of Blackwater Lake (Mississippi River) about three miles from Cohasset, Minnesota.

The plant site is on the flood plain of the Mississippi River with soil conditions dictating pile foundations. The first units constructed on the site (Units 1 and 2) are supported on timber piles. Unit 3 is supported on Union Metal Monotube piles with a design capacity of 60 tons (535 kN) and vary in length from 55 to 95 feet (16.8 to 29m).

Unit No. 4, located immediately north of Unit 3, is also supported on monotube piles. A total of 5,457 piles support the main power plant structures including the turbine-generator, boiler, air quality control system, a 600 foot (183m) chimney, the coal handling system, the cooling tower pump structure, and various tanks and support structures. A general location plan for the plant site is presented as Figure 1.

[1] Senior Soils Engineer, Ebasco Services Incorporated, Norcross, Georgia

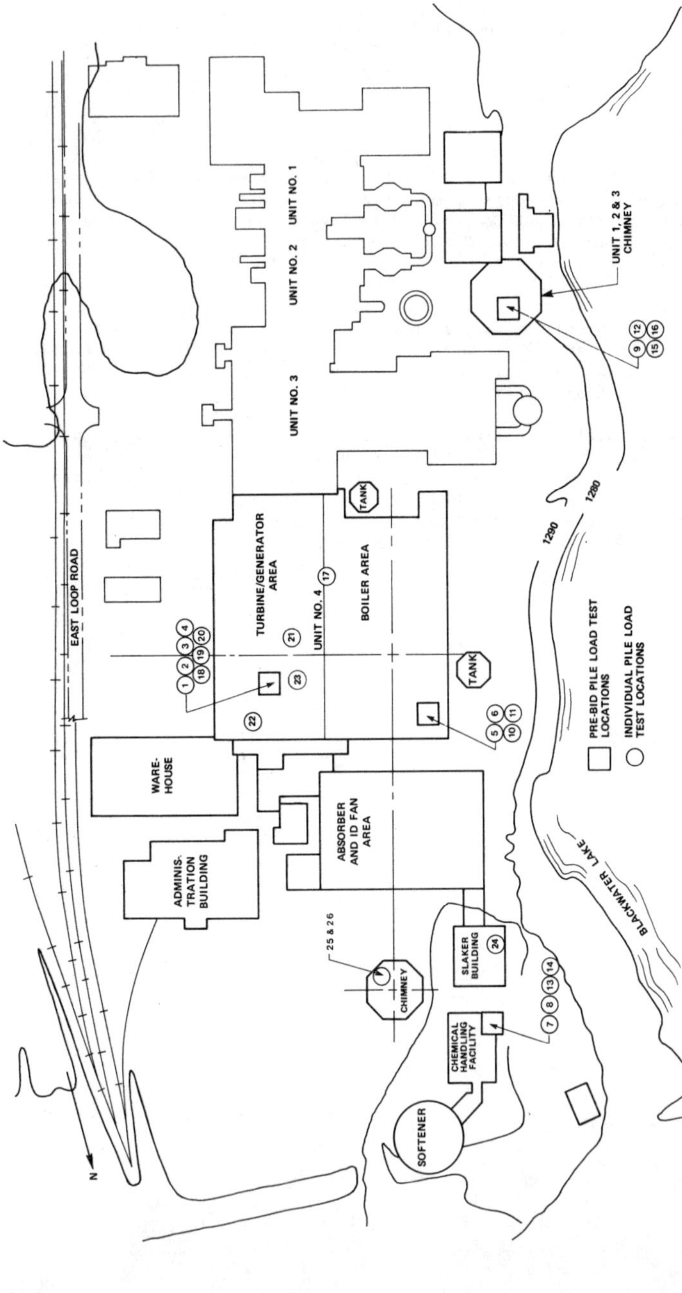

Figure 1
LOCATION PLAN

FOUNDATION INVESTIGATION

A foundation investigation was performed by an independent foundation consultant (Ref. 5) with a review by Ebasco. The investigation included a determination of the subsurface soil conditions and engineering properties, as well as recommendations for pile types, lengths, and driving criteria.

Subsurface Conditions - A generalized soil profile is shown on Figure 2. A fine sand and silty fine sand fill has been placed over the site area ranging in depth from 1 to 7 feet (.3 to 2.1m). The natural soil below the fill consists of sandy silts and silty sands that extend to depths ranging from 1 to 16 feet (.2 to 4.9m) below ground surface. A layer of medium stiff to stiff clay is present below the silts and sands and extends to depths ranging from 17 to 26 feet (5.2 to 7.9m) below the ground surface. The clay layer is underlain by a thick stratum of interbedded silts and silty fine sands extending to depths ranging from 61 to 97 feet (18.6 to 29.5m). The silts and silty sands are soft to medium stiff and loose to medium dense, respectively. Below the interbedded zones are fine to medium sands which become progressively denser with depth. The sands are medium dense to dense and extend to bedrock (granite) 265 feet (80.8m) below ground surface. These lower sands form the bearing stratum.

A shallow water table is present, perched on the clay layer. The permanent water table was recorded at 21 feet (6.4m) below the ground surface and is at approximately the same level as the water level in the Mississippi River.

Study of Pile Types - The original design load for the Unit 4 piles was 100 tons (890 kN). With a safety factor of 2.0, the required ultimate capacity of the piles would be 200 tons (1780 kN). The 100 ton (890 kN) design load represented a 66 percent increase over the 60 ton (535 kN) piles used for Unit 3. The increased pile capacity resulted in increased depth of penetration into the bearing sand.

The following types of piles were considered:

1. Steel H-piles (12 BP 53).
2. Concrete filled steel pipe - 10.75 inch (27.3 cm) O.D. with 0.25 inch (0.6 cm) wall thickness.
3. Concrete filled Raymond Step-Taper Pile - 00# Boot - 9-1/2 inch (24.1 cm) tip section with bottom two sections 12 feet (3.7 m) in length.
4. Concrete filled Union Monotube Pile - 8 inch (20.3 cm) tip section; 12 inch (30.5 cm) butt; and 30 foot (9.2m) tapered section.
5. Concrete filled Armco Hel-Cor Pile - 14 inch (35.6m) O.D.

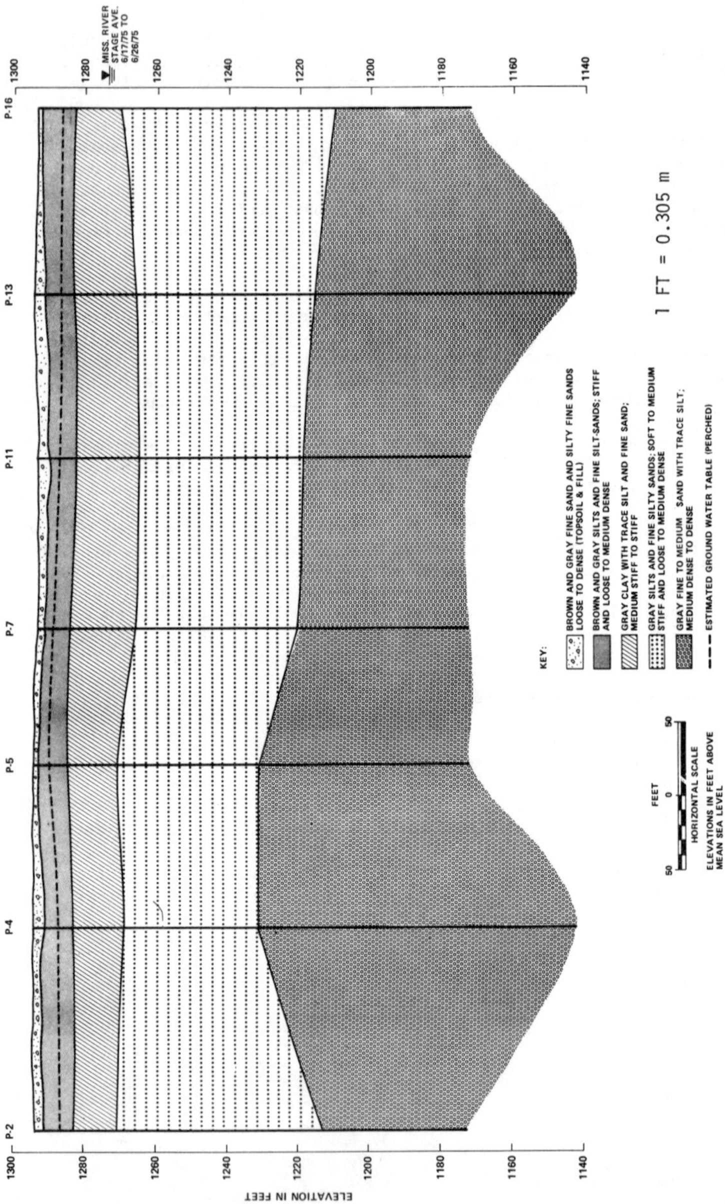

Figure 2
GENERALIZED SOIL PROFILE

Based on the results of the static analyses, it was determined that the Step-Taper pile, Monotube pile and Hel-Cor pile were the more economically competitive pile types for the site soil conditions.

Dynamic (Wave Equation) Analyses were performed on the Step-Taper Pile and the Monotube Pile. It was assumed that the Hel-Cor pile would perform similarly to the Step-Taper Pile. From these analyses it was determined that the three pile types would be suitable for use on the project. Although the Monotube Pile was indicated marginal, inclusion of this pile type was considered justified because pile load tests performed during construction of Unit No. 3 showed that the average ultimate capacity of the pile was greater than the ultimate dynamic resistance computed by the wave equation analysis. This could be due to either (1) inaccurate assumptions used in the analysis, (2) soil freeze, (3) other factors which are not fully understood, or (4) a combination of the above. A comparison of the wave equation results vs. load test results for the monotube pile is presented in Figure 3.

Preliminary Pile Driving Criteria and Procedures - Pile driving criteria and procedures were developed based on the soil conditions and the results of the static and dynamic analyses. The piles were to be preaugered to a depth of 50 feet (15.3m) to minimize heave (due to displacement of soil by the piles), decrease the driving resistance in the upper strata, and reduce the possibility of negative skin friction. The driving criteria were based on a combination of minimum driving resistance for the last two inches (5.1 cm) of driving and a minimum of 15 feet (4.6m) of penetration into the bearing stratum.

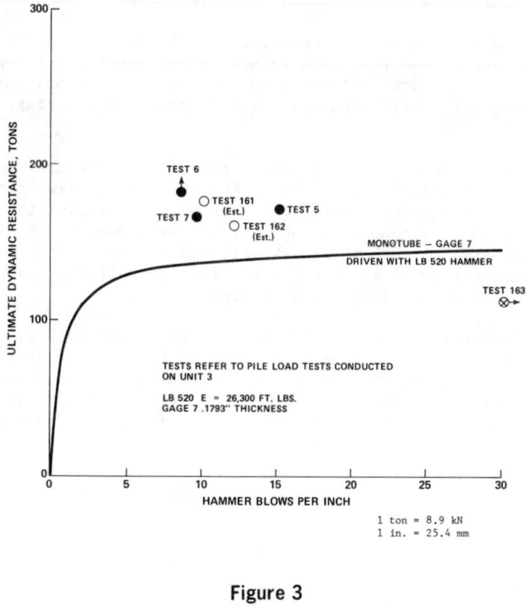

Figure 3
ULTIMATE DYNAMIC RESISTANCE
MONOTUBE PILE (Unit 3)

PILE LOAD TESTS

Prebid Test Program - A pile load test program was conducted prior to the release of the pile bid proposal. The purpose of the test program was to:

1. Compare the performance of various pile types.
2. Obtain information related to the required pile lengths and depth of penetration into the "bearing stratum".
3. Develop production pile driving criteria.

The test program information was to be utilized to design the most economical pile system. Additional savings was realized by providing the pile load test information to the contract bidders (allowing the contractors to reduce the contingency portion of their bid).

A total of 14 piles, 7 step-taper piles and 7 monotube piles, were driven and load tested at 4 locations in the plant area. The driving criteria varied for both pile types resulting in tip elevation ranging from 10 feet (3.1m) to 35 feet (10.7m) into the bearing stratum.

Pile driving was erratic and unpredictable. Piles stopped at the various depths depending upon blow count, embedment in the deep sand layer, and retap criteria. Only seven of the piles successfully carried the 200 ton (1780 kN) ultimate load with the two pile types exhibiting similar load-carrying capabilities. The failing load tests were attributed to the different soil conditions at the various locations. The data for all the Unit 4 pile load tests is summarized in Table No. 1.

Preconstruction Test Program - The monotube pile was chosen for Unit 4 based on the prebid load test program and the performance of the Unit 3 foundation. Pile driving criteria were revised and a second load test program initiated to verify the new criteria prior to the start of production pile driving.

Four pile load tests were planned prior to production pile driving. Two of the tests (No. 15 and 16) were made in the area of the new Units 1, 2 and 3 Chimney, an area which experienced failing tests during the prebid testing program. Another test (No. 18) was performed in the Unit 4 turbine pedestal area which had also experienced failing tests. The fourth test (No. 17) was performed between the turbine and boiler areas at a location which had not previously been tested.

The revised criteria required augering to a depth of 35 feet (10.7m). The piles were driven to a minimum blow count of 164 blows/foot (5.4 blows/cm) for tip grades at or above elevation 1185 (361.4m) and 82 blows/foot, below elevation 1185 (361.4m). Piles could be driven with the Vulcan 80C or Conmaco 115 having a driving energy of 24,500 ft.-lbs (33 kJ).

TABLE 1

TEST PILE DATA

Load Test No.	Date Driven	Pile No.	Length (ft.)	Blows/ft for last foot	Ultimate Capacity (tons)	Auger Length (ft.)	Hammer
1	2/23/76	R-1-A	102	12/68 redrive	135	32	Vulcan-80C
2	2/23/76	R-2-A	88	20	84	32	Vulcan-80C
3	2/24/76	M-1-A	89	4	104	44	Vulcan-80C
4	2/24/76	M-2-A	102	13/25/9"	122	44	Vulcan-80C
5	2/24/76	M-1-C	107	64	255	44	Vulcan-80C
6	2/24/76	M-2-C	117	180	285	44	Vulcan-80C
7	2/24/76	M-1-D	81	200/6"	250	44	Vulcan-80C
8	2/24/76	M-2-D	83	302	235	44	Vulcan-80C
9	2/25/76	M-1-B	102	20	135	44	Vulcan-80C
10	2/25/76	R-1-C	107	13	240	32	Vulcan-80C
11	2/25/76	R-2-C	117	35	280	32	Vulcan-80C
12	2/26/76	R-1-B	102	15	90	32	Vulcan-80C
13	2/26/76	R-1-D	107	23	240	32	Vulcan-80C
14	2/26/76	R-2-D	82	16	130	32	Vulcan-80C
15	6/ 3/76	H-1	129	102	170	35	Vulcan-80C
16	6/ 4/76	D-1	123	110	150	35	Vulcan-80C
17	6/ 9/76	988	88	210	300	38	Vulcan-80C
18	6/10/76	38	118	88	110	38	Vulcan-80C
19	6/24/76	36	170	90	150	17	Vulcan-80C
20	6/23/76	85A	173	186	170	35	Vulcan-80C
21	6/25/76	194	89	175/4"	280	35	Vulcan-80C
22	8/ 2/76	397	189	19/2"	300	35	Conmaco 115
23	9/ 2/76	178	112	200	160	0	Vulcan-80C
24	11/14/77	4287	93	32/3"	160	0	Vulcan-80C
25	11/22/77	4571	104	61	140	0	Vulcan-80C
26	11/22/77	4571	104	61	200	0	Vulcan-80C

1 ft = 0.305 m   1 ton = 8.9 kN

Piles driven in the turbine pedestal area and the Unit 1, 2 and 3 Chimney Area again failed to obtain the required ultimate capacity of 200 tons (1780 kN). The pile driven between the turbine and the boiler areas was loaded to the jack capacity of 300 tons (2670 kN).

Two more test piles (tests No. 19 and 20) were then driven and tested in the turbine area. These piles were driven to varying criteria and resulted in pile lengths of 170 feet (51.9m) as compared with lengths of 88 to 118 feet (26.8 to 36m) for the previous test piles driven in the area. Load testing of the longer piles failed to meet the 200 ton (1780 kN) ultimate load. A third test pile (No. 21) was driven in the turbine area approximately 60 feet (18.3m) from Test No. 18 and 19. This pile was driven to refusal at 89 feet (27.1m) and was load tested to an ultimate load of 280 tons (2490 kN).

From the test programs, it was evident that for at least a portion of the turbine area, it would not be economically reasonable to obtain 100 ton (890 kN) design piles. In this area, the design pile was reduced to 75 tons (670 kN) and additional piles added.

Based on the preconstruction test program, criteria for the production pile driving was defined as follows:

1. Preauger to elevation 1250 (381m).
2. Drive to 164 blows/ft. (5.4 blows/cm)) for tip grade above elevation 1125 ft. (343m).
3. Drive to 82 blows/ft. (2.7 blows/cm) for tip grade elevation 1125 ft. (343m).
4. Use continuous driving to establish blow count resistance.
5. Reduce pile design capacity to 75 tons (670 kN) in areas where piles meet the 82 blows/ft. criteria at, or below, elevation 1125 (343m).
6. Use driving energy of 24,500 ft.-lbs. (33 kJ)

PRODUCTION PILE DRIVING

Production pile driving began for the Warehouse foundation, northwest of the turbine area. At the contractor's request the first two piles were driven with and without preaugering. Little difference was noted in the pile length and the driving characteristics and augering remained as part of the driving criteria. Pile length in the Warehouse area was approximately 92 feet (28.1m). Driving was relatively easy until the last 3 to 4 feet (.9 to 1.2m), when blow counts significantly increased. Refusal was usually obtained in the same foot that the criterion of 164 blows/foot was reached.

The Warehouse area had been excavated to elevation 1288 (392.8m). As the driving approached the turbine area excavation, elevation 1277 (389.5m) the piles began to increase in length to approximately 160 feet (48.8m).

After driving several piles in the turbine area, a pile was driven next to test pile No. 21. The test pile had been driven and tested prior to the 17 foot (5.2m) excavation. It had been preaugered 35 feet (10.7m), driven to refusal at 89 feet (27.1m) and tested to an ultimate load of 280 tons (2490 kN). The new pile was driven 163 feet (49.7m) before the 75 ton criterion (670 kN) was met.

As driving continued in the turbine area, pile lengths continued to increase. The longest pile driven was 215 feet (65.6m). The rate of pile driving was less than three piles per rig per shift. At the same time, settlement of the ground surface and cracks in the soil around the excavation were observed.

A monitoring program was established to record ground and pile movement. Temporary benchmarks on individual piles and pile groups indicated up to 3 inches (7.6 cm) of dragdown on previously driven piles. The monitoring program was used to establish a safe distance between pile driving and placement of the concrete pile caps.

The potential impact on the construction schedule, cost and the integrity of the foundation were of major concern. The driving records, load test results, boring logs, static and dynamic analyses and all other related information were again reviewed and evaluated.

DISCUSSION AND ANALYSIS OF PILE DRIVING DATA

The design and analysis of the pile capacity were based on four main considerations:

1. A study of pile load-carrying capacities by static analyses.
2. A study of pile types and driving equipment by means of dynamic pile-driving analyses.
3. Test pile driving and load test results.
4. In situ soil conditions

Static Analyses - The theoretical pile bearing capacity was calculated by static formula based on the soil type and laboratory test results.

The results of the static analyses are presented in Figure 4. Also plotted on this figure are the actual load test results. The various lines were drawn using different soil parameters. The lines marked 30°, 35° and 40° include end bearing and frictional resistance along its entire length. The dotted line marked 30° neglects overburden effects due to the upper 70 feet (21.4m) based on the discussion presented under the subtitle "Soil Conditions". The shallow piles that successfully carried design loads are in areas where medium dense to dense sand occur. The actual test results generally plot within the 35° to 40° zone with full support. The deeper piles with poor results generally plot along the 30° line neglecting the effect of overburden pressure in the upper 70 feet (21.4m). An explanation for this is the "void space concept" presented in the "Soil Condition" section.

Dynamic Analyses - The second consideration is dynamic pile-driving formulas. Many dynamic pile-driving formulas have been proposed and used. The basic assumption is that the ultimate load-carrying capacity is equal to the dynamic driving force. The formulas have been modified to account for energy losses in elastic compression of the pile cap, pile and ground, and for hammer efficiency. The two formulas used for the dynamic analyses were the Engineering News Formula and the Hiley Formula.

The wave equation was previously discussed in the Foundation Investigation Section. Some modifications would have been necessary due to the change in hammer type and cushion material. However, the results obtained using the wave equation were no better than the less complicated equations and the data was not correlated.

The dynamic pile-driving formula indicates a scatter of results, which can be attributed to the varying density of the bearing sand layer.

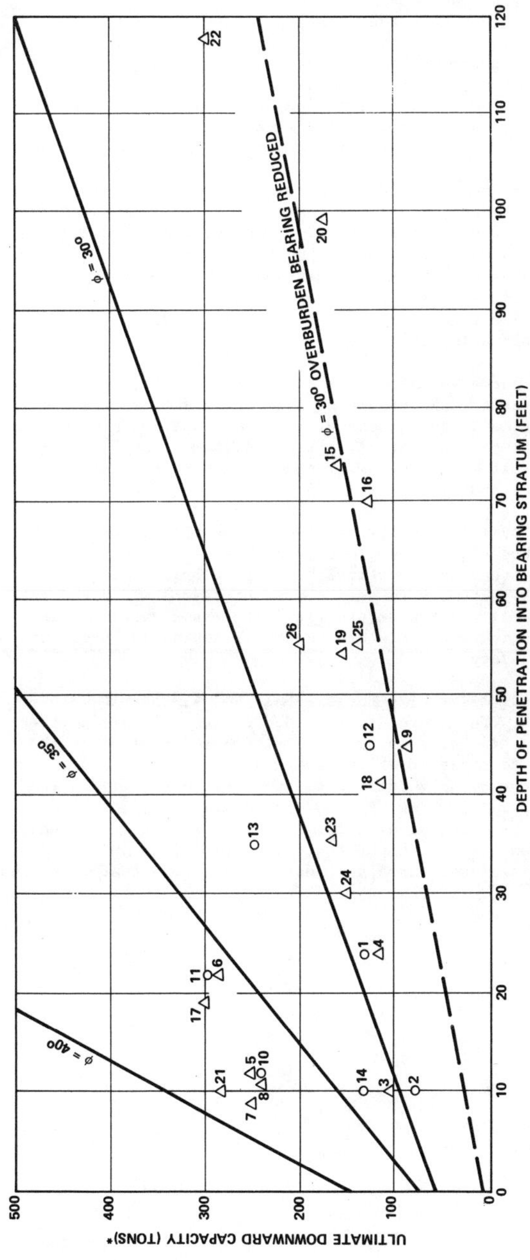

NOTE:

- ○ DENOTES RAYMOND STEP – TAPER PILES
- △ DENOTES UNION METAL MONOTUBE PILES
- * FROM LOAD TEST

1 ft = 0.305 m
1 ton = 8.9 kN

Figure 4
ULTIMATE DOWNWARD CAPACITY VS DEPTH OF EMBEDMENT USING STATIC ANALYSIS

Load Testing - The most important consideration for selecting the load carrying capacity of piles was pile load testing.

Movement at the pile head caused both by elastic deformation of pile and soil and plastic deformation of soil was measured. Plastic deformation is the most significant value obtained from load tests in order to guard against undue settlement of structures. The distribution of total movement and plastic movement is illustrated in load test Figures 5, 6, and 7.

Loading and unloading was used to determine whether the test load was being carried by the stratum selected, or if the center of resistance was at a higher point. The curve of plastic deformation was obtained by removing the load from the pile several times during the process of adding loads by increments, and observing and plotting the rebounds. The points where the rebound lines intersect the settlement axis are projected horizontally to points under the respective loads removed. These points are then connected by a smooth curve (Plastic Deformation Curve).

The plastic deformation curve was studied for determination of ultimate pile capacity. By subtracting the values of this curve from corresponding points on the total settlement curve, the elastic deformation curve was plotted and compared with the theoretical elastic deformation line of the pile. Typical elastic settlement curves are shown in the load tests depicted in Figures 5, 6, and 7.

The ultimate pile capacity is typically defined as the load beyond which the pile will begin to break into the ground, that is, when the tangent to the load-settlement curve becomes "vertical." In pile engineering practice, however, it is customary to select a point beyond which the rate of change increases markedly or the settlement increases excessively in comparison with the load increment, to denote this point as failure, and then apply a factor of safety.

From the pile load test data for Clay Boswell Unit No. 4, the ultimate failure point is obvious. Therefore, in accordance with pile engineering practice, the safe pile load has been selected at the point at which the gross settlement begins to exceed 0.05 inches (.13 cm) per ton (8.9 kN) of additional load, or at the point where the plastic settlement begins to exceed 0.03 inches (0.08 cm) per ton (8.9 kN) of additional load. These values are then divided by a factor of safety of 2 for static load. While the static and dynamic analyses were theoretical or empirical in their selection of pile capacities, the load test data gives the actual safe load for each pile.

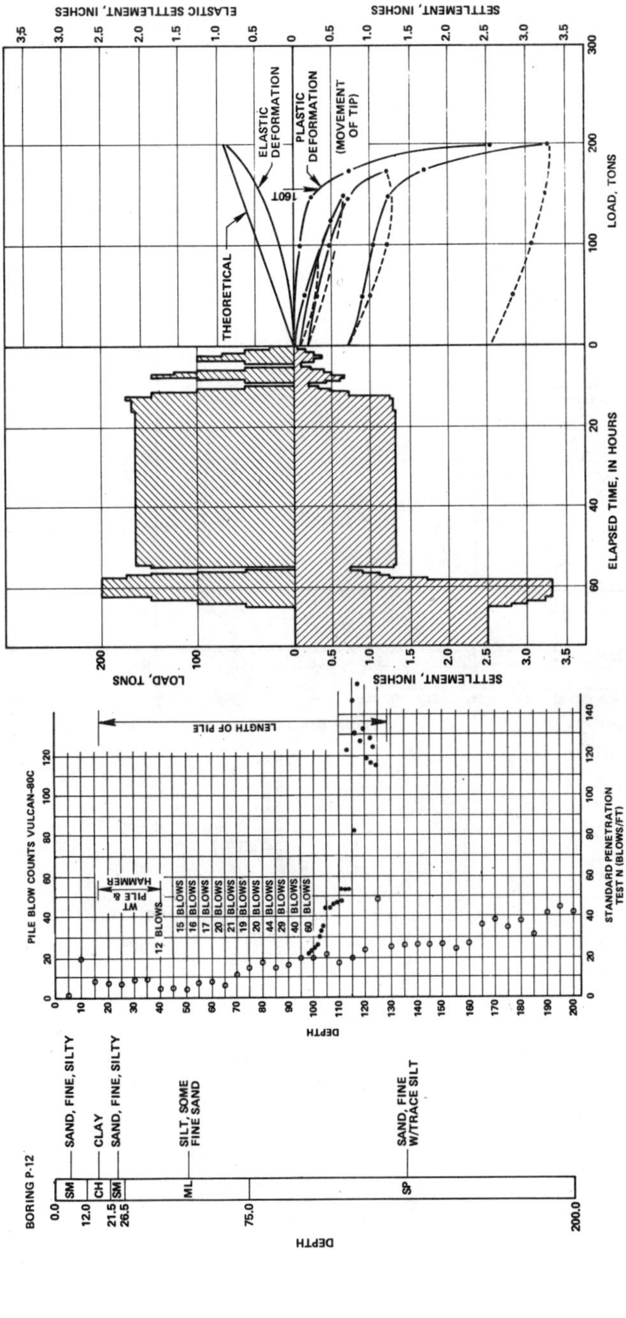

1 in. = 25.4 mm
1 ft = 0.305 m
1 ton = 8.9 kN

Figure 5
PILE LOAD TEST NO. 23
PILE NO. 178 SEPT. 10-13, 1976

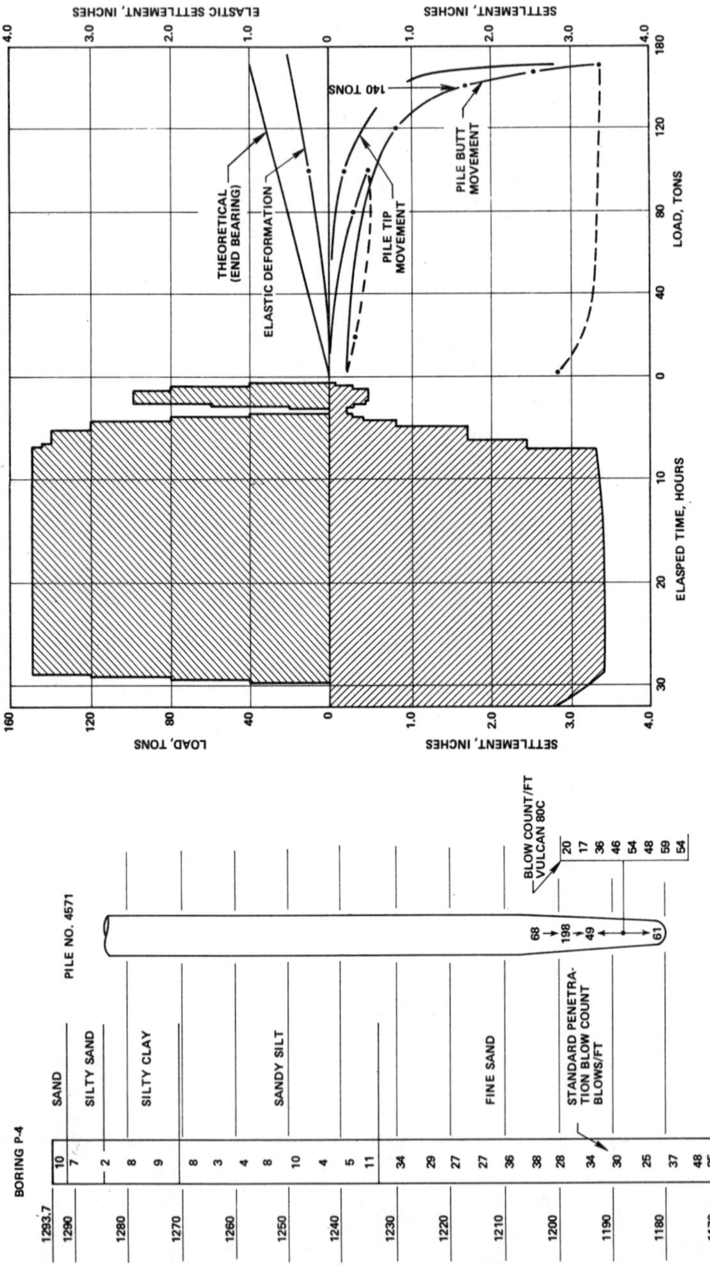

1 in = 25.4 mm
1 ft = 0.305 m
1 ton = 8.9 kN

Figure 6
PILE LOAD TEST NO. 25
PILE NO. 4571 NOVEMBER 19-20, 1977

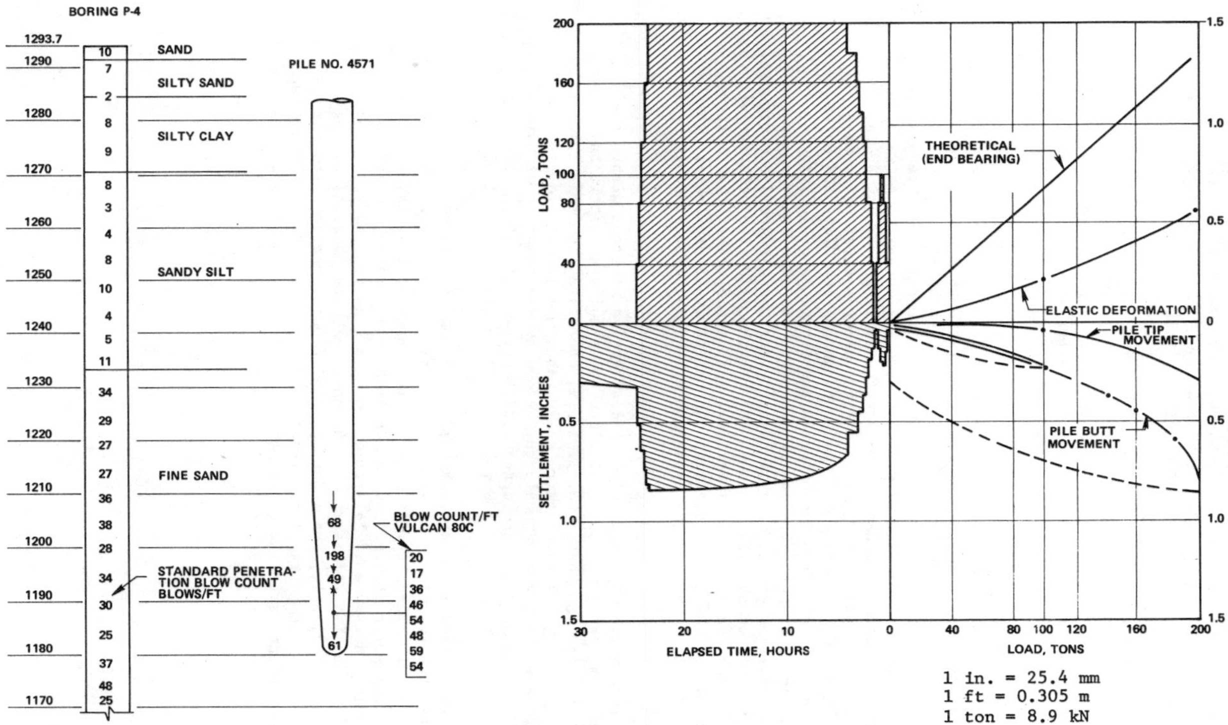

Figure 7
PILE LOAD TEST NO. 26
PILE NO. 4571 DECEMBER 16-17, 1977

Soil Conditions - After driving the test and reaction piles for load tests Nos. 19 and 20 in the turbine area, boring P-12A was made to provide additional data. During the drilling of Boring P-12A, a loss of drilling mud was noticed at a depth of 72 feet (22m). This coincides with the interface of the silt and clay strata and sand stratum. A loss of drilling mud in coarse materials is not unusual. However, in the fine sands and silts, it is unlikely unless a void exists. This same loss of drilling mud was encountered in Boring P-37. Grout from the void grouting repair of Unit 3 intake area foundations, appeared in holes being augered 50 feet (15.3m) away, also indicating that void space exists.

Using the blow-count data from Boring P-12 and the full overburden pressure, a relative density of 50 to 60 percent is obtained using the Gibbs and Holtz relationships. At this relative density, vibration from shocks due to pile driving could cause the sand to settle. This settlement could result in a void space between the silt and sand strata if the silts arched rather than settled with the sand. Figure 8 compares blow-counts from the standard penetration test from Boring P-12 and P-12A. P-12 was drilled prior to driving piles and P-12A after several piles were driven for the turbine pedestal. Again, if we use the Gibbs and Holtz relationship, the data from Boring P-12A and assuming no overburden pressure to occur at the interface of the silts and fine sands due to arching, we obtain the same relative density for these reduced blow-counts. Thus, the reduced overburden pressure reduces the effective stresses within the sands thereby reducing the frictional resistance within that stratum. It is to be noted that the blow-counts for P-12A are approximately one-half of those for P-12. This can be explained by the void space theory discussed above.

The void space would not only reduce the effective stresses within the bearing stratum, but would also allow the upper soils to settle and cause negative drag on the piles. It would, therefore, be impossible for the piles to successfully carry a 100 ton (890 kN) safe design load in areas where the sand would settle due to the dynamics of pile driving. In other areas where dense sand occurred the piles fetch up and they are able to carry a 100 ton (890 kN) safe design load.

A second explanation for the erratic pile driving is related to the ground water conditions at the site. The lower ground water table is a 240 foot (73.2m) aquifer confined by bed rock and the clay layer. Pile driving in the silty stratum and in the fine sand stratum can develop high pore water pressures. This condition can lead to such phenomena as "pile run" and "soil freeze". Both phenomena were observed during pile driving. Basically, the increase in pore water pressure reduces the effective stress in the soil, resulting in "pile run". After driving has stopped and the excessive pore water is allowed to dissipate, the effective stress increases and "soil freeze" is observed.

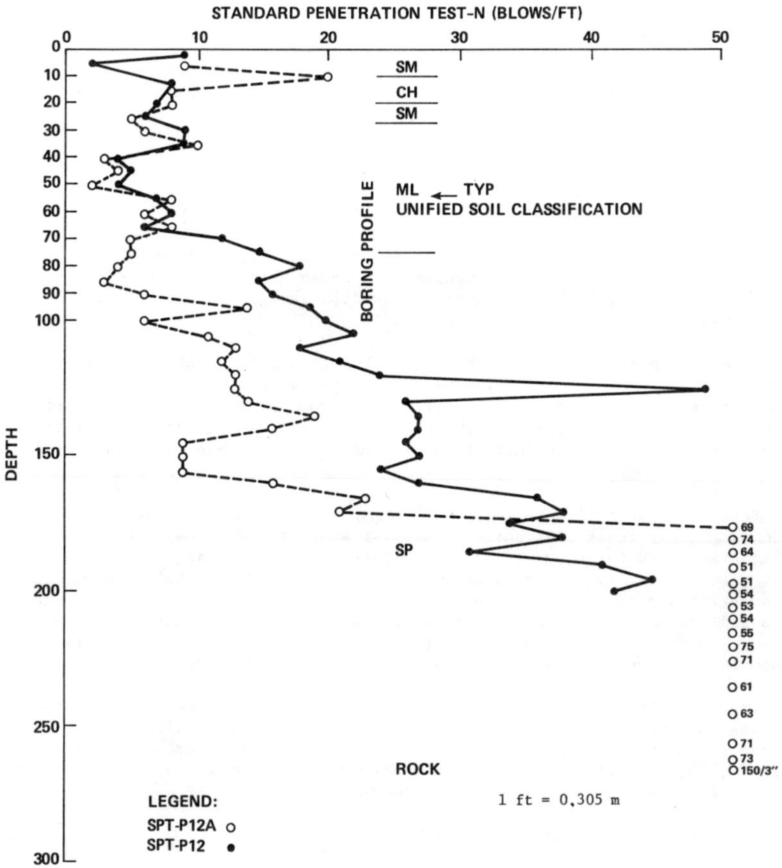

Figure 8
COMPARISON OF BORINGS P12 & P12A

With this explanation, settlement would occur as the result of soil consolidation as the pore water pressure disipated. Grout migration and possibly the loss of drilling mud could be contributed to hydraulic fracturing. Erratic pile driving would reflect varying pore water pressures resulting from local variations in density of the sands and the number of piles driven in the area.

The significance of pore water pressure phenomena is that piles may be driven into the bearing stratum with significantly less energy than predicted by the dynamic analyses. After dissipation of the excessive pore water, redriving the pile or, more importantly, load testing the pile will demonstrate the actual load carrying capability.

FINAL PILE DRIVING CRITERIA AND VERIFICATION OF PILE CAPACITY

Final driving criteria was developed based on the review and analyses. Production pile driving in the turbine area began driving to the new criteria. To verify the pile capacity resulting from the new criteria a pile was selected for pile load test No. 23.

The pile selected was driven without any preaugering since it was thought that if the upper clays and silts were displaced downward rather than removed, they would settle with the sand rather than arch. Without the void space being developed, the full overburden pressure of the overlying silts and clays would allow the friction along the pile to develop and a higher load could be supported.

The Test Pile was loaded to 165 tons (1470 kN) and held for 42 hours, since at higher loads continuous settlement would occur. The pile was designated an 80 ton (710 kN) design pile. This pile was selected as a test pile since: 1) it was located in the turbine pedestal pile group adjacent to deep piles and could be considered in a transition area between deep and shallow piles and 2) it required two retaps before obtaining 164 blow per foot criteria. The majority of piles driven to the new criteria reach 164 blow per foot during driving or on the first retap.

The total actual load for each pile in loose sand areas on the project was reviewed and compared to the design load available. Where piles were driven to the new criteria and require retaps, they could, as with test pile No. 23, support 80 tons (710 kN) with a factor of safety of at least two. Pile load test No. 23 is presented in Figure 5.

The final driving criteria was presented as follows:

1. Limit preaugering to a depth of 5 feet (1.5m).
2. Drive piles to a depth of approximately 95 feet (29m) unless a blow-count of 164 blows/ft. criteria is obtained at a higher elevation. A series of three or more piles should be driven before the first pile in the series is retapped to allow a minimum of 30 minutes before a pile is retapped.
3. If a blow count of 164 blows/ft. is obtained on the retap, driving is completed.
4. If the pile is restarted with less than 164 blows/ft., the pile should be driven in 10 foot (3.1m) increments, allowing a minimum of 30 minutes before retapping. Driving is completed when 164 blows/ft. is obtained, either during driving or on the restart.

Piles driven to the above criteria were designated 80 ton (710 kN) capacity pile for design purposes.

As production pile driving continued in the turbine and boiler area, with only a few exceptions, the driving criterion was obtained during the initial driving or on the first retap.

Phase II pile driving began in the absorber and ID fan area. Piles were driven to essentially the same criteria as above. Driving resistance was lower than had been experienced during the Phase I driving. Two pile load tests were performed to verify the driving criteria. Load Test No. 24 (Figure No. 5) was made in the foundation for the scrubber building. This test verified the 80 ton (710 kN) design pile. In the chimney area, load test No. 25 (Figure No. 6) resulted in a 70 ton (625 kN) design pile. Since the chimney foundation had been designed for 100 ton (890 kN) piles, an additional 92 piles were added to the foundation. After driving in the chimney area had been completed, load test No. 26 (Figure No. 7) was performed on the same pile tested in load test No. 25. The pile was loaded to the 200 ton (1780 kN) capacity of test jack, verifying a design capacity of 100 tons (890 kN). The increase in the carrying capacity of the pile was contributed to densification of the area, soil freeze, or a combination of the two. Densification would result from driving the 377 displacement piles in the chimney foundation.

CONCLUSIONS

The design analysis of the pile capacity was based on static formula, dynamic analysis, pile load tests, and the soil conditions. At the Clay Boswell Unit 4 site, the most important tool used for the development and verification of the driving criteria was the pile load test. The varied and unusual soil conditions of the flood plain soils required a total of 26 pile load tests.

The soil conditions were analyzed together with the reaction to deep penetration into the bearing stratum. Pile driving and capacities were adjusted to accommodate the most practical combination. Modifications to the driving criteria were required during production driving. The modifications were made, based on all available information and verified by pile load tests. The results of this effort was an early completion of the pile driving schedule at a reasonable cost.

The performance of the foundation has been monitored during the construction of the plant. At the time of this writing, approximately 90 percent of the dead loads have been placed on the more heavily loaded foundations. Settlement in these areas has been on the order of 1/2-inch (1.3 cm). This can be compared with the 3/4 to 1 inch (1.9 to 2.5 cm) of total settlement recorded for the existing Clay Boswell Unit 3.

REFERENCES

1. Chellis, R.D., Pile Foundations. Second Edition, McGraw-Hill Book Company, Inc. (1961).

2. Terzaghi, K. & Peck, R.B., Soil Mechanics in Engineering Practice, Second Edition, John Wiley and Sons, Inc., N.Y. (1967).

3. Department of the Navy, Design Manual, Soil Mechanics, Foundations, and Earth Structures, DM-7, (1971).

4. Gibbs, H.I. and Holtz, W.G., Research on Determining the Relative Density of Sands by Spoon Penetration Testing, Proceedings, Fourth International Conference on Soil Mechanics and Foundation Engineering, Vol. 1, Page 35, 1957.

5. Dames & Moore, Report, Foundation Investigation Proposed Unit No. 4, Clay Boswell Steam Electric Plant For Minnesota Power & Light Company (1975).

TIMBER PILES IN FINE-SILTY SAND

Raj P. Khera[1], MASCE, Tonis Raamot[2], FASCE, and Morris Berkowitz[3], MASCE.

INTRODUCTION

This paper deals with the foundation installation of a reinforced concrete framed structure. The building is essentially square in plan and consists of three floors, including the ground floor. The design called for the columns to be supported by creosoted timber piles of 30 tons (267 kN) capacity. The contract specifications called for pile lengths of 25 ft (7.62 m). The tip diameter was specified to be a minimum of 10 in (254 mm) and the butt diameter to be a minimum of 12 in (305 mm). The number of piles per column varied between 3 and 6 with the majority of the columns resting on 5 piles.

The specifications called for a single- or a double-acting hammer with 15000 ft lbs (20340 N m) as its minimum rated energy. The Engineering News Formula was specified for determining the dynamic resistance of the piles. A total of 627 piles were to be driven. Among these, 5 piles had been designated for load tests to be performed in accordance with ASTM D-1143 method in order to verify and/or adjust, if necessary, the pile driving procedure.

In this paper the following items are considered:

1. The specified pile lengths of 25 ft (7.62 m) are compared with those obtained by using the static formula.
2. The driving resistance of piles is determined using different dynamic formulae.
3. The need for pile load tests is demonstrated.
4. The applicability of static and dynamic formulae, in determining pile lengths and driving criteria, respectively, is evaluated.
5. The need for giving due considerations to soil properties and soil response during pile installation is demonstrated.

SITE CONDITIONS

At the time when borings were made, the ground elevation varied between 125 ft (38.1 m) and 140 ft (42.7 m) above mean sea level. The soil data from the boring logs are shown in Fig.1 through Fig.5. The distance between the borings and the corresponding test piles varied between 5 ft (1.52 m) and 25 ft (7.62 m). Test boring numbers B1 and B3 extended to elevations of about 76 ft (23 m). The remaining three borings extended

1 - Professor, Department of Civil & Environmental Engineering, New Jersey Institute of Technology, Newark, New Jersey, presently Research Scholar, Bundesanstalt für Wasserbau, Karlsruhe, West Germany.
2 - President, Raamot Associates, New York, New York.
3 - Consulting Engineer, Trenton, New Jersey.

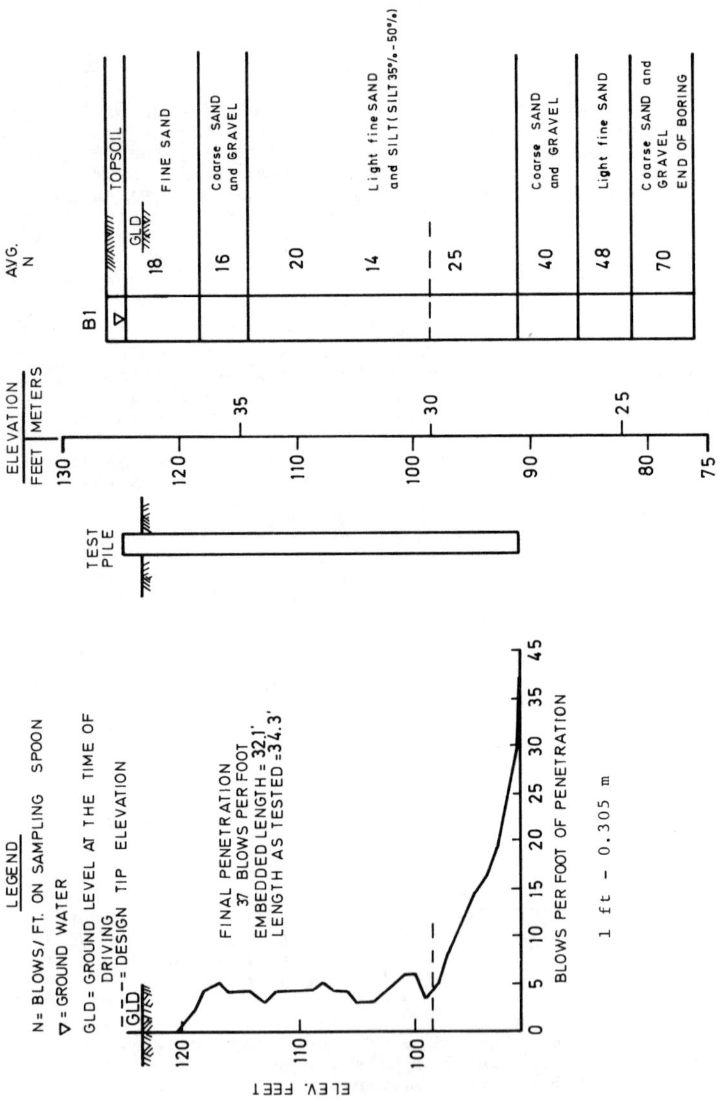

FIGURE 1

PILE DRIVING LOG AND SOIL DATA FOR BORING CLOSEST TO THE PILE FOR LOAD TEST NO.1

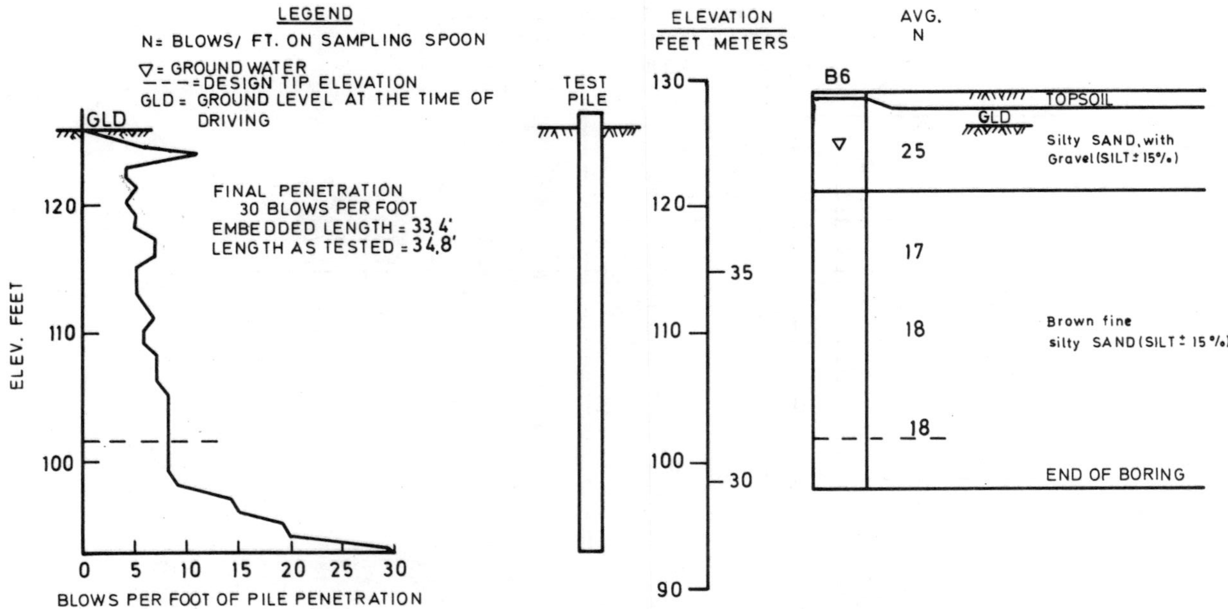

1 ft - 0.305 m

FIGURE 2

PILE DRIVING LOG AND SOIL DATA FOR BORING CLOSEST TO THE PILE FOR LOAD TEST NO. 2

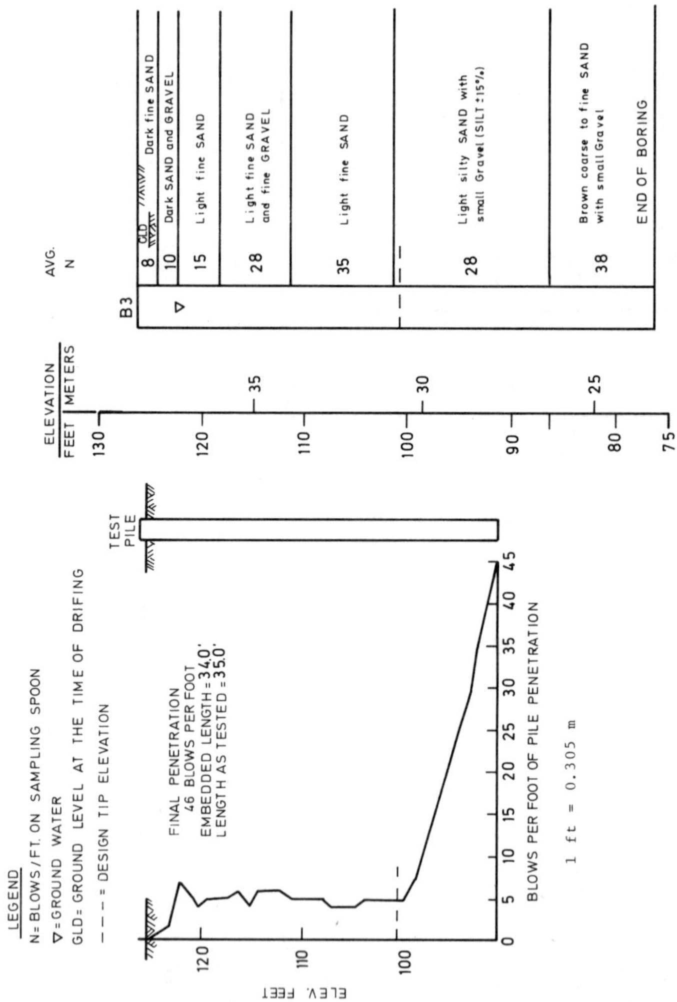

FIGURE 3
PILE DRIVING LOG AND SOIL DATA FOR BORING
CLOSEST TO THE PILE FOR LOAD TEST NO.3

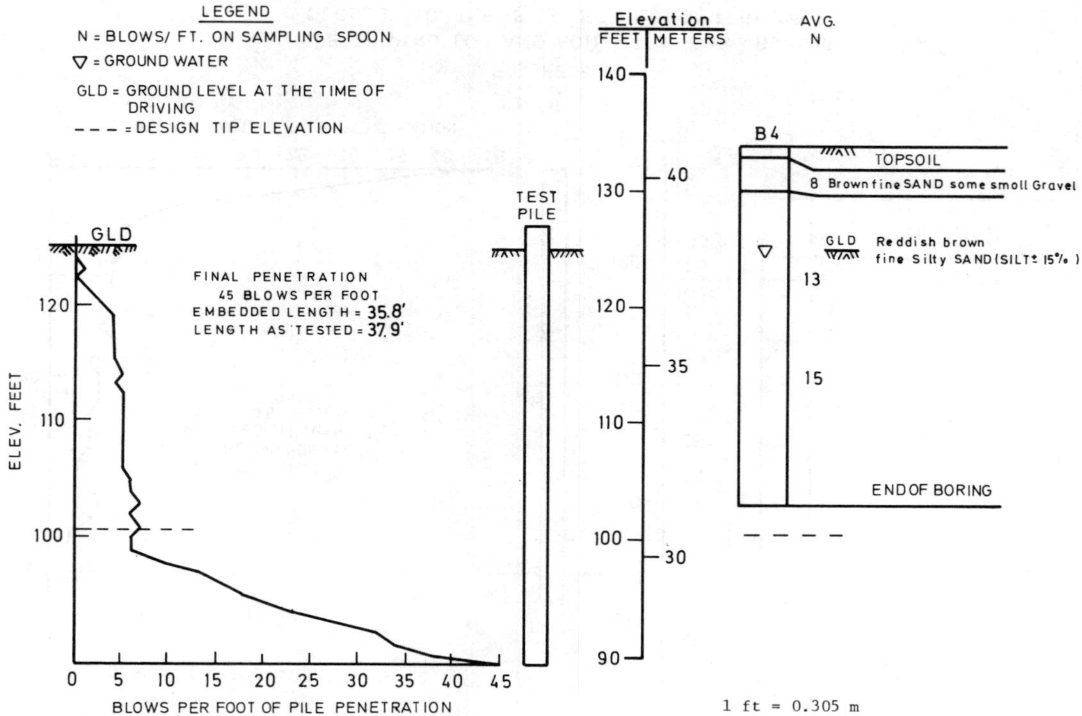

FIGURE 4
PILE DRIVING LOG AND SOIL DATA FOR BORING CLOSEST TO THE PILE FOR LOAD TEST NO. 4

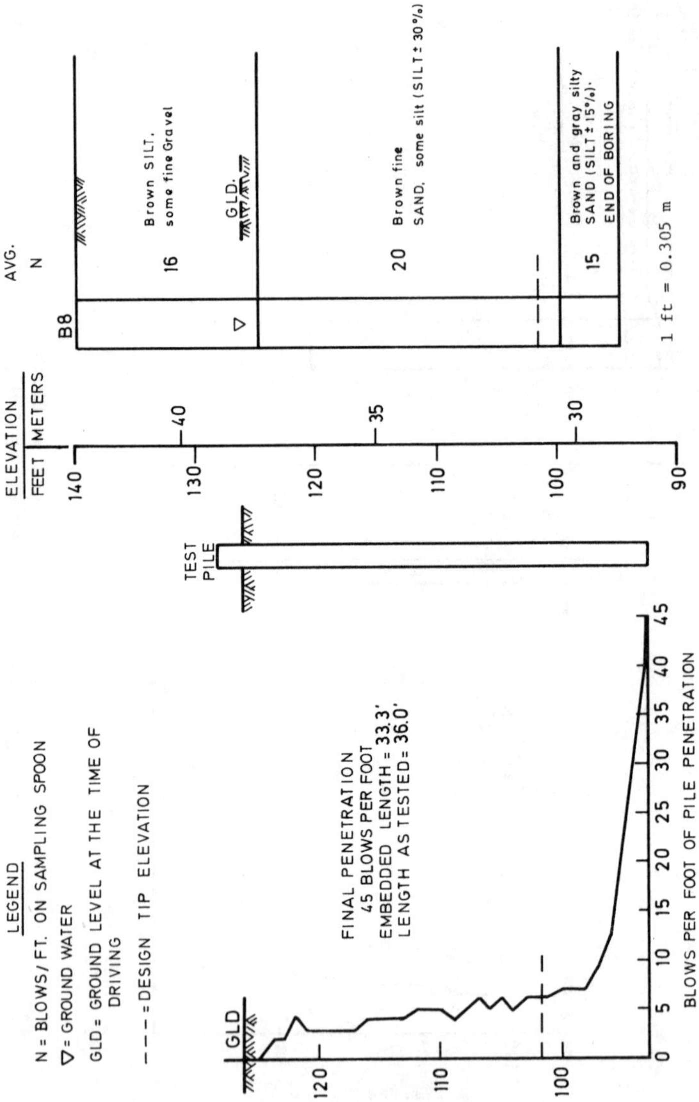

FIGURE 5
PILE DRIVING LOG AND SOIL DATA FOR BORING CLOSEST TO THE PILE FOR LOAD TEST NO. 5

to elevation of 95 ft (29 m) or greater.

Borings B6, B4 and B8 show the soil is either fine silty sand and/or fine sand with some silt, i.e. the fine sand contains approximately 15 to 30 percent silt. The standard penetration resistance (N) values show the soil to be of medium density with a tendency to be on the lower side of medium (B6 – N_{avg} = 20 blows per ft (per 0.3 m); B4 – N_{avg} = 12 blows per ft (per 0.3 m); B8 – N_{avg} = 17 blows per ft (per 0.3 m)). Boring B1 has a layer of gravel to an elevation of about 115 ft (35 m). Below it lie fine sand and silt mixtures to an elevation of about 92 ft (28 m) with an N_{avg} less than 20 blows per ft (per 0.3 m). Notice the similarity in the material characteristics between B1 and the previously mentioned three borings to elevations of 92 ft (28 m) and above. Below this depth for B1 the N value takes a jump to 40 and then increases to 70.

Boring B3 shows a sand and gravel layer at about the same elevation as B1 except that the thickness of this layer is greater and so is its density. N_{avg} for this boring to an elevation of about 87 ft (26.5 m) is near 30, i.e., the density is at the boundary between medium and dense. Like boring B1 the N value increases at greater depths and so do the gravel contents.

STATIC PILE LOAD CAPACITY

The ultimate static load carrying capacity of a pile is the sum of its base resistance and shaft resistance. For cohesionless soils, the static load on piles may be determined by the following empirical formula (Meyerhof, 1956):

Pile base resistance

$$q(b) = \frac{0.4ND(b)}{B} \quad \begin{array}{ll} \leq 4N & \text{for sand} \\ \leq 3N & \text{for silt} \end{array} \tag{1}$$

where q(b) = ultimate unit base resistance ($tons/ft^2$)

N = standard penetration resistance (blows/ft)

D(b) = depth of pile base (ft)

B = width or diameter of pile (ft)

Pile shaft resistance

$$f(s) = \frac{N}{50} \tag{2}$$

where f(s) = ultimate unit shaft resistance (tons/sq.ft)

N = the average standard penetration resistance within embedded length of pile (blows/ft)

Meyerhof (1976) notes that observed unit point resistance is generally in fair agreement with eq.1 except at greater overburden pressures at the point when q(b)/N decreases. If the N value corresponding to an effective overburden pressure of 1 tsf (100 kN/m^2) is taken as standard and other values near the point are corrected empirically as suggested by Peck et al (1974), the estimates of the point resistance will be improved somewhat.

As shown in Fig.1, Fig.4 and Fig.5, a 25 ft (7.62 m) pile as specified by the original contract will rest entirely in the material with N values ranging between 8 and 20. For B6 in Fig.2, except for the top 5 ft (1.5 m) where N=25, the remaining 20 ft (6.1 m) length of pile will also be resting in a material with N less than 20. It is well known that in developing resistance the upper 1/5 to 1/4 of a pile contributes very little to the overall pile capacity. Therefore, the higher N value for the top 5 ft (1.5 m) of the pile will not improve the pile carrying capacity significantly.

For B1, B6, B4 and B8 in the upper 25 ft (7.62 m) of soil N_{avg} = 16. This N_{avg} value will be used for computations of shaft resistance. Similarly, N varies between 14 and 20 at the expected pile tip elevation. N_{avg} for the pile base is also 16 blows per ft (per 0.3 m). Since the major portion of the material is fine sand with 15 percent to 30 percent silt, the limiting value of 3N for unit base resistance is considered more appropriate. Therefore, for N_{avg} = 16 from eq.1, q(b) = 48 tsf (4800 kN/m^2) and from eq.2 f(s) = 0.32 tsf (32 kN/m^2). For a pile with a tip diameter of 10 in (254 mm), a butt diameter of 12 in (305 mm) and a length of 25 ft (7.62 m), the ultimate resistance is 49 tons (357 kN). Using a factor of safety of 2.0, a safe load on a pile will be 24.5 tons (178 kN), which is less than 30 tons (267 kN), the specified pile capacity. Even lower values of the ultimate pile load capacity are obtained from the empirical method proposed by Nordlund (1963) or Terzaghi and Peck (1967) for a 25 ft (7,62 m) long pile.

If it is assumed that N_{avg} does not change with increasing depth, then for q(b) = 48 tsf (4800 kN/m^2) and f(s) = 0.32 tsf (32 kN/m^2), the length of the pile needed to support 30 tons (267 kN) with a factor of safety of 2.0 will be 36.7 ft (11.2m). Borings B6, B4 and B8 terminate above this depth, whereas borings B1 and B3 penetrate to a greater depth indicating the soil to have higher N values below 25 ft (7.62 m). The density of the materials in B1, Fig.1, to an elevation of about 90 ft (27.5 m) is similar to that of B6, B4 and B8. The density of the material in B3 is substantially higher than the rest of the borings through most of its depth. A similarity between B1 and B3 is observed at elevations below 88 ft (26.8 m) and 91 ft (27.7 m). Here the materials do not contain silt and the N values are 38 and greater. Therefore, the computed length of 36.7 ft (11.2 m) for the pile may not always materialize, depending upon where it meets the dense layer (N = 30-50) and how much it penetrates into it. Generally it is difficult for a timber pile to penetrate very deeply into dense sand and gravel.

According to Meyerhof (1976) if the properties of the cohesionless-bearing strata vary near the pile base as is the case here (medium dense fine sand and silt above the base, and dense coarse sand and gravel below the base), then the average values in the corresponding failure zone between about a four pile diameter above the base and a one pile diameter below the base should be used in determining the pile base resistance. Hence, in order to develop the desired base resistance, the pile point must penetrate one to two feet into the dense coarse sand and gravel layer.

DYNAMIC PILE LOAD CAPACITY

As permitted by the specifications, a Vulcan 1, single-acting hammer, with a rated energy of 15000 ft lbs (20340 N m), was selected by the contractor. For this hammer, the ram has a weight of 5000 lbs (22.25 kN) with an equivalent fall of 3 ft (0.92 m). The dynamic pile resistance for this hammer was determined by using the Engineering News Formula which is given by Equation 3.

$$P = \frac{2W(r)h}{s + 0.1} \tag{3}$$

in which

P = allowable load per pile (pounds)
W(r) = weight of ram (pounds)
h = drop height of ram (inches)
s = set per blow under final few blows (inches)

For a 30 ton (267 kN) capacity pile, the final penetration of the pile was computed from eq.3 to be 30 blows per foot (per 0.3 m). A safety factor of 6.0 has been implicitly assumed in eq.3. For comparison purposes, the Hiley formula and the Modified Engineering News Formula have also been presented. Chellis (1961) recommends the use of the Hiley formula which for a single-acting hammer is given by:

$$Pu = \frac{e_f W(r)h}{s + 1/2(C_1+C_2+C_3)} \times \frac{W(r) + e^2 W(p)}{W(r) + W(p)} \tag{4}$$

where
Pu = ultimate load per pile (pounds)
e_f = efficiency of hammer
C_1 = temporary allowance for pile head and cap (inches)
C_2 = temporary compression of pile (inches)
C_3 = temporary compression of ground (inches)
e = coefficient of restitution
W(p) = weight of pile (pounds)

From eq.4, for the design load of 30 tons (267 kN) and a factor of safety of 3.0, using Vulcan 1 hammer with an efficiency of 0.75, the final driving resistance of the pile was determined to be 55 blows per foot (per 0.3 m).

Based on extensive testing of piles and pile-driving hammers, a modified version of the Engineering News Formula (1965) was presented in the following form:

$$Pu = \frac{1.25\ e_f}{s + 0.1} \times \frac{W(r) + e^2 W(p)}{W(r) + W(p)} \qquad (5)$$

Using eq.5 and the suggested factor of safety of 6.0, a pile must be driven to a final penetration resistance of 45 blows per foot (per 0.3 m).

Thus, both these formulae (eq.4 and eq.5) yield a driving resistance substantially higher than the Engineering News Formula (eq.3).

TEST PILES AND LOAD TESTS

All load tests were performed in accordance with ASTM D-1143 specifications. Load was applied to the pile by hydraulic jack acting against a reaction beam welded to two anchor piles. Uniform load increments of 6 tons (53.4 kN) were added every hour. The final load of 60 tons (534 kN) was maintained for 24 hours, except in cases where the pile failed before. Unloading was carried out in decrements of 6 tons (53.4 kN) with each load maintained for one hour.

As called for in the specifications, the driving of piles was governed by the Engineering News Formula. The first test pile was driven to a depth of 32 ft (9.8 m) to a final resistance of 37 blows per foot (per 0.3 m), which is somewhat higher than the required value of 30. A very rapid development of the resistance to driving was the reason for the higher blow count which had to be averaged for the last foot of penetration. An examination of the pile driving log in Fig.1 shows that in the last couple of feet the driving resistance increased from less than 20 blows per foot (per 0.3 m) to 37 blows per foot (per 0.3 m). In the same figure, the boring log for B1 shows that near the base of the pile, the soil changes from fine sand and silt to coarse sand and gravel. Also, the standard penetration resistance jumps from an average value of 25 to 40 blows per foot (per 0.3 m). It should be noted that even though the pile was penetrating a soil (fine sand and silt) of medium density (N = 14 - 25) for almost three-fourths of its length, the driving resistance was less than 5 blows per foot (per 0.3 m).

Results of Load Test 1 and Load Test 2 are plotted in Fig.6. For Load Test 1 before the application of the final load, the pile had reached a total settlement of over 1 in (25 mm) and was deforming at a rate of about 0.0002 in per minute (0.005 mm per minute). Upon the application of the final load increment, continuous pumping of the jack was required to maintain the load to 60 tons (534 kN). About 1/2 in (13 mm) deformation had taken place under the final load, when it became very difficult to maintain the load. At this time, the test was terminated by reducing the load to zero and taking

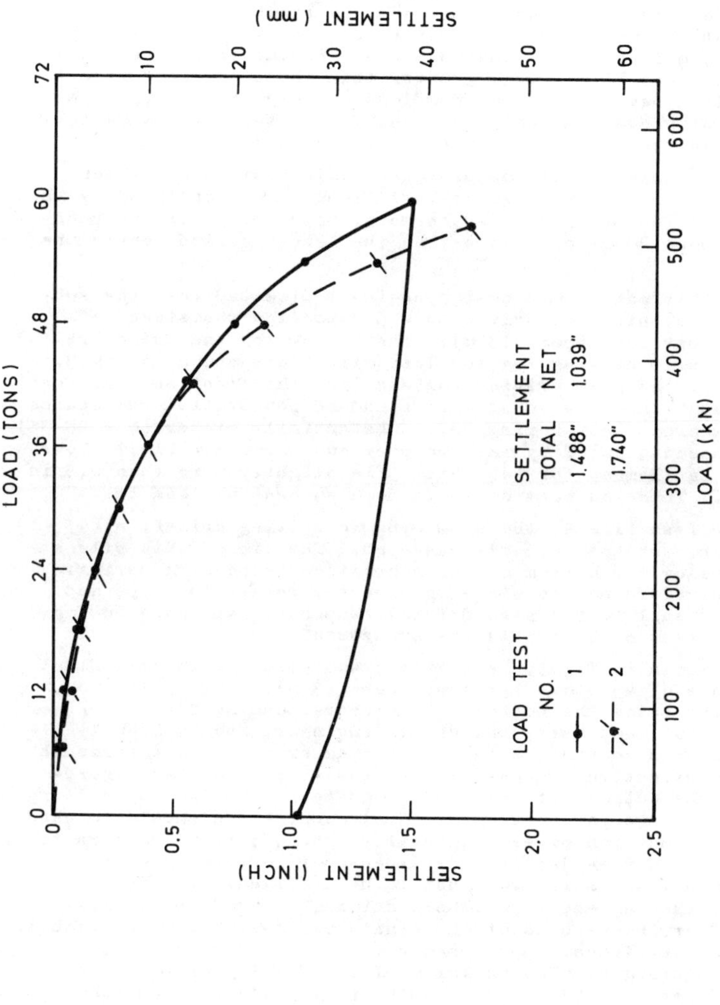

FIG. 6 SETTLEMENTS FOR LOAD TESTS NO. 1 AND NO. 2

the final reading.

Upon the direction of the engineer responsible for the design, Test Pile 2 was driven to a final resistance of 30 blows per foot (per 0.3 m) and it penetrated to a depth of 33 ft (10 m). Once again, the pile drove rather easy for the first several feet, and the final resistance developed quickly. As shown in Fig.2, the underlying soils for Test Pile 2 to a depth of about 30 ft (9.1 m) are similar to that near Test Pile 1. Results of the first two load tests, as shown in Fig.6, indicate that the measured settlements for the two piles were similar up to a load of 36 tons (320 kN). Beyond this, the settlements for Test Pile 2 were greater than those for Test Pile 1. Test Pile 2 developed a maximum load of only 58 tons (516 kN) under continuous pumping of the jack. The test had to be terminated as the pile had settled more than 1.7 in (44 mm).

The inadequacy of the design dynamic resistance, which was obvious by the failure of Test Pile 1, was confirmed by the failure of Test Pile 2. Furthermore, both the piles had penetrated about 30 percent more than the length called for in the design.

At this point, the design engineer directed that the subsequent test piles be driven to a penetration resistance of 45 blows per foot (per 0.3 m). The boring log and the corresponding pile driving log for Test Pile 3 are shown in Fig.3. As before, the pile driving was easy for the first several feet, even though the blow counts for standard penetration resistance in the nearest test boring were substantially higher (N = 28-35) than those indicated by the two previous borings. Fig.7 shows the net settlement for Test Pile 3 is slightly more than 0.5 in (13 mm). This was considered to be a successful load test.

For Test Pile 4, the same dynamic driving criterion (45 blows per foot (per 0.3 m)) was used. The tip of this pile extended below the bottom of the nearest corresponding boring. It is therefore not possible to directly relate the type and nature of soil to the pile driving response. As shown in Fig.7, the load test on this pile was successful too.

As shown in Fig.5, Test Pile 5 was also driven to a final resistance of 45 blows per foot (per 0.3 m). This pile, too, penetrated below the bottom of the corresponding boring. Prior to the final load test, the design engineer, who apparently had developed confidence in the driving criteria as the result of two consecutive successful load tests, stated that regardless of the outcome of the final load test, the driving criterion of 45 blows per foot (per 0.3 m) penetration should be used for the production piles. Naturally, the question arose as to how one could then justify the expense of performing another test, when its results were not to be utilized. The design engineer then agreed to withhold driving of production piles until after the results of the final load test became available. The total settlement approached 2 in (51 mm) for Test Pile 5 as shown in Fig.8. The validity of the driving criterion of 45 blows per foot (per 0.3 m) which previously yielded two successful load tests, became doubtful. Out of a total of five pile load tests, three had failed, and no clear-cut dynamic

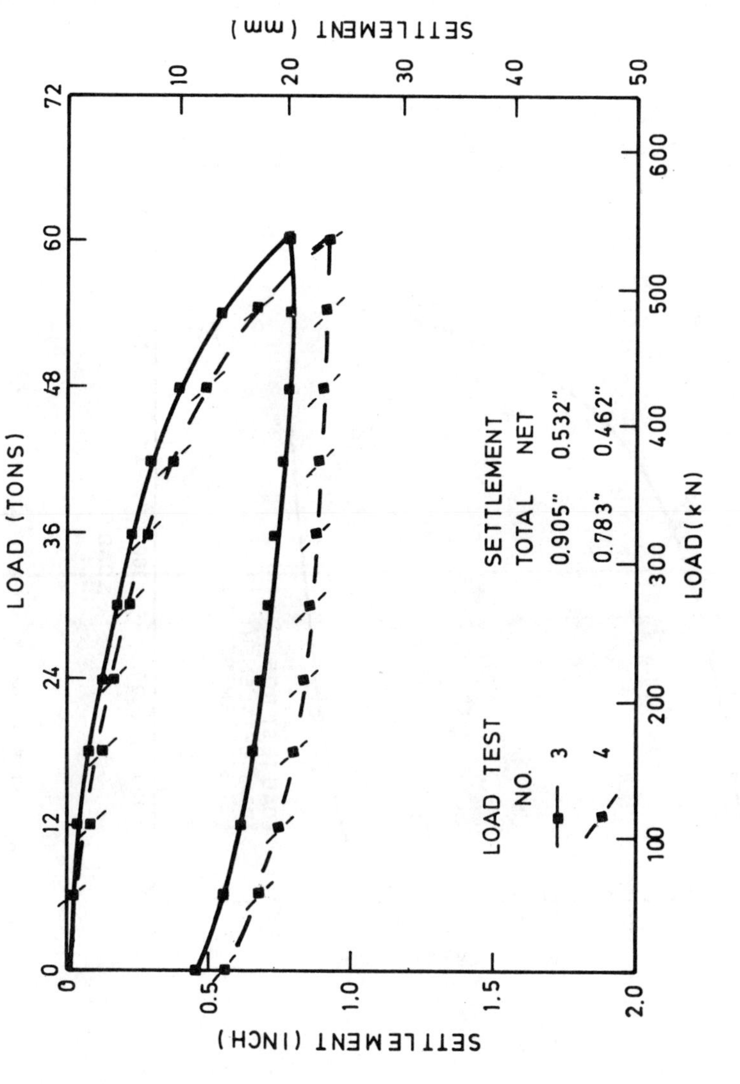

FIG.7 SETTLEMENTS FOR LOAD TESTS NO. 3 AND 4

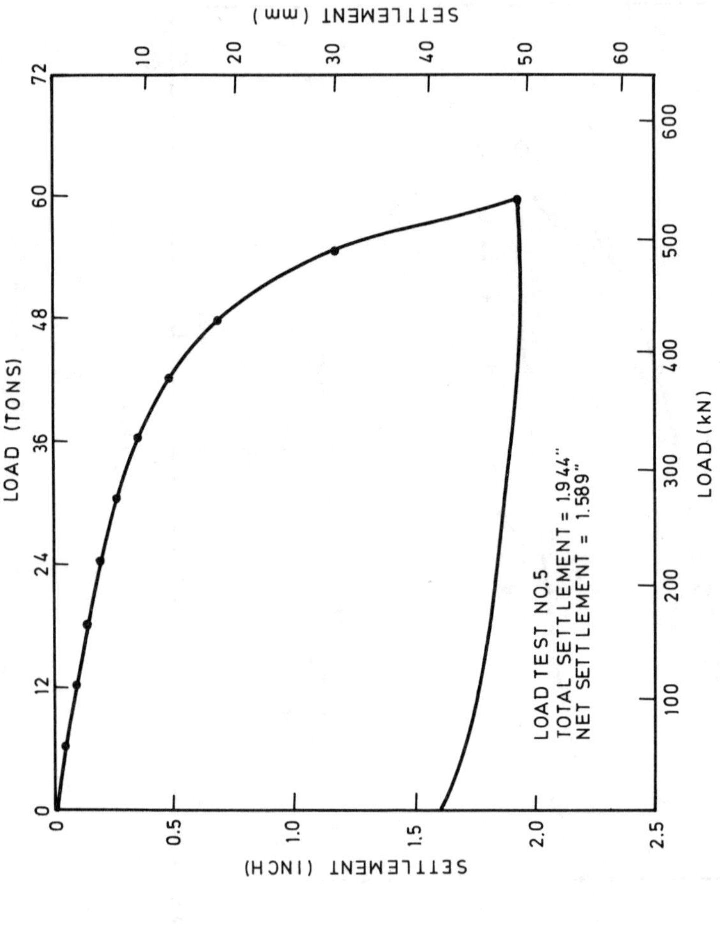

FIG. 8 SETTLEMENT FOR LOAD TEST NO.5

driving criterion had emerged.

ANALYSIS AND DISCUSSION

At this point, the data from the entire testing program was critically reviewed. All of the test piles penetrated to a depth of between 33 ft and 36 ft (10 m and 11 m) even though they were driven to different penetration resistances. Typically, these piles encountered a low resistance of 4 to 6 blows per foot (per 0.3 m) to driving for the first 25 ft (7.6 m). The required resistance was developed in the last 2 ft to 5 ft (0.6 m to 1.5 m) of penetration. Specific conditions and variations were as follows:

(a) For Test Pile 3, boring B3 (Fig.3) shows that the soil in the vicinity of this pile is fine sand, and of much higher density than in any of the remaining four borings. The pile tip is indicated to be embedded more than four times its diameter into the soil with an average N of 28. Therefore, using N_{avg}=23 for shaft resistance and N = 28 for base resistance from eq.1 q(b) = 84 tsf (8400 kN/m^2) and from eq.2 f(s) = 0.46 tsf (46 kN/m^2). With the pile base diameter of 10 in (25.4 mm), the butt diameter of 12 in (30.5 mm) and embedded length of 34 ft (10.4 m), it has an ultimate capacity of 90 tons (800 kN). For Test Pile 4 the corresponding boring (Fig. 4) terminates about 15 ft (4.5 m) above the pile base. The lack of soil information does not permit computations of the pile capacity with the static formula.

Since both piles 3 and 4 did not fail, a comparison of their driving characteristics reveals a remarkable similarity. The driving resistance for the first 25 ft (7.6 m) is around 5 blows per foot (per 0.3 m) even though N for B3 (pile 3) is more than twice that of B4. Since B3 shows the soil to be of medium density (N = 10 - 30), with a tendency to be on the higher side, the low blow count for pile 4 may be explained from the fact that B3 is about 25 ft (7.6 m) from this pile and is, therefore, not a true representative of the soils that may be present around the pile. During the final 5 ft (1.5 m) of driving, the resistance to penetration for both the piles increased from 20 blows per foot (per 0.3 m) to 45 blows per foot (per 0.3 m). Observe that for pile 3 this increased resistance came as the pile passed from a fine sand layer into the layer consisting of silty sand and gravel, but for pile 4, soil data is lacking to make such observation. Since both pile load tests were satisfactory, their success is attributed to the similarity in their response to the final 5 ft (1.5 m) of driving. It will be shown in the subsequent section that such driving response was not observed for the piles that failed.

(b) Like piles 3 and 4, the penetration resistance of piles 1, 2, and 5 was low for the first 25 ft (7.6 m). But, unlike piles 3 and 4, these latter piles fetched very rapidly and their resistance built up from less than 20 blows per foot (per 0.3 m) to the required 30 or 45 blows per foot (per 0.3 m) in a distance of 2 ft (0.6 m) or less. Therefore, the contribution from the lower portion of the pile shaft to the total pile resistance was less than that for pile 3 or pile 4. For pile 1, the reason for the rapid increase in driving resistance is attributed to the high N value of 40 in coarse sand and gravel deposit immediately below the pile tip. For the other two cases, soil data in the immediate vicinity of the pile were not available as the corresponding borings had terminated several feet above the bases of these piles. However, from the remarkable similarity in the driving response of the three piles that failed, it may be inferred that soil conditions near their tips were very similar.

On the basis of the total analysis of soil characteristics, pile driving response, pile load tests, etc., the following driving criteria was established:

1. Piles were to be driven to a penetration resistance of 45 blows per foot (per 0.3 m) for the last foot, provided they had developed a minimum penetration resistance of 25 blows per foot (per 0.3 m) in each of the preceding 4 ft (1.2 m).
2. If the driving resistance developed in a shorter distance of penetration than a total of 5 ft (1.5 m) (see item 1), then the piles were to be driven to a resistance of 60 blows per foot (per 0.3 m) for the last foot.

During the installation of production piles, it was found necessary to drive about one-fourth of the piles to the final resistance of 60 blows per foot (per 0.3 m).

Observe that the driving criterion established by 1 is slightly more stringent than that for Test Piles 3 or 4 which were tested successfully.

The validity of the criterion established by 2 (60 blows per foot (per 0.3 m)) was not checked by load tests. However, any driving criterion more severe than this has the potential of damaging the piles and is generally not recommended.

The Engineering News Formula is considered to have been developed from friction piles in cohesionless soils. The pile driving records for this project clearly indicate that these piles were more like end bearing piles rather than friction piles. The Modified Engineering News Formula (45 blows per foot (per 0.3 m)) gave somewhat better results but still, in one instance (Load Test 5), the error was on the unsafe side. The Hiley formula yielded a driving criterion of 55 blows per foot (per 0.3 m). Piles driven to a resistance of 55 blows per foot (per 0.3 m) would have, in all probability, yielded successful load tests. However, had all the piles been driven to Hiley criterion, error would have been on the safe side, though the cost would have been more. As has been pointed out in the past by many investigators, the applicability of any dynamic formula is a matter of chance, since the very assumptions on which these formulae are based are questionable.

CONSEQUENCES OF CHANGED CRITERIA

The contractor renegotiated the prices for pile installation for the following reasons:

1. The new lengths of piles were between 35 ft to 40 ft (10.5m to 12 m) as opposed to the original contract lengths of

25 ft (7.6 m). The cost per unit length was more for longer piles than for shorter piles.

2. The new driving criteria (45 to 60 blows per foot (per 0.3 m)) was much more stringent than the original design called for (30 blows per foot (per 0.3 m)). It required more time to drive the piles to their final resistance.

The final procedure for payment was quite involved. The contractor did not accept any responsibility for a damaged, out-of-plumb, or out-of-location pile, once it had reached a penetration of 25 ft (7.6 m) or had developed a driving resistance of 30 blows per foot (per 0.3 m). Also, beyond the original depth of penetration (25 ft (7.6 m)) or driving resistance (30 blows per foot (per 0.3 m)), a different unit price had to be paid. As one would expect, the new prices were appreciably higher than those in the original contract. Needless to say, there was considerable delay in the completion of the job.

CONCLUSION

Based on several load tests on wooden piles, driven into silty fine sand and fine sand deposits overlying dense layers of sand and gravel, the following conclusions are made:

1. Empirical static formulae due to Meyerhof (1976) gave reliable estimates of pile capacity and length.
2. In silty fine sand or fine sand, the dynamic pile resistance determined by the Engineering News Formula is not safe.
3. The blow count of the final one foot (0.3 m) of driving is not in itself sufficient to assure the safety of a pile. The pile tip must penetrate adequately into the bearing stratum to develop the needed base resistance.
4. Piles that developed their final resistance with a minimum of 20 blows per foot (per 0.3 m) for the final 5 ft (1.5 m) showed adequate capacity, when driven to a final resistance of 45 blows per foot (per 0.3 m) required by the modified Engineering News Formula.
5. If a pile increased in its driving resistance from 20 blows per foot (per 0.3 m) to 45 blows per foot (per 0.3 m) in less than 2 ft (0.6 m), it failed under the applied load. For this type of pile driving response, the modified Engineering News Formula did not yield safe results.
6. The Hiley formula yielded the most conservative dynamic driving criterion. Had all the piles been driven to this criterion, both the cost and time of installation of piles would have increased.

REFERENCES

"A Performance Investigation of Pile Driving Hammers and Piles," Joint Study by the Bureau of Public Roads, U.S. Department of Commerce, the Michigan Road Builders Association, Wayne State University, and Rep. Hammer Manufacturers, Michigan State Highway Commission, Lansing, Michigan, March, 1965.

Meyerhof, G. G., "Penetration Tests and Bearing Capacity of Cohesionless Soils," Journal of the Soil Mechanics and Foundations Division, ASCE, Vol. 82, No. SM 1, January, 1956, pp. 1-19.

Meyerhof, G. G., "Bearing Capacity and Settlement of Pile Foundations," Journal of the Geotechnical Engineering Division, ASCE, Vol. 102, No. GT 3, March, 1976, pp. 195-228.

Nordlund, R. L., "Bearing Capacity of Piles in Cohesionless Soils," Journal of Soil Mechanics and Foundations Division, ASCE, Vol. 89, No. SM 3, May, 1963, pp. 1-36.

Peck, R. B., Hanson, W. E., and Thornburn, T. H., Foundation Engineering, John Wiley and Sons, New York, New York, 1974.

Terzaghi, K., and Peck, R. B., Soil Mechanics in Engineering Practice, Second Edition, John Wiley and Sons, New York, New York, 1967, p. 530.

The Behavior of Steel Rocket Shaped Pile

By Peter Y. Lee[1], M. ASCE and Lawrence W. Gilbert[2]

INTRODUCTION. The Louisiana Power and Light Company (LP&L) is charged with the responsibility for the design, construction, and operation of many energy related structures. Among these structures are high voltage transmission lines. Traditionally, these lines, when extending over long distances, were supported by steel trussed spaced along the route. These structures were usually very large, requiring wide rights of way and blending poorly with the surrounding landscape. Due to the increased cost of land and the concern for preserving the environment, utility companies recently have begun to use single-pole steel structures in large transmission systems. These structure foundations are generally simpler and smaller than those of trussed towers and consist of 30 to 50 feet (9.2 to 15.3 meters) long open ended octagonal shaped steel piles with four wings. Due to limitations in the size of pile driving equipment that can operate effectively in the marsh, the piles are usually fabricated in two sections and are joined together in the field with a male-female connection at midpoint.

The southern Louisiana area provides one of the worst combination of loading and soil conditions for the design of these single-pole foundations. Since this region is subject to frequent fierce hurricanes, wind loads are so severe that lateral load requirements generally govern the design of the foundation. Soil conditions in this area consist typically of very soft peat and organic clay underlain by very soft to soft normally consolidated clays.

The design of single-pole foundations to resist lateral loads can be subdivided into two basic steps. First, determine the lateral load applied to the superstructure that would tend to cause its failure. Second, size a pile sufficiently large to provide resistance to these loads. While our understanding of the intensity, duration, variability and cyclic nature of applied loads generated by hurricane force winds is far from complete, this paper will concentrate on predicting foundation resistance to lateral loads. Although a number of theoretical and semi-empirical approaches exist to calculate the resistance of a pile to lateral loading, their direct application to the single-pole foundation pile is not clear. The interrelationship among the degree of subsurface investigation, number and type of laboratory (or field) tests, methods of analysis, and type pile used for a specific problem cloud the predicted value of lateral capacity. Reliable assessments of a design approach to predict behavior under load can best be made by evaluating well instrumented lateral pile load tests.

Two such full-scale laterally loaded pile tests were conducted during 1977-1978 at Tulane University, sponsored by LP&L. Both test piles were 40 feet long. However, the geometry of the two piles were different, resulting in piles having differing stiffness and resistability to lateral loads. One pile was fabricated without wings, the other, the rocket shaped pile, had wings attached to the upper 20 foot (6.1 meter) length.

---

[1]Professor of Civil Engineering, Tulane University, New Orleans, LA

[2]Lect. in Civil Engineering Department, Tulane University

The objective of the initial phase of the pile load test program was to develop a methodology for LP&L that could be applied to future designs of single pole transmission tower foundations. The methodology was developed based on the performance of the pile without wings when subjected to lateral loading. Details of this methodology were previously reported (Lee and Gilbert, 1979). Developed procedures are presented in summary form later in this paper. The second phase of the program was to verify the developed methodology by using it to predict performance of another pile with different geometry (the rocket pile) and to compare predicted performance to observed behavior. This paper presents the results of this second phase of the pile test program.

CURRENT ENGINEERING PRACTICE. Significant advances on the subject of resistance of piles to lateral loads have been made within the past decade at the University of Texas, Austin. The computer program entitled, "Analysis of Laterally Loaded Piles by Computer" (Reese, 1977) is widely used to predict the performance of piles subjected to lateral loading. This program solves the following differential equation derived on the assumption that the pile is a linearly elastic beam and that the soil reaction may be represented as a line load;

$$EI \frac{d^4 y}{dx^4} - P_x \frac{d^2 y}{dx^2} - p = 0 \text{ - - - - - - - } (1)$$

In which P_x = axial load on pile; y = lateral deflection of the pile at point x along the pile length; p = soil reaction per unit length; and EI is the flexural rigidity of the pile.

The behavior of the soil surrounding the laterally loaded pile can be described in terms of p-y curves. These curves relate soil resistance to pile deflection at various depths below the ground suface. In general, these curves are non-linear and depend on depth below ground surface, stress-strain properties of the soil, and the number of load cycles applied to the soil. Recommended procedures for developing p-y curves for piles in soft clay have been proposed by Matlock, (1970). The procedures are semi-empirical and rely heavily on pile load tests conducted at Lake Austin and Sabine, Texas. A simplified summary of the procedures was presented by Reese (1975) as follows:

A. Short Term Static Loading:
   1. Obtain the best estimate of the variation of shear strength and effective unit weight with depth. Also obtain the value of ε_{50}, the strain corresponding to one-half the maximum principal stress difference.
   2. Compute the ultimate soil resistance per unit length of shaft, p_u, using the smaller of the values given by the equations below:

$$p_u = (3 + \frac{\gamma' x}{c} + \frac{0.5x}{b})\, cb \text{ - - - - - - - } (2)$$

$$p_u = 9cb \text{ - - - - - - - - - - - - - - - - - - } (3)$$

When γ' = average effective unit weight from ground surface to p-y curve; x = depth from ground surface to p-y curve; c = shear strength at depth x; and b = width of shaft. The value of p_u is computed at each depth where a p-y curve is desired, based on the shear strength at that depth.

3. Compute the deflection, y_{50}, at one half the ultimate soil resistance from the equation:

$$y_{50} = 2.5\ \varepsilon_{50}\ b \text{ - - - - - - - - - - - (4)}$$

4. Points describing the p-y curve are now computed from the relationship:

$$p/p_u = 0.5\ (y/y_{50})^{1/3} \text{ - - - - - - - (5)}$$

B. Cyclic Loading:

1. Construct the p-y curve in the same manner as for short term static loading for values of p less than 0.72 p_u.
2. Solve equations 2 and 3 simultaneously to find the depth, x_r, where the transition occurs.
3. If the depth to the p-y curve is greater than or equal to x_r, the p is equal to 0.72 p_u for all values of y greater than $3y_{50}$.
4. If the depth to the p-y curve is less than x_r, then the value of p decreases from 0.72 p_u at $y = 3y_{50}$ to the value given by the expression given below at $y = 15y_{50}$:

$$p = 0.72\ p_u\ \left(\frac{x}{x_r}\right) \text{ - - - - - - - - - - -(6)}$$

The value of p remains constant beyond $y = 15y_{50}$.

It should be emphasized that the procedures recommended by Matlock are heavily influenced by the results of pile load tests conducted at Lake Austin and Sabine, Texas. Certain problems should be recognized before the above decribed procedure is used for the LP&L pile in Louisiana marshlands.

1. The diameter of the LP&L piles at the ground surface is substantially larger than those piles tested by Matlock.
2. Soil conditions at Lake Austin consisted of silts and clays with the upper layer overconsolidated due to desiccation. At Sabine the soil conditions appeared to be typical of slightly overconsolidated marine deposits. Procedures based on tests performed at these sites may not be applicable for piles driven in Louisiana marsh environments.
3. The constructed p-y curve is clearly dependent on the stress-strain properties of the soil. The type of soil test used to obtain these soil properties should be critically selected and its ability to predict performance verified locally.
4. Tolerable deflections of LP&L service piles during hurricane loading are generally much larger than those of most other piles. The validity of the p-y curve at large values of deflection has not been widely tested. Also, the lateral load required to develop a complete foundation failure is rarely determined by pile load tests.
5. The behavior of piles under cyclic loading is function of soil sensitivity. Procedures developed by Matlock may not be appropriate for the Louisiana marsh deposits.
6. Hurricane winds acting on a transmission tower system generate significant ground line moment in addition to horizontal shear load. For best results the load system applied to a test pile should duplicate as nearly as possible the load system applied to the service piles.

In response to the above problems, a design methodology was tailored for the specific problem faced by LP&L. The methodology developed (Lee and Gilbert, 1979) was based on the following premises:

1. The computer program, "Analysis of Laterally Loaded Piles by Computer", (Reese, 1977) and procedures for developing p-y curves for piles in soft clay (Matlock, 1970) are widely used to predict performance of laterally loaded piles. No changes in the computer program itself were made. However, for reasons previously mentioned, a new method for constructing p-y curves for the very soft peats and very soft, highly plastic clays encountered in Louisiana marshes is necessary.
2. Soil parameters required for the construction of p-y curves are obtained primarily from unconfined compression tests. This type strength test has been performed routinely for other LP&L foundations designs. Since equations proposed for p-y curves are largely semi-empirical, a more sophisticated testing program was not considered.
3. The methodology was developed from the results of the pile load test performed on the test pile without wings. This pile was loaded laterally until it underwent a complete foundation failure (in the soil).

The construction procedures for developing p-y curves for piles in very soft peat and very soft highly plastic generally follow the above procedures proposed for soft clay. Parametric studies were required to yield a procedure that accurately models the behavior of the test pile under both short term static and cyclic loading. Values of design shear strength ($S_u = c$) and density presented in Figure 1 and design ε_{50} values of 4 percent in the peat and 1.5 percent in the clay were used in the parametric studies. These values of ε_{50} were obtained from unconfined compression test stress-strain curves. A summary of the resulting construction is outlined below:

A. Short-Term Static Loading:
   1. Develop a design profile of c, γ' and ε_{50}. These values should be based primarily on unconfined compression tests.
   2. Calculate the trend of P_u with depth as the smaller of the values obtained by the equation:

$$p_u = (3 + \frac{\gamma' x}{c'} - \frac{0.5x}{b})\, c'b \qquad (7)$$

$$p_u = 9\, c'b \qquad (8)$$

   where c' = 2.6 times the value of design shear strength determined by unconfined compression tests.
   3. For the depth at which a p-y curve is desired, calculate

$$y_{50} = 0.1\ \varepsilon_{50}\ b \qquad (9)$$

   3. Points describing the p-y curve for short term static loading can be obtained by solving:

$$p/p_u = 0.5\ (y/y_{50})^{1/3} \qquad (10)$$

B. Cyclic Loadings:
   1. Construct the p-y curve as described above for short term static loading for values of p less than 0.91 p_u, corresponding to a y value of $6y_{50}$.

2. The value of p remains constant beyond $6y_{50}$.

SITE CONDITIONS AND SOIL PROPERTIES. The Chalmette site is located within a brackish-fresh water marsh deposit that developed between the natural levees of the present day Mississippi River and of abandoned distributatires that formed during the lengthening process of the Mississsippi River. Typically, these marsh deposits consist of a surface layer of very soft coarse to medium texture fibrous peat, underlain by very soft to soft, highly plastic clay silt layers. The ground-water at the test site is slightly higher than the natural ground surface and fluctuates slightly due to tidal and wind effects. Two undisturbed borings were made for the pile load test. These borings indicated that the surface peat layer at the test site is 15 feet (4.6 meters) thick. The underlying clay deposit extends to a depth of 54 feet (16.5 meters), at which depth an 8 foot (2.44 meters) thick layer of nearshore gulf sand is encountered. Overconsolidated clays of Pleistocene origin underly this fine sand and extend to the maximum depth penetrated by the borings. Representative samples recovered from the undisturbed borings were used in a testing program consisting of: visual classification and moisture content determination (w_n), Atterberg Limit determination (AL), total unit weight determination (γ_t), unconfined compression (U) and triaxial (UU) tests. Figure 1 presents w_n and AL data and bands of the average γ_t and of the undrained shear strength (S_u) as measured by U, UU, and LV strength tests. The results of S_u measured by these three laboratory tests yielded values in the same range. Also shown on this figure are values of γ_t and S_u values selected for design. Field vane tests were not performed at the test site. Since the foundation contained numerous roots, pieces of wood, and other obstructions, field vane tests would be expected to give biased results. A field self-boring presuremeter (The University of Texas Model) was used to perform soil tests at the alternate test site in Chalmette during December, 1977. In general, the undrained shear strengths (S_u) obtained from this field test were substantially higher than the values of S_u obtained from laboratory unconfined compression tests (ranging from 230% to 320% higher). Since this field test was not directly performed at the pile test site, it merely gave a good indirect comparison between the field undrained shear strength and the design values of S_u selected from the laboratory tests (U,UU, LV).

The design values of S_u selected for the clay layer below a depth of 15 feet correspond to a ratio of S_u to vertical effective stress ($\bar{\sigma}_{vo}$) equal to 0.25. This strength trend is considered reasonable for normally consolidated, highly plastic clay. Foott and Ladd (1972) reported values of $S_u/\bar{\sigma}_{vo}$ ranging between 0.22 and 0.26 for similar type soils. Results of the consolidation tests performed indicate that the soil above the Pleistocene clay is normally consolidated.

PILE SELECTION, INSTRUMENTATION AND TESTING. The pile test was designed to duplicate as nearly as possible the loads experience by a typical service pile during a storm of hurricane intensity. A combination of horizontal shear load and ground line moment was applied to the test pile with a cable (approximately 300 feet (91.5 meters) long) attached to a pull pole at 59 feet (18 meters) above ground. A tension load was applied through this cable by drawing on the cable with a crane located on a shell service road constructed by LP&L on the marsh surface. The cable formed an angle of approximately 10 degrees with respect to the horizontal when tensioned.

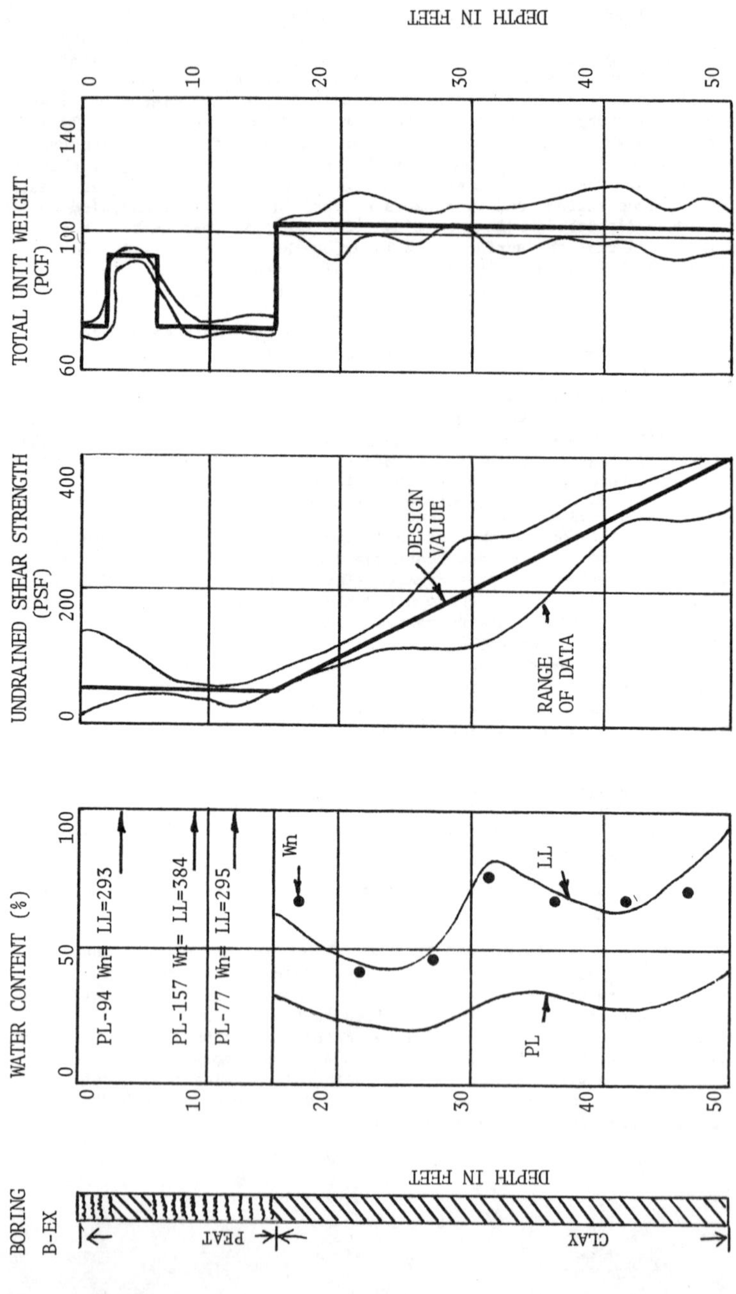

(1 ft = 0.305 m; 1 psf = 47.9 N/m^2; 1 pcf = 0.157 kN/m^3)

FIGURE 1. SOIL PROPERTIES AT CHALMETTE TEST SITE

A plan and profile of the test site is shown in Figure 2. The rocket shaped pile tested in this program was similar to those commonly selected by LP&L to support their transmission towers. The rocket pile was a 40 feet (12.2 meters) long, octagonally shaped steel pile with four wings. It was open-ended and tapered from 31 inches (0.79 meter) at the top of 24 inches (0.61 meter) at the bottom. Details of this test pile are shown in Figure 3. Pile driving records for the rocket pile are shown in Figure 4.

The performance of the test pile under lateral loading was monitored with the following measurements:

1. Load Measurements: The load applied to the test pile by the attached cable were measured with an hydraulic load cell linked to the cable near the loading crane. By using a special blocking arrangement, the crane operator was able to sustain the desired load to a precision of ± 50 pounds (222 N). The load cell was monitored continuously during the test program.
2. Deflection Measurements: The deflected shape of the test pile under load was measured with a Digitilt Model 50306 slope inclinometer. The inclinometer casing was fastened to the tension side of the pile for its entire length. Measurements were taken at the end of each loading increment.
3. Strain Measurements: The strain at various points along the tension side of the test pile was measured with AILTECH Weldable Strain Gages. Gages were located at 2 feet (0.61 meters) below ground line and at 5 feet (1.53 meters) increments down for the entire pile length.
4. Ground Line Deflection: Ground line deflection was measured with a scaled device attached to the pile two feet above ground line. The device was hinged and equipped with a bubble level on the upper section, allowing for horizontal leveling before each reading. Ground line movement readings were taken with a transit located approximately 50 feet (15.3 meters) from the pile.

Under hurricanewind loading, conventional transmission towers are subjected to an equivalent horizontal load centered at about 59 feet (18 meters) above ground line. This increases in magnitude gradually over a period of 7 to 10 hours and is characterized by successive increases and decreases in load due to gusting winds (Figure 5). The general philosophy of load application followed in this test program was to statically, then cyclically, load the test pile over a period of from 7 to 10 hours. The pile was then unloaded to determine the net permanent deflection. This was followed by a gradual reload and, then, continued load of the test pile until complete foundation failure was observed. Each static load increment was applied for a minimum period of 1 hour or until horizontal movement at ground line had essentially ceased. After sustaining the static load, the pile was then subjected to 25 cycles of alternately loading between the static load increment of 50 percent of that increment (e.g., cycle between 10 kips (44.5 KN) and 5 kips (22.25KN) cable load). The cycle load was applied at a rate of approximately 4 cycles per minute. After the cyclic loading, the load was again held constant for at least 30 minutes before advancing to the next load increment.

The pile load test program for the rocket pile was carried out in one day. The procedures generally followed in the program are outlined below:

Phase I: Static load applied at increments of 5 kips (22.25 KN), 10 kips, (44.5 KN) and 20 kips (89 KN).

Phase II: Cyclic load applied at the end of the 10 kips, (44.5 KN) 15 kips, (66.75 KN) and 20 kips (89 KN).

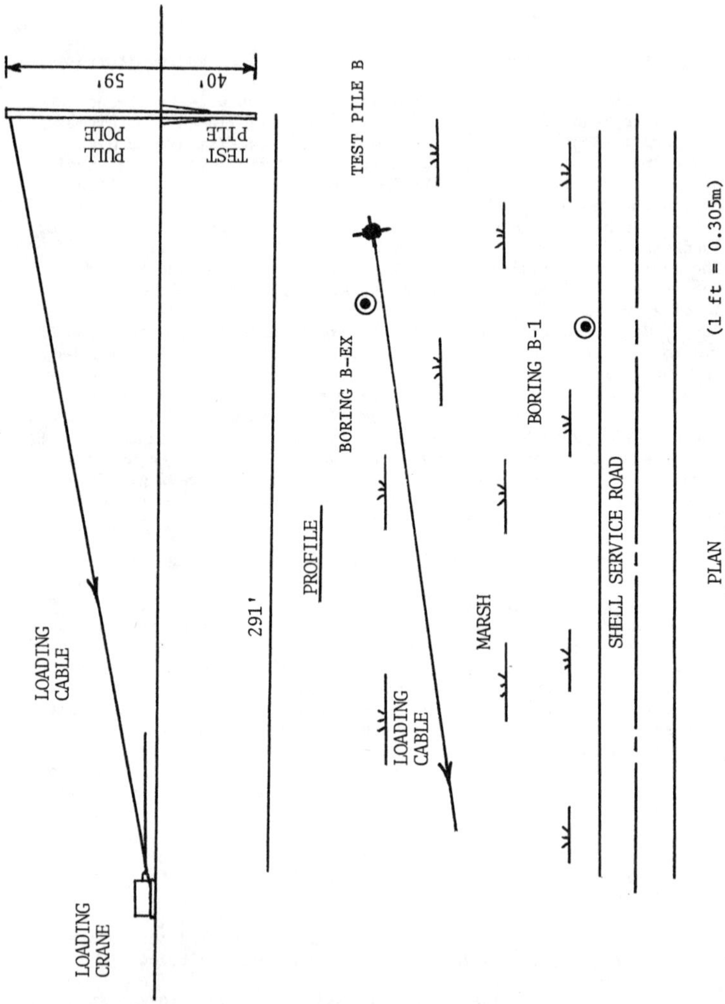

FIGURE 2. CHALMETTE TEST SITE

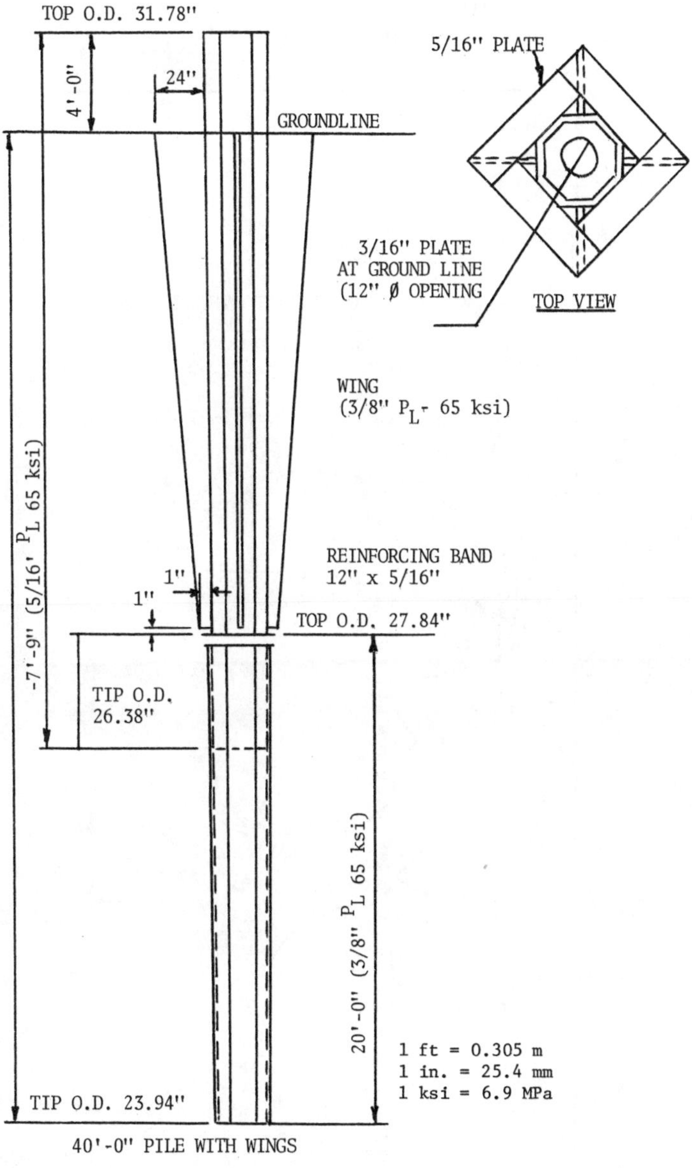

FIGURE 3. DETAILS OF TEST PILE (PILE WITH WINGS)

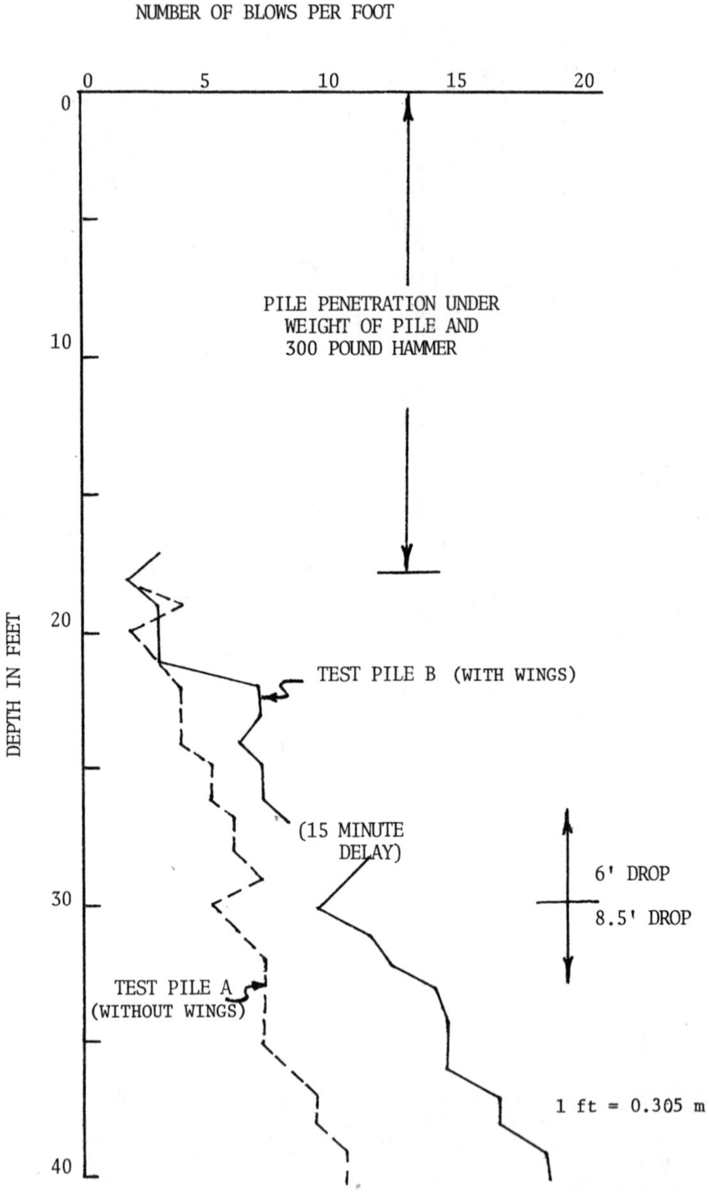

FIGURE 4. PILE DRIVING RECORD

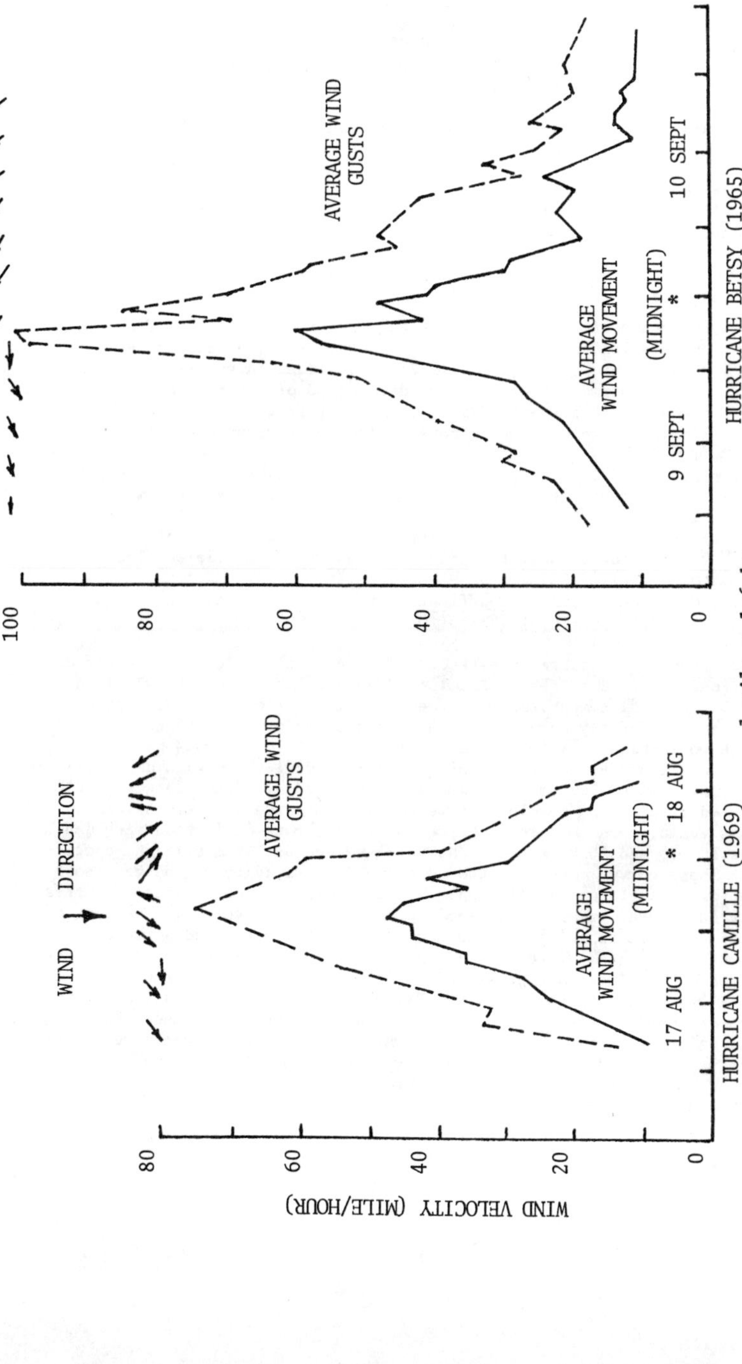

1 mile = 1.6 km

FIGURE 5. WEATHER STATION DATA -- PORT SULPHUR, LA

Phase III:	Unload the test pile
Phase IV:	Reload at increments of 5 kips (22.25 KN), 10 kips, (44.5 KN), 15 kips (66.75 KN), 20 kips (89 KN), and 25 kips (111.25 KN).
Phase V:	After loading the test pile to failure, then unload the test pile.

A summary of the load-deflection performances of the rocket pile is shown in Figure 6. Curves of deflection with depth for the test pile subjected to the static and cyclic load increments and for the unload pile are presented in Figure 7. Foundation failure was achieved at a cable load of 25 kips(111.25 KN). The rocket pile failed structurally after sustaining 25 kips(111.25 KN) for 35 minutes (yielded in steel).

Some general observations that can be made regarding the data presented in Figure 6 and 7 are:

1. The shape of the load deflection curve resembles that of an axially loaded pile. A "plunging" type failure occurred at a cable load of 25 kips (111.25 KN)(yielded in steel).
2. The net permanent ground line deflection of the test pile loaded to 80 percent of its ultimate or failure load are within the tolerance of transmission tower structures (± 2") (± 5.1 cm).
3. The ground line deflection of the pile reloaded to its previous maximum load returned essentially to its previous maximum deflection.
4. Measured deflections with depth indicate that very small deflections occurred near the pile tip even when the pile was loaded to a level of 80 percent of failure.
5. Additional groundline deflection resulting from cyclic loading was not significant (5% to 10%).

Families of p-y curves constructed for test pile B for both static and cyclic loading based on the methodology developed at Tulane are shown in Figure 8. General comparisions between the pile performances predicted with these sets of p-y curves and the measured performances of the test pile B are shown in Figures 9 through 11. The curve of load versus ground line deflection (Figure 9) and curves of load versus deflection with depth (Figure 10) for the applied increment loads show very good agreement between deflections predicted by the development method and measured deflections. Figure 11 presents measured and predicted curves of load versus deflection with depth for the 10, 15 and 20 kips (44.5 KN, 66.75 KN and 89 KN) static and cyclic loads. These curves emphasize the relative insensitivity of the very soft foundation soils to cyclic loading. Measured ground line deflection increased by only 10 percent under cyclic loading under a load level of 80 percent of the ultimate pile capacity. These curves also show very good agreement between predicted and measured horizontal deflections. Curves of moment versus depth for static loads are presented in Figure 12. Because the wings of test pile B lack stiffness, the rigidity (EI) values of the upper part of the wings were questionable. The measured moments and the predicted moments based on 100% EI values were not matched. If an EI value of 36% had been used at the top of the wings and increased gradually to the 100% EI value used at the bottom of the wings, a general agreement between measured and predicted moments would have been achieved.

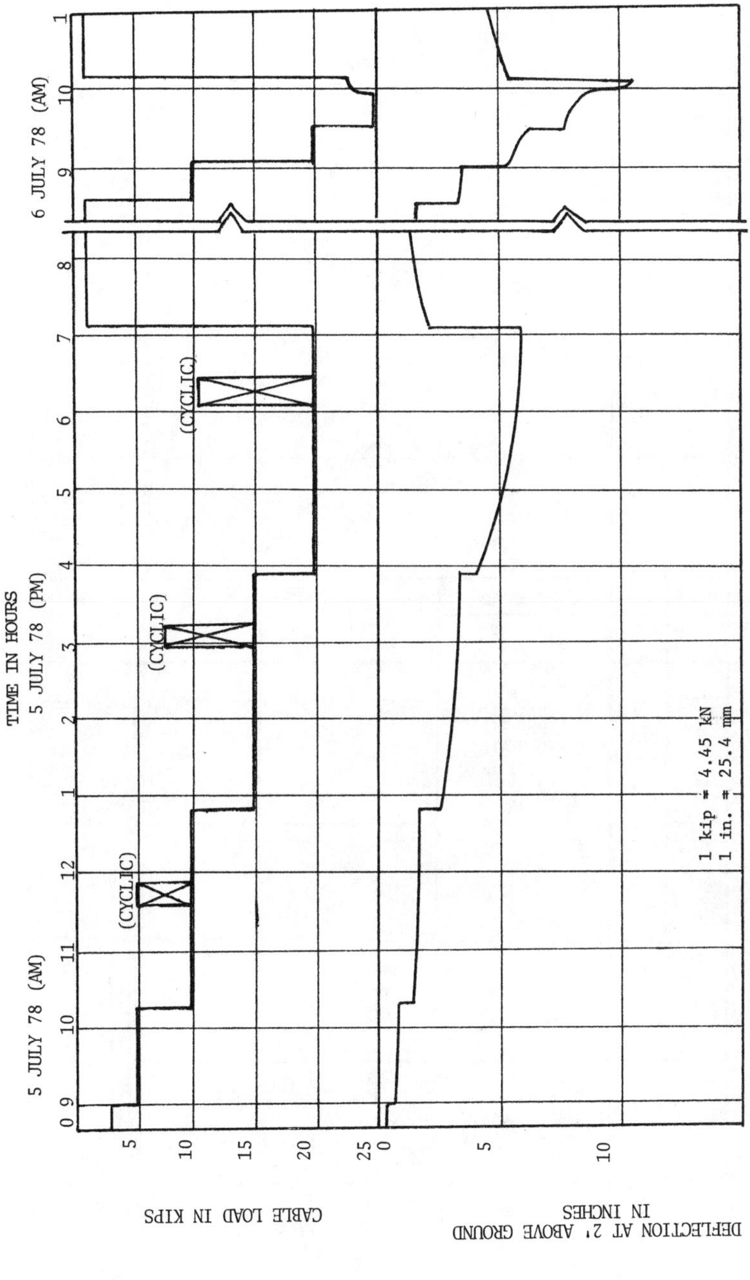

FIGURE 6. LOADING SCHEDULE, PILE WITH WINGS (TEST PILE B)

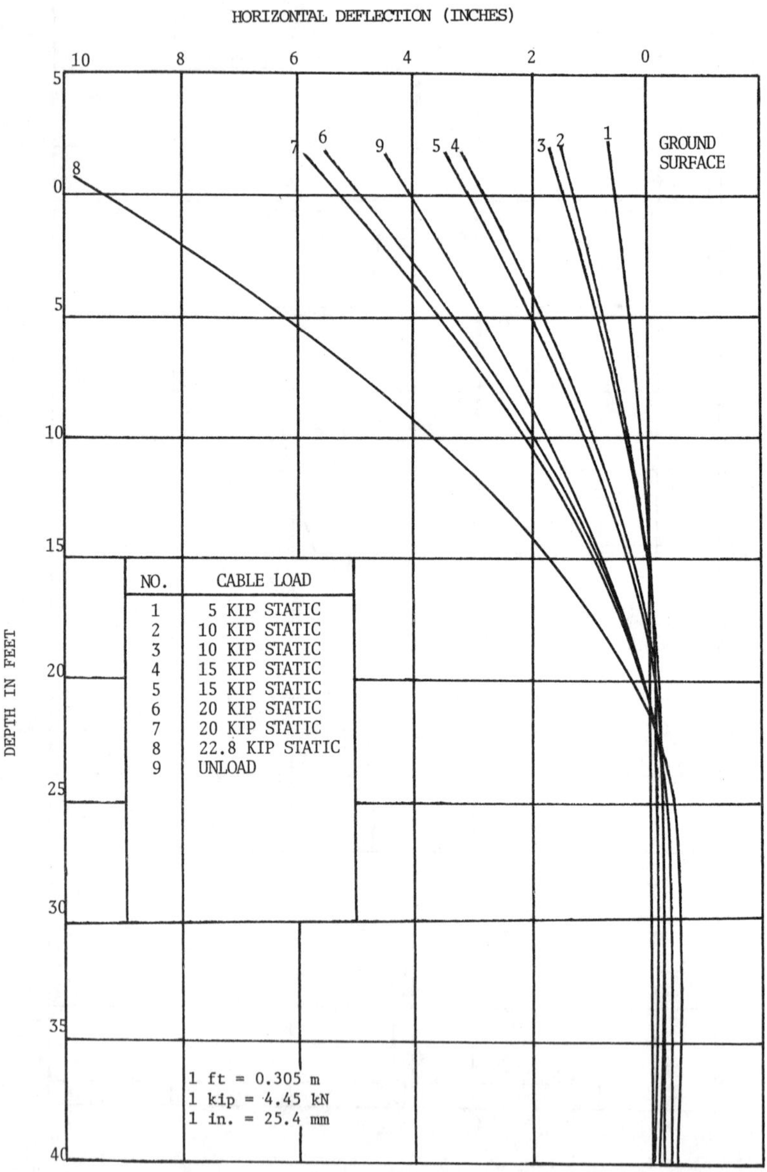

FIGURE 7. MEASURED DEFLECTION WITH DEPTH (1 ft = 0.305 m) (PILE WITH WINGS, TEST PILE B).

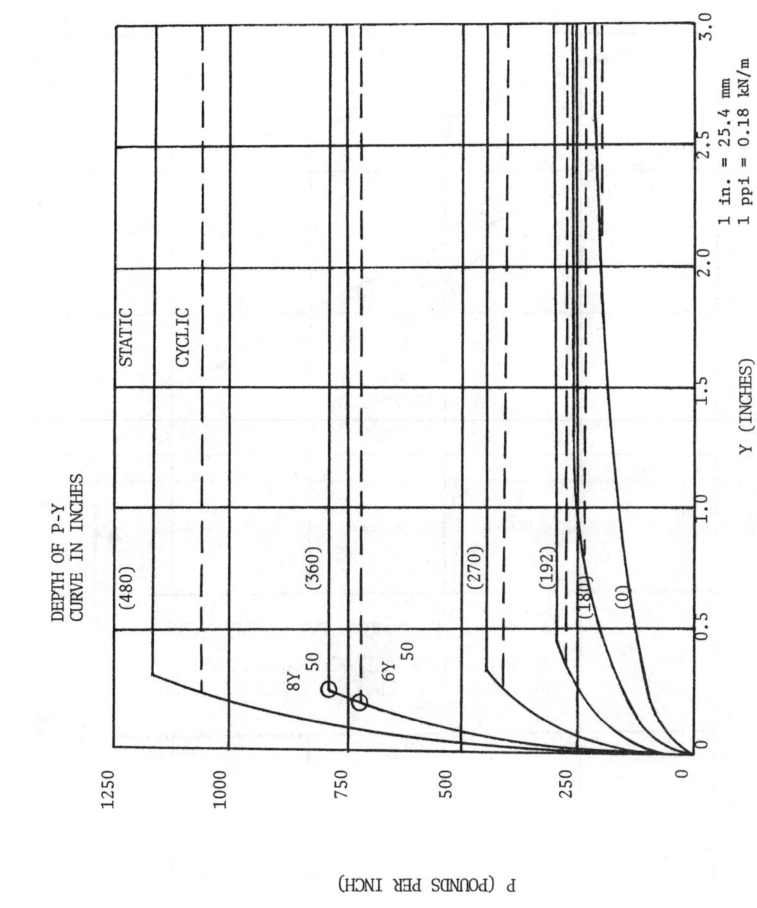

FIGURE 8. SHORT TERM STATIC AND CYCLIC P-Y CURVES.
(PILE WITH WINGS, TEST PILE B)

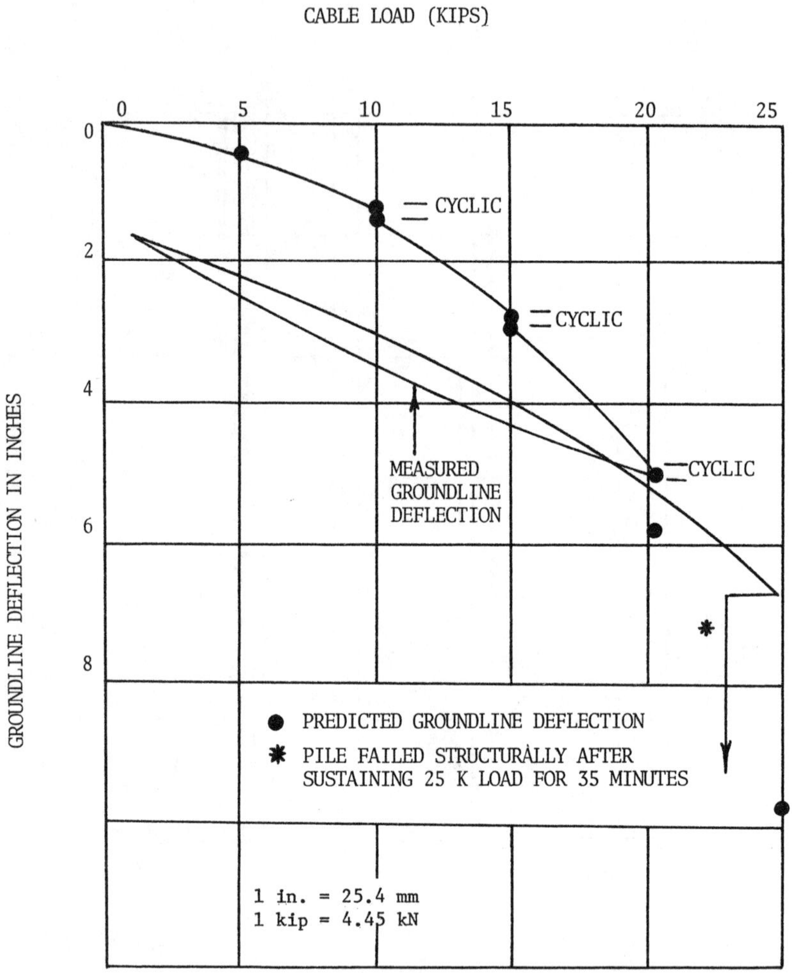

FIGURE 9. MEASURED AND PREDICTED GROUNDLINE DEFLECTION (PILE WITH WINGS, TEST PILE B)

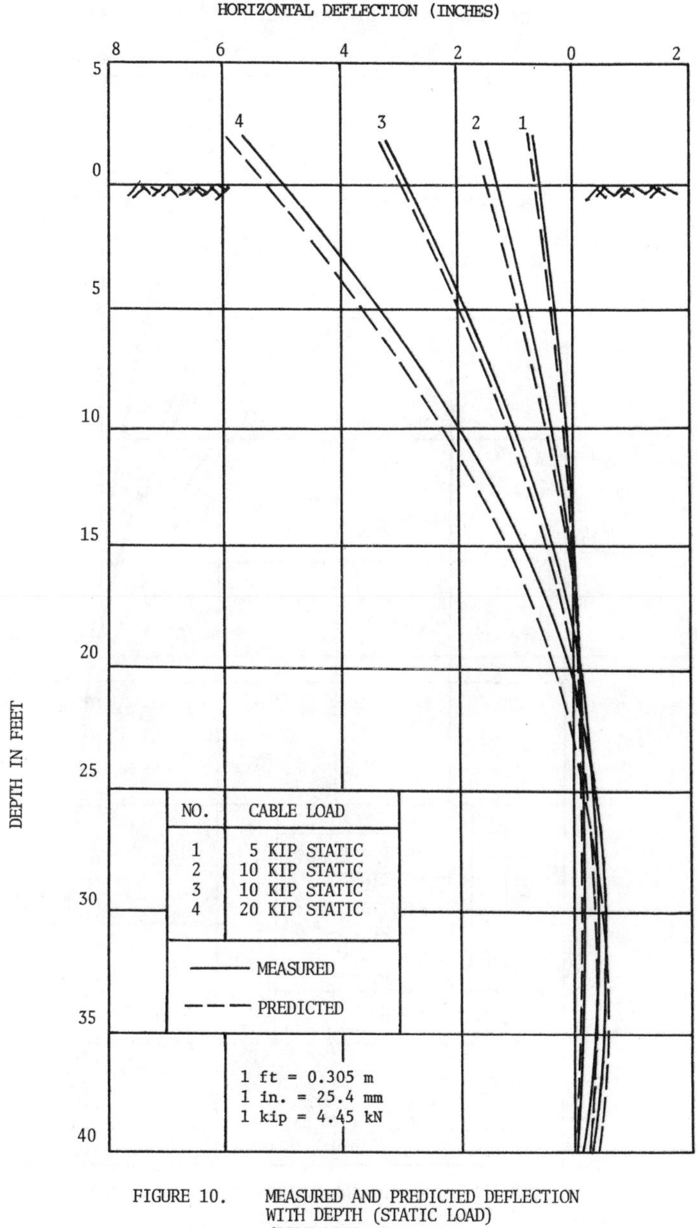

FIGURE 10. MEASURED AND PREDICTED DEFLECTION WITH DEPTH (STATIC LOAD) (PILE WITH WINGS, TEST PILE B)

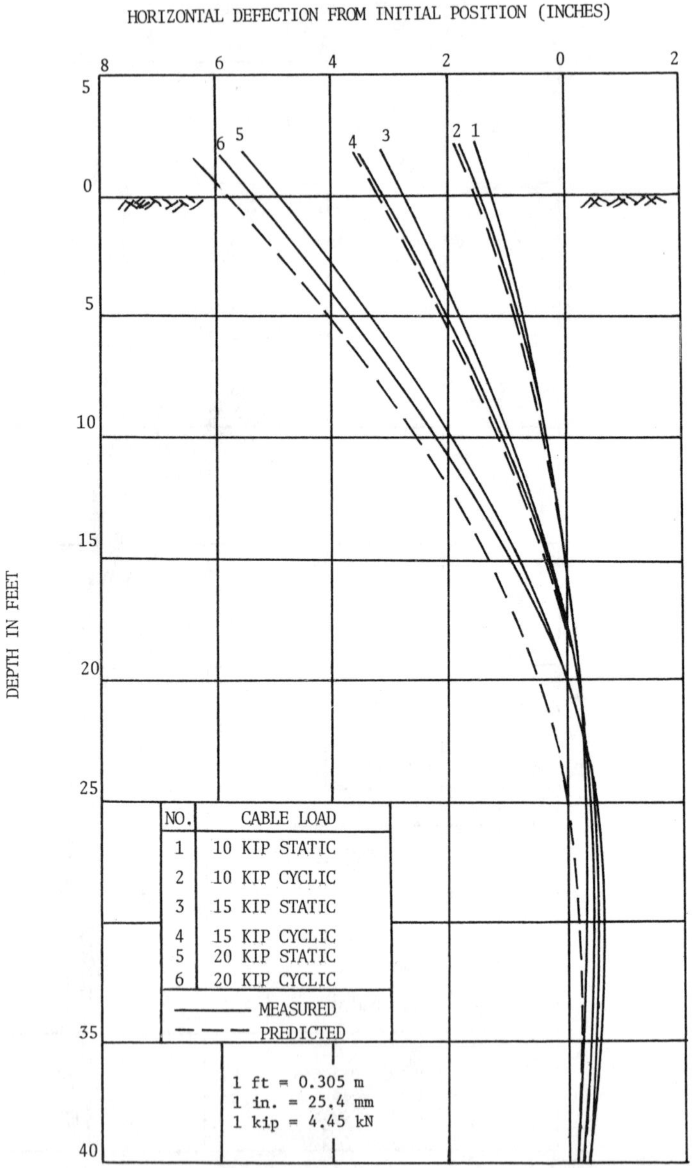

FIGURE 11. MEASURED AND PRECICTED DEFLECTION WITH DEPTH (CYCLIC)
(PILE WITH WINGS, TEST PILE B)

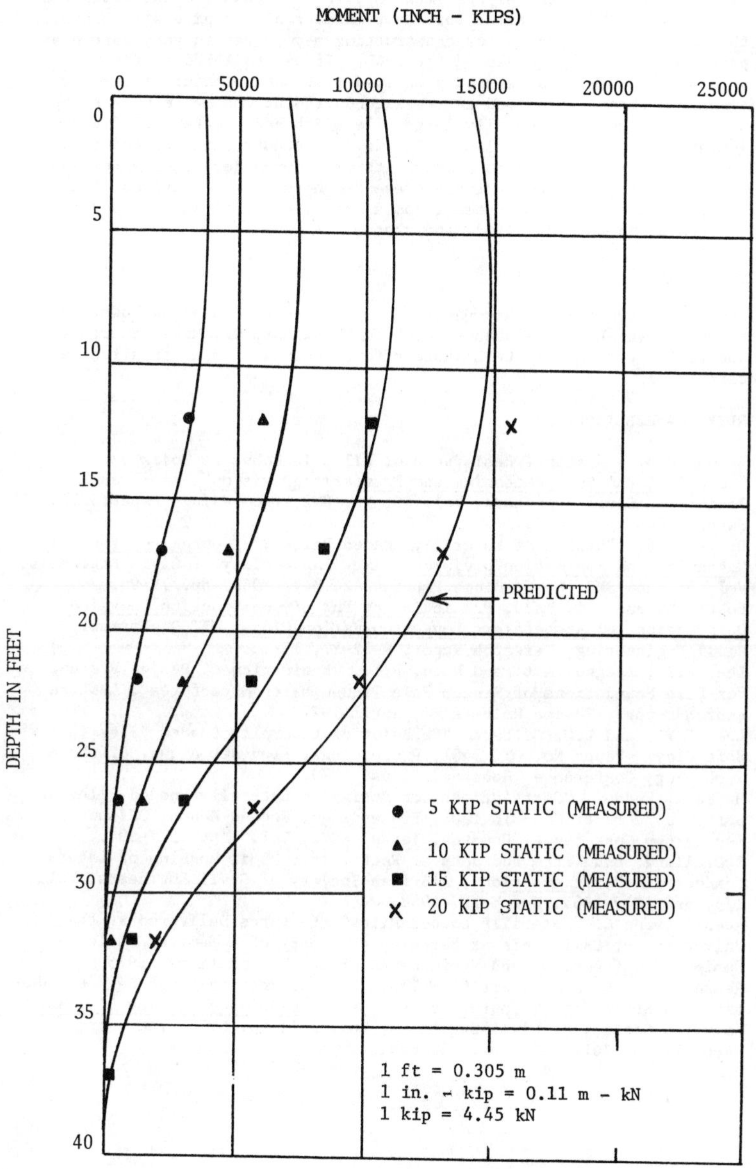

FIGURE 12. MEASURED AND PREDICTED MOMENT WITH DEPTH (PILE WITH WINGS, TEST PILE B)

Figure 12 shows a general agreement between the measured and predicted moments based on 100% EI values for the remainder of the pile with wings.

CONCLUSION. A new procedure for constructing p-y curves in very soft clay and peat was previously developed for LP&L. It was tailored to fit the measured performance of a tapered pile load test conducted at Chalmette, Louisiana. A second pile load test was conducted at Chalmette on a rocket shaped pile having substantially large pile width and greater stiffness, particularly near the ground line. The new p-y construction procedure was applied to this rocket pile to predict its behavior under lateral loading. The results of this second test are presented herein. Comparison of computed and measured lateral movement and moment with depth indicates very good agreement in both magnitude and shape.

ACKNOWLEDGEMENTS

The research presented in this paper was sponsored by Louisiana Power and Light Company (LP&L). The authors wish to thank both Mr. Mike Bauer, Jr. and the field personnel of LP&L whose efforts helped to make this test a successful one.

APPENDIX I - BIBLIOGRAPHY

1. Broms, B.B., "Lateral Resistance of Piles in Cohesive Soils", "Jounal of the Soil Mechanics and Foundation Division", American Society of Civil Engineers, Vol. 90, No. SM2, Proceedings Paper 3825, March 1964, pp. 27-63.
2. Broms, B.B., "Design of Laterally Loaded Piles", "Journal of the Soil Mechanics and Foundation Division", American Society of Civil Engineers, Vol. 91, No. SM3, Proceedings Paper 4342, May, 1965, pp. 79-99.
3. Foott, R. and C.C. Ladd, Prediction of End of Construction Undrained Deformations of Atchafalaya Levee Foundation Clays, MIT Department of Civil Engineering, Research Report R-72-27, 1972.
4. Lee, P.Y., Avegno, A.B. and Hunn, D.T., "Evaluation of Design Procedures for Pile Foundations of Single Pole Transmission Structures", LP&L research report, Tulane University, April 1977.
5. Lee, P.Y., and L.W. Gilbert, "Behavior of Laterally Loaded Piles in Very Soft Clay, Paper No. OTC 3401, Proceedings, Eleventh Annual Offshore Technology Conference, Houston, Texas, 1979.
6. Matlock, Hudson, "Correlations for Design of Laterally Loaded Piles in Soft Clay", Paper No. OTC 1204, Proceedings, Second Annual Offshore Technology Conference, Houston, Texas, 1970, Vol. 1 pp. 577-594.
7. McClelland, Bramlette and John A. Focht, Jr., "Soil Modulus of Laterally Loaded Piles", Transactions, American Society of Civil Engineers, Vol. 123, pp. 1049-1063, New York, 1958.
8. Reese, Lymon C. "Laterally Loaded Piles", Lectures Delivered at the University of California at Berkeley as a part of a Seminar Series--Design, Construction, and Performance of Deep Foundations, 1975.
9. Reese, Lymon C. and Robert C. Welch, "Lateral Resistance of Deep Foundations in Stiff Clay", Journal of the Geotechnical Engineering Division, American Society of Civil Engineers, Vol. 101, No. GT7, Proceedings Paper 11456, July 1975, pp. 633-649.

10. Reese, Lymon C., "Laterally Loaded Piles: Program Documentation:, Journal of the Geotechnical Engineering Division, American Society of Civil Engineers, Vol. 103, No. GT4, Proceedings Paper 12862, April, 1977, pp. 287-305.
11. Simmons, J.M., "Prelininary Design and Testing of a Nonrigid Swamp Foundation for Steel Poles", LP&L, 1971.
12. Sullivan, William, R., Development and Evaluation of Unified Method for the Analysis of Laterally Loaded Piles in Clay, University of Texas, Austin Department of Civil Engineering, Masters Thesis, 1977.
13. Terzaghi, Karl, "Evaluation of Coefficients of Subgrade Reaction", Geotechnique, Vol. 5, No. 4, December 1955, pp. 297-326.
14. United States Corps of Engineers, Geology of the Mississippi River Deltaic Plain, Southeastern Louisiana, Technical Report no. 3-483, Vol. 1, U.S. Army Engineer Waterways Experiment Station, Vicksburg, Mississippi, July,1958.

APPENDIX II - NOTATION

b = width of pile shaft

c = shear strength of soil at depth x

c' = 2.6 times undrained shear strength

EI = flexural rigidity of pile

P_x = axial load on pile

p = soil reaction per unit length

p_u = ultimate soil resistance per unit length of pile

S_u = undrained shear strength

x = depth from ground surface to p-y curve

x_r = depth from ground surface to transition (?)

y = lateral deflection of pile

y_{50} = lateral deflection of pile at one-half the ultimate soil resistance

ε = strain

ε_{50} = strain corresponding to one-half maximum principal stress difference

γ_t = total unit weight of soil

γ' = average effective unit weight of soil from ground surface to p-y curve

ω_η = moisture content

PREDICTION AND VERIFICATION OF FRICTION PILE BEHAVIOR

Michael W. Montgomery[1], Member ASCE

INTRODUCTION

In early 1976 a major expansion was planned for a large paper board mill in eastern Georgia. Proposed structures included a Board Machine, which is very sensitive to differential settlement, and several large and heavy High-Density Stock Chests. A thorough geotechnical investigation revealed that piles would be required for the Board Machine and Stock Chests. Piles would derive most of their capacity from side friction due to the firm sands through which they would be installed and in which they would terminate.

In order to allow use of the most economical, technically adequate type of pile, displacement-compatibility pile analyses were used to predict design capacity as a function of length for four types of piles--thin-shell mandrel-driven, closed-end pipe, steel H-sections, and augered cast-in-place. Realistic pile length predictions were particularly important, since lengths would vary for a given design capacity due to different efficiencies in transferring load along the pile shaft, and since a distinct bearing stratum did not exist. The designer requested that the geotechnical consultant estimate pile lengths for the various pile types as the basis for bidding by pile contractors. Obviously, realistic design lengths were required to assure a fair competitive basis for interested contractors and to achieve the most economical foundation.

On the basis of competitive bidding and the estimated lengths, augered cast-in-place piles were selected. Four full-scale load tests were performed on piles varying in length from 25 to 40 ft (7.6 to 12.2 m). Analysis of the test results revealed the rate of load transfer to the soil with depth and provided a method of verifying the displacement-compatibility analysis used in the design phase. Comparison of load test results and the analyses shows good agreement, as described in this paper.

PLANT FACILITIES

The expansion consists of a 660-foot-long (201 m) Board Machine Room with the Board Machine centered in it longitudinally, Finishing Area, Shipping and Storage, Stock Preparation, Pulp Mill, and various High Density Stock Chests (see Figure 1). Column loads for the buildings range up to 300 kips (1330 KN). The building columns are supported by shallow spread footings.

---

[1] Project Manager, Law Engineering Testing Company - Atlanta, Georgia

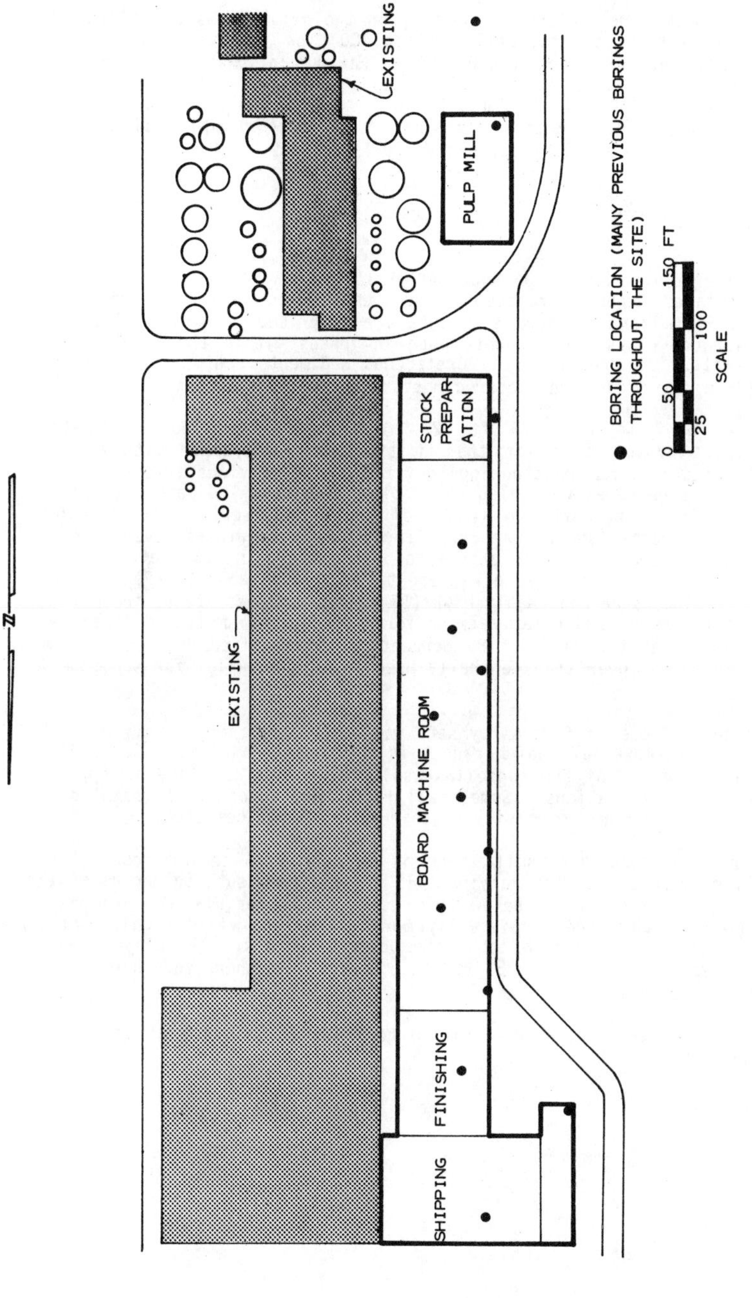

FIGURE 1
SITE PLAN (1 FT = 0.305 M)

The Board Machine is pile supported, on two column rows 24 ft (7.3 m) apart. Column loads are typically 200 to 400 kips (890 to 1780 KN) with maximum loads of 600 kips (2670 KN). Maximum allowable differential settlement is 0.2 in (0.5 cm). The pile-supported High Density Stock Chests are 70 ft (21 m) high and 42 ft (13 m) in diameter. The lower one-third of the Chests tapers to a bottom diameter of 26 ft (7.9 m). The contact pressure at the base level is approximately 7 ksf (340 KN/m2).

SUBSURFACE CONDITIONS

The site is located in the South Atlantic Coastal Plain. The soils are Undifferentiated Upper Cretaceous sediments (previously termed the Tuscaloosa Formation). These sediments are comprised primarily of clayey sands, although clays are present, both in lenses within the sand and in massive units. In the area, the Cretaceous sediments are in the range of 150 ft (46 m) thick and overlie Precambrian gneisses and schists.

Fifteen borings up to 75 ft (23 m) deep were drilled during the geotechnical investigation for the additions, in accordance with ASTM D 1586. The data from numerous borings drilled during previous investigations for existing structures were also available. The data indicated generally similar conditions beneath the length of the Board Machine and the High Density Stock Chests (see Figure 2). Clayey sand fill placed several years earlier to proposed site grade exists to depths averaging 10 feet. At the extreme south end of the area of interest the fill is absent. Penetration resistance values were reasonably high (15 to 29) and consistent, indicating controlled placement and relatively uniform compaction. Measured densities were, however, in the range of 85 percent of the Modified Proctor maximum density, which is lower than generally used in this material for structural fills.

A layer of loose to firm silty sand was encountered immediately beneath the fill. This material, which ranges in thickness from 3 to 5 ft (0.9 to 1.5 m), is comprised of the topsoil/subsoil layer left in place during the previous filling operations. Some small roots were found in the samples of this material; however, no heavily organic material was detected.

Beneath the topsoil/subsoil layer is a zone of materials consisting primarily of firm sands with varying silt and clay content, to depths of 12 to 25 ft (3.7 to 7.6 m). Below this zone, the materials are primarily sands; however, they are consistently more clayey and yield significantly higher penetration resistances than the upper materials. The clayey sands extend to depths of from 42 to 53 ft (12.3 to 16.2 m), where very hard sandy clays are encountered.

Groundwater varies in depth from 10 to 16 ft (3.0 to 4.9 m) below the ground surface.

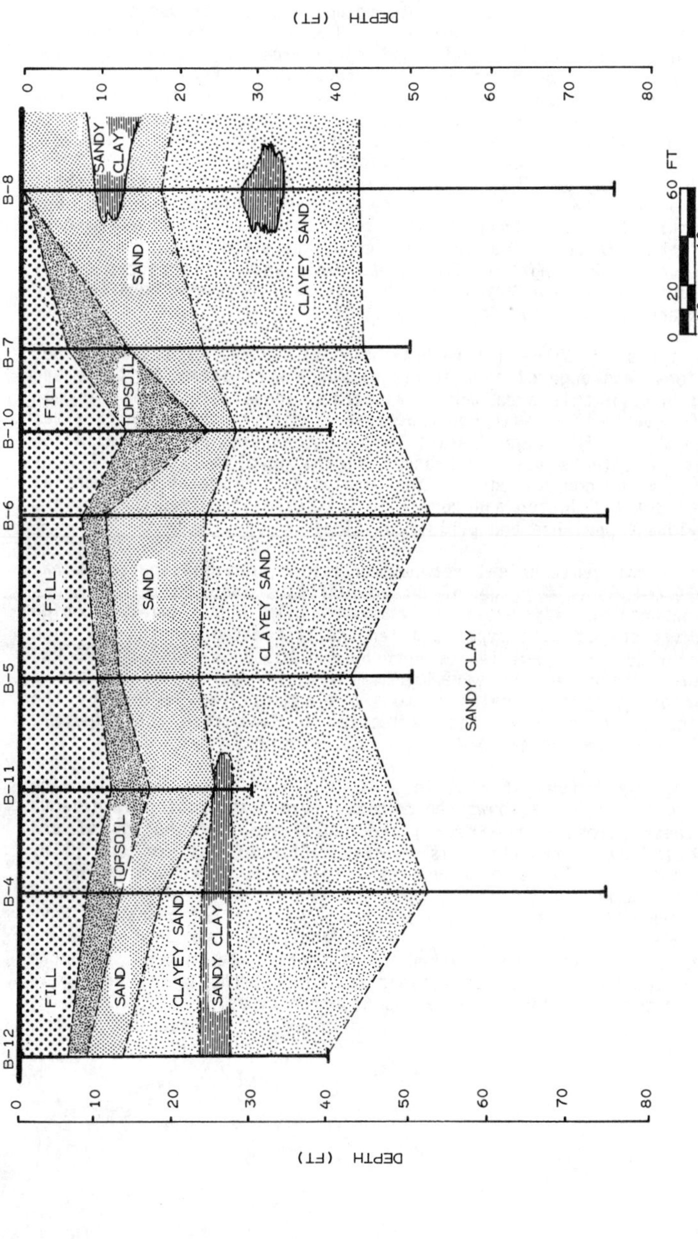

FIGURE 2
SUBSURFACE CONDITIONS (1 FT = 0.305 M)

FOUNDATION ANALYSES AND RESULTS

Conditions at the site were determined to be suitable for use of shallow spread footings for all building columns, on the basis of data from direct shear tests and consolidation tests and experience in the area. The sensitivity of the Board Machine to differential settlement and the magnitude of the Stock Chest loads dictated pile foundations for these structures. Direct shear data was used in a displacement-compatibility analysis (1) to analyze a wide variety of piles. Use of this analysis procedure results in computed load-deflection curves, utilizing a model which considers both end bearing and load transferred through shaft friction. Soil parameters utilized include not only ultimate strength, but the relationship between strength and deformation. In this manner, shortening of the pile under load is considered, as is the variation in soil deformation with depth along the pile shaft. This analysis produces curves of load vs. deformation for any point on a pile. Long-term behavior and group effects are not addressed by this analysis.

The four types of piles (thin-shell mandrel-driven, closed-end pipe, steel H-sections, and augered cast-in-place) which were deemed suitable and practical for use in this area were evaluated. These pile types vary in their rate of load transfer with depth because of differences in shaft size, shaft roughness, shaft shape, shaft modulus, taper, and installation procedure (as it affects normal stresses on the in-place pile). All of these factors were considered in the analysis. Parameters such as coefficient of shaft friction and normal soil stresses required assumptions based on previous experience and published values.

At the time the geotechnical report was issued, design capacities had not been selected, but a range of 50 to 100 tons (445 to 890 KN) was considered reasonable. By examining computed pile butt load-deflection curves for a variety of pile types and lengths, estimated lengths for each type of pile for design capacities of both 50 and 100 tons (445 and 890 KN) were selected. These lengths were minimum values required to achieve a safety factor of at least 2 relative to the ultimate load defined by the load-deflection curve and to limit deflection to less than approximately 0.25 in (0.63 m) at the design load.

Typical computed curves of butt load vs. deflection for two piles are shown in Figure 3. Figure 4 shows the computed load distribution with depth for one of these piles, demonstrating the negligible contribution of end bearing. This load distribution is representative of most of the piles analyzed. Estimated pile lengths presented to the designer are shown in Table I. This Table indicates the computed variation in pile length of various pile types for two design capacities. Similar relationships have been demonstrated on projects where similar analyses have been used and several pile types have been tested. It should be noted that two contractors proposing different pile types stated that they felt their pile length was unreasonably estimated relative to other types.

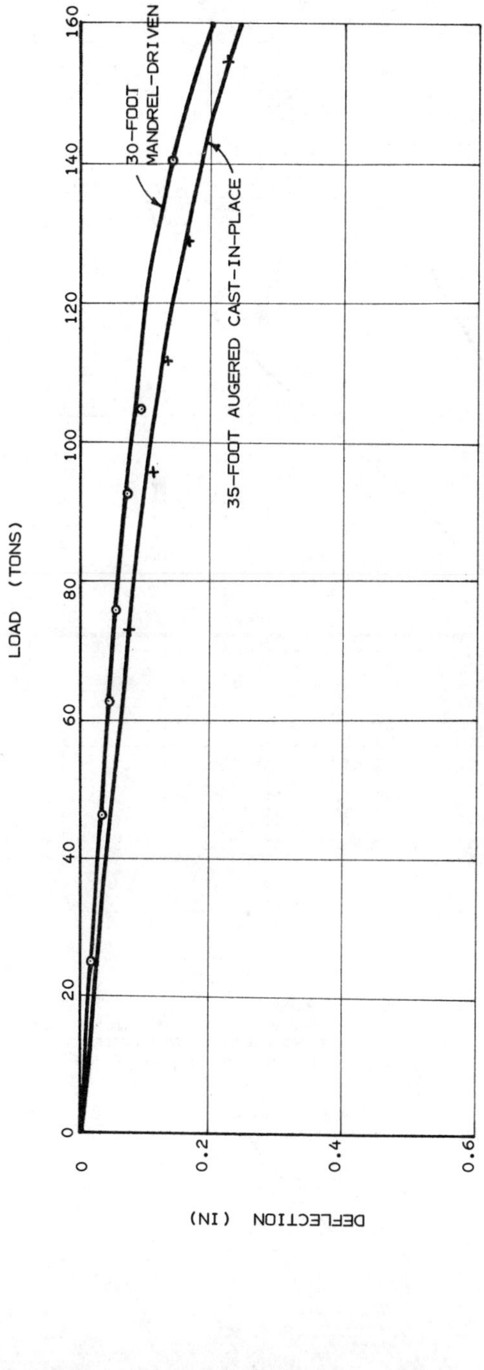

FIGURE 3
TYPICAL COMPUTED LOAD VS. DEFELCTION
CURVES (1 IN = 2.54 CM; 1 TON = 8.896 kN)

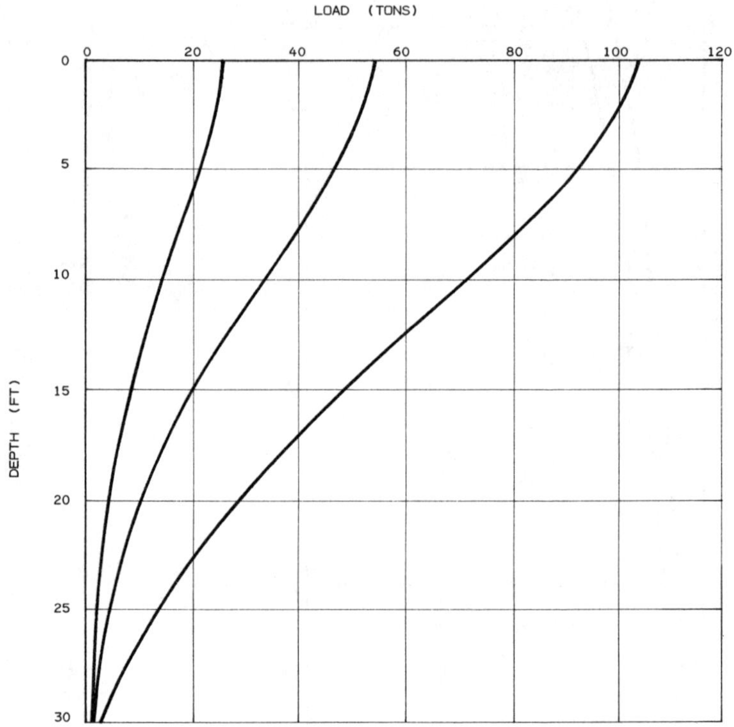

FIGURE 4
TYPICAL LOAD DISTRIBUTION WITH DEPTH
30-FOOT MANDREL-DRIVEN PILE
(1 FT = 0.305 M; 1 TON = 8.896 kN)

TABLE I - ESTIMATED PILE LENGTHS

	ESTIMATED PILE LENGTH (FT)		
PILE TYPE	50-TON *	100-TON *	75-TON **
Thin-shell Mandrel-driven cast-in-place			
9 1/2"	25	--	--
11 3/8"	--	35	--
10 3/8"	--	--	32
Augered cast-in-place			
14"	30	--	--
16"	--	40	37
H-Piles			
HP10X42	35	--	--
HP14X89	--	45	--
HP12X53	--	--	39
Pipe			
12", 3/8" wall	40	--	--

* Estimated by geotechnical consultant

** Estimated by AE by interpolation

Note: 1 ton = 8.896 kN; 1 ft = 0.305 m; 1 in = 2.54 cm

During final design, a 75-ton (670 KN) design capacity was selected by the designer. The piles were designed for negligible lateral and uplift loads with the exception of 36 tension piles (25-ton or 220-KN capacity) and 18 battered piles. Because 75-ton (670-KN) piles had not been addressed specifically in the geotechnical report, the designer interpolated from the first two columns of Table I to determine pile quantities on which to receive bids (last column of Table I). Such interpolation was a reasonable approach, but it was complicated by choice of intermediate pile sizes for the thin-shell mandrel-driven piles and steel H-sections. Closed-end piles were eliminated from consideration because of the geotechnical consultant's recommendation that they be limited to a 50-ton (445 KN) design capacity. Bids were based on those lengths for 1440 piles for the Board Machine only. Bid quantities therefore varied from 45,600 to 56,520 ft (13,910 to 17,250 m). Augered cast-in-place piles (intermediate bid quantity) were selected on the basis of the bidding. Inspection of Table I indicates that the estimated length for 16-inch (41 cm) augered cast-in-place piles should have been approximately 30 ft to 33 ft (9.1 m to 10.0 m).

TEST PILE PROGRAM

At this stage of the project, the designer contacted the geotechnical consultant so that a full-scale test pile program could be jointly developed.

Four full-scale load test were specified. This number exceeds that which would have been required for driven piles, since installation of augered piles was based solely on a depth criterion (no driving record). Because subsurface conditions are relatively uniform over the length of the Board Machine, pile lengths ranging from 25 to 40 ft (9.6 to 12.2 m) were selected. In this manner, the maximum amount of information on pile behavior (rate of load transfer with depth and relationship between load and deformation) could be obtained, and the minimum length required for adequate safety at the design load could be determined. This approach required the assumption that subsurface conditions over the length of the Board Machine are constant. This was considered a reasonable assumption within practical limits, on the basis of the detailed geotechnical investigation.

The test piles were installed by the equipment and operators to be used in the production piling operation, and in the same manner. A continuous flight auger was rotated into the ground to the specified depth. Cement-sand grout was pumped down the hollow auger stem, ejecting a plug at the bottom prior to auger withdrawal. Grouting continued as the auger was withdrawn, at a rate such that the grout was always in contact with the bottom of the auger.

Large steel beams spanning between two sets of three reaction piles at each test location were used as reactions for the hydraulic jacks which applied the test loads. The tests were performed essentially by the procedure described by ASTM D 1143-75, "Standard Method of Testing Piles Under Axial Compresssive Load". The Standard Loading Procedure was followed with three exceptions:

1. Unloading decrements were maintained for one-half hour.

2. The test load was maintained on Test Pile C for only 21 hours.

3. Test Piles B and C were loaded to only 1.8 times the design load because of limited jack capacity.

Loads were determined from hydraulic gauges on the jacks and recent calibration curves. Deflections were measured at the top of each pile by a pair of dial gauges graduated to 0.001 in (0.002 cm) and by a pair of wire-mirror-scale devices.

TEST PILE RESULTS

Test results are shown in graphical form on Figure 5. The vertical line at the test load represents movement in the interval between one hour after application of the final increment and 24 hours. In each case, the rate of movement was negligible at the end of the full test load increment. The unloading cycle is not shown. In each case except the 25 ft (7.6 m) pile, approximately half the total movement was recovered on unloading.

The relationship between the four test piles is essentially that expected, with the longer piles showing less deflection at a given load. The single exception is the 40 ft (12.2 m) pile whose behavior was similar to that of the 35 ft (10.7 m) pile, and not better as might be expected. This minor discrepancy may be attributable to a delay during installation caused by malfunctioning equipment, which resulted in an open hole for 30 minutes, which was subsequently reaugered. Disturbance due to stress reduction and groundwater effects probably occurred.

All piles except Test Pile D (25 ft or 7.6 m long) show essentially linear load-deflection relationships to 1.8 or 2 times the design load. Except for Test Pile D, deflections at the design load were 0.15 in (0.38 cm) or less, and at the full test load were 0.42 in (1.1 cm) or less. All three of these piles recovered about one-half their deflection on unloading. Test Pile D had a permanent net set of almost 0.9 in (2.3 cm), and showed indications of impending failure if loaded to slightly more than 150 tons (1330 KN), although the rate of deflection at the full test load was negligible at 24 hours.

The test pile program gave results which were in good agreement with pile behavior anticipated on the basis of the displacement-compatibility analysis. All piles except Test Pile D exhibited satisfactory performance for a 75-ton (670 KN) design capacity. Test Pile D was marginal as a 75-ton (670 KN) pile. Since there are no driving resistance records for augered cast-in-place piles to supplement other installation criteria (in this

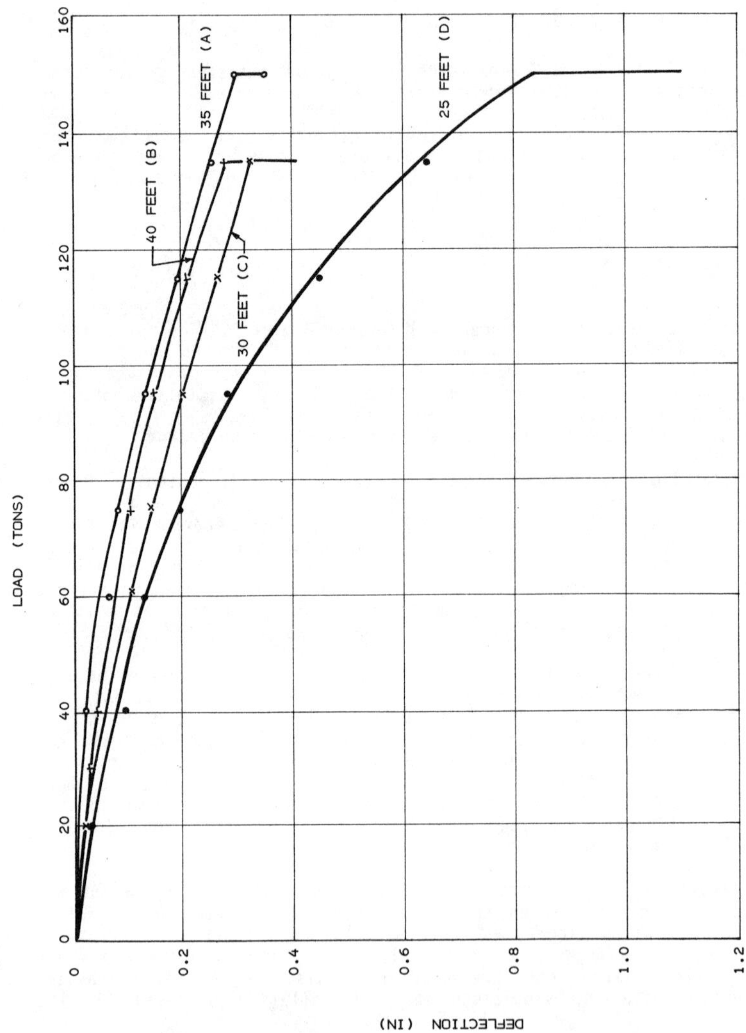

FIGURE 5
LOAD TEST RESULTS (1 IN = 2.54 CM; 1 TON = 8.896 kN)

instance, depth), a 25-foot (7.6 m) depth could not be recommended. A 30-foot (9.1 m) installation criterion was recommended with confidence.

COMPARISON OF COMPUTED AND MEASURED PILE BEHAVIOR

An indication of the good agreement between computed and measured pile behavior is the fact that the final installation depth criterion is the same as that which was anticipated during the geotechnical investigation. Although 75-ton (670 KN) piles were not specifically considered during the investigation, examination of Table I demonstrates that for a 16-inch (41 cm) pile, 30 to 33 ft (9.1 to 10.0 m) would be anticipated as the design length. Further evidence of the validity of the analysis is shown by Figure 6, which superimposes the measured butt deflection vs. load for a 35-foot (10.7 m) pile on that computed for a 16-inch (41 cm) by 35-foot (10.7 m) pile during the geotechnical investigation. Reasonably good agreement is evident up to twice the design load, and deflections at the design load are extremely close.

CONCLUSIONS

Several conclusions can be drawn from this case history:

1. For projects where friction piles will be utilized, reasonable estimates of length are critical to selection of the optimum pile length (shortest length that will have adequate safety factor against failure and will exhibit tolerable deflection at the design load).

2. Where several types of friction piles are technically feasible, close estimates of their depth are very important in preparation of bid documents. Different piles may require different lengths for the same capacity due to differences in rate of load transfer with depth. This makes it unreasonable to have every contractor use the same bid quantity. Therefore, a fair selection process demands close depth estimates to avoid unfairly penalizing a particular pile type.

3. The displacement-compatibility solution utilized in this project is shown to predict the behavior of augered cast-in-place piles at this site quite closely. By inference, the predicted lengths of other types of piles are thought to be reasonable, thus establishing an appropriate relationship between pile lengths for this particular project. Such predicted relationships have been verified by more extensive monitoring and testing of various pile types at other projects.

4. The approach utilized is a practical analytic tool which requires no more extensive data than that required for a conventional static analysis (although strength tests must consist of direct shear tests). By allowing rational consideration of pile shortening and stress-strain characteristics of the soil, the behavior of a pile can be described analytically, and agreement between observed and predicted performances is improved.

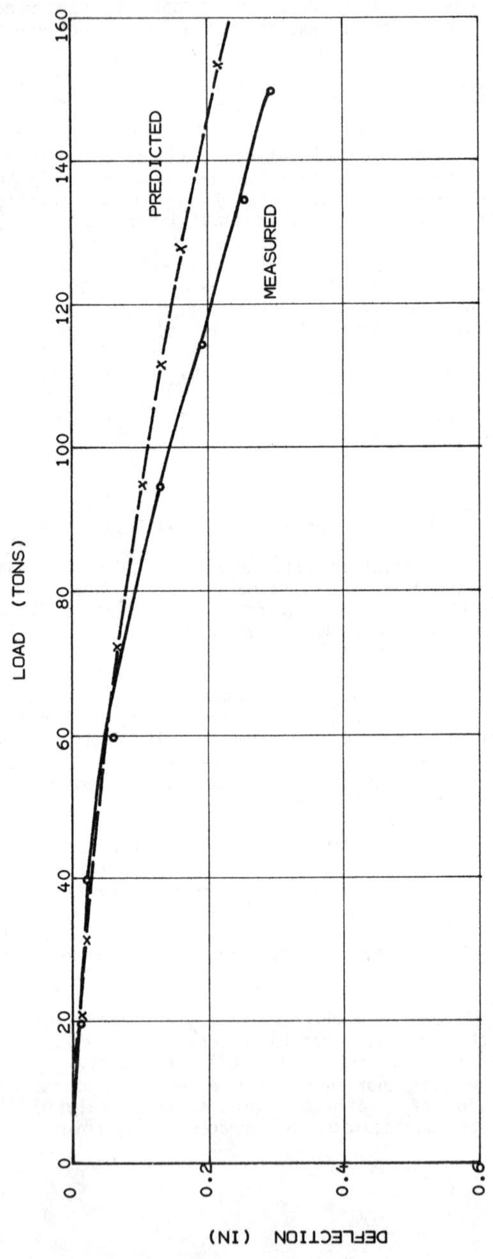

FIGURE 6
COMPARISON OF PREDICTED AND MEASURED
BEHAVIOR OF 35-FOOT PILE
(1 IN = 2.54 CM; 1 TON = 8.896 kN)

REFERENCES

1. Montgomery, M.W., "Pile Load Tests to Evaluate Load Transfer Mechanisms", Behavior of Deep Foundations, ASTM STP 670, R. Lundgren, ed., 1979.

CAPACITIES OF STEEL H PILES SUPPORTED IN SHALE

By Larry G. O'Dell,[1] M. ASCE, Alan N. Stilley,[2] A.M. ASCE

Introduction

The Powder River Basin of Northeast Wyoming and Southeast Montana contains the largest deposits of coal in the United States, in terms of economically mineable reserves. This Basin extends from Douglas, Wyoming, north to Glendive, Montana, and is about 100 miles (161 km) wide. Development of western coal resources has stimulated the mining activity. Campbell County Wyoming is known as the coal capital of the world, where seven new mines have begun operations within the past four years. Most of the coal is mined, processed, and shipped by rail to coal fired steam generating plants in the Midwest or South.

The Carter Mining Company has developed two mines near Gillette, Wyoming. This paper discusses the Caballo Project, where the coal processing and storage facilities consist of:

- Truck Dump-Hopper
- Primary Crusher
- Conveying and Sampling System
- Train Load-out Facility
- Two Concrete Storage Silos

The critical structures are the two concrete coal storage silos, 70 feet (21.35 m) in diameter and 200 feet (61 m) high. They each store 12,500 tons of coal, and have a total weight of 25,000 tons, (1 ton = 907.2 kg).

Plant facilities, for economic reasons, are not located over mineable coal reserves. Structures are generally located in valleys where the coal has been eroded or on deposits of baked shale, (areas where the coal has burned). Foundation conditions are generally poor, and deep foundation systems are normally required.

This paper provides a case history of the steel H pile foundation for the coal storage silos at the Caballo Mine.

Purpose

During the geotechnical investigations, the normal field and laboratory tests were made, including coring the bedrock stratum and performing strength and compressibility tests. Unconfined compression and unconsolidated-undrained triaxial shear tests were performed in the laboratory, and pressuremeter tests were performed during the field investigations.

Prior to construction, 16 test piles were driven with a Raymond Model 3/0 single acting steam hammer, with a rated energy of 40,625 ft-lbs. (55,088 joules), and monitored using the Case-Goble method of dynamic pile analysis. This method of analysis consists of measuring the force and acceleration during actual driving. Transducers and accelerometers transmit signals corresponding to force and acceleration into a field analyzer. Calculations

[1]Vice-president, Northern Testing Laboratories, Inc., Billings, MT.
[2]Geotechnical Area Manager, Northern Testing laboratories, Inc., Billings, MT.

are made and a printed record of the predicted ultimate bearing capacity and maximum compressive force is made.

Two piles were load tested in accordance with ASTM D1143, the cyclic method. The ultimate pile load capacity for 12 by 53 steel H piles driven into the shale stratum is compared using laboratory strength data and the Dynamic Analyzer. Calculated values are then compared with the pile load test results.

Material Properties

Geology and Material Characteristics - The mine site is located within the Wasatch Formation, which consists of interbedded layers of shale, sandstone and coal. Typically, a thin layer of alluvial or residual soil overlies the shale or sandstone.

In many locations, a material unique to this area, termed locally as "baked shale", is present. This material results when interbedded layers of coal ignite and bake the overlying shales and sandstones. The heat process changes the shale characteristics. It becomes a hard, granular porous material, with properties similar to angular gravel. After the coal has burned, the overlying materials collapse into the resulting void. This causes the deposit to have a very loose consistency, with many voids. Support of structure foundations directly in the baked shale is normally not feasible due to the loose consistency of the material and the presence of large voids. For large structures, it is necessary to transmit the loads through the baked shale to the unburned materials below. At some locations, where the heat from burning was very high, large "clinkers", or fused materials, often several feet in diameter, are formed. These clinkers make drilling or driving piling through the deposits extremely difficult. Installation of drilled caissons through the baked shale is nearly impossible, and non-displacement piling that can withstand difficult driving conditions are best suited. Steel H piles satisfy both of these criteria, and are used extensively.

The shales and sandstones are "soft rock" materials which typically show a weathered zone several feet thick near the surface or contact. After this zone is penetrated, the materials are normally competent and capable of supporting moderately high loads.

The topography consists of low rolling hills, separated by wide shallow drainages. The baked shale typically caps the hills at the higher elevations.

Subsurface Profile

Eleven test borings were made at the silo site to establish a typical subsurface profile. After final grading was completed, the profile consisted of clay, (CL), extending from near the surface to a maximum depth of 13 feet (4.0 m). Below the clay, baked shale extended to depths of about 16 to 25 feet (4.9 to 7.6 m). At several locations, the clay was not present and baked shale was exposed at the surface. Below the baked shale, sandstone or shale was encountered, and extended to the depth of the test borings, over 100 feet (30.5 m).

A typical subsurface profile through the silo area is shown in Figure 1.

Laboratory Test Results - Laboratory tests which were performed on the relatively undisturbed core samples of sandstone and shale, included one-dimensional consolidation tests, unconfined compression tests, and unconsolidated undrained, triaxial compression tests. In an attempt to simulate the load-unload cycles the silos would experience, a portion of the consolidation and triaxial tests were subjected to cyclic loading. The results of the laboratory strength tests are summarized in Table 1.

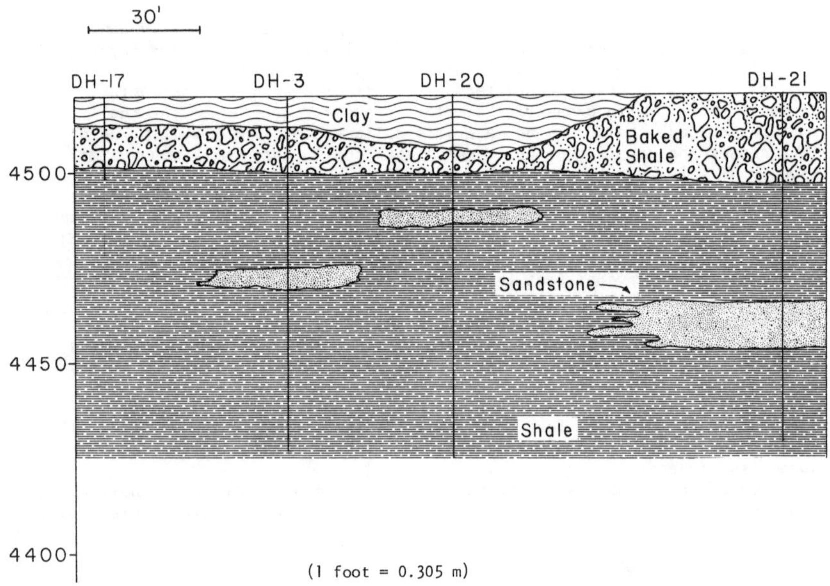

FIGURE 1 - Subsurface Profile, Silo Site

TABLE 1: Summary of Laboratory Strength Properties

Material Type (1)	Depth of test, feet (2)	Type of Test (3)	Moisture Content, percent (4)	Unit Dry Weight, pounds per cubic foot (5)	Test Results (6)
Shale	29.4 - 29.9	Unconfined Compression	11	127	qu = 12 ksf
Shale	45.3 - 45.8	Unconfined Compression	12	125	qu = 36 ksf
Shale	78.0 - 78.5	Unconfined Compression	14	119	qu = 13 ksf
Shale	78.5 - 79.0	Unconfined Compression	13	119	qu = 10 ksf
Shale	45.4 - 45.9	Unconfined Compression	13	118	qu = 47 ksf
Shale	53.5 - 54.0	Unconfined Compression	12	122	qu = 16 ksf
Shale	33.7 - 34.2	Unconfined Compression	2	148	qu = 470 ksf
Shale	43.0 - 43.5	Unconfined Compression	9	132	qu = 77 ksf
Shale	70.0 - 70.5	Unconfined Compression	9	138	qu = 95 ksf
Shale	75.5 - 76.0	Unconsolidated Undrained Triaxial Test, confining pressure = 15 ksf	11	125	Maximum Deviator Stress = 75 ksf
Shale	76.0 - 76.5	Unconsolidated Undrained Triaxial Test, confining pressure = 11 ksf	10	126	Maximum Deviator Stress = 68 ksf
Shale	99.5 - 100.0	Unconsolidated Undrained Cyclic Triaxial Test, confining pressure = 12 ksf	17	112	Maximum Deviator Stress = 17 ksf
Shale	37.7 - 38.2	Unconfined Compression	13	123	qu = 31 ksf
Shale	48.5 - 49.2	Unconfined Compression	10	129	qu = 15 ksf
Shale	43.5 - 44.0	Unconsolidated Undrained Triaxial Test, confining pressure = 12 ksf	8	137	Maximum Deviator Stress = 130 ksf
Shale	44.0 - 44.5	Unconsolidated Undrained Triaxial Test, confining pressure = 18 ksf	8	137	Maximum Deviator Stress = 98 ksf
Shale	45.0 - 45.5	Unconsolidated Undrained Triaxial Test, confining pressure = 18 ksf	8	132	Maximum Deviator Stress = 118 ksf
Shale	45.5 - 46.0	Unconsolidated Undrained Triaxial Test, confining pressure = 12 ksf	9	131	Maximum Deviator Stress = 95 ksf
Shale	65.5 - 66.3	Unconsolidated Undrained Triaxial Test, confining pressure = 6 ksf	11	126	Maximum Deviator Stress = 58 ksf

NOTE: 1 ft. = 0.305 m, 1 ksf = 4790 N/m^2, 1 pcf = 16.01 kg/m^3

Examination of the laboratory strength test results indicates a wide range of values. Visual examination of the test specimens indicated that many of the sample strengths were controlled by "slickensides", or planes of weakness in the shale. Slickensides are planes along which, apparently, movement has occurred in the past. These joints or planes of weakness occur at random locations and generally do not have any specific orientation.

Strength parameters used in design were selected by using typical values and disregarding tests that were apparently low due to incipient fractures.

CALCULATED PILE CAPACITIES

Pile load capacities were calculated for standard H pile sections by three methods. First, using static design methods and the strength properties determined from the laboratory tests, the depth of penetration into the bedrock required to develop maximum load capacity was calculated for various pile sections using the following formula:

$$Q_{ult} = Q_b + Q_s \quad (1)$$

where Q_{ult} = ultimate pile capacity, Q_b = base resistance, Q_s = shaft resistance.

For the shale stratum, controlled primarily by cohesion, the equation is rewritten as

$$Q_{ult} = CN_cA_b + Cpl \quad (2)$$

where C = cohesion, N_c = bearing capacity factor, A_b = area of pile base, p = pile perimeter, l = length of pile penetration into shale.

The average unconfined compressive strength is about 34 kips per square foot. Using a cohesion value of 17 ksf (1 ksf = 47,900 N/m^2), pile capacities are summarized in Table 2.

TABLE 2: Summary of Calculated Pile Capacities

Steel H Pile Section (1)	Penetration into Bedrock, feet (2)	Allowable Pile Capacity, Tons Safety Factor = 3 (3)
10 x 57	5	64
10 x 57	10	111.5
12 x 53	5	82
12 x 53	10	138*

*Exceeds allowable working stress in pile.
NOTE: 1 ft. = 0.305 m, 1 ton = 8.896 kN

In order to achieve 100 ton (890 kN) working load piles, 10 by 57 sections would require about 10 feet (3.0 m) of penetration and 12 by 53, about six to seven feet (1.8 - 2.1 m).

Second, pile driving criteria was computed from driving formulas, Table 3. These included the Hiley equation, the Janbu equation, and the Engineering News formula. These methods required the assumption of a pile-hammer energy, final driving resistance, and cushion characteristics. Final driving resistances and cushion characteristics were selected from values recommended in the literature (Ref. 2). Pile-hammer energy required to develop the maximum load capacity of the piles was then calculated.

TABLE 3: Summary of Pile Driving Formula Results

Type Pile (1)	Hammer Energy Foot-Pounds (2)	Hammer Type (3)	Final Blows/ inch (4)	Equation (5)	Factor of Safety (6)	Allowable Load, Tons (7)
10 x 57	26,300	Link Belt 520	20	Hiley	3	67
10 x 57	42,000	Vulcan 014	20	Hiley	3	90
10 x 57	41,000	Raymond 3/0	20	Hiley	3	85
10 x 57	26,300	Link Belt 520	30	Hiley	3	70
12 x 53	26,300	Link Belt 520	20	Hiley	3	60
12 x 53	42,000	Vulcan 014	20	Hiley	3	80
12 x 53	41,000	Raymond 3/0	20	Hiley	3	75
12 x 53	26,300	Link Belt 520	30	Hiley	3	62
10 x 57	32,500	Vulcan 010	20	Janbu	3	28
10 x 57	42,000	Vulcan 014	20	Janbu	3	50
Any	30,000	Any	20	EN	6	200
Any	40,000	Any	20	EN	6	266

NOTE: 1 ft-lb. = 1.356 joules, 1 in. = .0254 m, 1 ton = 8.896 kN

Third, pile capacities were calculated from results of the field pressuremeter tests. This test is performed in a drill hole and is a measurement of volume change with changes in pressure. The test results are normally presented as a plot of pressure versus volume change, with corrections applied for the test apparatus characteristics. From this plot, it is possible to determine the lateral earth pressure at rest, the preconsolidation pressure, a creep pressure, which is the proportional limit of the material, and a limit pressure, which is the theoretical point at which shear failure occurs. From the straight line portions of this plot, a deformation modulus of the soil or rock can be calculated. Special procedures have been derived for the design of deep foundation systems using pressuremeter data.

The bearing capacity Menard (Ref. 3) uses for a concrete pile foundation is:

$$Q_{ult} = P_o + K(P_l - P_o)A_b + \frac{P_l - P_o}{5.5}\, pl \qquad (3)$$

Q_{ult} = ultimate pile capacity
P_o = lateral earth pressure at rest
K = a coefficient based on soil type and size of pile
P_l = limit pressure of soil
A_b = area of the base
p = perimeter
l = length of pile penetration into the shale.

Pile capacities from the three methods are presented in Tables 3 and 4.

The problem with specifying high energy hammers and hard driving resistance is that piles may be overstressed during installation. M. T. Davisson (Ref. 1) lists general criteria for matching hammer energy, driving resistance and pile type from wave equation analysis. Typical examples of these criteria for 12 by 53 steel H piles are presented in Table 5.

Dynamic Analysis - To aid in piling evaluation, the Case-Goble method of dynamic pile testing was performed on 16 piles (12 x 53 steel H) during the load testing program. These studies were performed by Soil Exploration Company (Ref. 4), and the results of the test pile analysis are shown in Table 6, which contains the pile number, depth of penetration below grade, maximum compressive force in the pile, final driving resistance and predicted ultimate static bearing capacity.

TABLE 4: Summary of Pile Capacities

Method (1)	Pile Section (2)	Safe Pile Capacity, tons (3)	Penetration into Bedrock, feet (4)	Hammer Size, foot-pounds (5)	Calculated Driving Resistance, Blows/inch (6)
1. Static based on laboratory tests	10x57 12x53	100 100	10 6	40,000 40,000	20 20
3. Pressuremeter	12x53	100	6		

NOTE: 1 ton = 8.896 kN 1 ft. = 0.305 m, 1 ft-lb. = 1.356 joules, 1 in. = .0254 m

TABLE 5: General Criteria for Hammer Energy and Driving Resistance

Pile Type (1)	Point Resistance (2)	Maximum Hammer Energy, foot-pounds (3)	Final Blows/inch (4)	Allowable Load, tons Factor of Safety=2 (5)	Pile Length feet (6)	Driving Stress, kips/square inch (7)	Yield Stress of Steel, kips/square inch (8)
12x53	100%	42,000	20	108	80	38	36
12x53	100%	32,500	20	111	20	36	36
12x53	100%	42,000	20	123	20	40	36

NOTE: 1 ft-lb. = 1.356 joules, 1 in. = .0254 m, 1 ton = 8.896 kN 1 ft. = 0.305 m, 1 ksi = 6900 kN/m^2

The longer test piles were typically driven to a higher final driving resistance than the 25 foot long (7.6 m) test piles. Higher forces in the shorter piles can be partially attributed to less energy loss due to bending. When driving the 40 foot (12.2 m) piles, about 20 feet (6.1 m) of piling extended above grade. This portion was free to deflect slightly during driving, thus absorbing some hammer energy.

The predicted ultimate pile load capacity varied from 150 to 250 tons (1.34 - 2.22 MN). Only two piles had calculated capacities less than 200 tons, which was twice the design working load. The two piles which were less than 200 tons were damaged during driving.

From the dynamic testing, ultimate capacities of 200 tons were generally achieved, and could be used for construction. This test program also indicated with the pile hammer and pile section selected, production piles would require close supervision to prevent overdriving and damage. The shorter pile lengths tended to be more susceptible to damage. Damage, when it occurred, was evident as bending of the flanges aboveground.

TABLE 6: Dynamic Analyzer Results

Pile Number (1)	Depth Below Grade, feet (2)	Depth to Shale, feet (3)	Maximum Compressive Force, kips (4)	Final Driving Resistance, Blows/inch (5)	Initial Pile Length, feet (6)	Predicted Ultimate Static Bearing Capacity, tons (7)
D-7	16.5	15	560	9	40	200
D-2	17.8	16	550	16	40	---
A-7	17.2	15	525	9	25	---
A-3	17.0	16	560	10	25	---
C-5*	17.1	16	570	10	40	250
B-36	21.5	15	545	8	25	230
A-37	20.8	16	515	8	40	235
D-43	21.5	16	480	8	40	175**
C-57*	20.0	15	590	11	25	230
B-89	19.2	14.5	530	9	40	210
B-82	14.3	14.5	570	8	25	210
B-75	18.5	13	500	4	25	150**
B-29	21.0	15	535	5	25	245
B-43	23.2	18	485	6	25	200
B-68	17.8	14	550	5	25	225
B-20	15.5	13	560	8	25	---

*Static Load Test.
**Damaged Pile
NOTE: 1 ft. = 0.305 m, 1 kip = 4.45 kN, 1 in. = 0.0254 m, 1 ton = 8.896 kN

Comparison of Calculated Capacities with Load Test Results

Load Test Results - Two of the test piles which were instrumented with the Dynamic Analyzer were selected for static load tests. For each of these piles, a system of eight anchor piles was driven around the test pile in order to apply the load to the test pile. All of the test piles were 12x53 steel H piles, driven without tip protection using the Raymond 3/0 hammer. An aluminum micarta cushion was used in the hammer. Loads were applied with a hydraulic jack system, and deflections were measured with a system of dial gauges, attached to an independent reference frame. The loading procedure used was in accordance with ASTM D1143, the standard or slow maintained load procedure. The load-settlement curve for each of the test piles is shown in Figures 2 and 3.

From the load settlement curve for test piles 1 and 2, it can be seen that the piling supported loads of twice the design capacity, without excessive deflection. Test pile C-5 was not actually failed. At a load of about 240 tons (2.14 MN), the anchor piles began pulling out of the ground. Test pile C-57 failed at a load of about 230 tons, where continued load application resulted in excessive settlement. At the maximum load of 230 tons, the stress in the steel is about 30 ksi (1 ksi = 6900 kN/m^2). This value is approaching the yield strength of the steel.

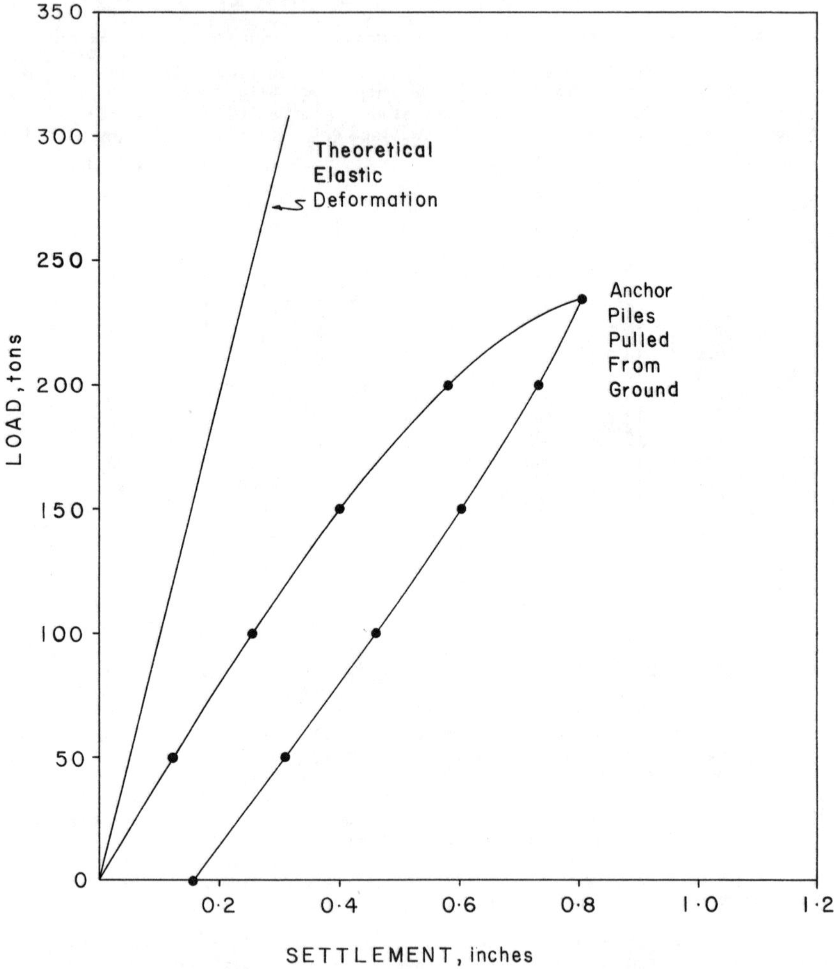

(1 ton=8.896 kN, 1 inch=0.0254 m)

FIGURE 2 - Load Settlement Curve for Test Pile C-5

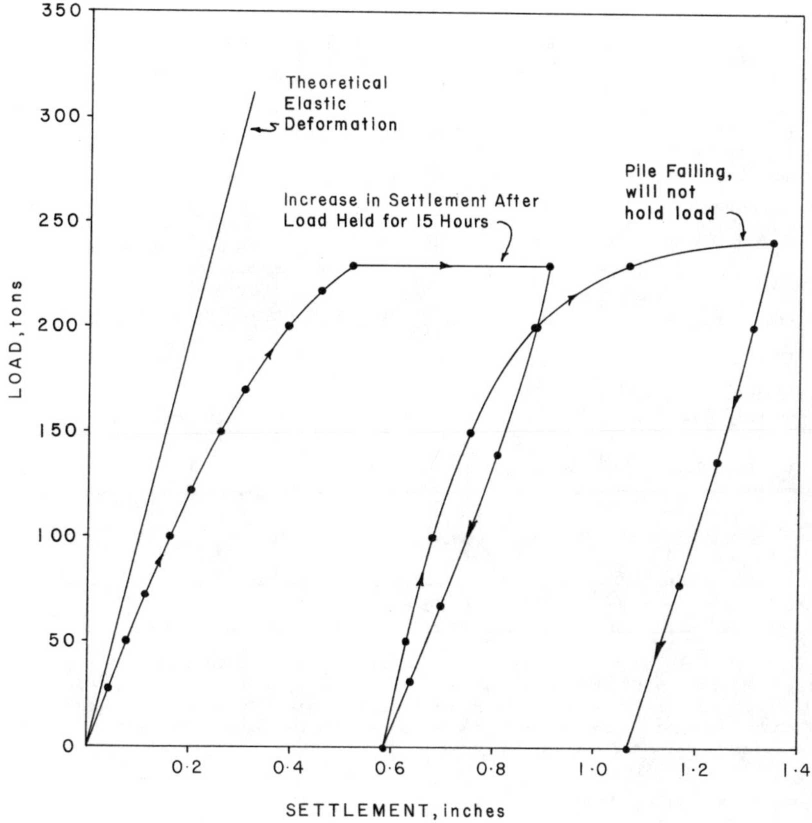

(1 ton=8.896 kN, 1 inch=0.0254 m)

FIGURE 3 - Load Settlement Curve for Test Pile C-57

Comparison of Calculated Pile Capacities with Load Test Results - An allowable single pile load of 100 tons for the 12x53 steel H pile was selected as a working load, based on the allowable stress in the steel. If a failure load of about 230 tons is selected from the load test, this results in a factor of safety of about 2.3. Table 7 summarizes the calculated allowable pile capacities.

TABLE 7: Summary of Pile Capacities

Method of Determination (1)	Factor of Safety (2)	Allowable Pile Load, tons (3)	Remarks (4)
Static Method Using Laboratory Strength Values	3.0	82	Assuming 5 feet penetration into shale
Hiley, Dynamic Formula	3.0	80	Assuming final driving resistance of 20 blows per inch with Vulcan 014 hammer
Pressuremeter	3.0	100	Assuming about 6 feet penetration into shale
Field Instrumentation by Dynamic Analyzer	2.4	100	Average value for Test Pile C-5 and C-57
Static Load Test	2.3	100	C-5 = 1.1 feet penetration into shale; C-57 = 5 feet penetration into shale

NOTE: 1 ton = 8.896 kN, 1 ft. = 0.305 m

Examination of ultimate calculated pile capacities shows that values ranged from a low of 240 tons with the Hiley Dynamic formula, to a high of 300 tons with the pressuremeter. These values compared to an ultimate load of about 240 tons (2.14 MN) measured with the Dynamic Analyzer, and 230 tons from the static load test.

Summary & Conclusions

A case history is given for a deep foundation system supported in a clay shale stratum. Preliminary design pile capacities were developed using field and laboratory strength parameters. In order to achieve 100 ton (890 kN) working load piles, penetration into the bedrock of at least six feet (1.83 m) was anticipated. Since hard driving was expected, a hammer rated between 35,000 to 40,000 ft-lbs. was specified (47.5 - 54.2 kJ). Overstressing the steel H piles was a concern and Dynamic Analyzer tests were performed in conjunction with the pile load testing program.

The results of the various methods of analysis indicated the following:

1. Calculated pile capacities from field and laboratory tests compared fairly well with pile load test data.
2. Penetration required to achieve the 100 ton (890 kN) working load piles was estimated at six feet (1.83 m). Typical penetrations were generally in the range of three feet (0.915 m).
3. Piles were driven with a Raymond 3/0 single acting steam hammer with a manufacturer's rated energy of 40,625 foot-pounds (55 kJ). The Dynamic Analyzer tests indicated piling were generally stressed near the ultimate stress allowed in the steel, and damage was observed on two piles during the test program. During production driving, close supervision was required in order to achieve the design pile capacity without pile damage. A 30,000 ft-lb. hammer would have been preferable, but was not available during construction.
4. From calculations made with pressuremeter data, the pile load capacity for steel H piles compared favorably with the static calculations.

Appendix I. - References

1. Davisson, M.T., "Pile Load Capacity," Proceedings, Design, Construction and Performance of Deep Foundations, ASCE - U. of California, August 1975, Berkeley.

2. Leonards, G.A., Ed., Foundation Engineering, McGraw-Hill Book Company, New York, NY, 1962, pp 651-658.

3. Menard, Louis, "Rules for Calculation and Design of Foundation Elements on the Basis of Pressuremeter Investigation of the Ground," Sols-Soils, No. 5, June 1963, pp 1-18.

4. Soil Exploration Company, "Dynamic Pile Test Near Gillette, Wyoming," April 1977.

Appendix II. - Notation

The following symbols are used in this paper:

A_b = area of pile base
C = cohesion
K = coefficient used in pressuremeter calculations
N_c = bearing capacity factor
l = length of pile embedment
p = pile perimeter
P_o = lateral earth pressure at rest
P_l = limit pressure of the soil
Q_b = base resistance
Q_s = shaft resistance
Q_{ult} = ultimate pile capacity
qu = unconfined compressive strength

AUGER-PLACED GROUT PILES IN GRAVEL

By Larry G. O'Dell,[1] M. ASCE, John M. Pool,[2] A.M. ASCE

Introduction

Strength parameters and consolidation characteristics of coarse granular soils are difficult to determine accurately. Often a considerable amount of judgment or past experience within an area is used to develop design criteria.

A case history is presented of a moderate size building supported on auger-grout piles in an alluvial gravel deposit. Ultimate pile capacities and estimated settlements were calculated using "N" values from penetration tests. Field pile load tests were conducted on two test piles and the allowable pile capacity determined by load test data is compared with analytical methods.

Project Description

The Phase II expansion project at the Billings Deaconess Hospital consists of a two-story addition approximately 170 feet (52 m) by 380 feet (116 m) in dimension. It has a basement within the center third of the addition which extends about 12 feet (3.7 m) below grade. The basement area is defined by the dashed lines on Figure 1. The eastern and western thirds are slab on grade construction. The addition spans over North 28th Street, creating a tunnel for vehicular traffic. The hospital and expansion project are joined at the second floor elevation and by a tunnel connecting the basement levels.

The expansion project is rigid steel frame construction with metal panels. Concrete floor slabs are constructed on metal decking. Structural loads supported on columns range from 175 to 500 kips (1 kip = 4.45 kN).

Subsoil Conditions

The site is located in the Yellowstone River Valley which is several miles (1 mile = 1.61 km) wide in this area. In the geologic past, the river eroded through the Eagle Sandstone and into the underlying Colorado Shale. The resistant cliff forming sandstone is a prominent geographic feature which surrounds the downtown area. This erosional valley was subsequently filled with alluvial deposits, and most of Billings has developed on the relatively flat floodplain.

At the proposed expansion site, 16 test borings were drilled to depths ranging from four to 33 feet (1.2 to 10.1 m) as shown on Figure 1. The soils profile consists of clay, overlying a sandy gravel, and then the Colorado Shale formation. The clay stratum contains a high percentage of silt and classifies as a CL, according to the Unified Soils Classification System. Lenses of plastic clay (CH) were encountered in two borings. The clay is soft to firm, and extends to depths ranging from nine to 17.5 feet (2.7 to 5.3 m) at the boring locations. The gravel stratum underlies the clay and extends to depths of about 26 feet (7.9 m). This deposit is generally dense, but occasional loose sand lenses were encountered. The sandy gravel is fairly well graded and contains cobbles up to 10 inches (25 cm) in size. The

[1]Vice-President, Northern Testing Laboratories, Inc., Billings, MT.
[2]Project Engineer, Northern Testing Laboratories, Inc., Billings, MT.

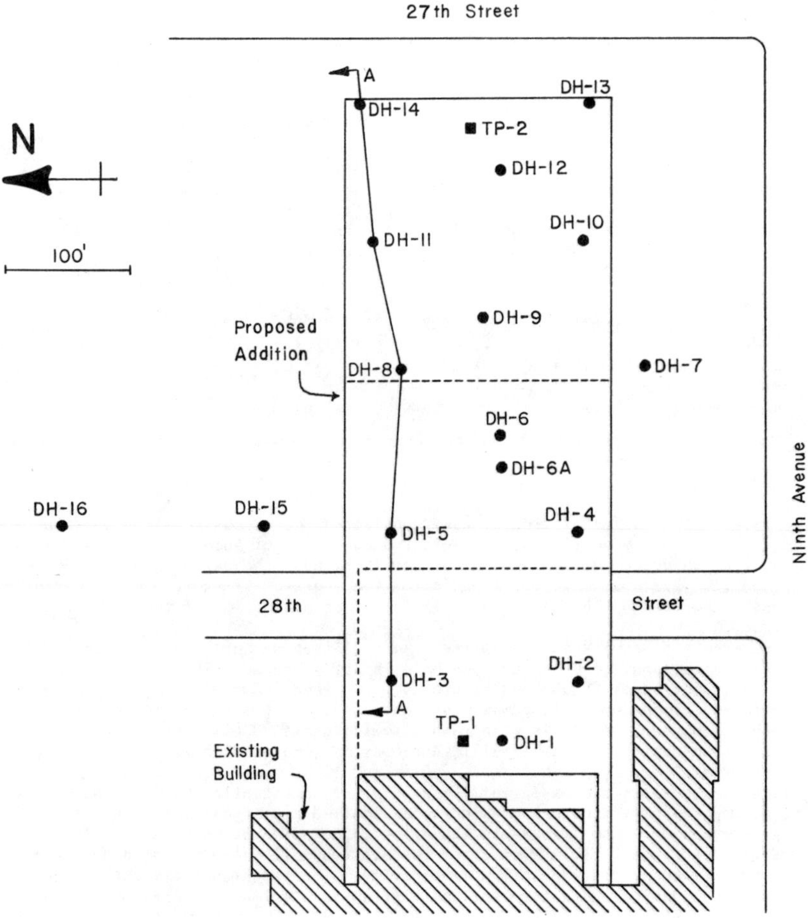

FIGURE 1 - Boring Locations (1ft.=0.305 m.)

particles are subrounded to rounded.

The underlying shale is competent and capable of supporting heavy structural loads. Groundwater was encountered 10 to 13 feet (3.1 to 4.0 m) below the existing grade, which is slightly above the gravel contact. Physical and engineering properties of the soils encountered are categorized as shown in Table 1.

A simplified geologic cross section of the soil profile is presented in Figure 2.

TABLE 1: Typical Field and Laboratory Test Values

Soil or Material Type (1)	Range of "N" Value Blows/foot (2)	Liquid Limit (3)	Plasticity Index (4)	Saturated Internal Friction Angle, Degrees (5)	Saturated Cohesion, Kips/ Square Foot (6)	Average Compressi-bility in Percent/ksf (7)
Silty Clay (CL)	2 - 21	44	22	30	0	1 - 1.5
Sandy Gravel (GW)	17 - 100+	-	-	-	-	-
Shale, Claystone	100+	33	16	-	-	-

Foundation System

Several foundation alternates were considered and discussed with the structural engineer. The foundation system selected consisted of auger-grout piles beneath the slab on grade portion and spread footings under the basement area. Both foundation systems were supported in the sandy gravel stratum. This paper discusses the design and subsequent pile load testing portion of the foundation system.

Auger-grout piles are cast in place like a caisson rather than being driven. Unlike a caisson, however, an open hole is not produced. The hole is advanced using a continuous flight hollow stem auger. When the required penetration is achieved, cement grout is pumped through the auger stem to the bottom of the hole. The auger is slowly withdrawn allowing grout to fill the void. The foundation contractor was Turzillo Corporation, who used their patented Auger-cast Piles.

Auger-grout piles are essentially a zero displacement pile and for load capacity, calculations are considered as small diameter caissons. Methods are presented in the literature (2,3) for calculating caisson capacities and settlement ranges. Selection of representative soil shear strength parameters is the key to predicting soil-pile behavior. The techniques presently used for load capacity and settlement calculations are primarily based on correlations with standard penetration test results or on the theory of plasticity, soil cohesion and friction angle. Relatively undisturbed samples of clays and sands are normally easy to obtain and tests for establishing representative shear strength parameters can be performed. However, in the case of coarse grained gravels with cobbles, undisturbed samples are nearly impossible to obtain and the use of N values from standard penetration tests requires considerable judgment since particle sizes affect the values. A typical N value of 40 was selected based on the standard penetration tests which did not encounter cobbles.

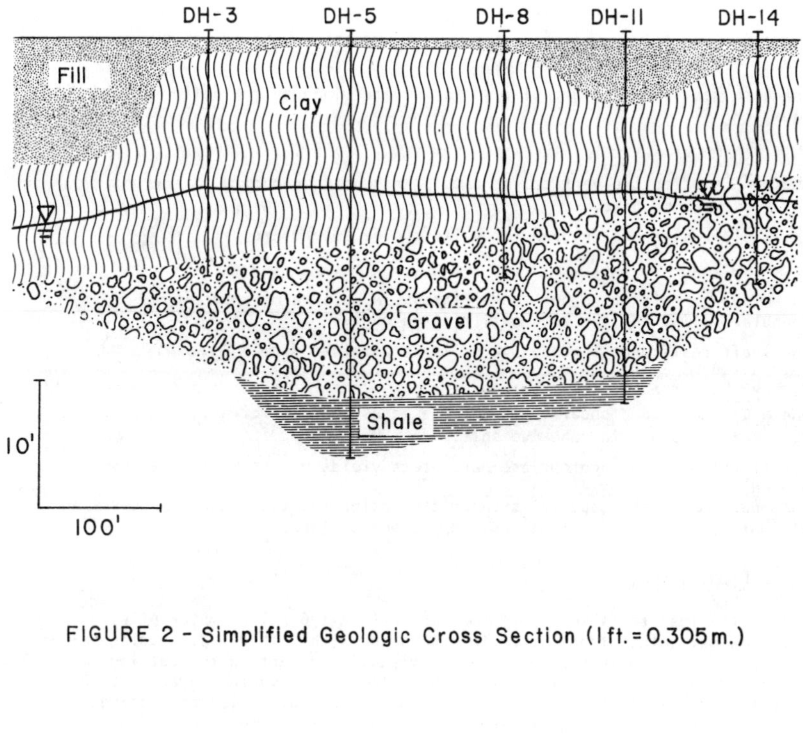

FIGURE 2 - Simplified Geologic Cross Section (1 ft. = 0.305 m.)

Based on experience with this gravel deposit and correlation with the typical N value, an internal friction angle of 45 degrees was selected for design calculations. Caisson load capacity correlations based on N values have been developed for granular soils (2) and are reproduced in Figures 3 and 4.

Using an N value of 40 for calculating the ultimate capacity yields an end bearing of 26 tsf and shaft resistance of 1.25 tsf in the gravel (1 tsf = 95.8 kPa). Using 0.6 tsf for shaft adhesion in the clay, the ultimate capacity of a 14 inch (36 cm) intrusion pile with seven feet of penetration into gravel through a clay layer 12 feet (3.7 m) thick becomes 86 tons (765.4 kN).

Calculating the ultimate capacity using the theory of plasticity is performed using the equation

$$Q_{ult} = Q_b + Q_s \quad (1)$$

where Q_{ult} = ultimate caisson capacity, Q_b = ultimate base capacity, and Q_s = ultimate shaft capacity.

The base resistance in granular soils is determined by the following equation:

$$Q_b = \bar{\sigma}_v N_q A_b \quad (2)$$

where $\bar{\sigma}_v$ = effective vertical stress, N_q = bearing capacity factor, and A_b = area of base.

The shaft resistance in granular soils can be determined by

$$Q_s = \bar{\sigma}_h A_s \mathrm{Tan}\,\phi \quad (3)$$

in which $\bar{\sigma}_h$ = effective horizontal stress, A_s = area of shaft in granular soils, and ϕ = internal friction angle.

The shaft resistance in cohesive soils is obtained from

$$Q_s = C\alpha A_s \quad (4)$$

where C = undrained shear strength, α = adhesion coefficient, and A_s = area of shaft in cohesive soil.

Substituting in the appropriate parameters yields Q_{ult} = 92 tons (1 ton = 8.9 kN).

Comparison of the capacities calculated using the two methods suggests an ultimate capacity of 90 tons (801 kN) is appropriate.

Pile Load Tests

Two pile load tests were performed on 14-inch (36 cm) diameter piles using the slow maintained load method in accordance with ASTM D1143. During installation, the borings were logged by visually inspecting the cuttings and observing the augering difficulty. The piles were advanced seven feet (2.1 m) into the gravel stratum and allowed to cure seven days prior to testing. Test pile locations, TP-1 and TP-2, are shown on the site plan in Figure 1.

Both piles were loaded to failure after the standard test was complete. Working loads of 30 and 40 tons (1 ton = 8.9 kN) were used to determine incremental load applications. The test piles were loaded in 25 percent increments to 200 percent of the design load. The 200 percent design load was maintained for 24 hours, then rebounded to zero in decrements of 50 percent of the design load. After 12 hours, the 40-ton pile was reloaded to 80 tons in 20-ton increments, then in four ton increments, until failure. The 30-ton pile was loaded to 60 tons, then loaded in 10-ton increments to failure. The load settlement plots are shown in Figures 5 and 6.

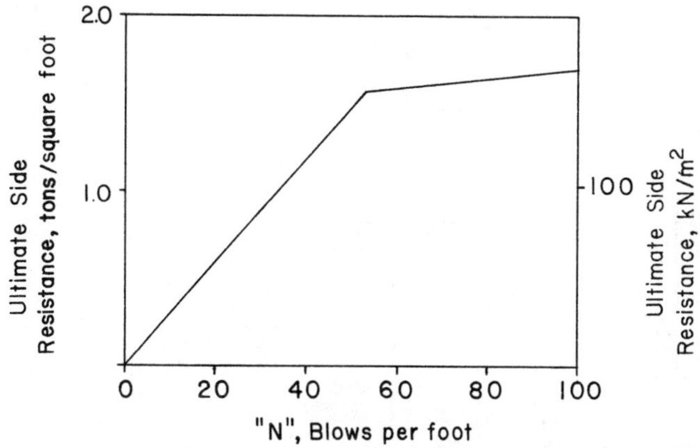

FIGURE 3 - Ultimate Side Resistance Versus Blow Count

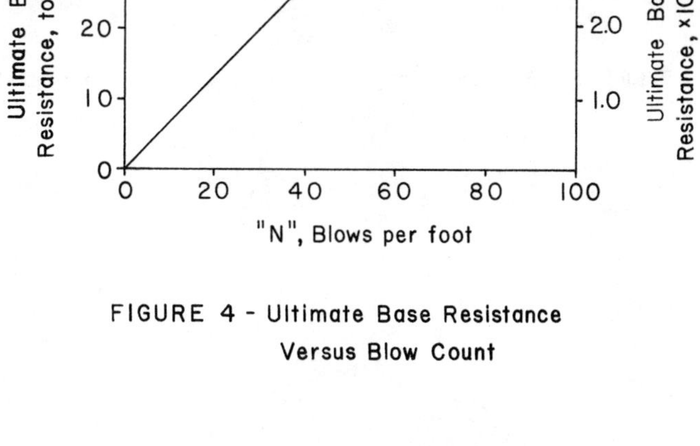

FIGURE 4 - Ultimate Base Resistance Versus Blow Count

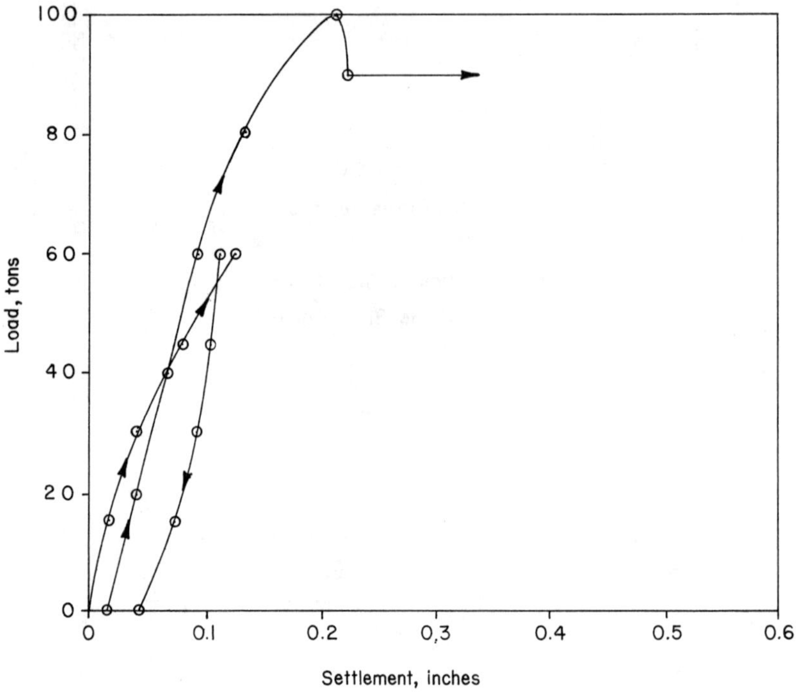

FIGURE 5 - Pile Load Test - 30 Ton Working Load (1 ton = 8.9 kN, 1 inch = 2.54 cm)

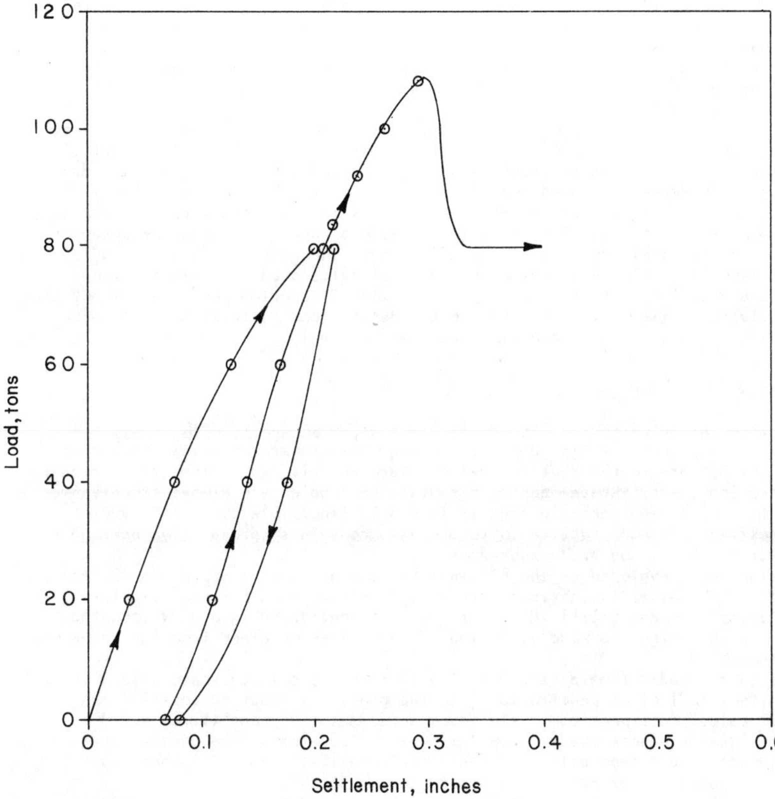

FIGURE 6 - Pile Load Test - 40 Ton Working Load (1 ton = 8.9 kN, 1 inch = 2.54 cm)

The determination of an ultimate pile load capacity from load testing is subject to some disagreement, as several methods are available for determining capacities (1). Both test piles reach a peak load, then experience a plunging failure. Continual plunging is noted at what is termed a "residual load", and is less than the peak load. Peak failure loads of 100 and 108 tons with residual loads after failure of 86 and 80 tons were recorded for test piles 1 and 2, respectively (1 ton = 8.9 kN).

In construction of augercast piles, grout is placed under pressure. The authors suspect the grout migrates into voids in the gravel stratum creating a fairly irregular soil-pile interface. The difference between peak and residual loads is believed to be a combination of shearing and reorientation of particles along the pile-soil interface, particularly in the gravel stratum. The long term ultimate pile capacity is suspected to be between the peak and residual values.

Using average values from the two test piles, and a design load of 40 tons, the factor of safety is 2.6 and 2.1 for peak and residual loads, respectively The load-settlement characteristics were calculated using emperical methods presented by Reese (2) which correlate the ratio of mobilized to ultimate shaft and base resistances to settlement. This method represents the results of numerous load tests. The probable settlement range is bounded by an upper and lower load limit for any given deflection. The results of these studies are presented in Figure 7, which is a plot of calculated settlement values with the actual load test values from test pile 2 superimposed. Settlement of an individual pile at the 40 ton (356 kN) design load is less than 0.1 inch (0.25 cm) from both calculated (2) and load test values.

Summary and Conclusions

A presentation of a case history is given of an augercast pile foundation system in a coarse granular deposit. Design values were calculated based on what the authors regraded as typical or representative "N" values from standard penetration tests. Gravel deposits containing cobbles are always troublesome to geotechnical engineers in that it is nearly impossible to obtain samples for testing. Common practice is to use assumed values, presumptive bearing pressures, or rely on past experience.

Design data developed on the Billings Deaconess Hospital Expansion Project based on "N" values is considered reliable. Correlation between calculated pile capacities and pile load test results is considered in a very acceptable range, considering the wide variation of "N" values obtained from the gravel stratum.

The calculated ultimate capacity of a 14-inch (35 cm) augercast pile with seven feet (2.1 m) of penetration into the gravel is about 90 tons (801 kN). Actual values obtained from field tests were 83 to 104 tons (1 ton = 8.9 kN) for residual and peak loads, respectively. Calculated settlements and actual settlement values from pile load tests indicate less than 0.1 inches could be anticipated at the design load of 40 tons (356 kN).

Correlation between calculated settlement and load test data is considered good.

Acknowledgement

The writers are grateful to the Administrators of Billings Deaconess Hospital, who kindly released the test data contained in this report.

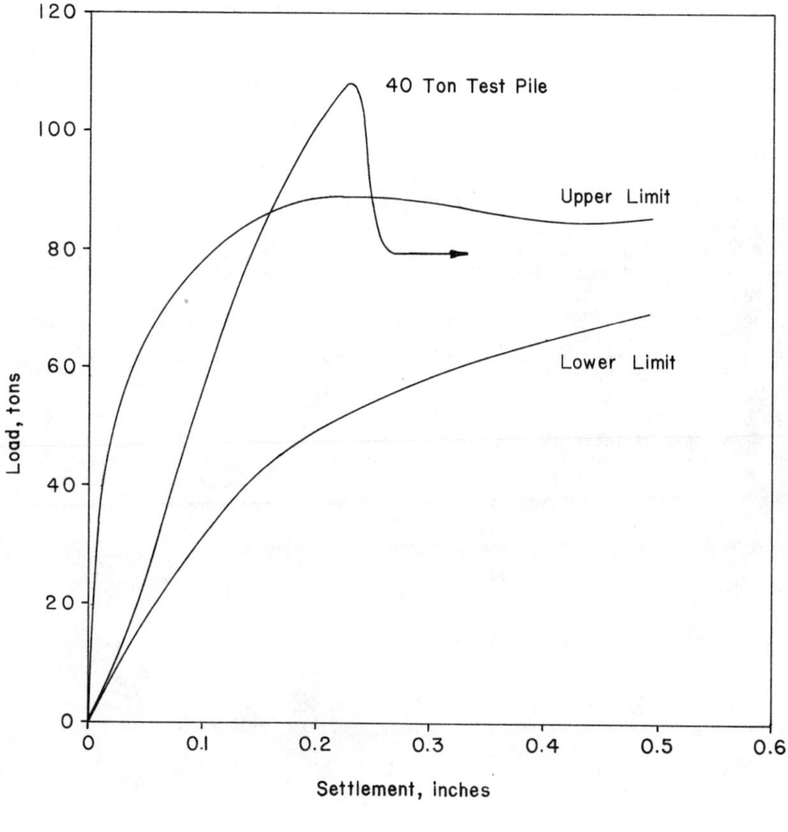

FIGURE 7 - Calculated Load versus Settlement Curves with 40 Ton Load Test (1 ton = 8.9 kN, 1 inch = 2.54 cm)

Appendix I. - References

1. Fellenius, B. H., "Test Loading of Piles and New Proof Testing Procedure," Journal of the Geotechnical Engineering Division, GT9, 1975.

2. Reese and Wright, "Drilled Shaft Design and Construction Guidelines Manual," Vol. 1, U.S. Department of Transportation, Federal Highway Administration, Offices of Research and Development, Implementation Division.

3. Woodward, Gardner, and Greer, "Drilled Pier Foundations," McGraw-Hill, Inc., New York, 1972.

Appendix II. - Notation

The following symbols are used in this paper:

A_b = area of pile base
A_s = area of pile shaft
C = undrained shear strength
N_q = bearing capacity factor
Q_b = base resistance
Q_s = shaft resistance
Q_{ult} = ultimate bearing capacity of pile
α = adhesion coefficient
$\bar{\sigma}_h$ = effective horizontal stress
$\bar{\sigma}_v$ = effective vertical stress
ϕ = angle of internal friction

A STUDY OF DOWNDRAG ON A CAST-IN-PLACE CONCRETE PILE

By Antal Partos* M. ASCE

In 1972/73 field testing at the Philadelphia International Airport confirmed the feasibility of preloading soft subsoils by lowering the piezometric elevation of the groundwater. Locations of test area and the proposed layout and design draw down elevations of a dewatering system for Apron E construction are shown on Fig. 1. Typical soil information and layout of dewatering test and the pile load test location are given in Fig. 2.

Airport structures and other structures located within the influence of proposed dewatering are supported on piles (cast-in-place concrete, timber, or composite (concrete-wood) piles). Because of concern about downdrag forces caused by the consolidation of the compressible silts and clays long term (5-8 months) pile load tests were conducted in 1974/1975. The pile load tests

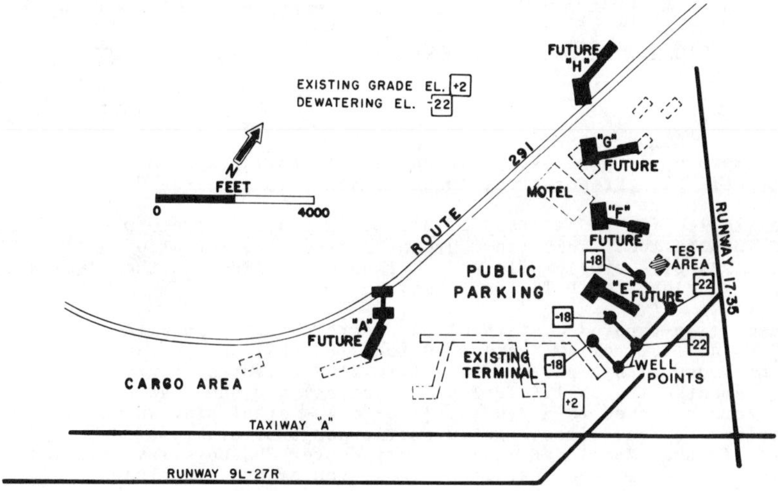

FIG. I. LAYOUT OF DEWATERING SYSTEM

* Antal Partos, Vice President, Chief Geotechnical Engineer, Site Engineers, Inc., Philadelphia, Pennsylvania

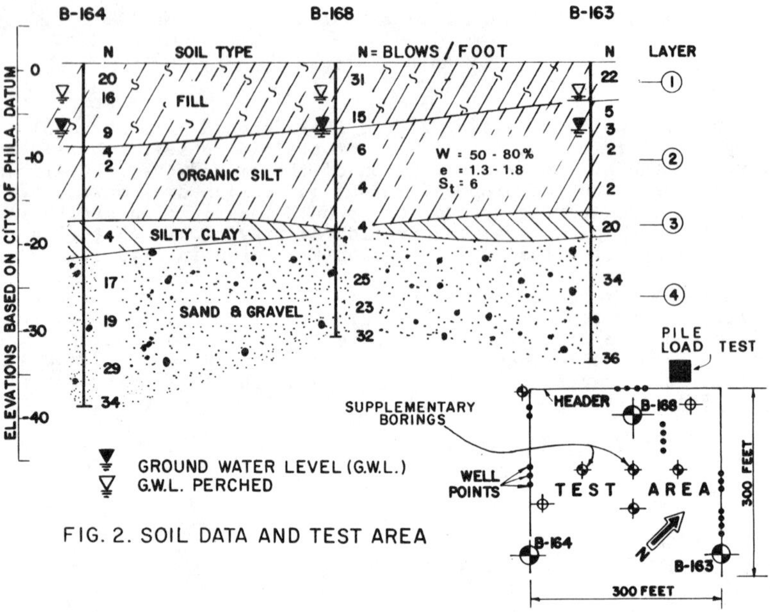

FIG. 2. SOIL DATA AND TEST AREA

were employed as part of the surcharge monitoring system of Contract: "Site Preparation, Apron E, Work No. A-582-SW."

A single pile and a group of 3 cast-in-place concrete piles were load tested. This paper presents general information on load tests of a single pile and a group of 3 piles and the results of the single pile (P-1) load test.

Each step-tapered cast-in-place concrete test pile with a 10-3/8 in. (26.4 cm) ϕ tip and 60 tons (534 kN) design capacity was provided with 4 pairs of electric strain gauges (Micro-measurements - EA 06-125-TB-350 T rosettes) and one load cell. They were attached to a 2.0 in. (5.08 cm) ϕ steel pipe which also contained 20 lead wires. An automatic gauge strain recorder system (Vishay-Model V/E 220) collected over 70,000 strain readings during the test. The test loads consisted of building loads (steel ingots) and surcharge loads of a 6 ft (1.83 m) thick stone fill and the lowering of the piezometric elevation of the ground water by 11 ft (3.36 m). The load test plan is shown in Fig. 3 and the flow of information is shown in Fig. 4.

DESIGN OF LOAD TEST

Since the purpose of the pile load tests was to provide data for predicting the behavior of the existing pile supported structures, cast-in-place concrete test piles were selected. The difficulty with instrumenting cast-in-place concrete piles using

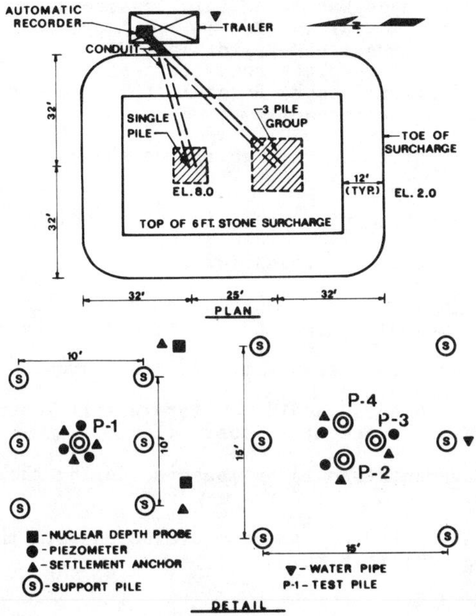

Fig. 3 LOAD TEST PLAN

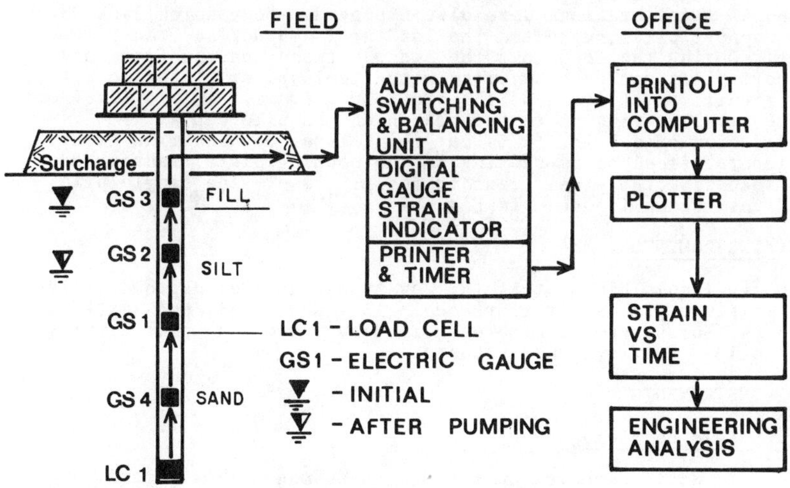

FIG. 4. FLOW OF INFORMATION

electric strain gauges was taken into consideration. Crandall in 1948 (3), Mohan, et al in 1963 (7), D'Appolonia and Hribar in 1963 (4), and Koerner and Partos in 1971 (6), reported their experience with cast-in-place concrete test piles using electric transducers. Publications by Bozozuk (1,2) and York et al (8) are also given in the Appendix.

The following criteria were governing the design of the load tests:

1. Maintain static pile loads for 5 months.

2. Maintain surcharge loads for 5 months.

3. Surcharge loads should affect only the below ground portion of the test piles (which is the case during dewatering).

4. Automatic information retrieval for 8 months.

5. Soil instruments should monitor subsoil behavior immediately adjacent to test piles.

6. Pile movements should be measured during the full testing period.

7. The load tests had to be coordinated with the schedule for Contract Work No. A-582-SW.

TEST SETUP

A steel frame platform and calibrated steel ingots were selected for incremental and long term loading. To provide stability and safety, support piles, horizontal steel beams, and diagonal steel bracings were also employed. Four screw jacks on the support piles supported the loading frame during load placement. During the load test the space between loading frame and support piles could be adjusted with steel spacer plates to arrest any significant tilting of the steel load frame. A 1 in (2.54 cm) thick steel bearing plate was placed on the pile cap after application of a 1 in. (2.54 cm) thick layer of nonshrink grout. Horizontal steel beams resting on the bearing plate transferred the load from the steel frame to the pile cap. The single pile load test setup is shown in Fig. 5.

SOIL INSTRUMENTS

Heavy liquid piezometers (R. Joy type), nuclear depth density and moisture probes (Troxler Models 1351 and 1251), and settlement anchors (Boros type) monitored soil behavior during pile driving, static load testing, and surcharging.

PILE INSTRUMENTS

1. Pile Settlements

   Scales mounted on mirrors, a permanent bench mark, and a surveyor's level were used to measure pile butt settlements.

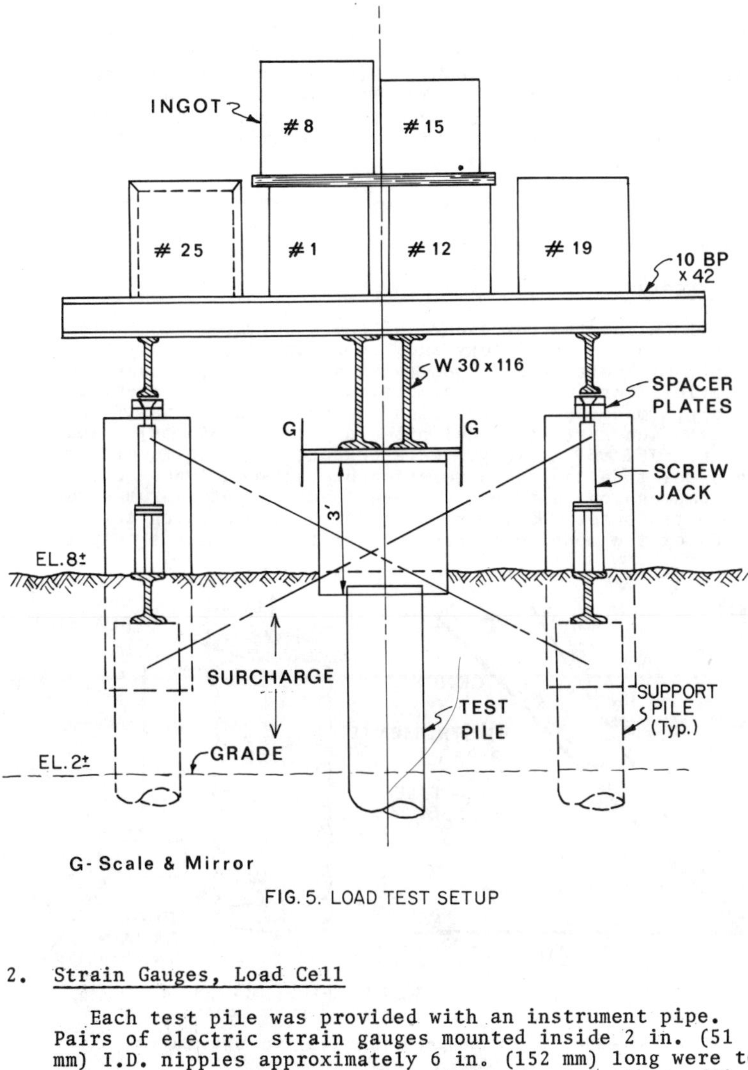

FIG. 5. LOAD TEST SETUP

2. Strain Gauges, Load Cell

Each test pile was provided with an instrument pipe. Pairs of electric strain gauges mounted inside 2 in. (51 mm) I.D. nipples approximately 6 in. (152 mm) long were to measure the vertical strain at different elevations. (Fig. 4). One pair of strain gauges was designed to monitor the pile strain in the upper fill (GS-3), two pairs of strain gauges to measure strain in the silt (GS-1, GS-2), and one pair of gauges measured strain in the lower sand layer GS-4), respectively. One calibrated load cell (LC-1) was designated to be placed at the tip of each pile. The 4 lead wires for each pair of waterproofed electric gauges and one load cell were contained in the instrument pipe which were carefully assembled with machine threaded and waterproofed connections. The instrument pipe was centered in the pile shell and then concrete was placed. The top of the instrument pipe was approximately 3 ft below pile cut-off elevation.

The 20 lead wires of each test pile were placed in a heavy duty rubber conduit and connected to the automatic gauge strain recorder.

3. Laboratory Tests

All calibrated components of the electric system were checked in the laboratory and, additionally, the elastic moduli of short steel pipes (with strain gauges), and the behavior (zero drift) of electric gauges in cast-in-place concrete model piles were also checked. Additionally compression tests on model piles of 14 in. (356 mm) ϕ and 30 in. (762 mm) in length were conducted in the laboratory in order to establish a relationship between loads measured by compression tests and between loads computed from strain gauge response. Fig. 6 shows that the compression test loads of the model pile are smaller than the loads computed

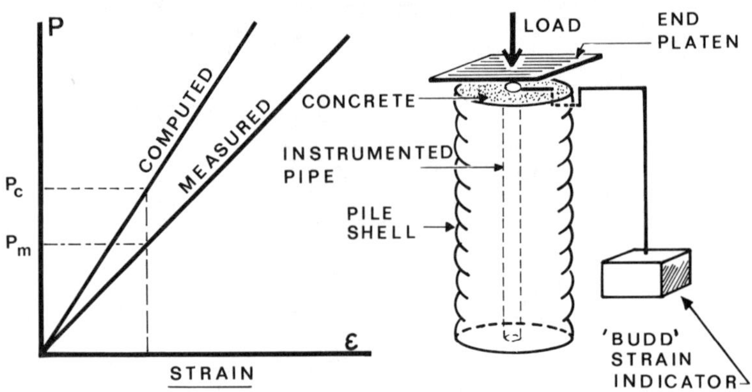

MODEL PILE: 14.25 in. ø
Length 30 in

P_m Load measured by compression testing equipment

P_c Load computed from response of strain gauges

$$F = \frac{P_m}{P_c} = 0.724$$

FIG. 6. LABORATORY COMPRESSION TEST ON MODEL PILE

from the strain gauge response. The ratio (F) of measured and computed loads is 0.724. This value is used as a correction factor in computing pile loads from the response of strain gauges. The value of F has not been checked for different concrete moduli and/or model test pile geometries.

PILE LOADS

In order to compute pile loads, all steel and concrete areas at the electric gauge locations were transformed into steel areas (A_t) by using

$$A_t = A_{steel} + \frac{A_{concrete}}{n} \text{ ; where}$$

$$n = \frac{E_{steel}}{E_{conc.}} = \frac{29\text{x}10^6 \text{ psi}}{4.0\text{x}10^6 \text{psi}} = 7.2$$

1. Pile Load at Electric Strain Gauge (GS) Location

$P_z = E_s \text{ x } (A_t)_z \text{ x } \varepsilon_z \text{ x } F = I \text{ x } (A_t)_z \text{ x} \varepsilon_z$; where $I = E_s \text{ x } F$

P_z = load at depth z

$E_s = 20 \text{ x } 10^6$ psi (modulus of elasticity of steel)

ε_z = gauge strain at depth z ("net" value)

$(A_t)_z$ = Area of concrete transformed to equivalent area of steel, plus all other steel areas

F = Correction factor from laboratory tests (0.724)

2. Pile Load at Load Cell (LC) Location

Load computed with the use of data from calibration certificates.

Load in kips (4.45kN) = $V_o = \frac{\text{Strain}}{2} \text{ x } 2.115 \text{ x } 10^{-3}$

CONSTRUCTION OF SINGLE PILE LOAD TEST SETUP

The test pile and 4 support piles were driven from Sept 11 to Sept 13, 1974. Test pile driving records are presented in Fig. 7. The 60 ton (534kN) capacity test pile was driven with a Raymond 65C hammer (19,500 ft-lbs rated energy) until the specified dynamic resistance was encountered. The final required set for the last 4 in. (10.16 cm) of pile penetration was 5 blows/in. The upper approximately 10 ft. (3.05 m) of fill was preaugered with a continuous flight auger of 10 in. (25.4 cm) ϕ. The average diameter of the pile shell is 15-3/4 in. (40.0 cm) in the fill stratum. Therefore, preaugering did probably not reduce the amount of pile load transferred to the fill.

The strain gauges, instrument pipe sections, and load cells were assembled on a platform and stored on a special storage rack, covered by canvas, at the test site. Three metal fins

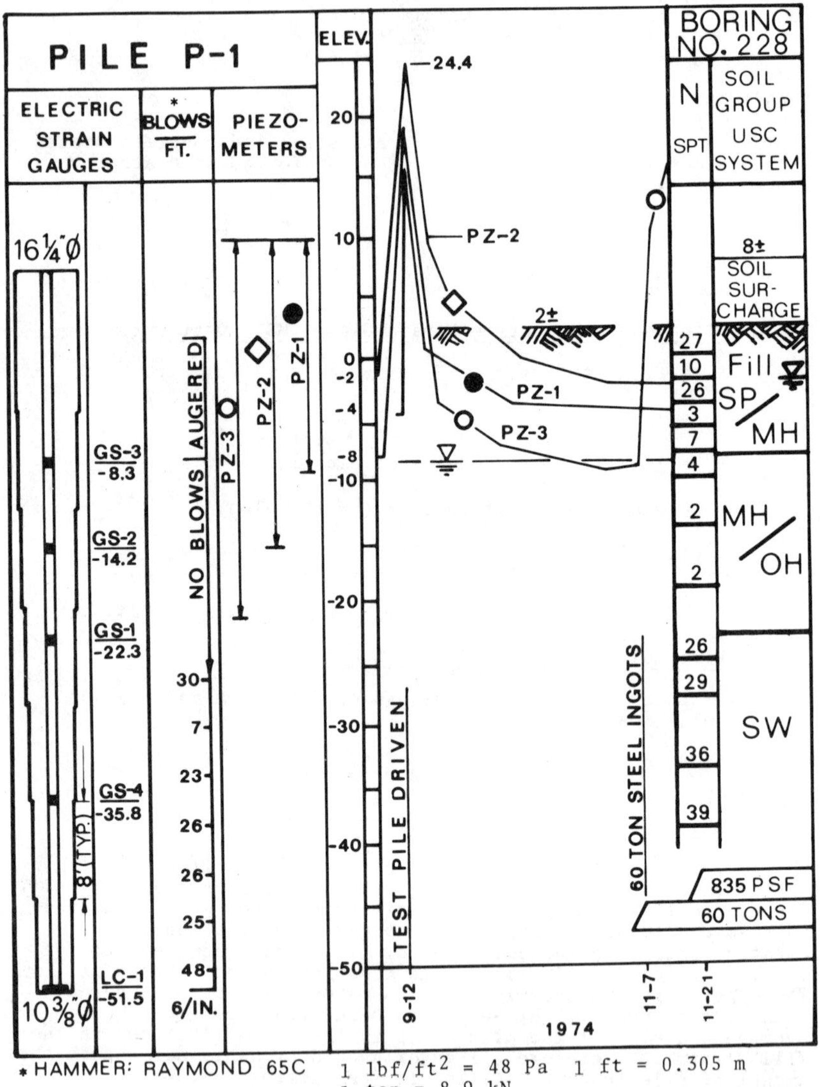

* HAMMER: RAYMOND 65C   1 lbf/ft^2 = 48 Pa   1 ft = 0.305 m
1 ton = 8.9 kN

FIG. 7. DRIVING RECORD AND PORE PRESSURES

(spacers) were fastened to the completed instrument pipe at four locations to aid in maintaining a vertical and centered position of the instrument pipe in the shell. All transducers were checked by a portable strain indicator and an ohmmeter. Hoisting of the instrument pipe from the storage rack (horizontal) to a midair position was done by a hydraulic crane with telescoping boom and a rope and pulley system. The rope was adjusted at its free end by the contractor's superintendent and the tip of the instrument pipe (load cell) was held and guided by hand while the instrument pipe was raised through an increasing angle with the horizontal until it reached its final free-hanging midair position. This method ensured no bending of the approximately 50 ft long instrument pipe.

After the instrument pipe had been lowered into the pile shell a small window was cut near the top of the shell and a short horizontal extension from the instrument pipe was used to connect the 20 lead wires of the 5 transducers to a terminal box. The system then was checked again for electric stability and signal intensity. On Oct 2, 1974 all lead wires were connected to the automatic gauge strain recorder and the entire system was checked for electric stability, scan, signal intensity, and laboratory calibration values. Instability of the load cell was observed on October 3. Therefore, the instrument pipe was removed from the pile shell, and the load cell was returned to the factory for check and repair. On October 5 the entire system was reassembled, checked again for electric stability, signal intensity, scan, and calibration values.

The influence of the environment (noise, vibration, temperature changes, time of day, stray electric currents from underground utilities, radio signals, etc.) on the electric gauges was checked from Oct 6 to Oct 8, 1974. It was observed that major temperature changes should be avoided. Therefore, the terminal boxes and the lead wire conduits were insulated and provided with automatic electric heating coils and the heating and air conditioning of the instrument trailer were improved. Temperature measurements in the pile shells indicated very little changes; they were generally 57°F. Outside temperatures varied from 50°F to 78°F; instrument room (trailer) temperature peaked at 92°F at 1425 hrs on October 7, 1974.

The concrete 28 day design strength was 5000 psi. ($3.45 \times 10^4 kN/m^2$). Shrinkage compensating cement (ChemComp) was used. The weights of each component of the platform and each steel ingot were measured and computations regarding ingot arrangement and a plan for the sequence of ingot placement were made. During the time of platform fabrication continuous "zero" readings were obtained and necessary adjustments in individual transducer circuits and scan control were made. At the beginning of November the transducer readings indicated electric stability and actual load testing began.

STATIC LOAD TEST, TESTPILE P-1

1. Test Load

The 60 ton (534kN) test load, consisting of the loading

frame and ingots was applied in 3 uniform increments from Nov 4 to Nov 5, 1974.

2. Pile Settlements

Three mirrors and scales were affixed to steel reinforcing bars welded to the north, south, and west sides of the base plate. Readings were made at 5 to 10 min intervals during load cycles and at greater time intervals thereafter. The last readings were obtained 47 hrs after full load application. The maximum average pile settlement was 0.17 in. (4.3 mm).

3. Pile Strain

Continuous readings (3 readings per min.) were obtained from all transducers during the incremental loading cycles and 30 min. to 1 hr. printout intervals were used between load application cycles and after the maximum test load was in place.

EARTH SURCHARGE

During the period from Nov 8 to Nov 19, 1974 the response of all soil and pile instruments was frequently checked so as to be able to determine when to initiate surcharge placement. A corrugated shell with a diameter several inches larger than the diameter of the test pile had been placed around the above ground portion of the test pile to isolate it from the surcharge. Protective casings were constructed for each soil instrument and the reference elevations for settlement anchors and piezometers were raised to facilitate readings during earth surcharging.

A 6 ft. (1.83 m) thick earth surcharge was placed from Nov 19 to Nov 21, 1974. The ground surface contact area was approximately 77 by 54 ft. (23.49 by 16.47 m) in size and the top of the surcharge occupied an area of approximately 60 by 40 ft. (18.30 by 12.20 m) in size. The surcharge was constructed in layers. Surcharge material was trucked to the test site and end dumped at peripherial locations. The weight of each truck load was certified. Surcharge adjacent to test pile and around instruments was placed carefully by hand and hydraulic backhoes. The total weight of surcharge material was 2178 tons (19384 kN) and it exerted an average 0.42 tsf (40.25 kN/m^2) pressure on the ground surface. The top and sides of the surcharge were covered by polyethylene sheets to maintain constant weight. The test area was enclosed by a steel wire fence and provided with an automatic floodlight system for safety and protection. No tilting of the load frame/ingot system was observed.

DEWATERING AT APRON E

From December 15, 1974 to January 2, 1975 the piezometric elevation of the ground water was lowered from Elev. -9 to Elev. -20, and held at approximately Elev. -20 until March 15, 1975. A 1 to 3 ft (0.31 to 0.92 m) temporary rise of the ground water piezometric level was measured during the last 2 weeks of March, 1975. The readings in April indicated that the water level was at Elev. -18, before dewatering operations were terminated on April 28, 1975.

UNLOADING OPERATIONS

Unloading of the total pile loads was done in 3 stages. The static loads (ingots) were removed on April 11, 1975. Earth surcharge removal started on April 15, 1975 and was completed by noon on the following day. The dewatering system was turned off on April 28. The last water level reading on June 10, 1975 indicated that ground water was at a piezometric level of Elev. -11, or approximately 2 ft (0.61 m) below the initial Elev. -9 as measured during October and November of 1974.

EVALUATION OF FIELD DATA FOR LOAD TEST MONITORING SYSTEM

1. Soil Instruments

Fig. 7 presents the driving records of Pile P-1, the elevations of strain gauges and piezometers, the soil profile, and the behavior of piezometers during pile driving and the placement of static pile loads (steel ingots). The water observation pipe, which was installed in the lower sand stratum underlying the silt layer indicated that the ground water level was at Elev. -8±. The water level in the fill located above the silt layer was approximately at El. -3. On Sept 12, 1974, piezometers PZ-1, PZ-2, and PZ-3 stabilized at Elev. -4, Elev. -2 and Elev. -8, respectively. During the driving of test pile P-1 excess pore pressures were measured in piezometers PZ-1, PZ-2 and PZ-3 at Elev. 16, Elev. 24 and Elev. 19 respectively. In other words the driving of test pile P-1 increased the initial pore pressures in piezometers PZ-1, PZ-2 and PZ-3 by 20 ft (6.10 m), 26 ft (7.93 m) and 24 ft (7.32 m), respectively. The undrained shear strength of the silt layer varies from 600 psf (28.7kN/m^2) to 800 psf (38.2kN/m^2). Therefore, the piezometer readings indicate that a plastic deformation (failure) occured in the silt to a distance of at least 2 ft (.61 m) away from the perimeter of the test pile.

The major portion of the excess pore pressures dissipated in approximately 12 hrs and all piezometer readings stabilized by the end of October 1974.

During static load test piezometer PZ-3 which was located at the bottom of the silt and near the top of the sand bearing stratum indicated a continuous increase in pore water pressure from approximately Elev. -8 to Elev. 15. The other two piezometers maintained their piezometric water levels. The application of a 0.42 tsf (40.25kN/m^2) surcharge load (Fig. 8) caused a rapid increase of piezometric elevations by 14 ft (4.27 m.), 6 ft (1.83 m) and 3 ft (0.92 m) for piezometers PZ-3 (bottom), PZ-2 (mid) and PZ-1 (top), respectively.

Shortly (2 weeks) after completion of soil surcharge placement the excess pore water pressures were reduced by 50% in all 3 piezometers and settlements on the order of 0.7 in. (1.78 cm) and 1.2 in. (3.05 cm) were measured by the settlement anchors with the lowest tip elevation (SA-6) and with the highest tip elevation (SA-2), respectively (Fig. 8). A

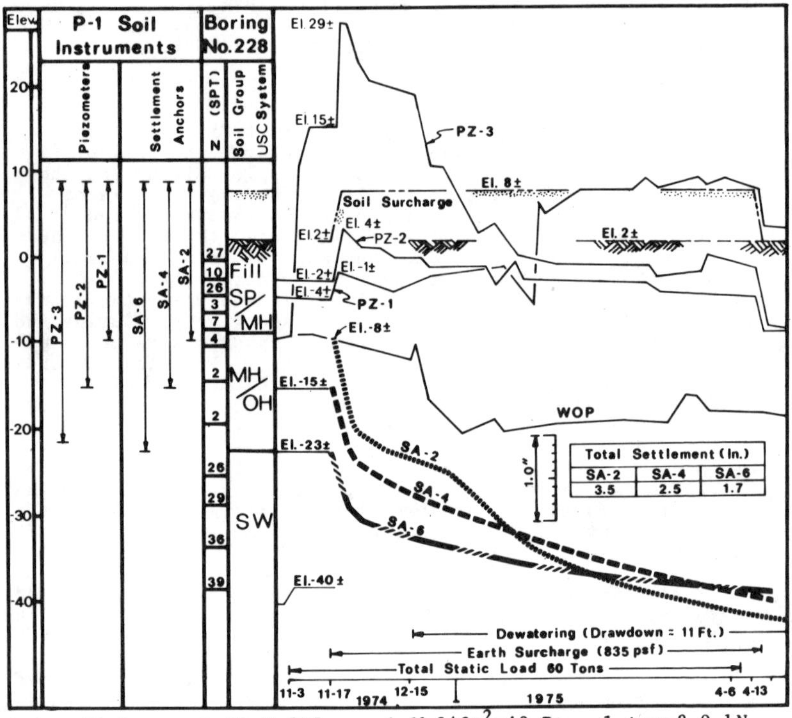

1 in.=25.4 mm   1 ft=0.305 m   1 lbf/ft^2=48 Pa   1 ton=8.9 kN

FIG. 8. PORE PRESSURES AND SETTLEMENTS

combined analysis of the piezometer readings and settlements reveals that the silt layer is not uniform.

At the start of dewatering operations the lowest piezometer, PZ-3, displayed a very substantial pressure loss. The 2 higher located piezometers indicated small changes in their elevations.

Settlements increased after full drawdown had been achieved (Jan 1975), and until the beginning of March. At that time the rate of settlement decreased greatly for SA-2 and SA-6 and moderately for SA-4 which was located in the center of the layer. The responses of piezometers PZ-2 and PZ-3 correlate well with the readings for corresponding settlement anchors SA-6 and SA-4. Piezometer PZ-1 developed problems on Jan 20 and repeated attempts failed to correct its malfunction.

During the removal of the earth surcharge, the decrease in piezometric elevation was approximately 3 ft (.92 m) at the middle and 6 ft (1.83 m) at the bottom of the silt layer. The maximum settlements measured by the settlement anchors with highest (SA-2), middle (SA-4), and lowest (SA-6) tip elevations were 3.5 in. (8.89 cm), 2.5 in. (6.35 cm) and 1.7 in (4.32 cm), respectively. Piezometer and water observation pipe readings were continued during the period of almost complete rebound of the piezometric elevations.

An evaluation of the data from all soil instruments confirms that the progress of consolidation of a non uniform compressible soil layer can be well monitored by installing several sets of 2 instruments (a piezometer and a settlement anchor) at different elevations.

2. Pile Instruments

The electric stability of all transducers for Pile P-1 was checked and, where necessary, adjusted on Oct. 28, 1974. All transducers had linear zero drifts from Oct. 28 to Nov. 4. The zero drift is an electric property of each transducer. It is the rate of gauge strain during load equilibrum conditions. The generally very small zero drift may be positive (upward) or negative, (downward). During the static load test the strains of all transducers rapidly increased and pile butt settlements were also rapid. The average maximum pile butt settlement was 0.17 in. (4.3 mm) at the end of the static pile load test.

Approximately 3 weeks after earth surcharge placement the transducers attained a near equilibrium condition. An average pile butt settlement of 0.23 in. (5.8 mm) was computed before the start of dewatering operations.

While the ground water piezometric level was maintained approximately 11 ft (3.36 m) below its original elevation the greatest increase in gauge strain was observed for GS-1 (bottom of silt), indicating a very significant load transfer to the pile due to consolidation of the silt caused by the weight of soil surcharge and an increase in effective unit weight of silt due to dewatering. GS-1 and GS-2 were the most active transducers during the combined earth surcharge and dewatering period ending April 11, 1975. The average maximum pile butt settlement was 0.52 in. (13.2 mm).

Upon removal of the static test load GS-1, GS-2 and GS-3 showed rapid relaxation and the rebound of the pile butt was 0.16 in. (4.1 mm).

No significant changes for the transducers were noted during earth surcharge removal and cessation of the dewatering operation. A final average net pile butt settlement of 0.27 in. (6.9 mm) was observed in May, 1975 before dismantling the test setup.

ANALYSIS OF PILE LOADS

The analysis uses the fundamental relationships between strain and elastic modulus, and stress to convert the electric gauge strains ($\mu\varepsilon$) into compression loads.

First the "total" gauge strain values which are the transducer inputs are plotted against time and loading conditions. Then the "net" gauge strain values are evaluated by adjusting "total" gauge strain values for the corresponding zero drift values. Zero drift values have been measured initially in the laboratory and in the field before and after concreting the testpile. During load test each electric gauge displayed a zero drift at equilibrium condition. The electric stability of the transducers was periodically checked during the load tests by obtaining internal calibration and zero readings manually. Those readings were reviewed during "net" gauge strain analysis. Pore pressure and settlement readings were also reviewed to assess the progress of consolidation and load transfer from soil to pile to select the dates when maximum "net" gauge strains under different loading conditions should be computed.

The procedures for computing a pile load at a strain gauge or load cell location are outlined in Section "PILE LOADS". The simplified formula which is used in computing pile loads at strain gauge locations is given below.

$P_Z = I \times (A_t)_z \times \varepsilon_z$; where

$I = E_s F$

The average load constant of the load cell obtained from the calibration certificate is 32,000$\mu\varepsilon$/1000 kips (4.45kN).

Fig. 9 shows the vertical soil profile, the gauge elevations, the vertical effective soil pressures and the pile loads. The vertical effective soil pressures are those of the overburden and surcharge (fill, dewatering). The soil pressures and pile loads are presented for 3 loading modes, as follow:

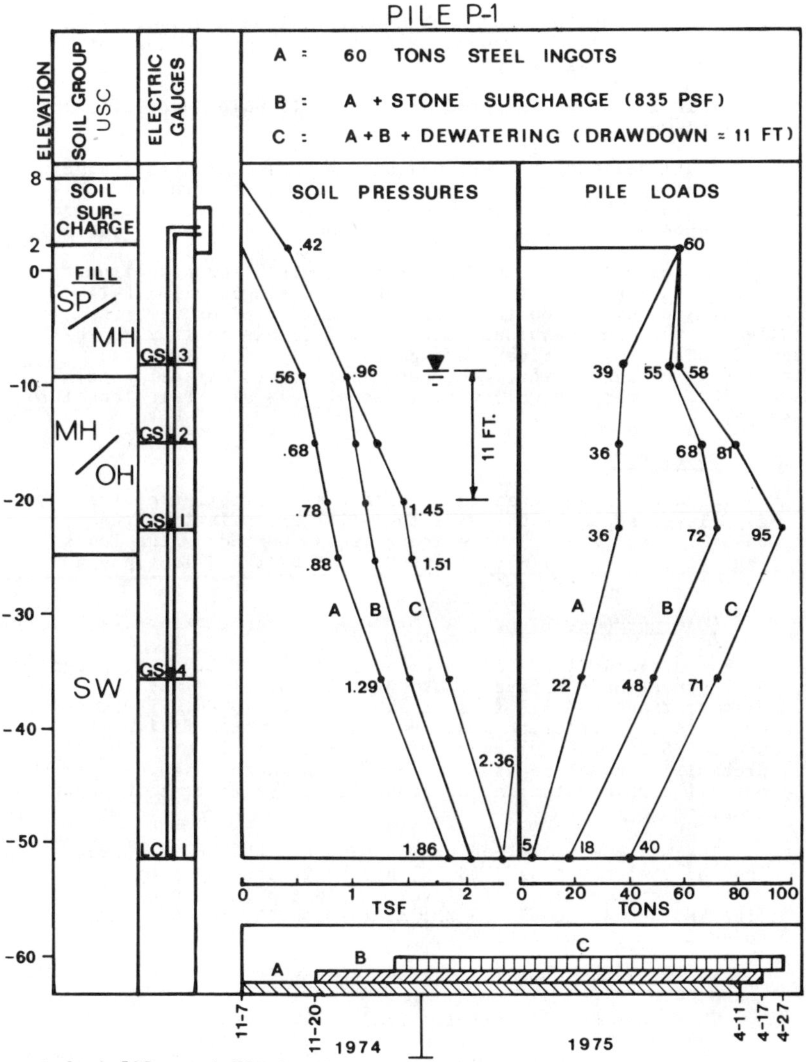

1 ft=0.305 m   1 TSF=96 kPa   1 ton=8.9 kN   1 PSF=48 Pa

FIG. 9. PILE LOADS

A. Static test load of 60 tons (534 kN) steel ingots.

B. Static test load and additionally 835 psf (40.25kN/m^2) surcharge fill pressure.

C. Static test load and surcharge fill load and additional surcharge load produced by 11 ft (3.36 m) drawdown of the ground water level.

The pile loads are computed at or near (loading mode B) equilibrium conditions.

Pile butt settlements were measured by a scale/mirror target and a surveying instrument. Pile tip settlements were not monitored. Measured butt settlements and computed values of elastic compression of the pile and the tip movement are given in Table 1.

Results of pile load tests completed in 1973 for new airport structures in the vicinity of the long term single pile (P-1) load test are summarized in Table 2. All test piles were installed in the same fashion. The upper fill was initially pre-augered and pile was driven through the fill and silt strata into the lower sand layer. The butt settlements of test pile 1-PE (Table 2) compare well with those of test pile P-1 (long term test).

1. Loading Mode A.

Figure 9 shows that the positive skin friction of the fill layer supports 60-39=21 tons (187 kN) pile load and 36-5=31 tons (276 kN) pile loads are supported by the lower sand stratum. Small loads are carried by the pile tip and the silt.

2. Loading Modes B and C

No dragdown load is indicated in the fill layer by gauge GS-3 since the positive friction of 21 tons (187 kN) is greater than 58-39=19 tons (169kN), but it is not likely that negative skin friction would not develop in the fill during up to 3.5 in. (89 mm) settlement of the silt layer. GS-3 probably is under registering because of some obstruction, or soil arching etc. in the generally loose fill at the gauge location.

The maximum pile load of 95 tons (846 kN) and downdrag load of 95-36=59 tons (525 kN) are registered by gauge GS-1 at the bottom of the silt layer. Pile tip loads are 18 tons (160 kN) and 40 tons (356 kN) at the conclusions of loading modes B and C, respectively.

3. Negative Skin Friction. (Loading Mode C)

Thickness of silt stratum (2H)	:15 ft (4.6 m)
Average pile perimeter (p)	:3.7 ft (1.1 m)
Pile shell length (L)	:15 ft (4.6 m)
Due to corrugation a 56% increase in shell length (ΔL)	:8.4 ft (2.6 m)
Adjusted pile shell length (L_c=L+ΔL)	:23.4 ft (7.1 m)

TABLE 1 - MEASURED AND COMPUTED PILE MOVEMENTS OF PILE P-1

LOADING MODE (Fig. 9)	MEASURED	COMPUTED	
	BUTT SETTLEMENT	ELASTIC COMPR.	TIP SETTLEMENT
	in. (mm)		
A	0.17 (4.32)	0.15 (3.81)	0.02 (.51)
C	0.52 (13.21)	0.24 (6.10)	0.28 (7.11)
Remove load A	0.36 (9.14)		
Remove all loads	0.27 (6.86)		

TABLE 2 - PILE LOAD TEST RESULTS

NO. DATE	TEST PILE DATA			LOAD TEST DATA	
	Length in Feet (Meters)				
	TOTAL	In Sand	Final Blows/in*	Load Tons (kN)	Pile Butt Settlement inches (mm)
1 F-BB May 73	33.0 (10.1)	16.0 (4.9)	4,4,4,4	60 /0 (534)/(0)	0.23/0.12 (5.84)(3.05)
				120 /0 (1068)/(0)	.52/ .28 (13.21)(7.11)
1-A TE May 73	38.0 (11.6)	24.0 (7.3)	5,5,5,5	60 /0 (534)/(0)	0.22/0.14 (5.59)(3.56)
				120 /0 (1068)/0	0.69/ .34 (17.53)(8.64)
1 PE May 73	68.5 (20.1)	37.5 (11.4)	4,5,5,5	60 /0 (534)/(0)	.20/ .10 (5.08)(2.54)
				120 /0 (1058)/0	.59/ .31 (14.99)(7.87)
P-1 L Nov 74-June 75	59.5 (18.1)	28.5 (8.7)	5,6,5,6	60 (534)	.17 (4.32)
				95 /0 (846)/(0)	.52/ .27 (13.21)(6.86)

*Hammer: Vulcan 65C, E(Ft.Lbs.) = 19,200; Capblock; Alu.-Mic
BB-Bridge B; TE-Terminal E; PE-Pavillon E; L-Long Term Test Site

a. Total surface (skin) area (As)

$As = L_c x p = 23.4 x 3.7 = 86.6\ ft^2\ (8.05 m^2)$

b. Average down drag load (P)

Load at gauge GS-3; 58-39=19 tons (169 kN)
Load at gauge GS-1; 95-36=59 tons (525 kN)

$$P = \frac{19+59}{2} = 39 \text{ tons } (347 \text{ kN})$$

c. Average negative skin friction (f_s)

$$f_s = \frac{P}{A_s} = \frac{78,000}{86.6} = 900 \text{ psf } (43.1 \text{ kN/m}^2)$$

$f_s = \beta \chi \bar{\sigma}$; where

$\bar{\sigma}$ = soil pressure at mid point of silt

β = coefficent (7)

$$\beta = \frac{fs}{\bar{\sigma}} = \frac{900}{2600} = .35$$

For silt the values of β vary from 0.25 to 0.35, as proposed by Garlanger (5).

CONCLUSIONS

Where piles are installed in compressible, cohesive soils the potential for downdrag should be evaluated. Consolidation of silts and clays after completion of pile driving would mobilize negative skin friction at small relative movement between pile and soil. In assessing the value of negative skin friction the gain in soil shear strength caused by consolidation should carefully be analyzed. A full scale load test, as presented in this paper is very helpful to observe the consolidation of the compressible soils and to measure downdrag forces.

Conclusions are presented below:

1. For monitoring consolidation of the subsoil the installation of sets of 2 instruments (a piezometer and a settlement anchor) at different elevations is very helpful.

2. Pile driving developed very high pore pressures in the silt layer. It exceeded the shear strength of the soil.

3. It is appropriate to study downdrag forces on a pile with actual service loads.

4. The use of electric gauges and an automatic strain recorder is a very convenient system. A backup system of instrumented model piles may be very helpful to measure gauge drifts under varying loading conditions.

5. Corrugated shells have a larger active skin area than smooth shells of the same pile geometry. Therefore, for the same value of negative skin friction a larger downdrag force is realized for a corrugated shell than a smooth shell of the same geometry.

6. This study confirmed that dewatering close to pile supported structures would cause a substantial increase in pile loads.

Acknowledgements

The foresight of the Department of Commerce, Division of Aviation, City of Philadelphia made it possible to conduct this full-scale field test. Special thanks to Messrs. Harold Taylor, Jr. and Jack Martins, Division of Aviation for their interest in this project. The contribution of Dr. Robert M. Koerner and Dr. Marty McCabe, Drexel University, Philadelphia, Pa., special consultants,is gratefully acknowledged. The ingenuity and unselfish cooperation of the contractor's special project superintendent, Jim Simpson, J. D. Morrisey, Inc., Philadelphia, greatly helped the completion of the load test setup.

The opinions and conclusions expressed are those of the author, and do not necessarily represent the views of the Department of Commerce, Division of Aviation, City of Philadelphia.

APPENDIX I - REFERENCES

1. Bozozuk, M., and Labrecque, A., "Downdrag Measurements on 270 Ft. Composition Piles," Performance of Deep Foundations, ASTM Spec. Publ. 444, 1966, pp. 15-40.

2. Bozozuk, M., "Downdrag Measurements on a 160-Ft. Floating Pipe Test Pile in Marine Clay. Canadian Geotechnical Journal, Vol. 9, 1972, p. 127.

3. Crandall, L.L.,'Electrical Resistance Strain Gauges for Determining the Transfer of Load from Driven Piling to Soil," 2nd ICSMFE, Vol. 4, Rotterdam, pp. 122-127. (1948)

4. D'Appolonia, E. and Hribar, J.A., "Load Transfer in a Step-Taper Pile," Journal SMFE, ASCE, SM6, Nov. pp. 57-77. (1963).

5. Garlanger, J. E., "Prediction of the Downdrag Load at Cutler Circle Bridge," M.I.T. (1973).

6. Koerner, R.M. and Partos, A., "Load Transfer Analysis of a Concrete Pile," Foundation Facts, Vol. VII, No. 1 (1971).

7. Mohan, D., Jain, G.S., Kumar, V., "Load Bearing Capacity of Piles," Geotechnique, Vol. 13, pp. 76-86 (1963).

8. York, D.L. Miller, V.G., and Ismael, N.F., "Long-Term Load Transfer in End-Bearing Pipe Piles,: Transportation Research Record No. 517, 1974, pp. 48-60.

APPENDIX II - BIBLIOGRAPHY

1. Berezantsev, V.G.; "Load Bearing Capacity and Deformation of Piled Foundations,:" Proceedings, Fourth Int. Conf. Soil Mechanics and Foundation Engineering, London, Vol. 1, pp. 283-286 (1961)

2. Clemente, Jr., F. M.; "Downdrag - A Comparative Study of Bitumen Coated and Uncoated Prestressed Piles," Pile Talk Seminar Papers pp. 49-71. New York City, March 1 and 2, 1979.

3. Kezdi, A.; "Colopol es Colopcsoportok Teherbirasa," Ep.ip. Kozl. Egyetem Tudomanyos Kozlemenyei, Budapest, 4, No. 3, pp. 25-87 (1958).

4. Ladanyi, B.; Expansion of a Cavity in a Saturated Clay Medium," Journal, SMFE, ASCE, SM4, pp. 127-161, (1963).

5. Partos A. and Koerner, R.M.;"Monitoring Consolidation Settlements of Soft Soils Induced by Dewatering," ASTM STP 584, pp. 112-126 (1975).

6. Terzaghi, K.; "Theoretical Soil Mechanics," New York, (J. Wiley & Sons) 1943.

7. Vesic, A.S.; "Bearing Capacity of Deep Foundations in Sand," National Academy of Sciences, National Research Council, Highway Research Record 39, pp. 112-153 (1963).

8. Vesic, A.S.; "Expansion of Cavities in Infinite Soil Mass," Journal, SMFE, ASCE, SM3, pp. 265-290, (1972).

PULL-OUT LOAD TEST OF A DRILLED PIER IN VERY STIFF BEAUMONT CLAY

By RICHARD A. PEARCE[1], M. ASCE
CARL L. BRASSOW[2], M. ASCE

The load transfer in a pile or pier-soil system and the movements associated with development of full skin friction have been studied in some detail in the last 10 to 15 years. Past field work from instrumented driven or bored pile load tests have presented recommendations for limiting load transfer or skin friction to values ranging from 800 (38.3 Kpa) to 1800 (86.2 Kpa) psf based upon the soil's undrained shear strength and the methods of installation. Past investigations have reduced the undrained shear strength by 30 to 80 percent to determine load transfer.

During preliminary analyses of a drilled pier foundation system for a project in Houston, Texas, total skin friction capacities of 2 to 3 ksf (95.8 - 143.7 Kpa) appeared feasible. In order to verify the design assumptions, a program to test a pier section was developed. The findings of that field study are provided in this paper.

General Site Geology and Soils Information

The site of the field test is located geologically in the Gulf Coast physiographic province. The soils in the area consist of Pleistocene age deposits which are part of the Beaumont Formation. In the Houston area, the Beaumont Formation is approximately 400 feet (122 m) thick and consists of ancient alluvium deposited in deltaic and marine estuarial environments.

The soils of the Beaumont Formation consist of medium to high plasticity, overconsolidated clays and silts which possess moderate to high expansive properties. The soils in the upper 15 to 20 feet (4.58 to 6.1 m) are generally fissured, slickensided and contain numerous deposits of calcareous nodules. Below depths of 20 feet (6.1 m), the soils become less fissured and contain more numerous lenses of clayey sands and clean fine sands. The Beaumont Formation in the Houston area, and specifically the study site, has previous consolidation pressures (P_c) of up to 8 ksf (383.2 Kpa).

From soil test borings drilled at the site which included one boring drilled immediately adjacent to the pier tested, an

---

1. Senior Engineer, Law Engineering, Houston, Texas
2. Chief Engineer, Law Engineering, Houston, Texas

idealized soil profile was developed which is shown on Figure 1. In addition, soil index properties, including standard penetration resistance, undrained shear strength, dry unit weights, natural moisture content and Atterberg limit tests are summarized on Figure 1.

As may be seen from Figure 1, generally three strata were identified at the site. The clay soils found at depths of 50 to 70 feet (15.3 to 21.4 m) were selected as the location of the test pier due to their relative uniformity and because the original assumption of the load distribution indicated that the majority of the total skin friction on a production pier approximately 100 feet in length would be developed in this stratum. The soil in the test zone consists of very stiff to hard reddish-brown silty clay containing interbedded seams of clayey fine sand which can vary in thickness from a few inches to several feet. The soil test boring drilled adjacent to the test pier location found consistent deposits of silty clay containing very few lenses of silt or fine sand in the proposed test zone.

The soils in the test zone are classified as moderately plastic clays (CL-CH) by the Unified Soils Classification System. Plasticity indices varied from 20 to 30 percent. The average value of undrained shear strength was approximately 4 ksf as determined by one-point unconsolidated-undrained (UU) triaxial tests with a confining pressure near or slightly above the effective overburden pressure. In order to gain further insight into the in-situ properties of the soils, a series of Menard pressuremeter tests were conducted in nearby test holes. The pressuremeter modulus to undrained shear strength (Em/c) ratio in the test zone was 100 to 250. The results of the pressuremeter tests are shown on Table I.

<u>Test Program</u>

During planning of the field study, several alternate types of tests were evaluated. The first test considered a full scale instrumented pier which would require testing a section approximately 4 feet (1.22 m) in diameter and 80 feet (24.4 m) in depth. The costs associated with such a procedure were prohibitive which led to tests performed on a selected pier section. Alternative test methods considered included a torsional test, a compressive test on a small scale pier section with a false bottom and a pull-out test. A pull-out test was selected for basically two reasons:

1) End effects normally associated with a compressive test could be minimized, and,

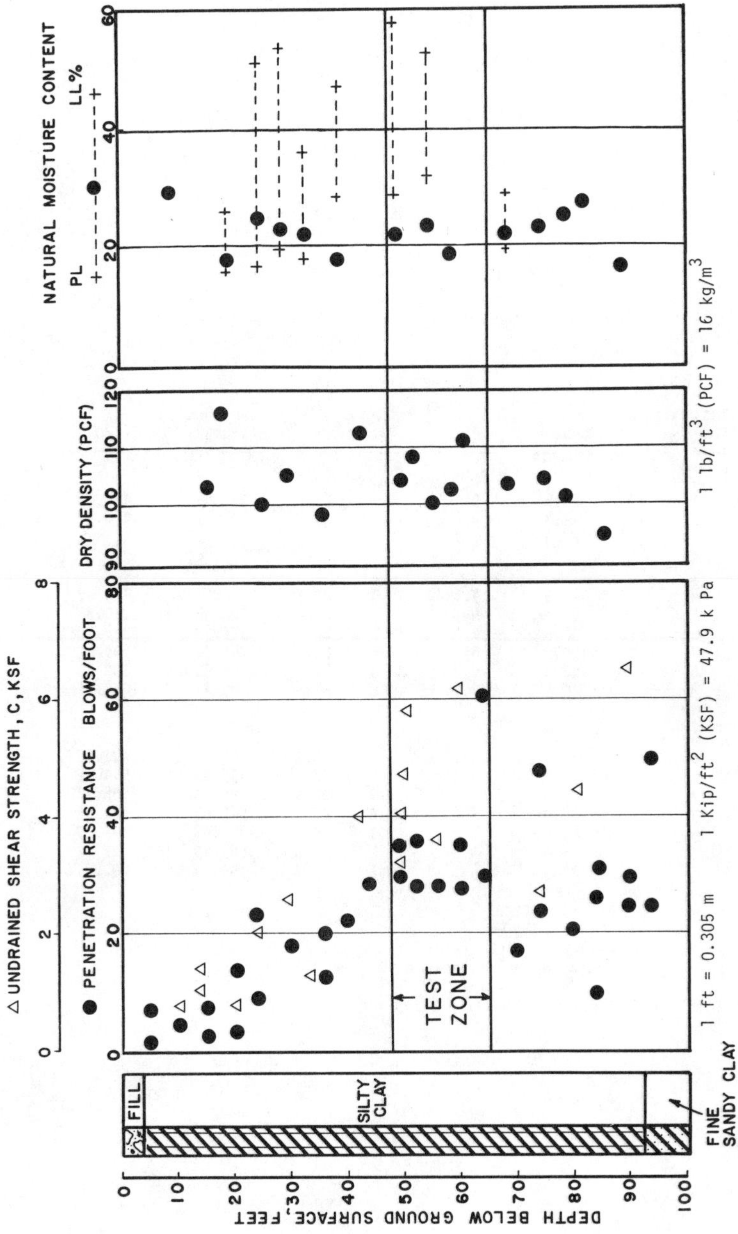

FIGURE I—SUMMARY OF SOIL CHARACTERISTICS

TABLE I

D A T A S U M M A R Y S H E E T

Menard Pressuremeter Tests

- - - - - - - - - - - - - - - - - TSF - - - - - - - - - - - - - - - - -

Test No.	Depth*	P_o	P_f	P	E(=0.33)	E(=0.5)
Boring D-1						
1	54.5	9.1	14.1	21.9	115.65	130.43
2	59.4	4.8	14.3	25.3	193.33	218.07
3	71.7	3.6	14.3	34.1	275.45	310.67
					196.25**	221.39**
4	80.5	7.6	13.1	20.5	161.77	182.47
5	83.8	4.8	11.9	22.4	186.55	210.39
6	91.5	6.4	13.2	28.1	335.43	378.30
7	100.5	6.1	12.6	34.1	745.30	840.57
8	106.5	4.7	22.3	41.4	312.35	352.35
9	120.5	7.1	19.6	56.3	949.71	1071.15
Boring D-5						
10	27.5	3.2	6.5	10.2	63.52	71.63
11	38.1	4.3	11.2	22.3	455.40	513.75
12	48.0	7.2	17.7	21.1	96.10	108.38
13	58.0	4.0	13.9	27.3	515.57	581.63
14	68.0	4.3	12.1	26.7	483.58	545.37
15	78.0	4.6	11.5	20.6	685.75	773.21
16	87.3	4.8	9.9	18.3	259.57	292.81

* Ground surface to probe center, in feet.
** Virgin modulus observed between 10.4 and 14.3 tsf

1 ft = 0.305 m   1 ton/ft^2 (TSF) = 95.8 kPa

2) A test section 3 feet (0.92 m) in diameter, which would be very close to the actual production pier size, could be used in the pull-out test and a specific soil stratum could be tested. This would not be true for either a torsional test or for a smaller scale compressive test on a pier with a false bottom.

The 3 foot (0.92 m) diameter section was installed by first drilling an upper 48 inch (121.9 cm) diameter shaft to a depth of 46 feet (14.0 m). Casing was then placed and seated in the bottom of the shaft to prevent sloughing into the open shaft. A bentonite drilling fluid was then introduced to model the actual proposed installation techniques. The following day, the shaft was advanced with a 36 inch (91.4 cm) diameter auger extending to a depth of 66.7 feet (20.3 m). The bottom of the hole was cleaned and a reinforcing cage was lowered to the base of the hole. The rebar cage was instrumented with two telltales. The configuration of the test section, rebar cage and telltales is shown on Figure 2. The telltales were placed near the anticipated top and bottom of the test section and were used to measure the deflection of the top and bottom of the pier during the test. Concrete was tremmied into place to a depth of 52 feet (15.9 m). Thirty (30) inch (76.2 cm) diameter belled piers (60 inch (152.4 cm) bell diameter) were drilled at a distance of 10 feet (3.05 m) (center to center) on both sides of the test pier to serve as reaction to the test load.

The reaction beams consisted of two wide flange sections bolted together and placed on two 300 ton (2670 KN) hydraulic jacks supported by the two reaction underreamed piers. Pressure gauges were used to read the applied load by each ram. The load tests were performed approximately seven days after the placing of the concrete. At the time of the load test, concrete cylinders made from representative material used for the construction of the test section had a compressive strength of approximately 2800 psi (19320 KN/M^2). A slump test was performed on a representative portion of the concrete in accordance with ASTM Designation C-143 and found to be 4 inches.

The procedure suggested by ASTM Designation D-3689 was followed using the optional quick load test method. The initial load cycle was conducted by applying ten ton load increments at 4-minute intervals until continuous jacking was necessary to hold the load. The deflections of the two telltales were measured by dial gauges while the reaction piers were monitored with string gauges. A measurement of deflection was taken directly after the load was applied and again at the end of the load increment.

After the pier had reached failure in the first cycle, the load was rebounded to zero and the pier was loaded a second time. For the second load cycle, the load was applied continuously until

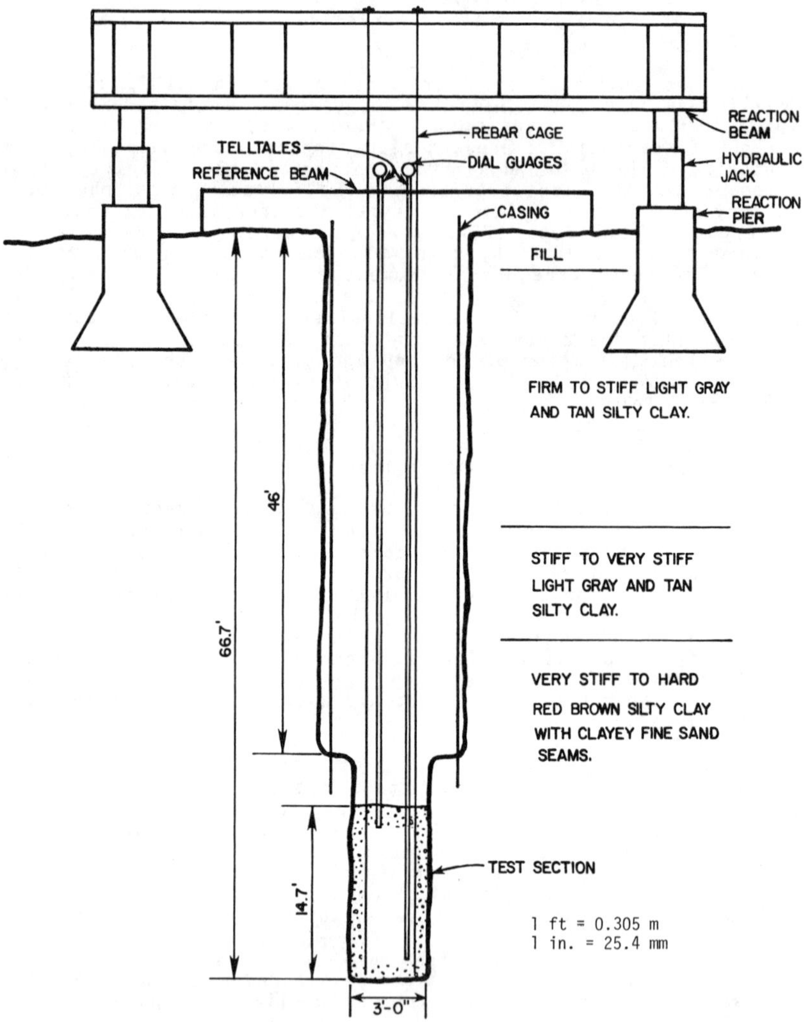

FIGURE 2— TEST PIER CONFIGURATION

further jacking did not increase the applied load. Measurements of load and deflection were taken at 2-minute intervals.

Interpretation of Test Results

The gross uplift movement versus load curve is shown as Figure 3. The movement shown on Figure 3 correlates with the top telltale readings in relation to the applied load. As may be seem in Figure 3 in the first load cycle, the load increased linearly up to approximately 130 tons (1157 kN). At that point, some deflection occurred between the initial readings and the final readings at each load increment. The ultimate load was interpreted as that load where continuous movement with no increase in load was achieved. This was achieved at a load of approximately 160 tons (1424 kN) and at a gross upper telltale movement of approximately 0.18 inches (4.57 mm). At this load, the bottom telltale movement was approximately 0.17 inches (4.32 mm).

In the second load cycle, the test pier was able to sustain a load of nearly 190 tons (1691 kN) before continuous jacking with no load increase occurred. This higher load in the second cycle is attributed to the quicker rate of loading where a continuous rate of uplift or a strain controlled procedure was used. The test pier was moved a total of approximately 2 inches (5.08 cm) during the loading procedure and no decrease in skin friction was noted.

Comparison with Laboratory Test Results

In the initial phases of this study, it was decided that the test results obtained from the test pier loading would be compared with the conventional undrained shear strength values determined in the laboratory and direct shear tests. The average undrained shear strength in the test zone was selected based on the results of one point undrained triaxial tests with confining pressures equal to or slightly above the effective overburden pressure. The results of these tests are shown on Figure 1 and an average cohesion of 4.0 ksf (191.6 N/m^2) was selected for design. The average skin friction developed over the length of the test pier was approximately 2.4 ksf (114.96 kN/m^2) which provides a ratio of load transfer to the design undrained shear strength of approximately 0.6.

Prior to performing the field test, load transfer curves were modeled by performing laboratory direct shear tests with concrete in the upper ring and soil in the lower ring. In addition, standard direct shear tests with soil in both rings were conducted. The concrete-soil shear tests were conducted by

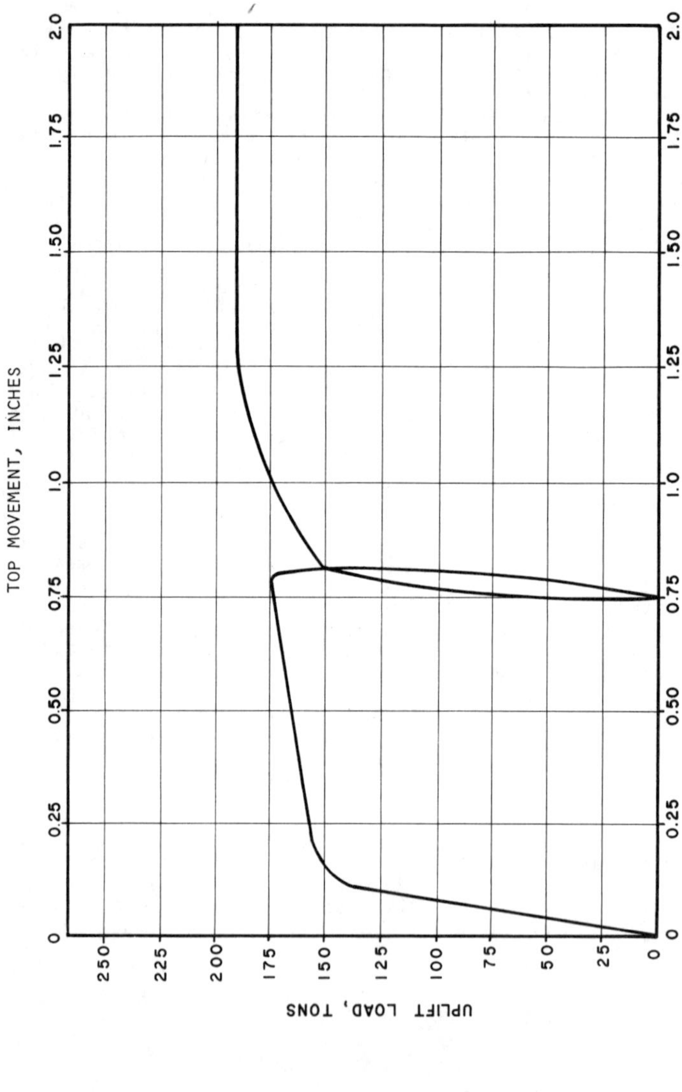

1 ton = 8.9 kN 1 in. = 25.4 mm

FIGURE 3 - UPLIFT VERSUS MOVEMENT CURVE

forcing a failure in the soil approximately 0.125 inches (3.18 mm) from the concrete-soil interface. Mortar mixed to the water-cement ratio expected in the field was placed in the upper ring of the direct shear device and soil was placed in the lower ring. The concrete soil interface was coated with bentonite slurry prior to placing the concrete. A normal pressure was placed on the concrete-soil test specimens which was equal to an effective lateral earth pressure using an at-rest earth pressure value, K_0, of 1.0. This pressure was held for a period of 7 days and then the sample was sheared.

In comparing the results of the direct shear tests and one point triaxial test to the field results, two different methods were used. First, the soil-on-soil direct shear tests were compared with the concrete-on-soil tests. It was found that, in both tests performed, a reduction in maximum developed shear stress of 30 percent occurred between the soil-on-soil sample and the concrete-on-soil sample. (The concrete-on-soil test was the lowest). This reduction in load transfer is probably due to moisture migration from the concrete and slurry placed on the sample, thus resulting in a lower shear strength due to higher moisture content[3]. The overall movement to maximum load transfer found in the direct shear tests ranged from approximately .1 to .15 inches (2.54 to 3.81 mm). This movement compares favorably with the .15 to .20 inches (3.81 to 5.08 mm) found in the field test.

The difference between the shear strength reduction in the laboratory tests (30 percent compared with the peak soil-on-soil direct shear values) and the field reduction (40 percent compared to average undrained shear strength) is probably due to the mechanics of pier installation whether it be remolding of the soils by the soil auger, some relaxation of the sides of the hole due to stress relief or other installation factors. The total shear stress developed in the laboratory direct shear tests were considerably higher than the actual average skin friction calculated from the field test. It is recognized that this may be a function of the magnitude of the consolidating normal pressure. However, it appears that the direct shear test procedure (soil-on-concrete) is appropriate for predicting the reduction in load transfer due to moisture migration or construction techniques and for estimating the total movement to develop ultimate load transfer. It is the authors' opinion that the procedure may not be appropriate for predicting the actual skin friction values developed along the pier. The shear strength profile developed from unconfined compression tests or unconsolidated undrained triaxial tests for fissured clays should be used with an appropriate reduction factor based on the direct shear test procedure. The relative values of reduction in skin friction versus deflection are shown in Figure 4. The figure

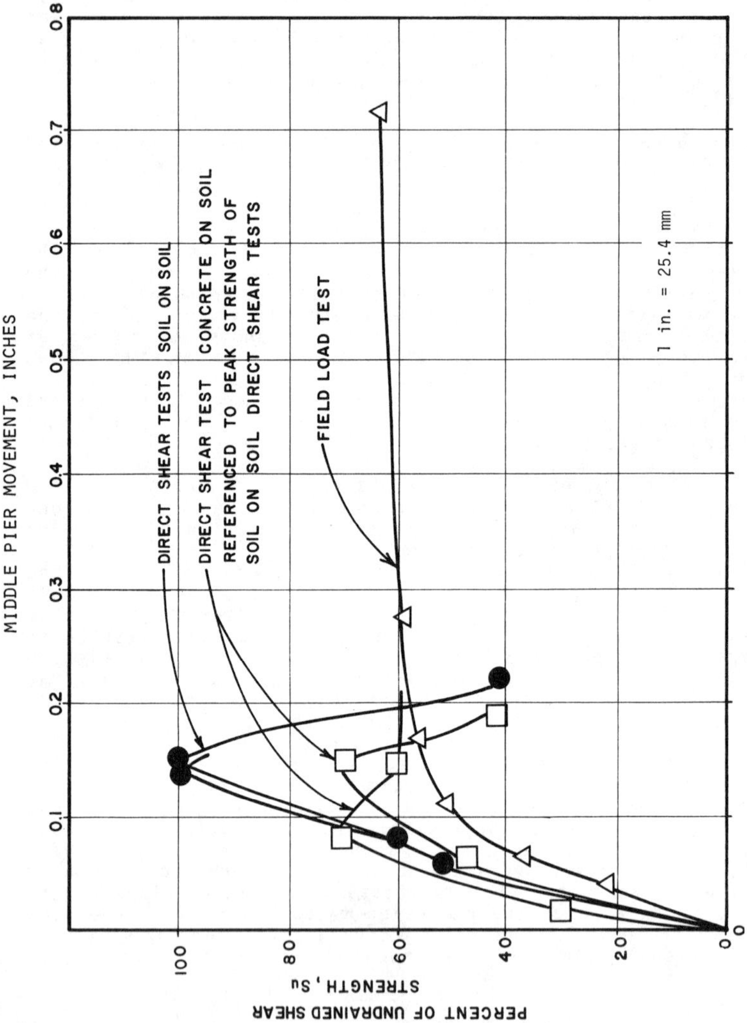

FIGURE 4 - LOAD TRANSFER VERSUS AVERAGE MOVEMENT OF PIER CENTER

provides load transfer information divided by peak load in the plots depicting the direct shear tests and load transfer versus undrained shear strengths for the field test procedure.

Conclusions

Several conclusions of this single load test can be made which may add to the state of knowledge concerning design of drilled piers in very stiff to hard clay soils. Our conclusions are as follows:

1) The reduction factors applied to undrained soil shear strengths that have been found in previous field tests[2,3] in Beaumont clays have been substantiated by the test discussed herein. However, the unit skin friction developed is somewhat higher than previously published limiting values of skin friction for either driven piles or drilled piers. This may be in part due to the lack of direct field information concerning load transfer in very stiff to hard clays. In designing drilled piers in very stiff to hard clays, the end effects within the lower 5 feet (1.53 m) of the drilled pier should be taken into account by an appropriate reduction in unit skin friction. The test results presented herein are considered an appropriate model for the middle section of a pier since the end effects were minimized by the test procedures.

2) Direct shear tests with concrete and soil compared to ordinary direct shear tests may be a useful laboratory procedure for predicting the amount of strength reduction at the soil pile interface since installation procedures can be modeled in the laboratory. Total movements in the laboratory direct shear tests were similar to those found in the field test performed as well as in tests performed by other[2,3].

APPENDIX I - REFERENCES

1. Coyle, Harry M. and Reese, Lymon C., "Load Transfer for Axially Loaded Piles in Clay", Journal, Soil Mechanics and Foundations Division, ASCE, SM2, March 1977, pp. 1-25.

2. O'Neill, Michael W., Reese, Lymon C., "Behavior of Bored Piles in Beaumont Clay", Journal of the Soil Mechanics and Foundations Engineers Division, ASCE, SM2, February 1972, pp. 195-213.

3. Reese, Lymon C., Touma, F.T., O'Neill, M.W., "Behavior of Drilled Piers Under Axial Loading", Journal of the Geotechnical Engineering Division, Geotechnical Engineering Division, ASCE, GT5, May, 1976, pp. 493-510.

4. Tomlinson, M.J. (1947), "The Adhesion of Piles Driven in Clay Soils", Proceedings, Fourth International Conference on Soil Mechanics and Foundation Engineering, London, England.

5. Woodward, Richard J. Lundgren, Raymond, Boitano, Joseph D., "Pile Loading Tests in Stiff Clays", Proceedings of the Fifth International Conference on Soil Mechanics and Foundation Engineering, Vol. II, Div. 3, Paris, France, 1961.

APPENDIX II - NOTATION

Note: The following symbols are used in this paper:

P_c = previous consolidation pressure
E_m = pressuremeter modulus
c = undrained shear strength
P_o = at-rest earth pressure
P_f = pressure at limit of elastic range
P = pressure at failure of soil
= poisson's ratio

PILE FOUNDATIONS FOR ONE MARKET PLAZA

by Edward E. Rinne[1], M. ASCE and Michael W. Stobbe[2], M. ASCE

INTRODUCTION

The One Market Plaza Office building complex was constructed between 1973 and 1975 at the foot of Market Street near San Francisco's waterfront. The project consists of two towers, 43 and 28 stories in height, which rise out of a 6-story structure featuring a covered mall and an architectural connection with the preexisting 8-story Southern Pacific building, a landmark on Market Street since 1917. Two levels of basement parking are provided beneath the complex. A Plot Plan is shown on Figure 1.

Foundation piles for this project are among the longest and most heavily loaded in San Francisco. The somewhat unique installation procedures and load testing program for these piles provide an interesting case history on the effects of installation techniques on the ultimate pile capacity.

[1]Partner, Dames & Moore, Los Angeles, California

[2]Project Engineer, Quinn and Associates, Lafayette, California; formerly Project Engineer, Dames & Moore, San Francisco

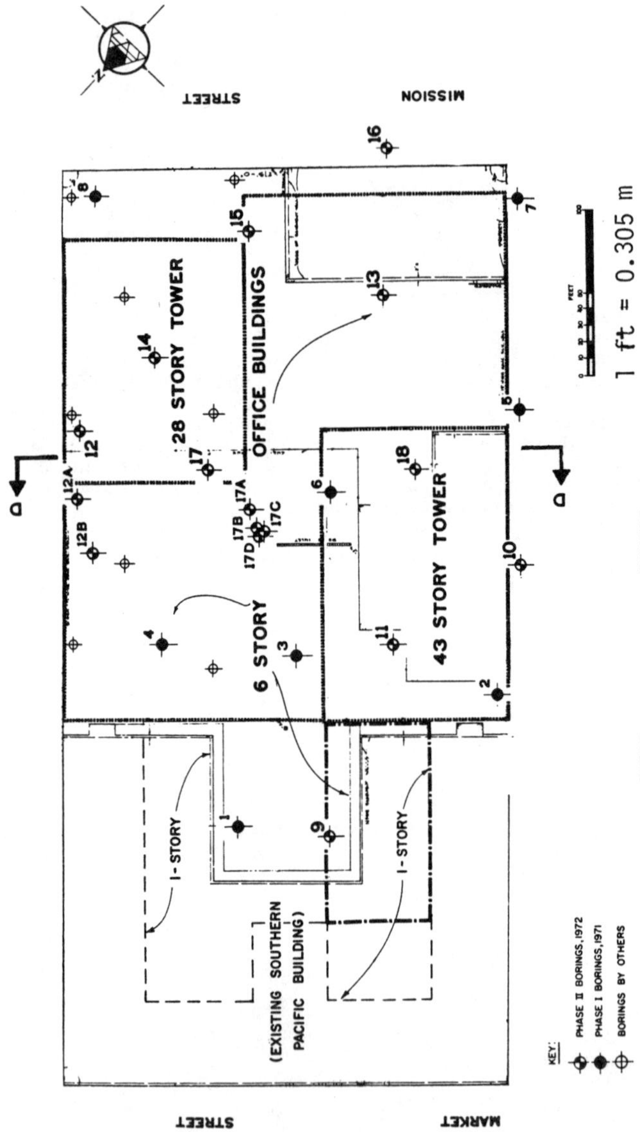

FIGURE I - PLOT PLAN

SOIL CONDITIONS

The building site is situated on filled land approximately 1,500 feet outboard from the original shoreline of San Francisco Bay. The site was filled approximately 100 years ago, and various timber pile-supported structures occupied the site prior to construction of the project.

A generalized subsurface section is shown on Figure 2. The street grade is at approximately Elevation 0 (San Francisco City datum). Borings drilled for the foundation investigation encountered approximately 20 feet (6m) of loose silty sand fill and rubble underlain by normally consolidated clay (Recent Bay Mud) to depths on the order of 100 to 110 feet (31 to 34m) below the street grade. The Bay Mud is underlain by medium to very dense sand and silty sands with interbedded stiff clay layers (Upper Sand). The Upper Sand generally extends to Elevation -135 to -140 feet (-41 to -43m), of which the lower 5 to 10 feet (1.5 to 3m) are significantly more dense than the soil above it, with standard penetration N-values on the order of 100 or greater. The Upper Sand is underlain by stiff to very stiff overconsolidated silty clay (Old Bay Clay). The Old Bay Clay extends as deep as Elevation -240 feet (-73m). A significant but discontinuous dense sand layer was encountered in the Old Bay Clay deposits at about Elevation -190 to -200 feet (-58 to 61m). This intermediate sand layer was found in the borings to range up to 24 feet (7m) in thickness, but also to pinch out completely beneath the center of the 43-story tower.

The Old Bay Clay deposits are underlain by dense to very dense clayey sand and gravel, and weathered shale and sandstone of the Franciscan Formation. Ground water was encountered in the fill at approximately Elevation -12 feet (4m).

A summary of undrained shear strength measurements performed on the Bay Mud and Old Bay Clay deposits is shown on Figure 3. Consolidation test data from the investigation for this project, as well as others in downtown San Francisco, indicate that the Recent Bay Muds are normally consolidated under the weight of the fill, and that the Old Bay Clay deposits are overconsolidated in excess of 3,000 to 4,000 pounds per square foot (144 to 192 kN/m^2) above the effective vertical stress existing prior to construction (Figure 4). The $C_c/1+e_o$ for these soils is on the order of 0.2 and $C_r/1+e_o$ ranges between 0.01 and 0.02.

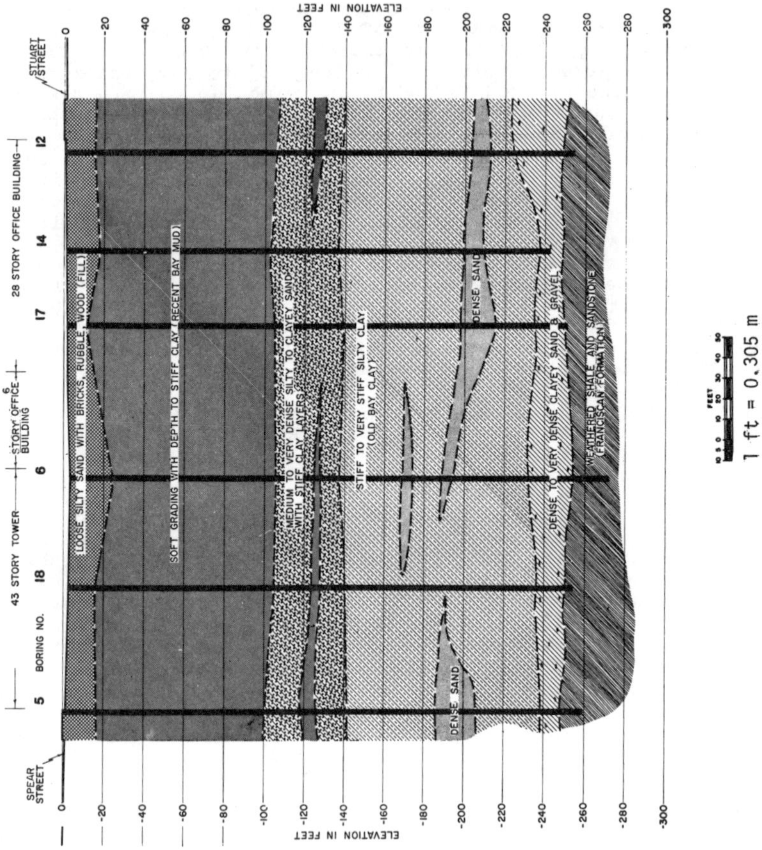

FIGURE 2. SOIL PROFILE THROUGH SECTION D-D

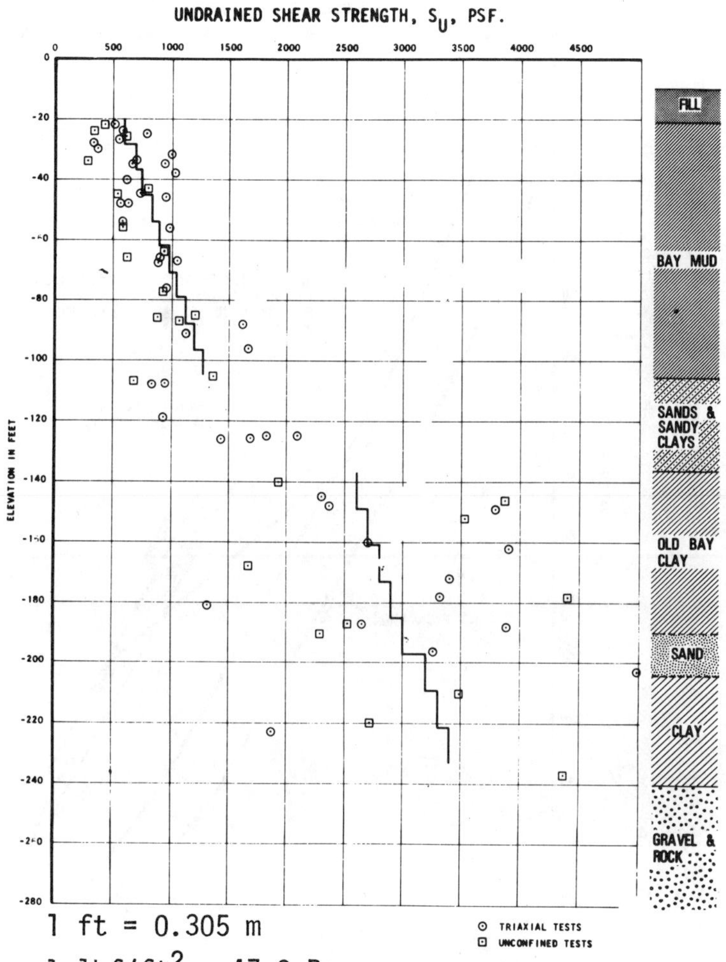

1 ft = 0.305 m

1 lbf/ft^2 = 47.9 Pa

FIGURE 3. - UNDRAINED SHEAR STRENGTH

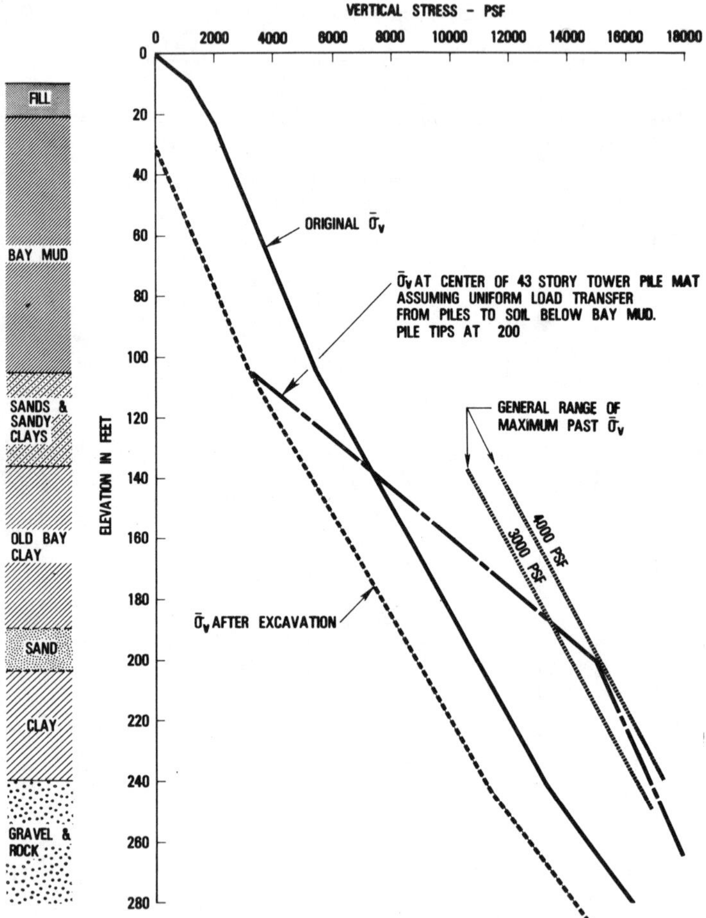

1 ft = 0.305 m

1 lbf/ft^2 = 47.9 Pa

FIGURE 4. - VERTICAL STRESS DISTRIBUTIONS

PREDESIGN DRIVING PROGRAM

During the course of an initial foundation investigation at the site, the design team of the architect, structural engineer, and foundation engineer arrived at the following foundation criteria for the project:

1. Support the high-rise structures on high-capacity piles in the range of 150 to 200 tons (1.3 to 1.8 MN) or more.

2. Support the 6-story structures on moderate capacity piles in the range of 60 to 80 tons (0.5 to 0.7 MN)

3. Limit differential settlements between the high-rise and the lower structures to allow them to be structurally connected; and

4. Provide protection against adverse settlement of the existing Southern Pacific building.

Based on these criteria and the conditions encountered in the initial investigation, long steel piles extending into the dense sands and gravels, overlying bedrock at Elevation -220$\pm$ feet (67m) or deeper, were recommended for support of the high-rise structures. These piles had recommended capacities of 205 to 220 tons and differential settlements between adjacent columns were estimated to be a fraction of an inch. It was recommended that the low-rise structures be supported by prestressed concrete piles extending into the upper sand layer directly underlying the Recent Bay Mud and designed for loads of 70 tons (0.6 MN).

A test pile program was initiated shortly after completion of the initial investigation. The primary purposes of the test pile program were:

1. To evaluate the vibration effects of pile driving on computer facilities located in the existing Southern Pacific building; and

2. To evaluate the feasibility and potential construction difficulties of driving very long piles for support of the proposed high-rise structures.

With respect to the feasibility of driving long piles, it was recognized from the soil data available at the time that difficult driving conditions could be expected in the intermediate sand layer interbedded in the Old Bay Clay deposits at approximately Elevation -200 feet (-61m). The steel test piles were driven at a location where the thickness of the sand layer was as great as could be anticipated with available data. A 14 HP 117 steel-bearing pile, and a 16-inch (0.40m) open-end pipe pile were driven at this location with a Kobe K-32 hammer and met practical refusal with 12- and 3-foot (4 and 1m) embedments respectively, into the intermediate sand layer. The test pile program also indicated a very high driving resistance for the prestressed concrete piles driven into the Upper Sands located immediately below the Recent Bay Mud.

PILE DESIGN

Results of the predesign driving program led to a more detailed evaluation of how the tower foundations would perform with pile tips extending only to Elevation -200 feet (-61m) (as driven in the test pile program). In addition, the design capacity of the concrete piles supporting the low-rise structure was reviewed.

The design capacity of the long piles was developed on the basis of adhesion of the supporting soils below the normally consolidated Bay Mud deposits. Previous pile load test data, collected in the site vicinity, indicated that this adhesion (approximately 2,000 psf (100 kN/m^2) average) would be capable of supporting twice the design load of 200 tons for either 14-inch (0.36m) HP sections or 16-inch (0.40m) pipe pile sections driven to Elevation -200 feet (-61m). End-bearing support from the intermediate sand layer was available in addition to this adhesion capacity over portions of the site. Due to the discontinuous nature of this intermediate sand layer, however, no additional capacity was considered in design other than permitting piles to stop short of the -200-foot (-61m) tip elevation provided suitable driving resistance was encountered in the intermediate sands.

The principal concern, however, was settlement; particularly the differential settlement between the 43-story tower and the adjacent 6-story structure. Settlements were computed on the basis of assumed uniform design load transfer along the piles from the bottom of the Bay Mud to the tip of the pile. The resultant effective vertical stresses using this assumption are plotted on Figure 4. The stress distribution was computed by cumulating the effects of a series of areal loads applied at various depths below the towers. The stress distribution was also computed using a Mindlin solution for each individual pile and superimposing the stresses from all adjacent piles having a significant effect on the stresses at the point of interest. Both methods gave similar stress distributions. When the computed stresses were combined with recompression moduli for the Old Bay Clays, based on both consolidation test data and analysis of settlement records for other buildings in downtown San Francisco, maximum building settlements were estimated to be on the order of 2 inches (50 mm). The differential settlements in the structural bays between the towers and the 6-story structures were estimated to range from 1 to 1-1/2 inches (25 to 38 mm). While these differential settlements did not satisfy the original design criteria, they were eventually accommodated in the structural design, and the project proceeded.

The low-rise building concrete pile capacities were raised from 70 to 80 tons (625 to 710 kN) on the basis of the driving test, except in one corner of the project where the Upper Sand was significantly thinner. At the latter location, the 70-ton (625 kN) design capacity was retained.

PILE INSTALLATION AND LOAD TESTING

The piling subcontractor chose to use 16-inch diameter pipe piles for the tower support, but proposed an alternative to the 0.5-inch (13 mm) wall thickness specified. He proposed using higher strength steel and a thinner wall thickness of 0.406 inch (10 mm).

Difficult driving was anticipated to penetrate the Upper Sands between about Elevations -100 and -140 feet (-30 to -43m). For this reason, the soils engineer recommended that if the subcontractor used pipe piles, they be driven open-ended. It was suggested, however, to reduce the amount of soil cleanout required before concreting, the subcontractor might consider placing a weak plug or cap in the pile tip. The plug was to keep some of the soft Bay Mud out of the pile during initial placement and then break upon commencement of driving (the piles typically penetrated the upper 60 to 80 feet (18m to 24m) under the combined dead weight of the pile plus hammer before driving was initiated).

The subcontractor, however, wished to completely avoid the time and expense of cleaning out the piles. To this end, he embarked upon an experimental driving program which began with an essentially closed-end pile and ended with the procedure shown on Figure 5. This arrangement was based on the fact that concrete filling was not structurally required in the bottom 35 feet (11m) of the pile due to load transfer considerations. The final configuration involved a relatively substantial breakaway plug located near the pile tip, and a solid 1-inch (25 mm) steel plate located 35 feet (11m) from the pile tip. This breakaway plug was to be stronger than the plug originally proposed by the soil engineers. The plug should be strong enough to withstand easy driving, but should break upon commencement of hard driving when the dense Upper Sands were reached. Thus, when working as planned, the pile would drive essentially closed-end through the softer Bay Mud deposits, the breakaway plug would break near the top of the Upper Sand stratum, the pile would drive as an open-end pile for the next 35 feet (11m) through the sand stratum, and finally drive essentially as a closed-end pile again through the Old Bay Clays. The pile was spliced in the field prior to the tip reaching the bottom of the Bay Mud and was driven the last 25 feet (8m) with a hollow pipe follower that was pulled after pouring the concrete through it to fill the pile.

The procedure shown on Figure 5 was evolved through a series of five test piles and load tests, and was that used for the majority of the piles driven for the project. Difficulties experienced were principally due to the reliance of the lower plate, which sometimes broke too soon or not at all.

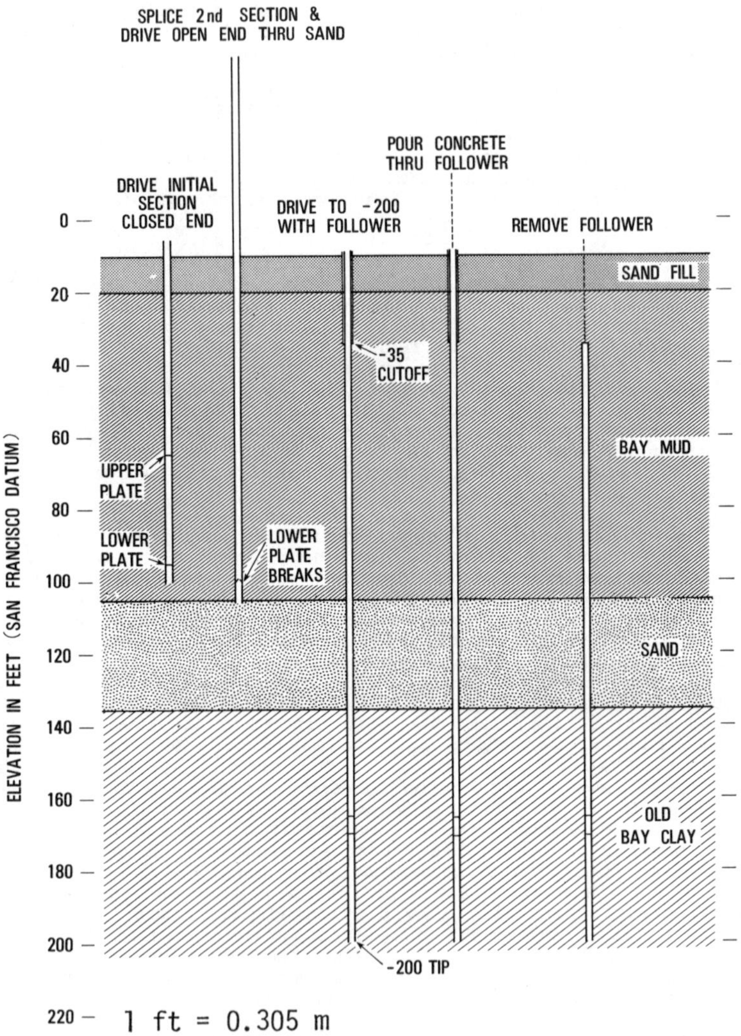

FIGURE 5.
SCHEMATIC SEQUENCE OF 200-TON PILE INSTALLATION

The first test pile (Pile B-49) was essentially a closed-end pile, having no lower plate and a 1-inch solid upper plate 35 feet (11m) from the tip. A small air vent was drilled through the pile wall just below the solid plate. The driving record for this pile and the log of the closest test boring are shown on Figure 6. As can be seen, extremely high driving resistance was encountered in the Upper Sand layer for this pile along with much higher than normal driving resistance in the Old Bay Clays below the sand layer. A wave equation analysis was performed on this pile, and the results are shown on Figure 7. For the final set, the wave equation indicated that the ultimate capacity, R_u, would be 300 tons (2.7 MN).

A static load test was performed on this pile 8 days after driving, and the results are shown on Figure 8. The pile failed at a test load of 300 tons (2.7 MN). This indicated a factor of safety far less than that desired. Comparison with the wave equation results indicates no setup in adhesional capacity during the 8 days between driving and testing.

This apparent lack of setup together with the anomalous high resistance in the Old Bay Clays, led the authors to believe that the pile had been damaged during driving. It was postulated that a "wrinkle" had developed at the pile tip which had prevented good pile-to-soil contact in the very stiff Old Bay Clays. This method of installation was thus both unsuccessful from the contractor's point of view, and unacceptable from the soil engineer's point of view.

The pile subcontractor then proposed a tack-welded end plug at the pile tip with a pair of heavy stop bars located 35 feet (11m) from the tip. This pile (Pile A-78) was driven with similar equipment to the previous pile and the driving record is shown on Figure 9. Here, only a few feet of fairly hard driving were encountered in the sand layer, the end plug broke as intended, and the pile tipped out with a driving resistance of 12 blows per foot. Referring back to Figure 7, the wave equation for this pile would indicate an ultimate dynamic resistance of only 200 tons (1.8 MN). A load test (Figure 10) was conducted on this pile 15 days after driving, and after successfully holding twice the 200-ton (1.8 MN) design load for 24 hours, the pile was loaded to failure at a constant rate of penetration to approximately 560 tons (5 MN). Following this successful test, the subcontractor began using this installation method for production piles.

The tack-welded plate procedure proved to be unreliable, however, and a large percentage of piles ran into difficulties due to either the plate breaking too soon or not breaking at all. Either way, the pile was essentially closed-end when it reached the Upper Sand and the subcontractor expended a great deal of effort in trying to advance these "hung up" piles. It appeared that the time spent in experimenting with the tack-welded plate was not paying off.

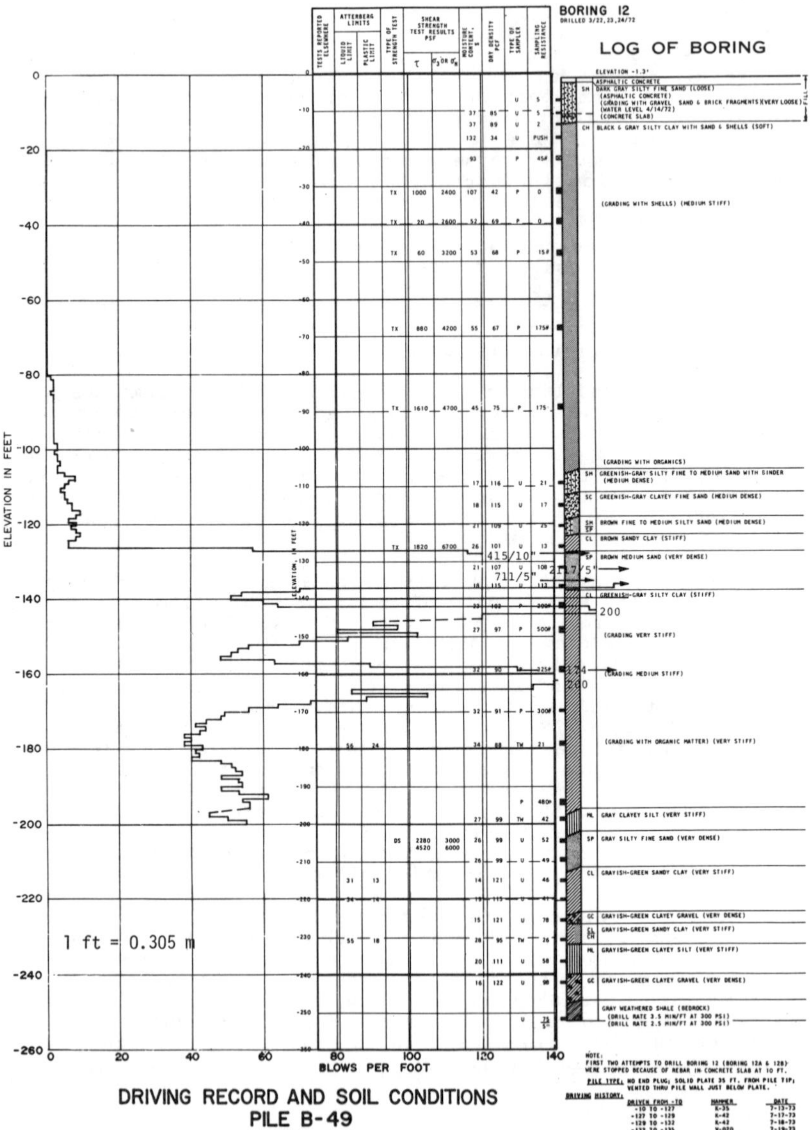

DRIVING RECORD AND SOIL CONDITIONS
PILE B-49
FIGURE 6

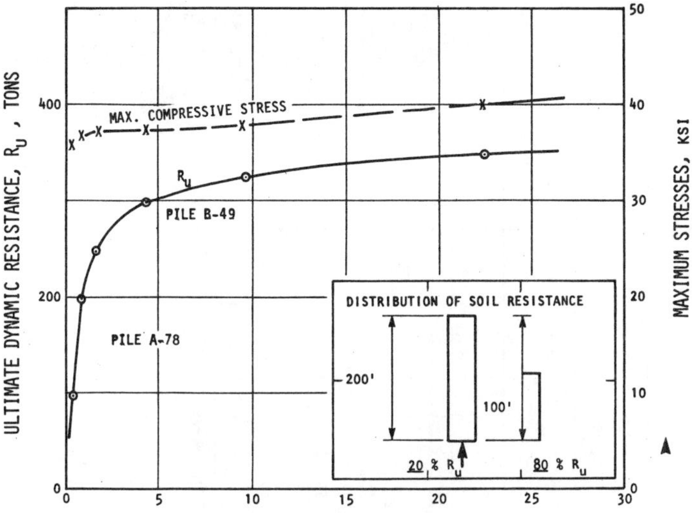

1 ton = 8.9 kN 1 in. = 25.4 mm 1 ksi = 6.9 M Pa

FIGURE 7. WAVE EQUATION ANALYSIS DIESEL 1 SIMULATION

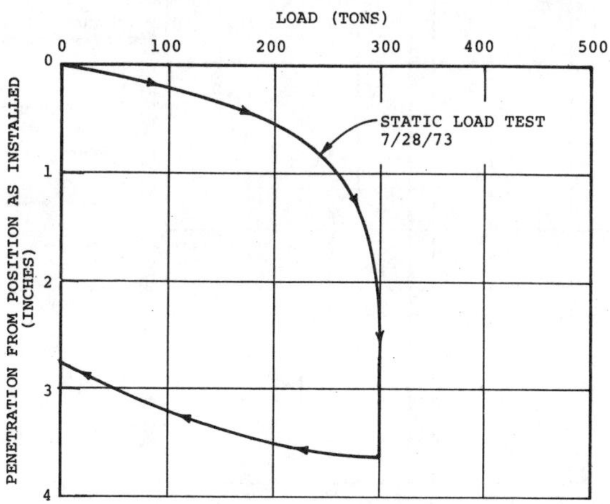

1 in. = 25.4 mm 1 ton = 8.9 kN

FIGURE 8. LOAD-SETTLEMENT RELATIONSHIP PILE B-49

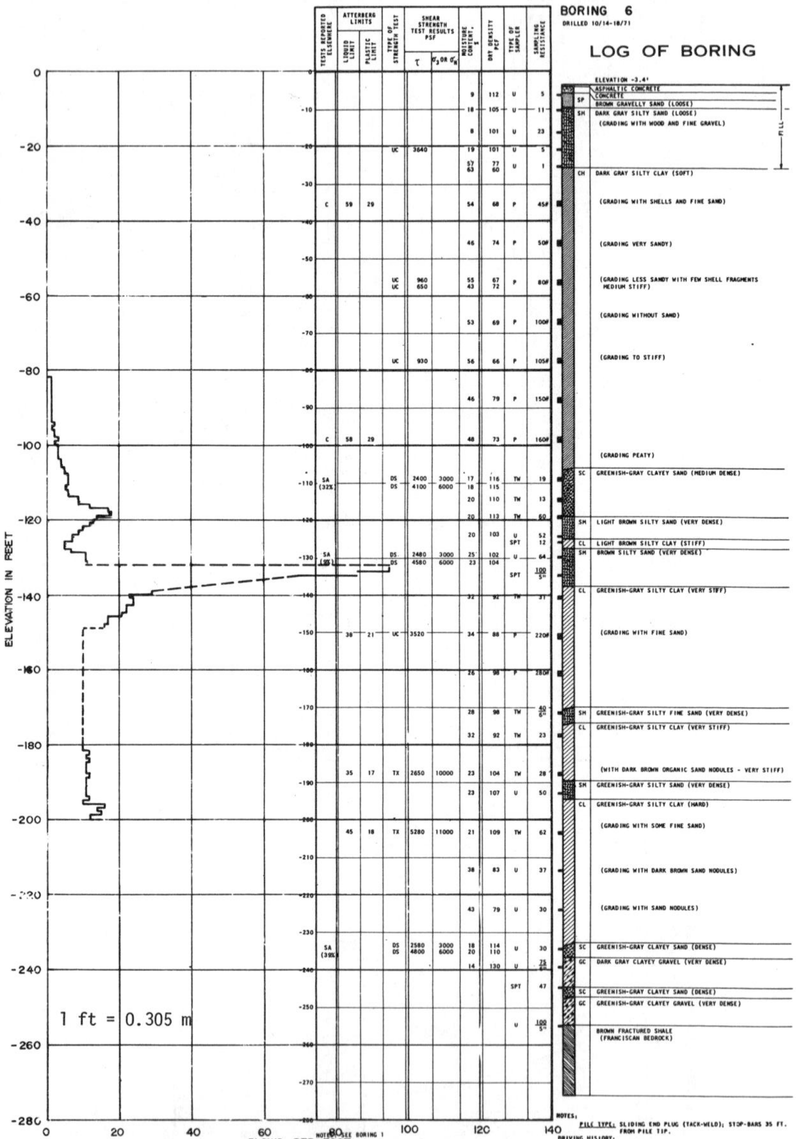

DRIVING RECORD AND SOIL CONDITIONS

PILE A-78

FIGURE 9

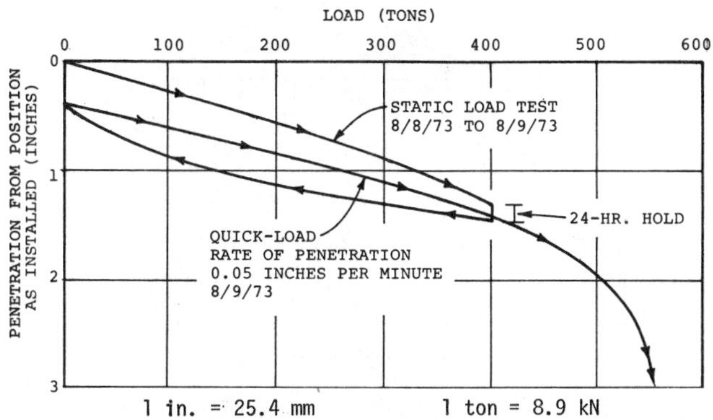

FIGURE 10. LOAD-SETTLEMENT RELATIONSHIP PILE A-78

At the suggestion of the soil engineers, an open-end pipe pile was driven, cleaned out and tested. The driving record for this pile (Pile B-72) is shown on Figure 11, and the load test data are shown on Figure 12. This pile was eventually loaded to 600 tons (5.3 MN) without failure (the load capacity of the testing equipment).

The pile subcontractor, however, elected to continue as much as possible with the breakaway plate concept. He replaced the stop bars with a solid 1-inch (25 mm) plate with a small air vent. The solid 3/8-inch (9.5 mm) plate at the pile tip was replaced with a plate that was split down the middle. With this arrangement, better control of the strength of the plate was anticipated by varying the amount of weld applied to the split. The first few piles driven with this arrangement were mostly unsuccessful, so the subcontractor decided to cut a large hole through the upper plate to enable him to find out, by means of a sounding line, when the lower plate was failing.

From these sounding line measurements, it was immediately apparent that for the next pile in which the breakaway plate failed as intended, the difference in hydrostatic pressure inside and outside the pile was causing appreciable blowing of the Upper Sands into the pile when the plate broke. It was the author's opinion that this significantly reduced the frictional capacity of the pile, and the method was considered unacceptable. Following this incident, the driving records of piles driven previously with some form of pile-tip breakaway plate were carefully reviewed for evidence of prior "blow-out" occurrances. (Test Pile A-78 was apparently unaffected.)

To balance the differential hydrostatic head, the subcontractor tried prefilling the piles with water, and a test pile was driven in this manner (Pile A-220). Filling the 16-inch pile with water was very tedious, however, and the subcontractor did

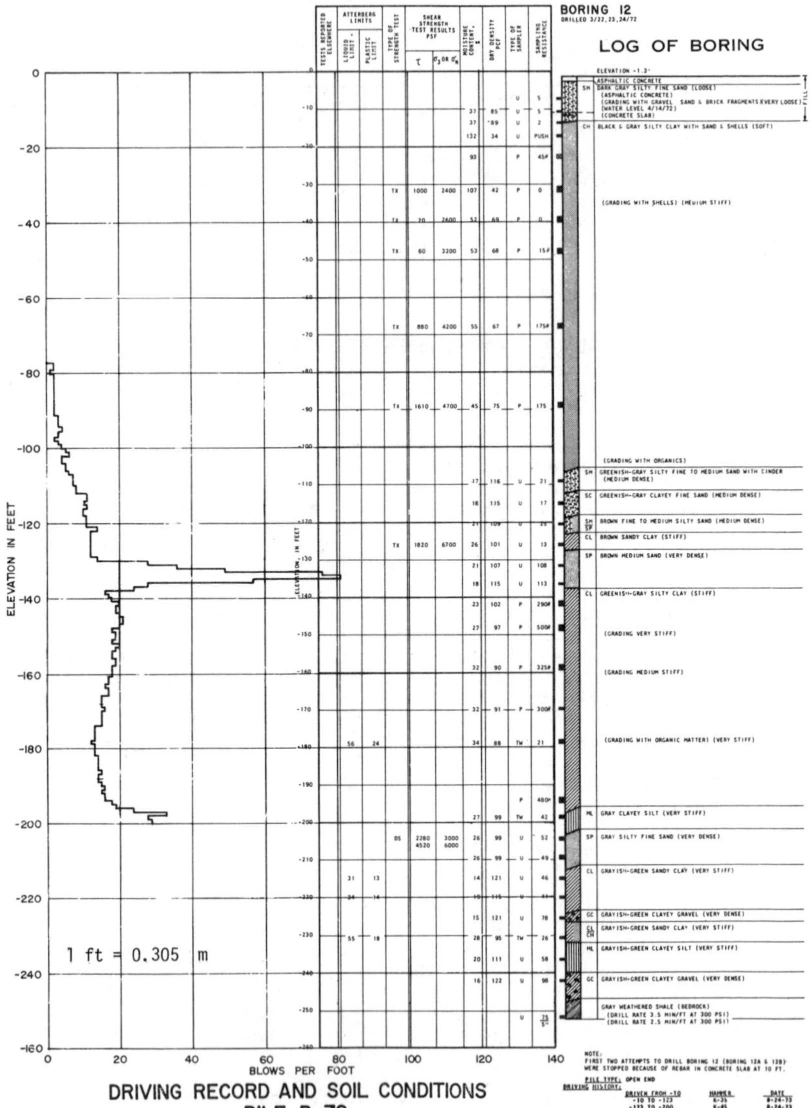

DRIVING RECORD AND SOIL CONDITIONS
PILE B-72
FIGURE 11

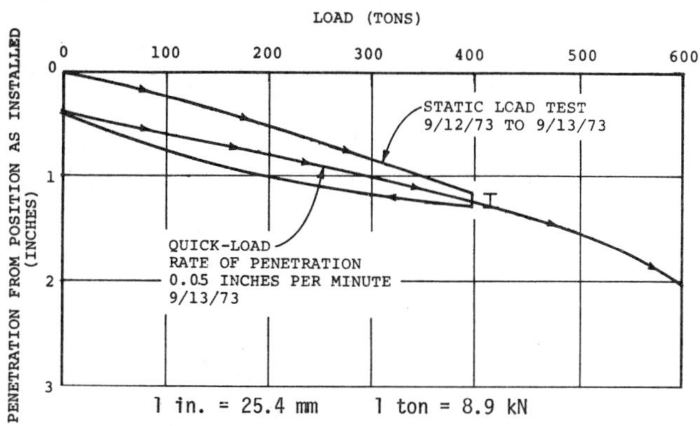

FIGURE 12. LOAD-SETTLEMENT RELATIONSHIP PILE B-72

not attempt to completely balance the expected hydrostatic head. The test pile was prefilled with 71 feet of water. When the plate broke, there was still an unbalanced head of about 56 feet (17m), and it was the author's opinion that a blow did occur. The driving record for this pile is shown on Figure 13, and the load test data are shown on Figure 14. This pile deflected excessively at twice the design load and failed at a constant rate of penetration at 460 tons (4.1 MN).

The authors then recommended that if the pile subcontractor still wished to use the breakaway plate concept instead of driving open-end, the plate be located approximately 5 feet (1.5m) from the pile tip to permit a plug of cohesive soil to impede the sand blowing problem. The fifth and final test pile (Pile A-246) was driven using this procedure, and the driving record is shown on Figure 15. There was no indication of a "blow" occurring with this pile. The load test results, Figure 16, show that this pile successfully held twice the design load for 24 hours and failed under a constant rate of penetration at 550 tons (4.9 MN).

BUILDING PERFORMANCE

Settlement points were established upon pouring of the basement floor. Maximum settlements, 16 months after the start of structural frame construction, were approximately 1-1/2 inches (38 mm). Maximum differential settlement between the 43-story tower and the adjacent 6-story structure was approximately 1 inch (25 mm). The new structures as well as the adjacent Southern Pacific building have performed well and there have been no signs of distress due to foundation settlements.

The observed settlements are generally in the range of those predicted on the basis of assuming a uniform load transfer from the piles to the soil. Had the conventional approximation been

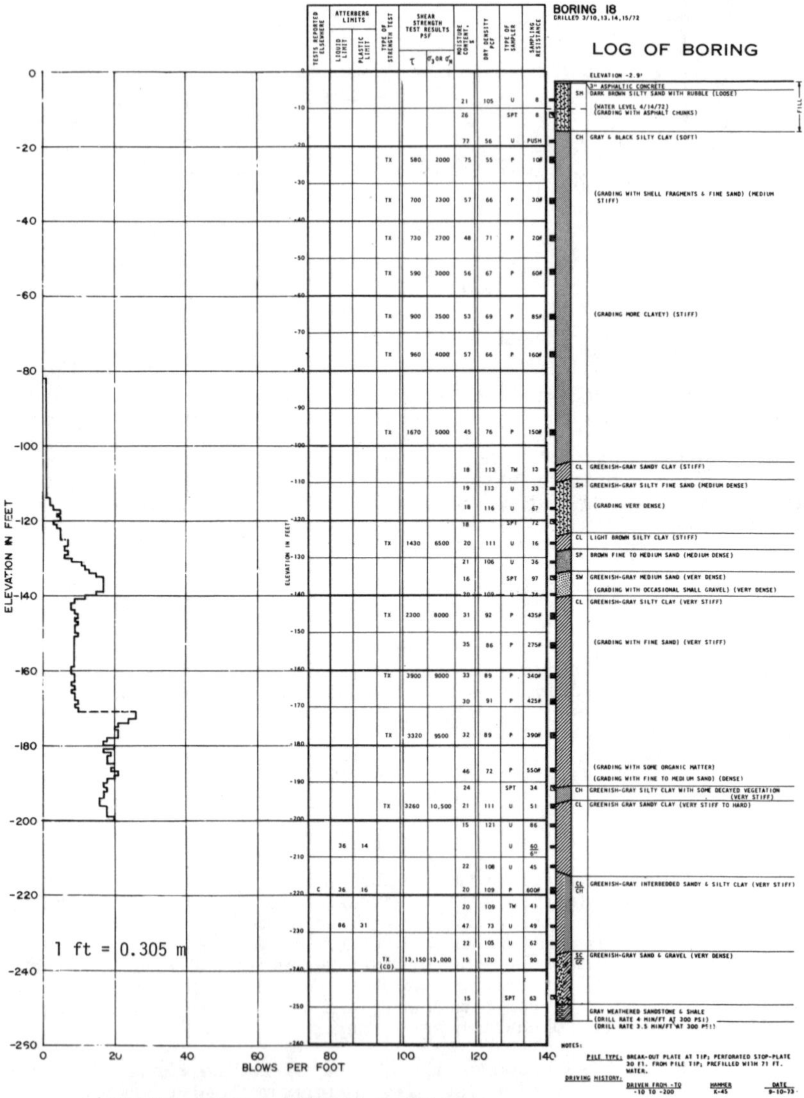

DRIVING RECORD AND SOIL CONDITIONS
PILE A-220
FIGURE 13

used in the settlement analysis, where an equivalent mat is placed at one-third the distance from the pile tip, significantly higher settlements would have been predicted since the computed vertical stresses would have tended to exceed the maximum past vertical stress.

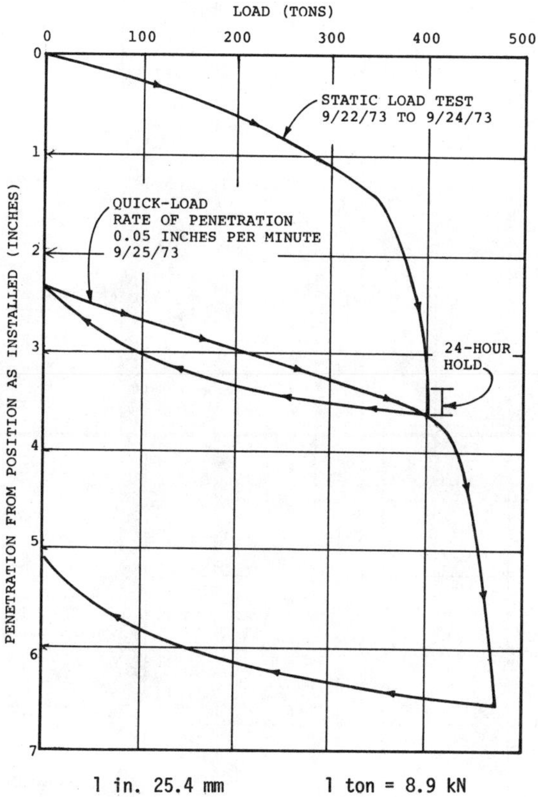

FIGURE 14. LOAD-SETTLEMENT RELATIONSHIP PILE A-220

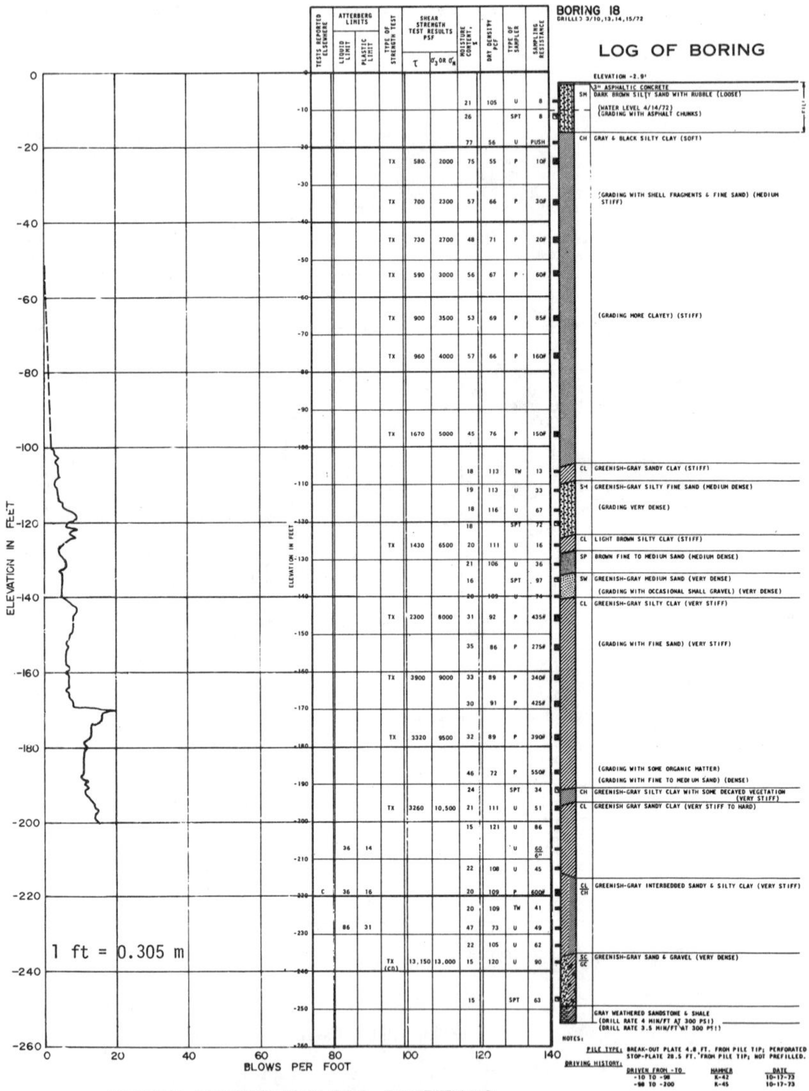

DRIVING RECORD AND SOIL CONDITIONS
PILE A-246
FIGURE 15

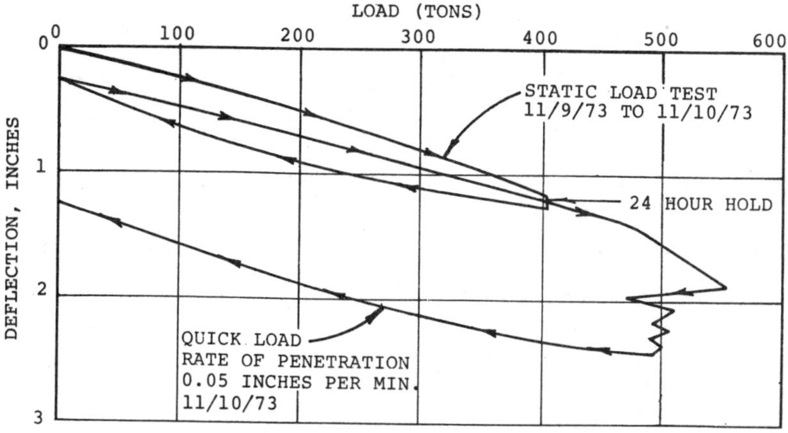

1 in. = 25.4 mm 1 ton = 8.9 kN

FIGURE 16. LOAD-SETTLEMENT RELATIONSHIP PILE A-246

CONCLUSIONS

This case history clearly shows the influence of pile installation techniques on the adhesion capacity of piles, even where jetting or predrilling is not a factor. A factor of two or more in the pile capacity due to installation techniques was apparent.

Where acceptable installation techniques were used, the pile load test results were consistent with calculated capacities based on previous load test programs in the local area. By comparison, computed pile capacities by various current methods (1,2,3) were generally consistent with these data. For this case, computed capacities using Meyerhof (1) tended to be somewhat higher than those computed using Tomlinson (2), and Vijayvergiya and Focht (3).

Observed building settlements were consistent with those computed using a uniform load transfer from the piles to the soil in the supporting strata.

ACKNOWLEDGEMENTS

The cooperation of the Southern Pacific Land Company, and The Equitable Life Assurance Society, owners of One Market Plaza, is gratefully acknowledged. The authors also wish to acknowledge the contributions of Walter H. Hensolt, Peter E. Quinn, J.P. Singh, and Robert D. Darragh to the foundation engineering for this project.

APPENDIX

REFERENCES

1. Meyerhof, G.G., "Bearing Capacity and Settlement of Pile Foundations," Journal of the Geotechnical Division, ASCE, Vol. 102, No. GT3, March 1976.

2. Tomlinson, M.J., "Some Effects of Pile Driving on Skin Friction," Proceedings ICE Conf.: Behavior of Piles, 1971.

3. Vijayvergiya, V.N. and Focht, Jr., J.A., "A New Way to Predict Capacity of Piles in Clay," OTC Paper 1718, 4th Offshore Technology Conference, 1972.

BEHAVIOR OF DRIVEN PILES ON SOFT LIMESTONE

By Curtis J. Roos, [1] A.M. ASCE and
E. W. Lingo, [2] M. ASCE

Introduction

Soft limestone formations are notorious for their erratic nature. These formations often consist of relatively thin, well-cemented calcareous layers separated by sands, clays, and sometimes voids. Although there has been some discussion of the difficulties experienced in driving piles into weak rock formations, few examples have been given regarding the behavior of piles driven in the soft limestones of the lower Atlantic Coastal Plain.

This paper presents a summary of the results of twenty load tests conducted to failure on a variety of driven pile types end bearing in soft limestone at ten sites within the region. The test data is presented in the form of load-deflection curves with the failure loads determined by a consistent graphical technique. Standard penetration test results and pile driving resistances are given for each test location. A range of compression moduli and ultimate bearing capacity are given for the soft limestone at the pile tips.

Piles driven into soft limestones are sometimes "overdriven" to seat the members into the bearing layer. Based upon the test results, a suggested method for hammer section and an empirical driving criteria for piles bearing on soft limestones are presented.

Soil Profiles

The locations of the pile load tests presented in this paper are shown on Figure 1. Standard penetration test results and pile driving resistances at the load test locations are shown on Figures 2 through 6. The test boring locations generally ranged from 2 to 50 feet (1 to 15 m) from the test pile locations.

The overburden soils generally consist of loose to dense sands, clayey fine sands, and some soft silt and clay layers.

---

1 Geotechnical Engineer, Law Engineering Testing Company, Jacksonville, Florida

2 Chief Engineer, Law Engineering Testing Company, Jacksonville, Florida

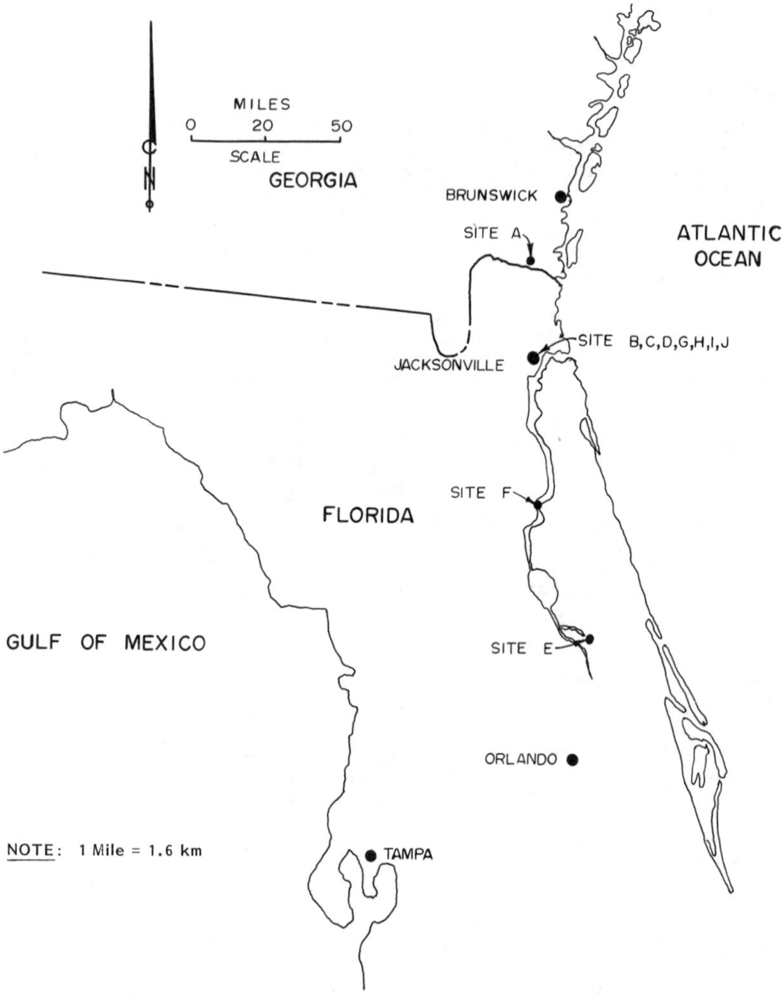

FIGURE 1. Location map

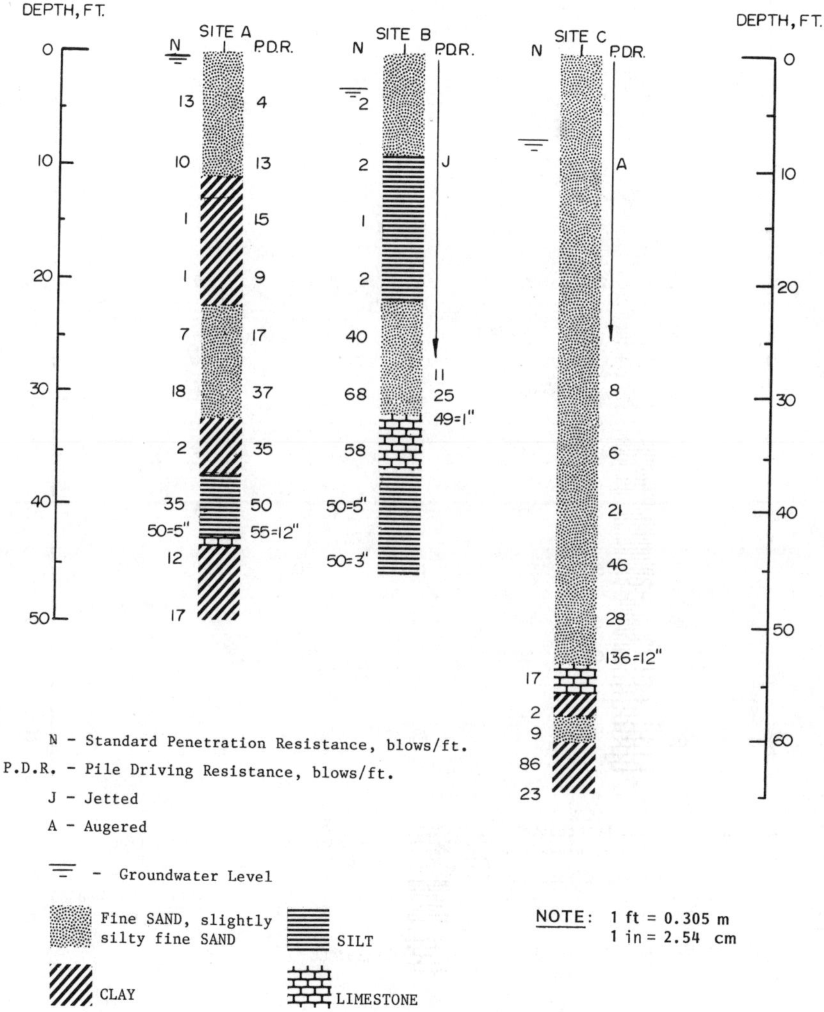

FIGURE 2. Boring and pile driving logs, Sites A, B and C.

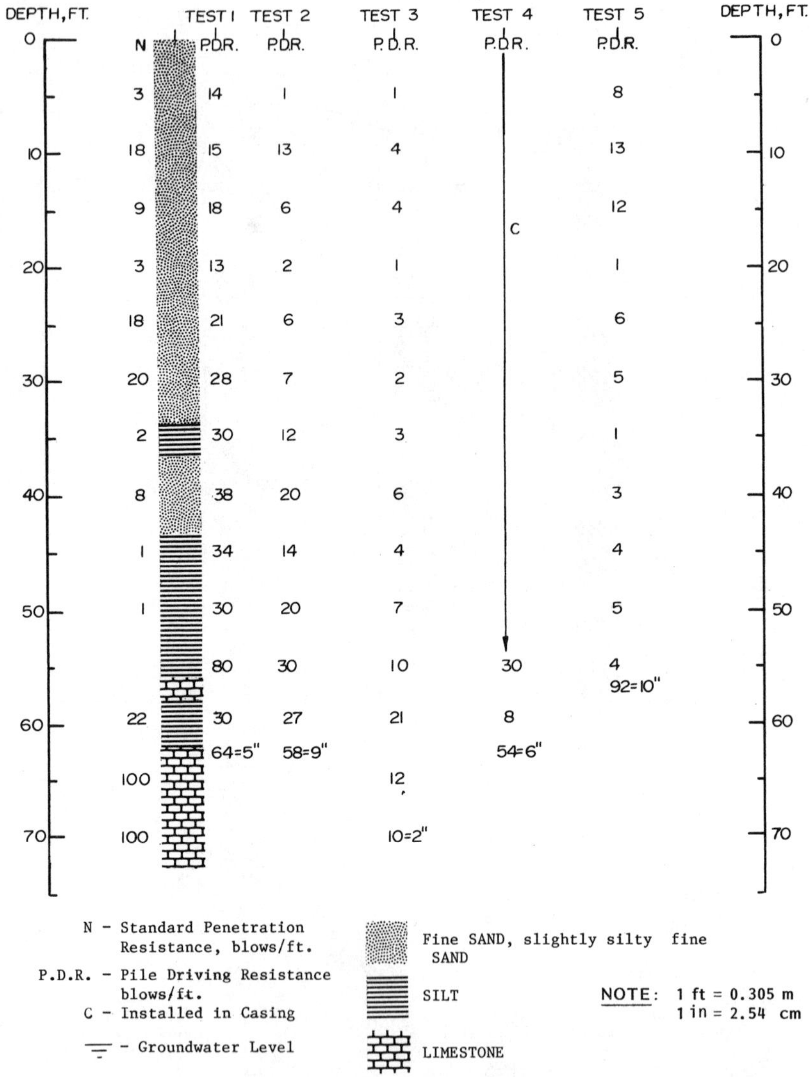

FIGURE 3. Boring and pile driving logs Site D

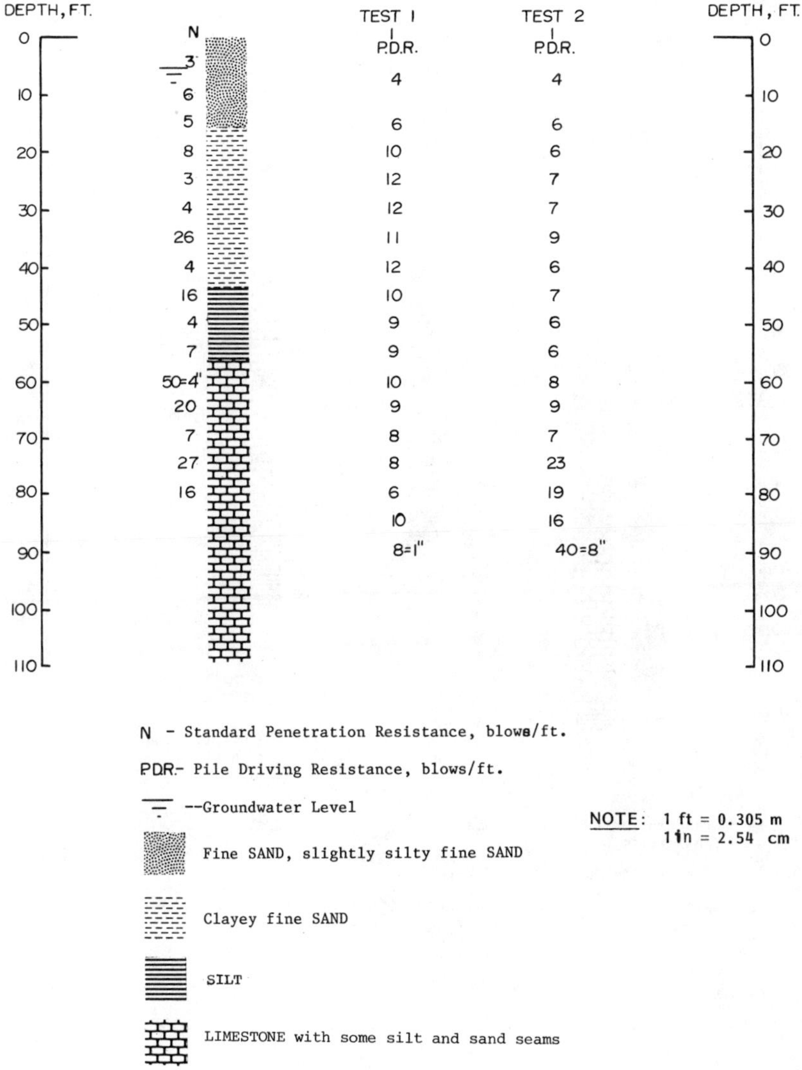

FIGURE 4. Boring and pile driving logs Site E

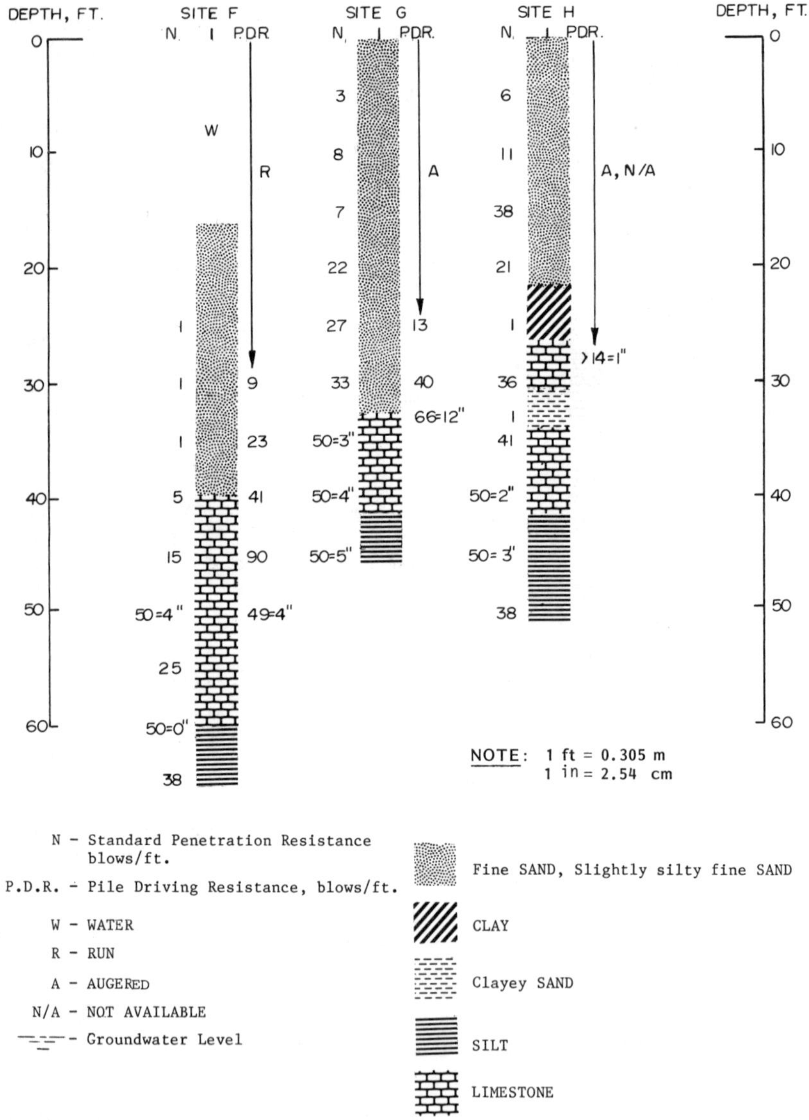

FIGURE 5. Boring and pile driving logs Sites F, G and H

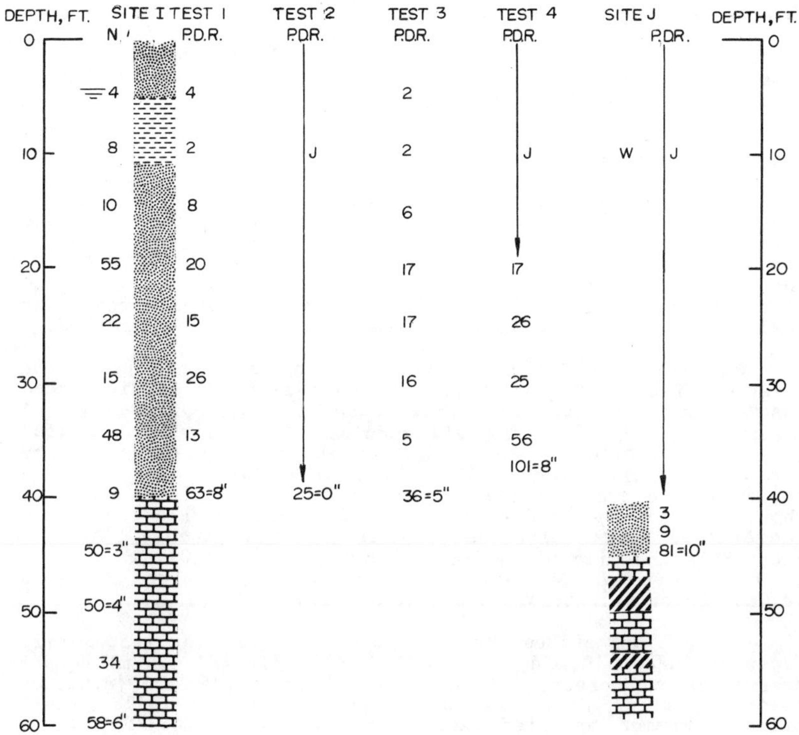

N- Standard Penetration Resistance, blows/ft.

P.D.R.- Pile Driving Resistance, blows/ft.

J - Jetted

W - Water

- Groundwater Level

Fine SAND, Slightly silty fine SAND

Clayey fine SAND

CLAY

LIMESTONE

NOTE: 1 ft = 0.305 m
1 in = 2.54 cm

FIGURE 6. Boring and pile driving logs Sites I and J

The soft limestone bearing layers ranged from less than one foot to more than 50 feet (0.2 to 15+m) thick. The soft limestone formations generally consisted of relatively thin, well-cemented calcareous layers separated by sand, silt, and clay seams at some locations. Although not indicated on the soil profiles, soil filled slots and solutioned zones within the limestone formations were common. Sowers (3) has discussed in detail the engineering implications of soft limestone formations.

Load Test Results

Figures 7 through 16 present the load - deflection curves for 20 test piles driven into soft limestone. The load tests were generally conducted in accordance with the standard loading procedure (slow, maintained load test) outlined in ASTM D1143. The data given in this paper shows the deflection one to two hours after each load increment was applied. Creep effects for load duration in excess of one to two hours have been excluded.

The failure load for each test was determined using the elastic offset criterion suggested by Davisson (2). This method assumes that the pile is a fixed base, free standing column of length (L), area (A), Young's modulus (E_p), and under a load (P). The elastic deflection (Δ) of the pile is calculated by $\Delta = PL/(AE_p)$. The scale of the deflection curve is then selected such that the elastic deflection line plots an angle of about 20° from the load axis. An offset line is then drawn from the elastic line a distance equivalent to 0.15 inches +(B/120), where B is the pile width (inches) or 0.38 + (B/120), where B is in cm. Failure is defined as the load at which the load deflection curve intersects the offset line.

Table I summarizes the load test results. The tested pile types included: 10, 14, and 18 inch (25, 35, and 45 cm) square prestressed concrete, 10 and 10-3/4 inch (25 cm) diameter concrete filled pipe, and step-tapered concrete filled shell piles. Hammer energies ranged from 15,000 to 43,750 ft-lbs (20 to 60 kN-m). Failure loads ranged from 50 to 250 tons (450 to 2200 kN).

Apparent Ultimate Bearing Capacity And Modulus

The apparent ultimate bearing capacity and compression modulus at the tip was calculated for the test piles with a constant cross-section that had standard penetration resistance information available in the overburden soils. Using static analysis, the approximate skin friction in the soils above the pile tip was determined. This value was subtracted from the failure load and the remaining load was divided by the tip area resulting in the apparent ultimate tip bearing capacity (Q_{ut}).

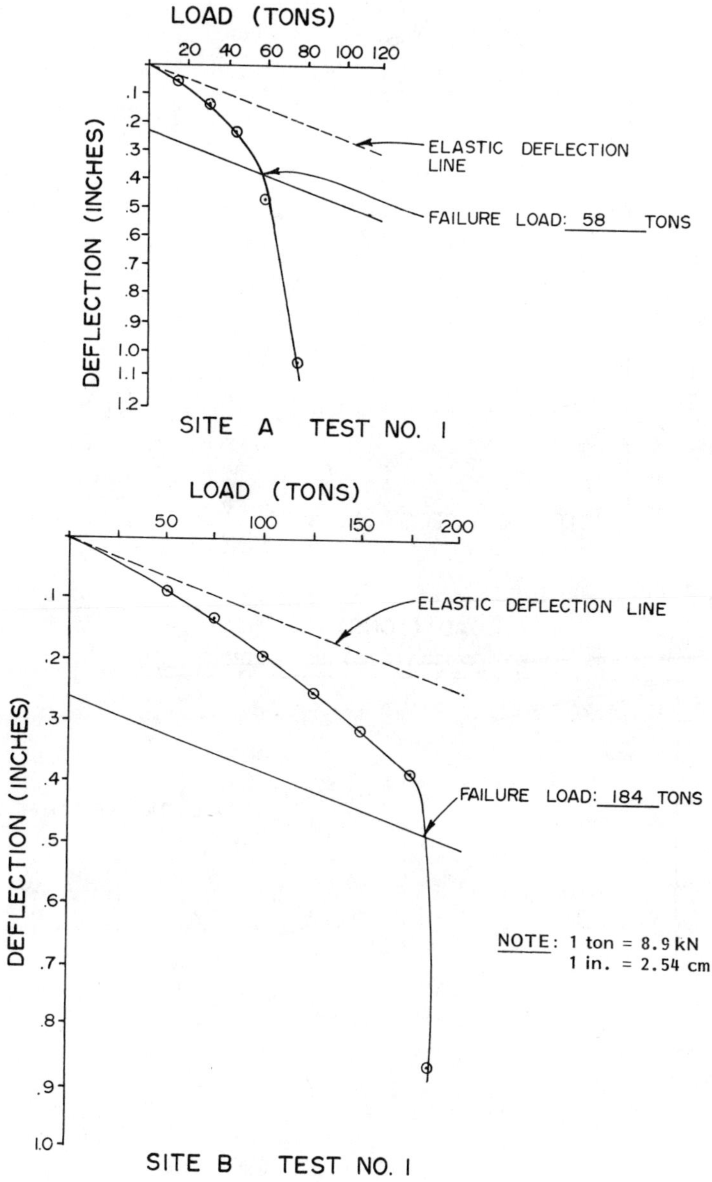

FIGURE 7. Load-deflection relationship Sites A and B.

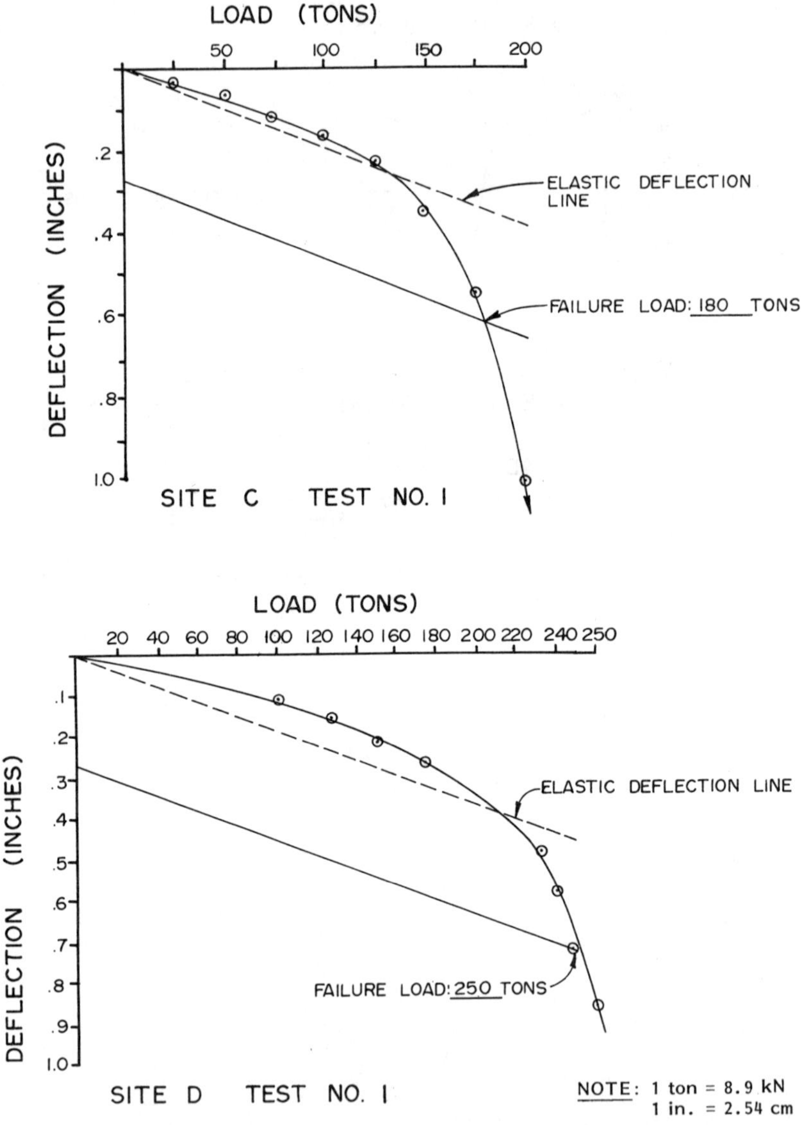

FIGURE 8. Load-deflection relationships Sites C and D

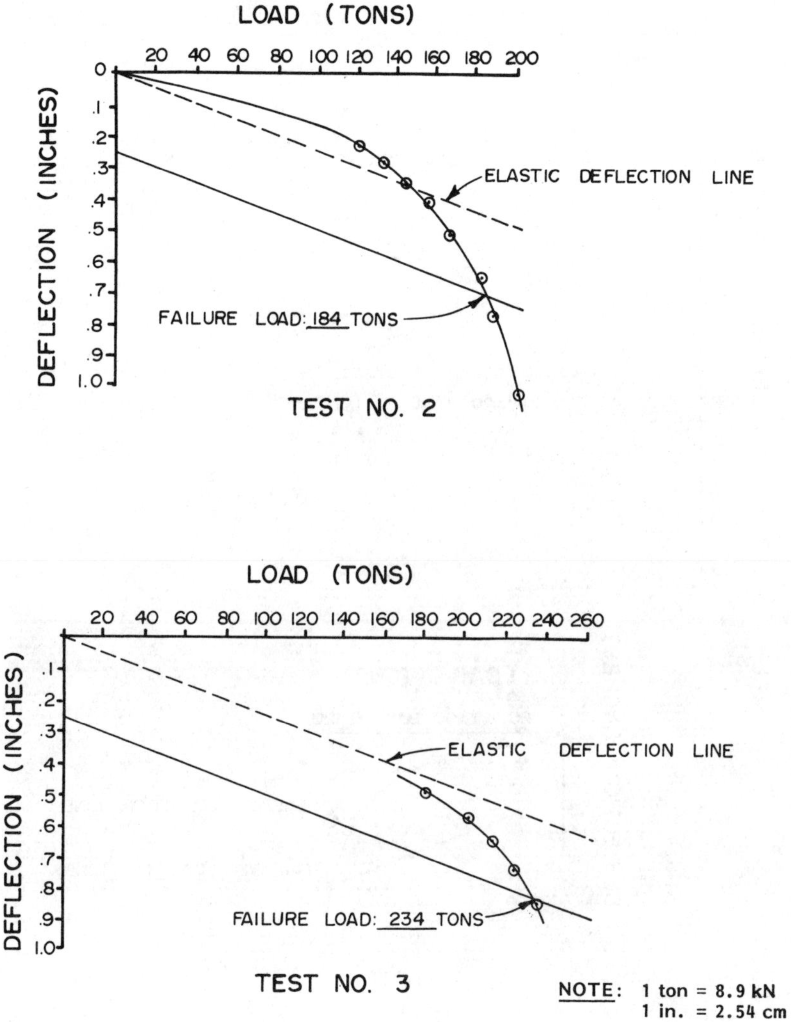

FIGURE 9. Load-deflection relationships Site D

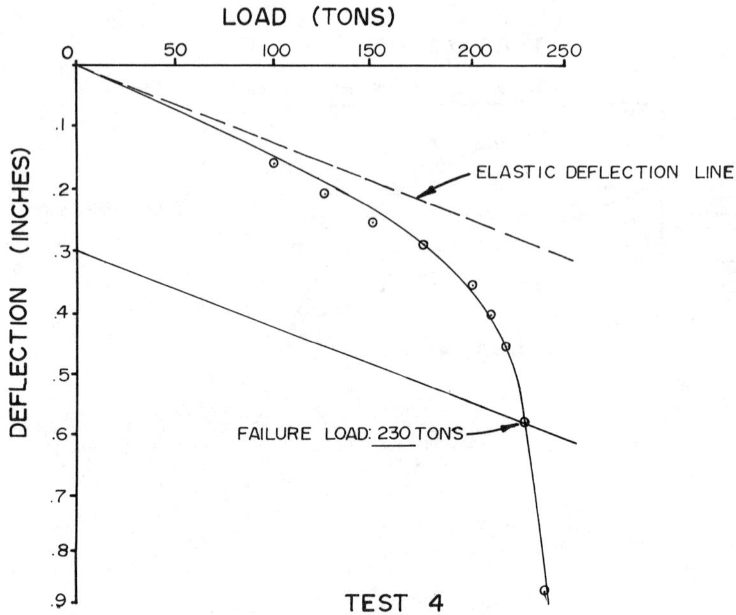

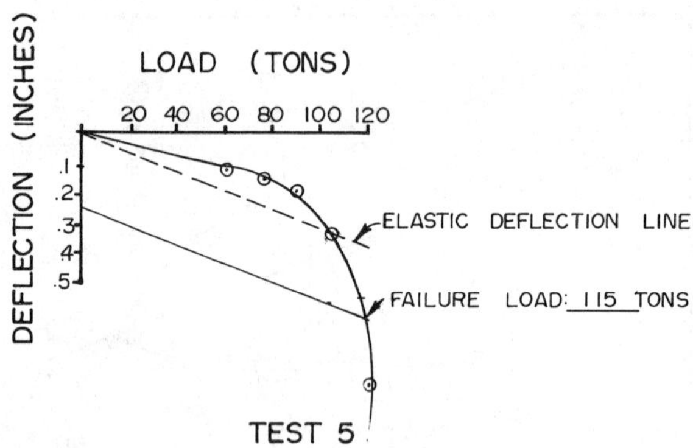

NOTE: 1 ton = 8.9 kN
1 in. = 2.54 cm

FIGURE 10. Load-deflection relationship Site D

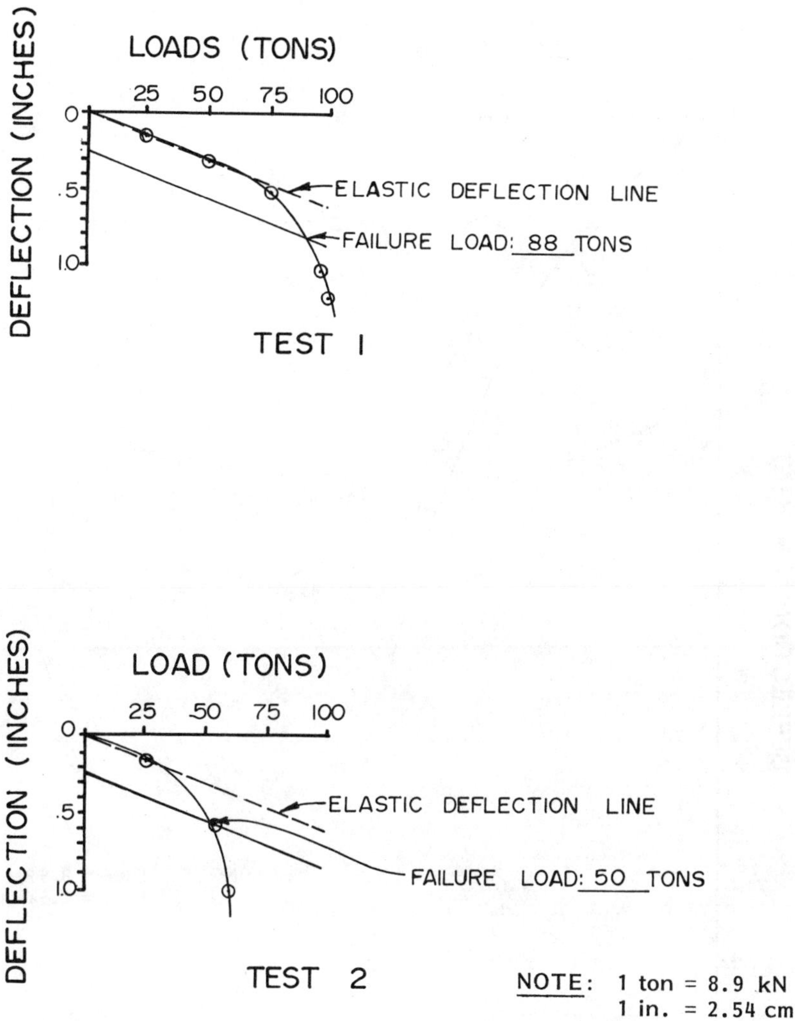

FIGURE 11. Load-deflection relationships Site E

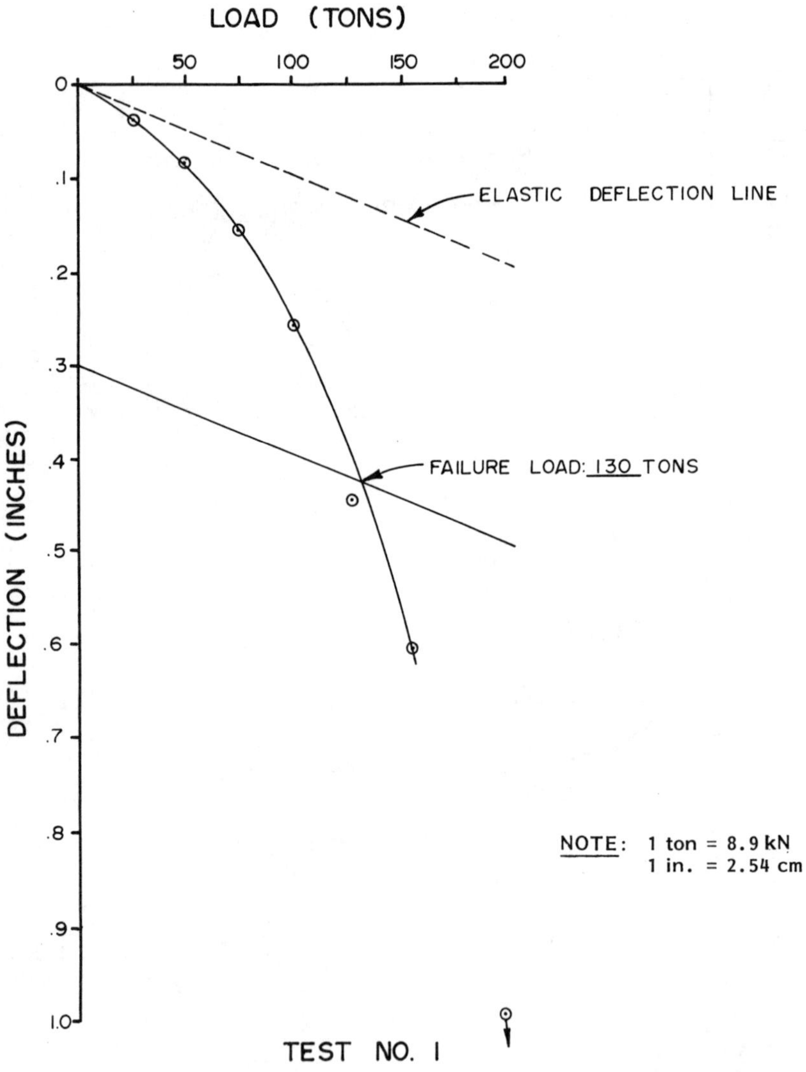

FIGURE 12. Load-deflection relationship Site F

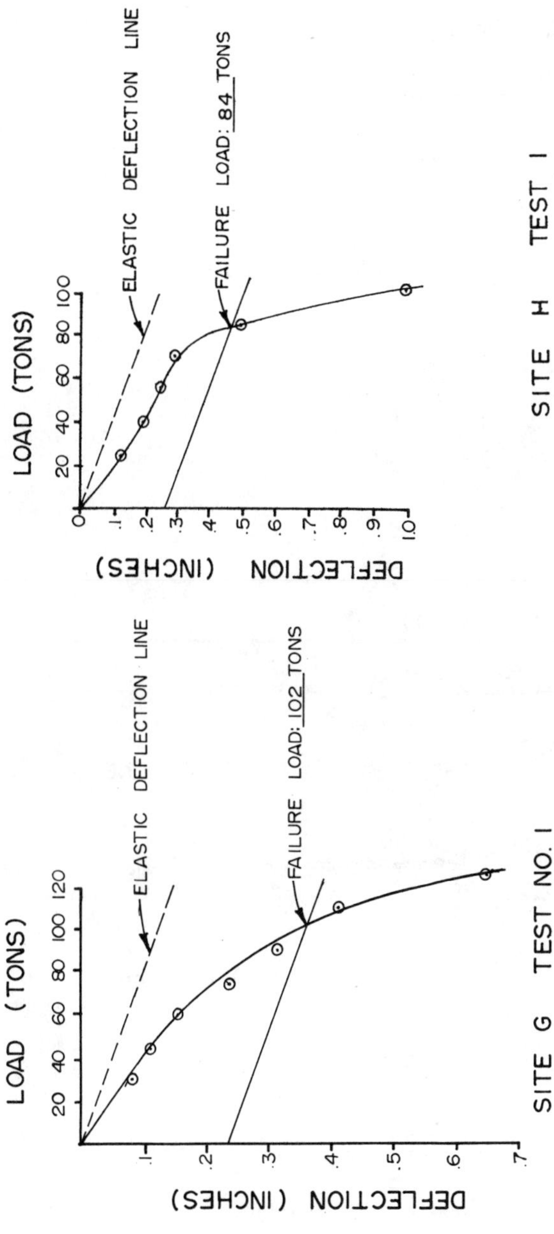

NOTE: 1 ton = 8.9 kN
1 in. = 2.54 cm

FIGURE 13. Load-deflection relationship Sites G and H.

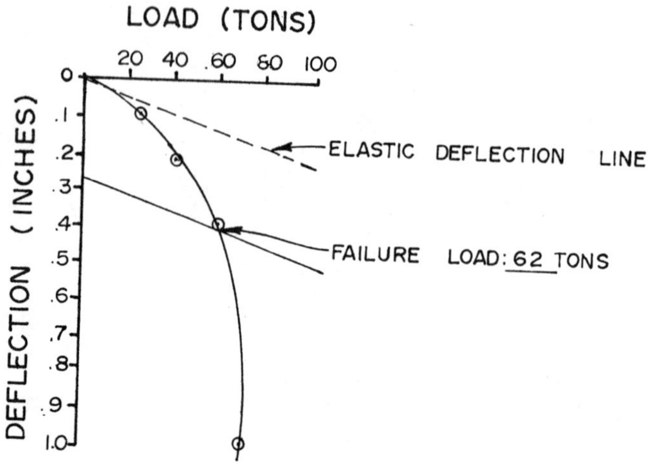

TEST 2

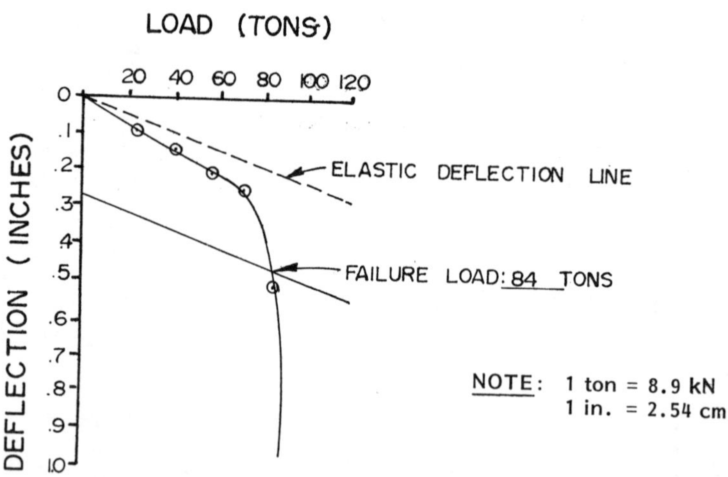

TEST 3

FIGURE 14. Load-deflection relationship Site H

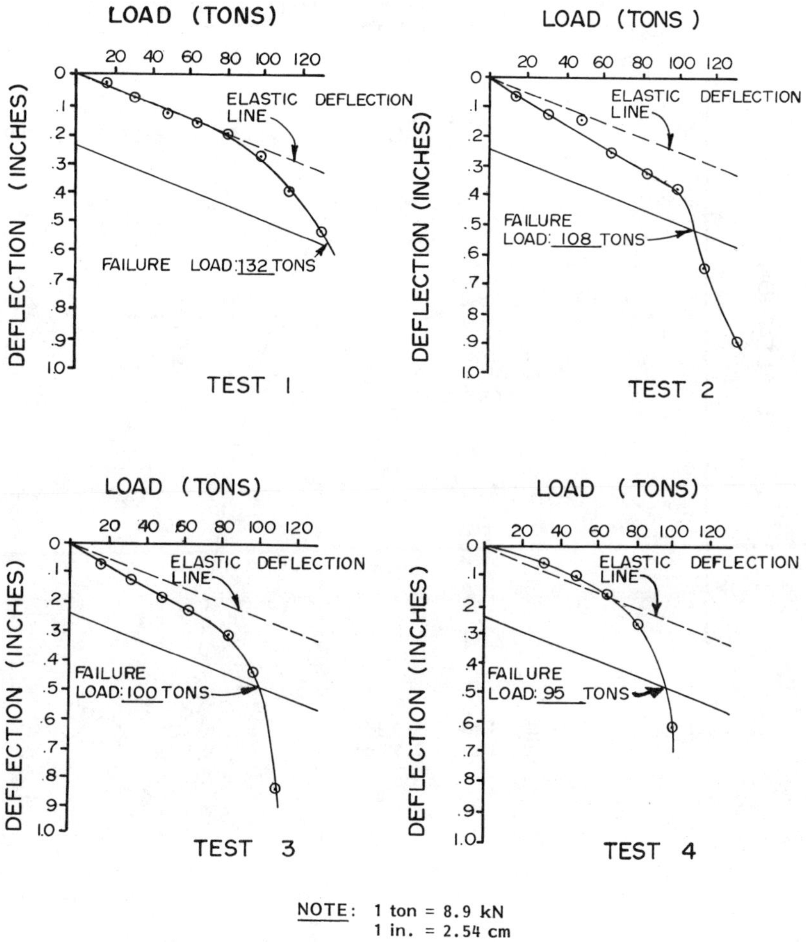

FIGURE 15. Load-deflection relationship Site I

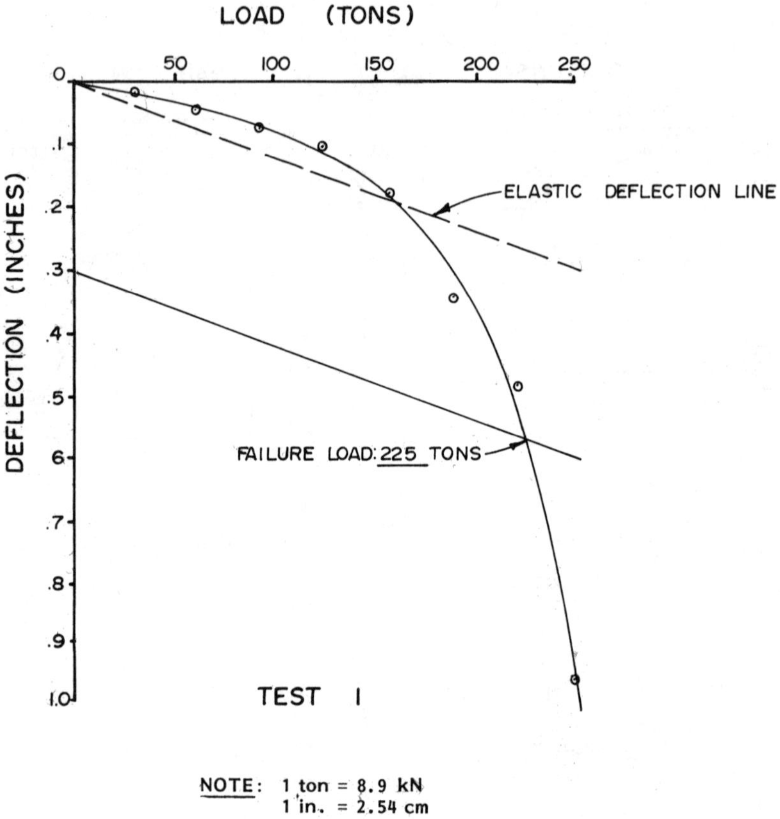

FIGURE 16. Load-deflection relationship Site J

TABLE I

Load Test Summary

Site	Test No.	Pile Type	Hammer Type	Energy (Ft-lbs)	Ram Wt (lbs)	Final Driving Resistance	Failure Load (Tons)
A	1	10" Square Prestressed Concrete	VULCAN 1	15,000	5,000	55=12"	58
B	1	14" Square Prestressed Concrete	LINKBELT 520	25,500	5,070	49=1"	184
C	1	14" Square Prestressed Concrete	VULCAN 010	29,000	10,000	136=12"	180
D	1	14" Square Prestressed Concrete	RAYMOND 65C	19,500	6,500	64=5"	250
	2	Step-tapered Concrete Filled Shell (8-5/8" dia. tip)	RAYMOND 65C	19,500	6,500	58=9"	184
	3	Step-tapered Concrete Filled Shell (10-3/8" dia. tip)	RAYMOND 5/0	43,750	17,500	10=2"	234
	4	18" Square Prestressed Concrete	RAYMOND 5/0	43,750	17,500	54=6"	230
	5	10-3/4" Dia. Concrete Filled Pipe w/0.25" Wall	RAYMOND 65C	19,500	6,500	92=10"	115
E	1	10-3/4" Dia. Concrete Filled Pipe w/0.25" Wall	CONMACO 65	19,500	6,500	8=1"	88
	2	10-3/3 Dia. Concrete Filled Pipe w/0.25" Wall	CONMACO 65	19,500	6,500	40=8"	50

Table I Continued

TABLE I (Continued)

Load Test Summary

Site	Test No.	Pile Type	Hammer Type	Energy (Ft-lbs)	Ram Wt (lbs)	Final Driving Resistance	Failure Load (Tons)
F	1	18" Square Prestressed Concrete	VULCAN 010	32,500	10,000	49=4"	130
G	1	Step-tapered Concrete Filled Shell (11-3/8" Dia. tip)	RAYMOND 65C	19,500	6,500	66=12"	102
	1	10" Dia. Concrete Filled Pipe w/0.188" Wall	VULCAN 1	15,000	5,000	>14=1"	84
H	2	10" Dia. Concrete Filled Pipe 2/0.188" Wall	VULCAN 1	15,000	5,000	>14=1"	62
	3	10" Dia. Concrete Filled Pipe w/0.188" Wall	VULCAN 1	15,000	5,000	>14=1"	84
	1	10-3/4" Dia. Concrete Filled Pipe w/0.219" Wall	VULCAN 1	15,000	5,000	63=8"	132
I	2	10-3/4" Dia. Concrete Filled Pipe w/0.219" Wall	VULCAN 1	15,000	5,000	25=0"	108
	3	10-3/4" Dia. Concrete Filled Pipe w/0.219" Wall	VULCAN 1	15,000	5,000	36=5"	100
	4	10-3/4" Dia. Concrete Filled Pipe w/0.219" Wall	VULCAN 1	15,000	5,000	101=8"	95
J	1	18" Square Prestressed Concrete	CONMACO 160	42,250	16,250	81=10"	225

The elastic compression of the pile was determined using the load distribution determined by static analysis and was then subtracted from the observed head deflection to give the tip movement (Δ_t). Assuming that the bearing formations were uniform within a depth of two pile diameters (2B) below the pile tip, the apparent limestone compression modulus (E_t) was calculated using:

$$E_t = \frac{B \quad Q_{ut}}{2\,\Delta t} \qquad (1)$$

The values of the apparent tip bearing capacity and tip compression modulus are given on Table II.

Figure 17 shows graphically the relationship between the standard penetration resistance and the apparent ultimate end bearing capacity and compression modulus at the pile tip. The shaded zone shows the range for soft limestone while the dashed lines show the influence of seams of soft materials within the limestone formation in the vicinity of the pile tip.

Design of Driven Piles of Soft Limestone

Subsurface Investigation Program

The purpose of a subsurface investigation program (for a site at which it is anticipated that driven piles on soft limestone may be a favorable foundation system) is to obtain information regarding the rock characteristics that will influence the driving and load carrying characteristics of the piles. Although continuous static or dynamic cone penetration tests, in-hole pressuremeter and geophysical methods may provide more data on certain characteristics of the soft limestone at a given location, the standard penetration test (whose limitations have been discussed in detail by others) is the most commonly used in-place test of the strength of soft limestone.

Continuous coring of the formation may not yield the desired information since seams of uncemented materials (ranging from soft clay to dense sand) may sometimes be washed away by the coring process yet they may have a significant effect on the strength of the limestone formation. Coring does, however, allow for a visual examination of rock texture as well as provide samples of the more cemented and intact portions of the limestone for laboratory testing.

If the standard penetration test is used to determine the characteristics of the soft limestone, sampling should begin as soon as the limestone formation is encountered. A cap of soft partially weathered limestone or a hard limestone seam may overlie a softer limestone. Standard penetration tests should

TABLE II

Apparent End Bearing Capacity and Modulus

Site	Test No.	Pile Type	Bearing Layer (N)	Apparent Ultimate End Bearing Capacity * (Q_{ut}) (tsf)	Apparent Modulus at tip* (E_t) (tsf)
A	1	10" Square Prestressed Concrete	Limestone & Clay (50=5") (12=12")	40	750
B	1	14" Square Prestressed Concrete	Limestone (58=12")	100	2700
D	1	14" Square Prestressed Concrete	Limestone (100=12")	140	5000
	4	18" Square Prestressed Concrete	Limestone (100=12")	90	3600
	5	10-3/4" Dia. Concrete Filled Pipe (0.25" Wall)	Limestone & Clay	140	1500
E	1	10-3/4" Dia. Concrete Filled Pipe (0.25" Wall)	Limestone w/silt seams	80	3500
	2	10-3/4" Dia. Concrete Filled Pipe (0.25" Wall)	Limestone w/silt seams	20	320
F	1	18" Square Prestressed Concrete	Limestone w/silt seams (50=4")	50	1100
H	1	10-3/4" Dia. Concrete Filled Pipe (0.188" Wall)	Limestone (36=12")	100	1900
	2	10-3/4" Dia. Concrete Filled Pipe (0.188" Wall)	Limestone (36=12")	60	1500
	3	10-3/4" Dia. Concrete Filled Pipe (0.188" Wall)	Limestone (36=12")	100	1900

TABLE II Continued

TABLE II (Continued)

Apparent End Bearing Capacity and Modulus

Site	Test No.	Pile Type	Bearing Layer (N)	Apparent Ultimate End Bearing Capacity * (Q_{ut}) (tsf)	Apparent Modulus at tip* (E_t) (tsf)
I	1	10-3/4" Dia. Concrete Filled Pipe (0.219" Wall)	Limestone (50=3")	170	5500
	2	10-3/4" Dia. Concrete Filled Pipe (0.219" Wall)	Limestone (50=3")	140	5700
	3	10-3/4" Dia. Concrete Filled Pipe (0.219" Wall)	Limestone (50=3")	120	3600
	4	10-3/4" Dia. Concrete Filled Pipe (0.219" Wall)	Limestone (50=3")	120	3500

* Includes only piles of constant cross-section and those having N values at tip.

NOTE: 1 in. = 2.54 cm
1 tsf = 96 kN/m^2

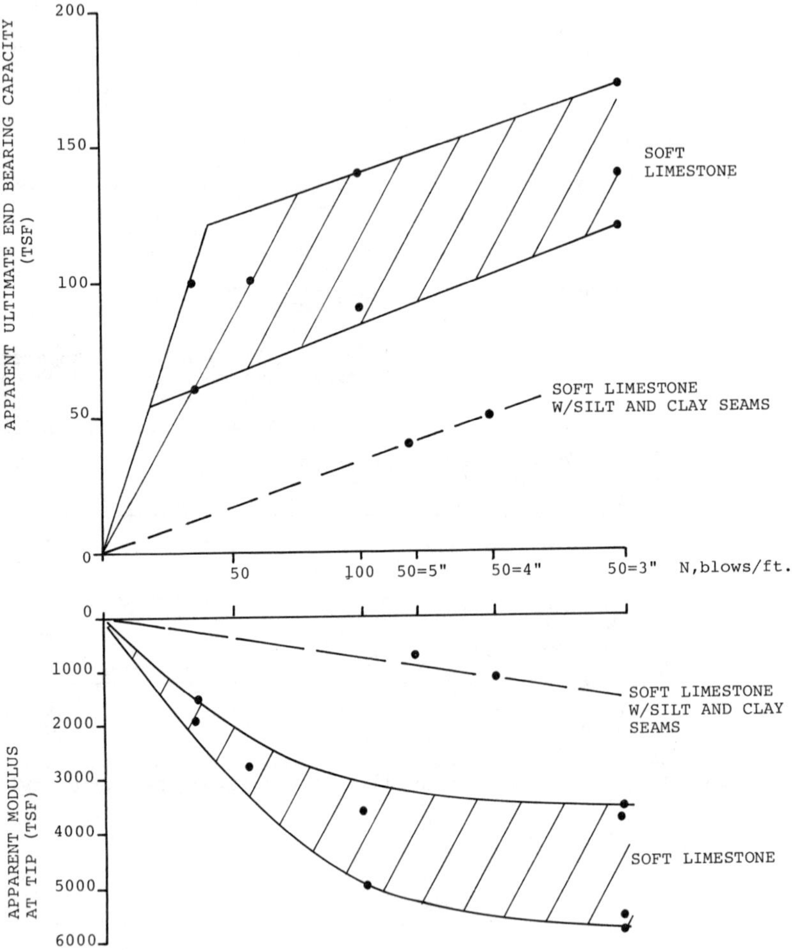

NOTE: 1 tsf = 96 kN/m^2
1 in = 2.54 cm

FIGURE 17. STP vs apparent ultimate capacity and compression modulus of limestone

continue on 2-1/2 foot (0.75 m) centers to about 10 feet (3m) below the probable pile bearing level. Each standard penetration test should be carried until 100 blows on the sampler are obtained as outlined in ASTM D1586.

Pile Capacity and Estimated Tip Elevations

In order to determine pile capacities in soft limestone formations, considerable engineering judgement and appropriate conservatism is required due to the non-uniform nature of the limestone formations. A common design approach is to perform a static analysis of the load carrying ability of the overburden soils and also determine the point resistance of the limestone at the pile tips.

A base movement of about 1/10 the pile width may be required to mobilize all the point resistance. Structure settlement considerations, however, generally limit the amount of capacity that may be assigned to the pile tip. The assigned tip capacity, therefore, should be based upon the values calculated from previous load tests in which settlement considerations have been included (i.e. Figure 17).

The product of the apparent ultimate bearing capacity at the pile tip, (Q_{ut}) and the area, (A), is the point resistance of the pile. A factor of safety of three is recommended for determining the allowable capacity in a soft limestone formation and a factor of safety of two is suggested for the capacity of the overburden soils.

Once the working capacity of the pile has been selected, the pile must be driven sufficiently hard to penetrate any thin well-cemented limestone seams that may overlie softer materials. Each pile then serves as a probe of the limestone formation at the pile location.

The presence of solutioned zones, cavities, slots, and voids complicates the analysis of the bearing capacity of the piles in soft limestones. The piles should be installed with a sufficient separation between the tips and the solutioned zones so that a punching shear failure does not occur. If the depths of limestone solutioning and the expected pile tip elevations are highly variable or if the site is in an area of active solutioning, the structure should be designed to tolerate the loss of some piles without catastrophic failure. An alternative in this case would be the consideration of other foundation systems. The mechanisms of limestone subsidence have been discussed by Sowers (3).

Tip elevations are very difficult to predict in soft limestone, even if pile locations are pre-drilled. If significant vertical slots or chimneys of non-cemented material occur within the limestone, the piles may penetrate the entire formation. Pile driving experience indicates that tip elevations at some sites may vary by 20 feet (6m) or more in the same cap. Test borings at every column location, therefore, will not necessarily enable the prediction of the pile tip elevations within a given cap.

Installation Criteria

In non-uniform bearing materials such as soft limestones, pile driving resistance (1/set) is a convenient method for determining when each pile has reached a satisfactory bearing depth. Should it be desired to examine the relative influence of all the parameters of the pile/hammer system or if these parameters assume extreme values, a wave equation analysis should be performed.

The following empirical relationship between pile capacity, hammer energy and driving resistance has been derived from the load test information presented in this paper:

$$\frac{1}{S} = \frac{1}{\frac{0.0015\ E}{Q_u} - 0.15} \qquad (2)$$

Where, $\frac{1}{S}$ = Driving Resistance, blows/in.

E = Hammer Energy, Ft-lbs.

Q_u = Ultimate Pile Capacity, tons

Note: 1 in. = 2.54 cm
1 ft.lb. = 0.00138 kN-m
1 ton = 8.9 kN

The test piles presented in this paper were driven using conventional pile driving equipment. The ratio of the weight of the pile or mandrel to the ram weights ranged from 0.4 to 2.0.

A plot of the measured vs the driving resistance predicted by equation 2 is given on Figure 18. A factor of safety of two is recommended.

It should be noted that the driving resistances obtained by equation 2 are generally higher than that obtained by the EN or Hiley Formulae. This should not be surprising since piles driven into soft limestones generally have higher driving resistances and it has been found that it is necessary to drive the piles

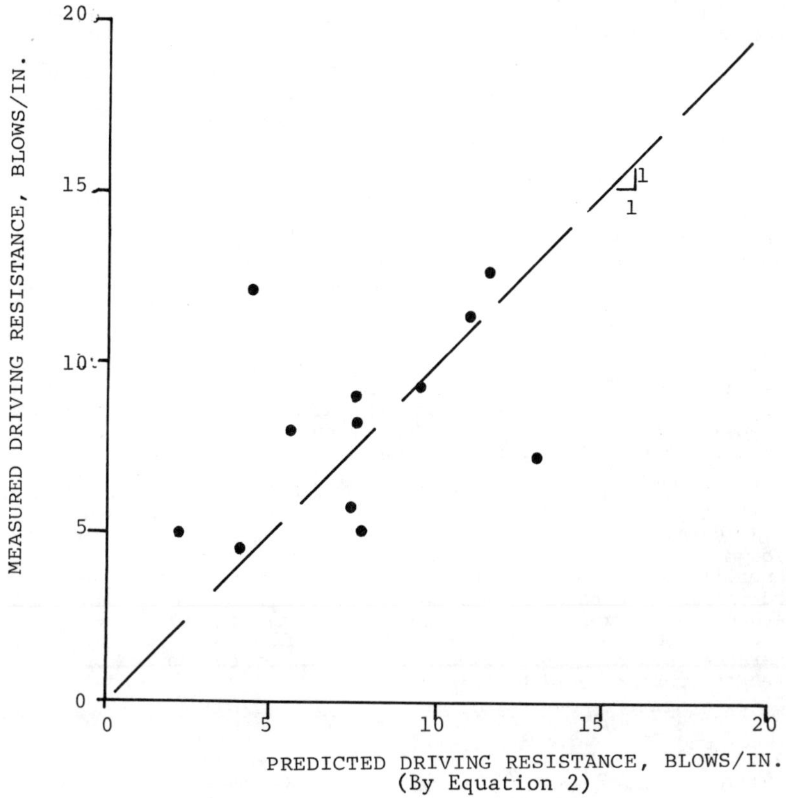

NOTE: 1 in. = 2.54 cm

FIGURE 18. Measured vs predicted driving resistance

sufficiently hard to penetrate any well-cemented limestone seams that may overlie a softer bearing layer within the formation.

It is suggested that the ratio of the weight of the pile or mandrel to the weight of the ram not exceed 2.0. It is also recommended that the piles not be stopped until sufficient penetration of the limestone occurs and the required driving resistance is obtained. Sufficient penetration into the limestone has generally been found to be 6 to 12 (15 to 30 cm) inches of penetration where the driving resistance is equal to or greater than that required by equation 2 and the driving resistance in the last inch (2.5 cm) is at least that given by equation 2. Therefore, if the required driving resistance was 8 blows per inch (2.5 cm) by equation 2 the required driving criteria would be at least 48 blows in the last 6 inches (5 cm) and 8 blows in the last inch (2.5 cm).

If a significant amount of hard clay is encountered within the limestone formation and the piles are expected to achieve a large percentage of their capacity from the clay, a dynamic formula such as equation 2 may not be appropriate due to the time dependent nature of the clay strength.

The driving resistances obtained from equation 2 should be verified by static load tests at each site. Quick maintained load tests (1) or constant rate of penetration tests are recommended for static load tests on soft limestone since they generally give a better definition of the load-deflection curve and result in signficantly shorter test times than the test methods that require a 24 hour or longer hold at twice the design load. If it is desired to determine the total creep characteristics of soft limestone, test durations on the order of weeks or months could be required.

Hammer and Pile Selection Criteria

The hammer should be selected and the pile should be designed so that the driving stresses do not structurally damage the pile. Davisson (2) has suggested limits on driving stresses. These limits for end bearing piles have been slightly modified to the form shown below:

Timber - Maximum hammer energy: 200 ft-lbs per blow for each square inch of pile tip.

Steel H, pipe pile (ASTM A36, A252, grade 2 minimum) - Maximum hammer energy: 2000 ft-lbs per blow for each square inch of steel in the pile cross section.

Prestressed concrete - Maximum hammer energy (ft-lbs per blow):

$$\frac{1}{25}\ (f'c - fp)\ B\ \ A$$

where:
f'c = concrete compressive strength, psi
fp = prestress force, psi
B = pile width, ft.
A = tip area, sq. in.

Note: 1 ft-lb = 0.00138 kN-m
1 in2 = 6.45 cm2
1 psi = 6.9 kN/m^2
1 ft = 0.305 m

Tension stresses during driving of prestressed concrete piles should be evaluated using the wave equation.

The hammer should be selected so that a driving resistance of not more than 10 blows per inch (2.5 cm) is required. Figure 19 graphically shows hammer pile selection criteria for end bearing piles. This figure indicates the maximum recommended hammer energies for several common pile types and the minimum hammer energy for a given design capacity based upon equation 2 with a maximum driving resistance of 10 blows per inch (2.5 cm).

Conclusions

Although soft limestone formations are erratic, it appears that there is a relationship between the standard penetration test resistance and the compression moduli and ultimate bearing capacity of the soft limestone at the pile tips. With proper selection of the pile type and hammer, piles driven into soft limestones may be driven sufficiently hard so that the piles penetrate any thin seams which may overlie more compressible layers within the limestone formation without structural damage to the pile.

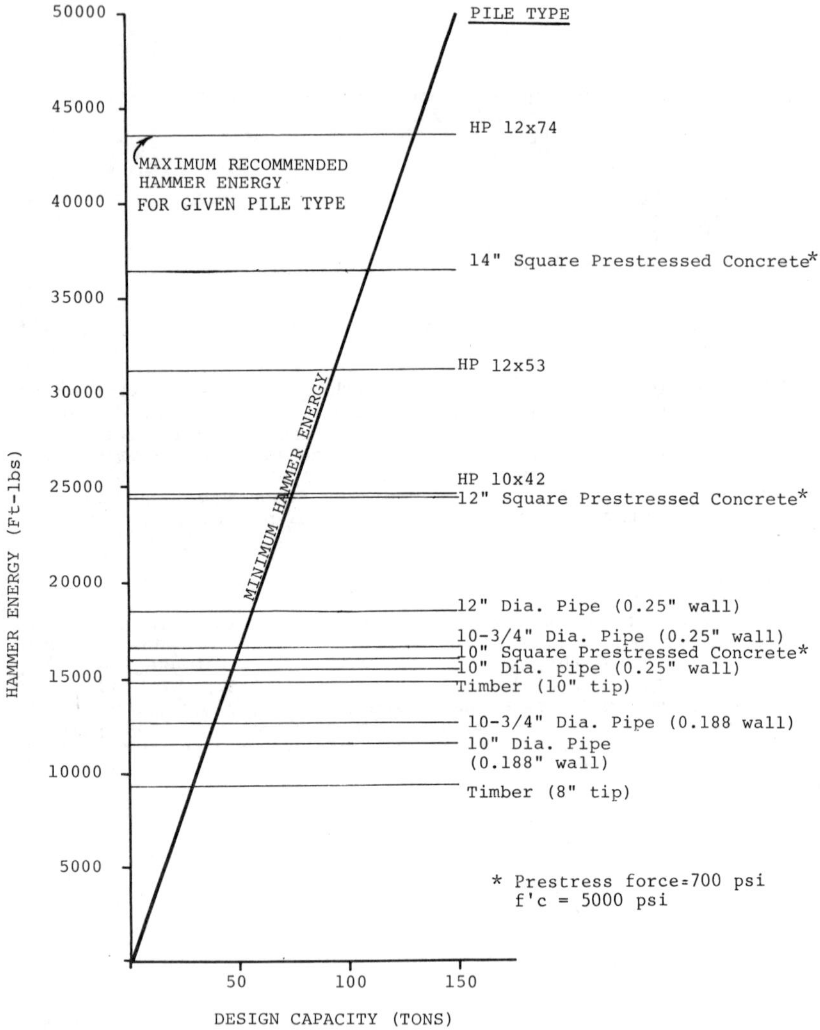

NOTE: 1in= 2.54 cm
1 psi = 6.9 kN/m^2
1 ft-lb = 0.00138 kN-m
1 ton = 8.9 kN

FIGURE 19. Hammer selection criteria for end bearing piles.

References:

1. Butler, H.D., and Hoy H.E., "Users Manual for the Texas Quick Load Method for Foundation Load Testing", Federal Highway Administration, Office of Development, Washington, 1977.

2. Davisson, M.T., "High Capacity Piles", Proceedings Lecture Series, Innovations in Foundation Construction, ASCE, Illinois Section, 1973, pp. 81-112.

3. Sowers, G.F., "Failures in Limestones in Humid Subtropics", Journal of the Geotechnical Engineering Division, ASCE, Vol. 101 No. GT 8, Proc. Paper 11521, August 1975, pp. 771-787.

Notation:

The following symbols are used in this paper:

A = crossectional area of pile
B = pile width
E = hammer energy
E_p = modulus of elasticity of pile
E_t = compression modulus of material at pile tip
$f'c$ = unconfined compressive strength of concrete
L = length of pile
P = load at top of pile
Q_u = ultimate pile capacity
Q_{ut} = bearing capacity at the pile tip
S = set

Δ = elastic deflection of pile
Δt = deflection of pile tip

PILE GROUP RESPONSE TO AXIAL AND LATERAL LOADING

By John B. Stevens[1], M. ASCE, D. Michael Holloway[2], A.M., ASCE, Yoshiharu Moriwaki[2], A.M., ASCE and Edward C. Demsky[3], M. ASCE.

INTRODUCTION

From November 1978 to March 1979, an extensive pile testing program was conducted for the St. Louis District of the US Army Corps of Engineers and the US Department of Transportation. This program was part of the larger investigation to evaluate various methods of rehabilitating Locks and Dam No. 26, Mississippi River, Alton, Illinois. As part of the pile testing program, three pile groups and a single pile were loaded to failure. This paper describes the site conditions, the pile groups and tests, and summarizes the test results.

TEST SITE LOCATIONS

The location of the test site, shown in Figure 1, was on the west bank of the Mississippi River, about one mile (1.7 km) downstream from the existing Locks and Dam No. 26. It is known locally as Ellis Island and is within the flood plain of the river.

SUBSURFACE CONDITIONS

Approximately 110 feet (34 m) of alluvial soils overlie the first rock unit at the test site. The upper rock formation is the St. Genevieve, an oolitic limestone. The overlying soil deposits are of glacial, alluvial and coluvial origin. Five distinct soil strata have been identified and are shown in Figure 2. These strata are, in descending order, flood plain deposits, recent alluvium, alluvial outwash (reworked alluvium), Wisconsinan outwash and Illinoian ice contact deposits.

The flood plain deposits are primarily soft to medium silty clays with occasional sand layers. The recent alluvium and alluvial outwash consist of fine to medium quartz sand with some silty sand and gravel layers. The sands are poorly graded. The Wisconsinan outwash and Illinoian ice contact deposits consist generally of coarse to medium sand with cobbles and gravel.

---

1. Asst. Prof., Civ. Engrg. Norwich University, Northfield, VT
2. Sr. Proj. Engr., Woodward-Clyde Consultants, San Francisco, CA
3. Civ. Engr., St. Louis District, US Army Corps of Engineers, St. Louis, MO

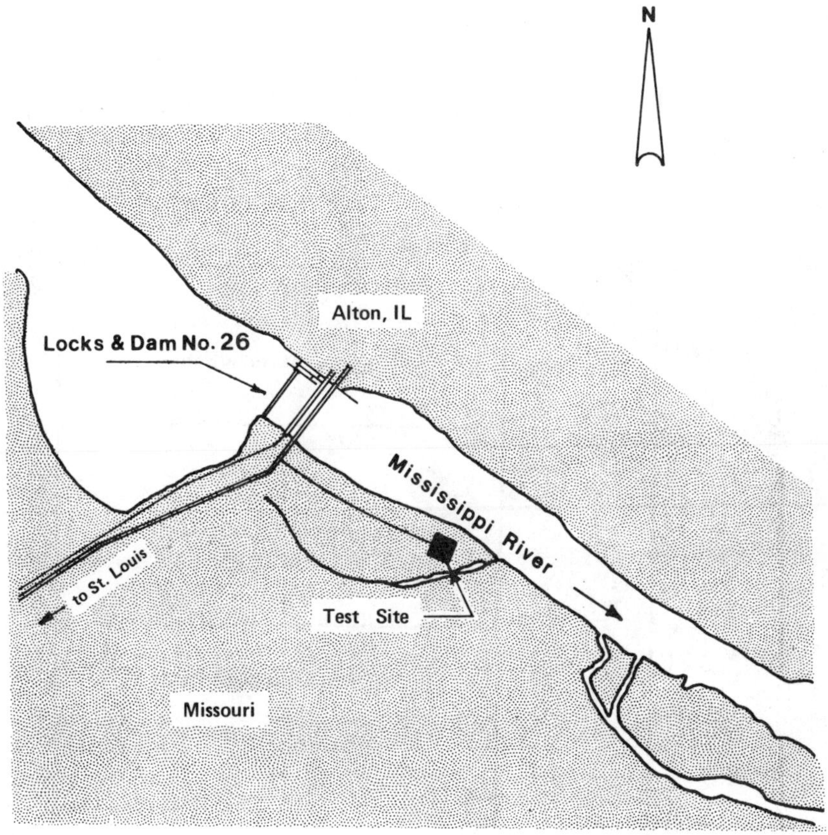

Figure 1 – TEST SITE LOCATION

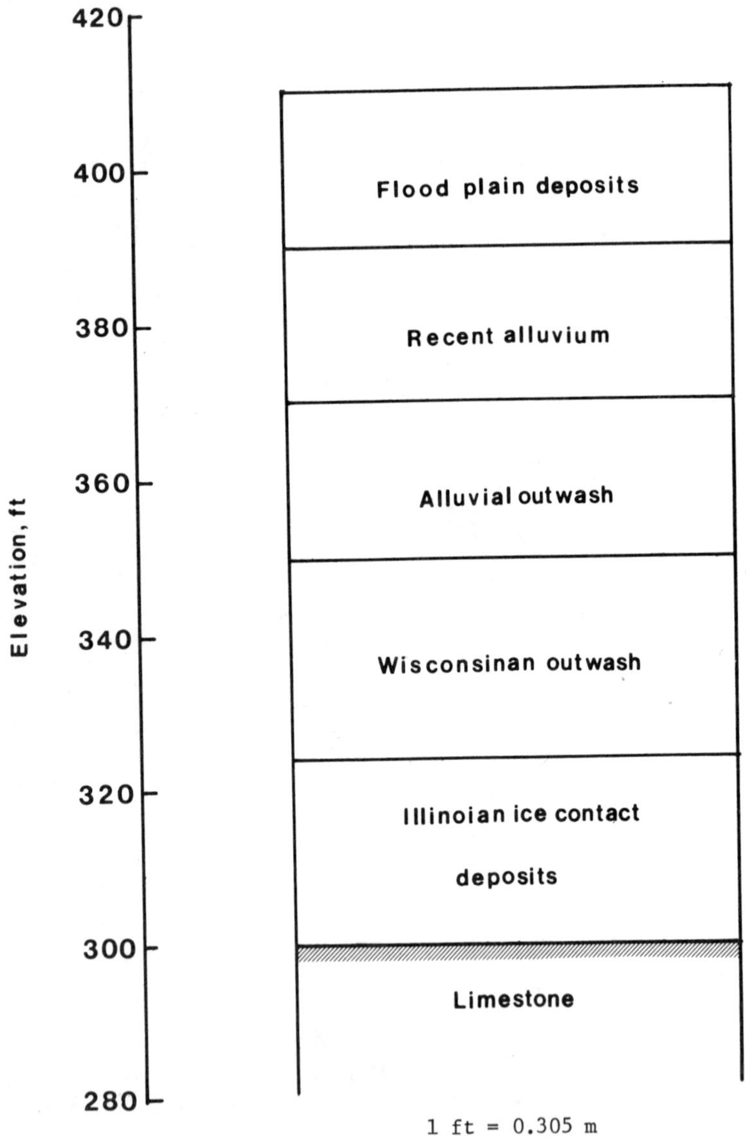

Figure 2 – GEOLOGIC PROFILE

The typical soil profile from 25 borings in the pile test area is given in Figure 3. Also presented in Figure 3 are the range of standard penetration blow counts and cone point resistance values. The cone had a 10-cm^2 (1.55 in^2) cross-sectional area and an angle of 60°. The cone was pushed into the soil at a constant rate of 4 ft/min (2 cm/sec).

From the standard penetration tests (SPT) and cone soundings, the ranges of relative density were estimated and are presented in Figure 4 with respect to depth. The relative density was estimated from the SPT results by the method presented by Gibbs and Holtz (1953) and from the cone soundings by the method presented by Schmertmann (1976).

The groundwater level at the site normally varied with the river level. To eliminate fluctuations during the test, the level was maintained between El. 389 ft (118.6m) and 390 ft (11.90m) (one foot (0.3m) below finished grade) by a dewatering system.

SOIL PROPERTIES

Prior to installation of the piles, the test area was lowered to El. 391 ft (119.3m) by excavation. This removed the flood plain deposits so that the piles were primarily installed in the alluvial sands. Therefore, the properties of the flood plain deposits are not presented.

A series of CID* triaxial compression tests were performed on samples obtained from the borings. The friction angle from these tests ranged from 38 to 41°. The dry unit weight from El. 391 to 350 ranged from 103 to 115 pcf (1.67 to 1.87 t/m^3). This is the zone in which the piles were installed.

DESCRIPTION OF PILES

Untreated green Douglas-fir piles were used for the test. The piles had butt diameters of 12 to 14 inches (30.5 to 35.5 cm) and tip diameters of 8.5 to 10.0 inches (21.6 to 25.4 cm). Initial lengths ranged from 43 to 45 feet (13 to 13.7 m).

Prior to installation the piles were instrumented to measure various aspects of their behavior in response to the applied loads. The instrumentation included:

1. Strain gages to measure axial and bending strains at seven to eleven locations along the pile.

---

*CID signifies isotropically consolidated, drained conditions.

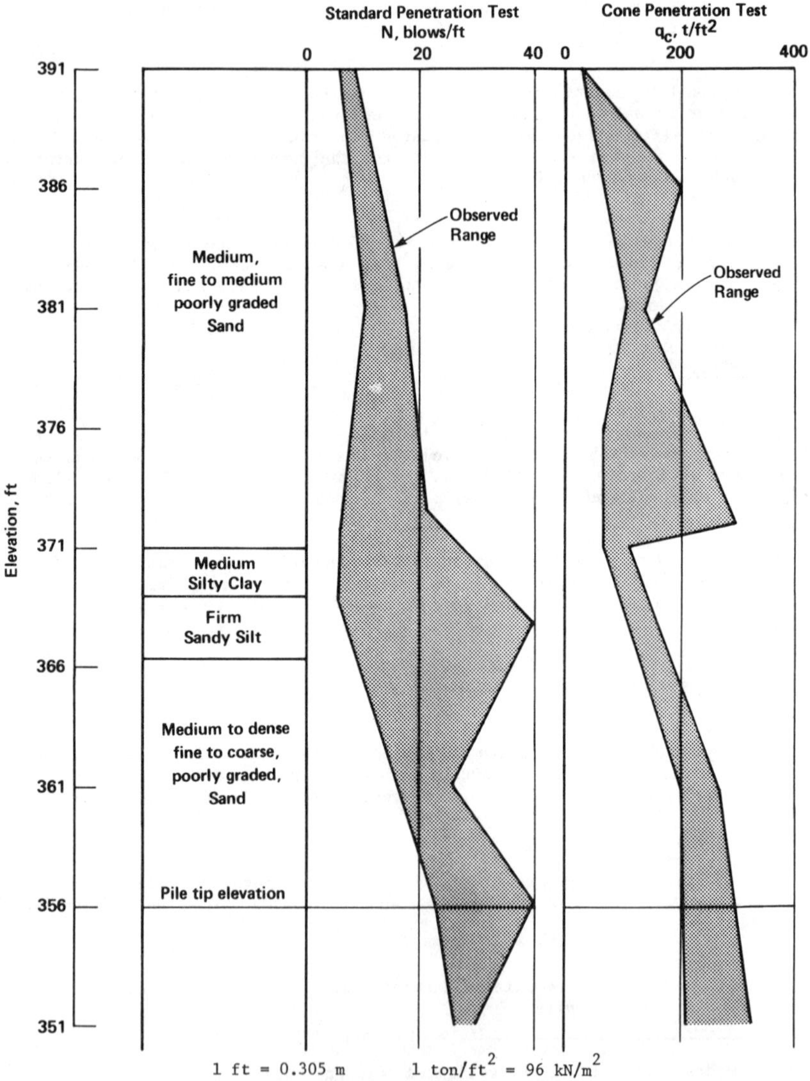

Figure 3 – GENERALIZED SOIL PROFILE AND PENETRATION TEST DATA

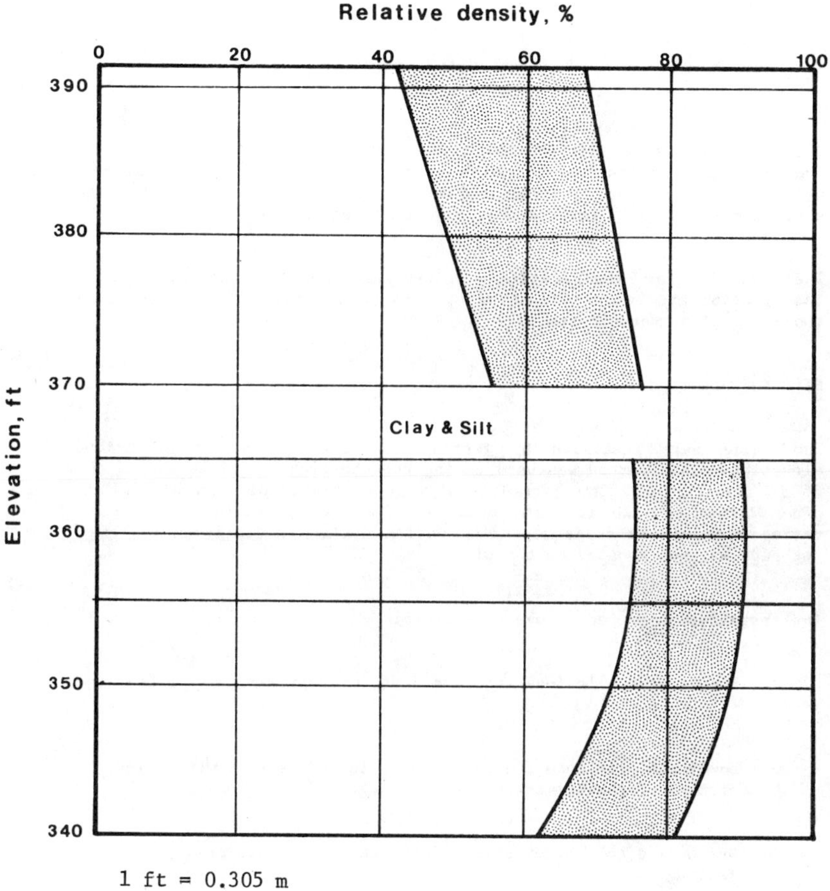

Figure 4 – RELATIVE DENSITY PROFILE

2. Telltales to measure the displacement of the pile tip and other points along the pile.

3. Slope inclinometer casings to measure lateral movement of the pile.

4. Thermocouples to measure temperature at various points of the pile.

Not all piles were instrumented alike. Of the 46 piles used in the test program, 32 were instrumented with strain gages, 46 were instrumented with telltales, and 7 were instrumented with thermocouples. The instrumentation schematic for a fully instrumented pile is given in Figure 5.

The details for the specific instrumentation used and its method of installation are beyond the scope of this paper, and are presented elsewhere (Woodward-Clyde Consultants, 1979).

PILE INSTALLATION

The piles were installed by jetting and driving to the prescribed tip elevation. This method was used during the construction of Locks and Dam No. 26 in the 1930's. The jetting equipment consisted of 2-in OD (5.1 cm) jet pipe with a 1-in ID (2.5 cm) nozzle. The pump pressure and rate of flow varied over a wide range depending on the jetting and driving results. The jet pipe was not mounted on the pile.

The installation procedure was as follows:

1. Prejet at a pile location to a depth of approximately 15 feet (7.6 m)

2. Lower the pile into the prejetted hole and seat it with light strokes of the hammer.

3. Drive pile to a penetration of 30 feet (9 m), assisted by jetting.

4. Drive pile to a final penetration of 35 feet (10.6 m) without jetting. This portion of the driving was monitored with a Pile Driving Analyzer System.

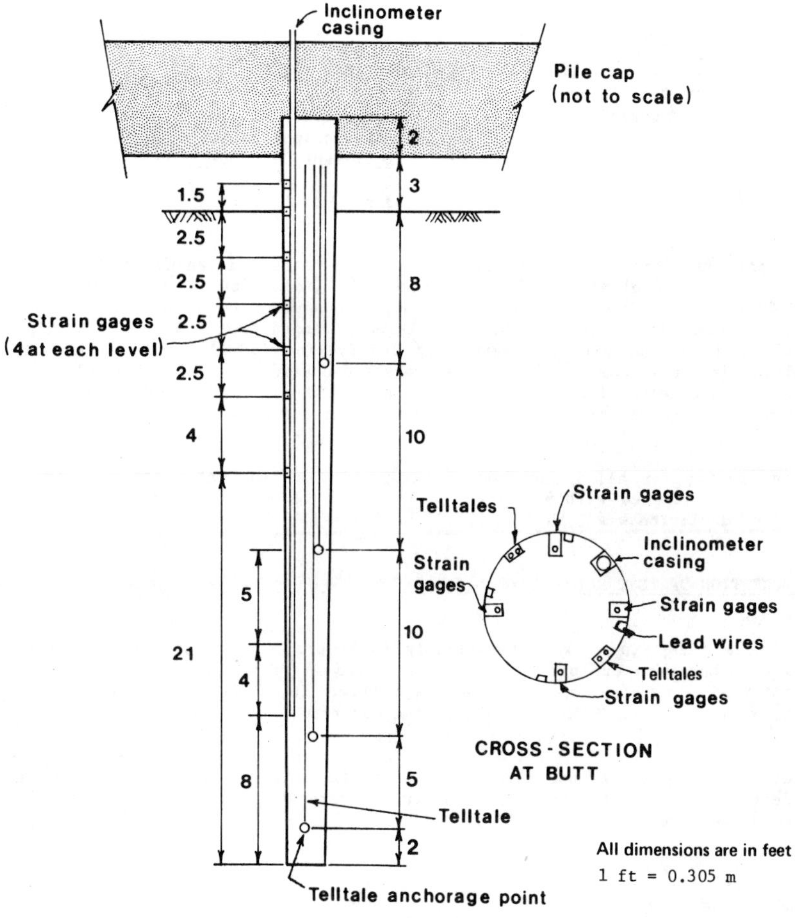

Figure 5 – PILE INSTRUMENTATION

5. Restrike pile after installation of all piles in a group.

A single-acting, Vulcan 1 air hammer was used to drive the piles. Details of the hammer/cap block assembly are:

Rated Energy, ft-lb (N-m)	15000 (20350)
Rated Speed, blow/min	50-60
Air Pressure at Hammer, psi (kPa)	80(550)
Weight of Ram, lb (kg)	5000 (2270)
Capblock Material	Duracrush (Mississippi Valley Equipment Co., St. Louis, MO)
Capblock Thickness, in (cm)	3 to 4 (7.6 to 10.2)

The blowcount for the final foot (0.3 m) of penetration ranged from 2 to 17 blow/in (1 to 7 blow/cm). Upon restrike, the final blowcounts ranged from 5 to 28 blow/in (2 to 11 blow/cm). From the dynamic measurements the maximum energy transferred per blow to the pile butt ranged from 1.4 to 5.5 ft-kips (1.9 to 7.5 kN-m) with an average of 2.8 ft-kips (38 kN-m) at the end of initial driving. For restriking the average decreased to 2.2 ft-kips (3 kN-m) and the range was 1.1 to 3.3 ft-ips (1.5 to 4.5 kN-m). In general, the higher blowcounts were associated with the lower measured hammer energies.

Pile spacing for all groups was a nominal 3 feet (0.9 m) on center. Most piles were located within 1 inch (2.5 cm) of the planned location, and the maximum deviation was 3 in (7.5 cm) in most cases.

DESCRIPTION OF TEST SETUP

The test setup is shown schematically in Figure 6. After the piles were driven, they were cut off leaving 5 feet (1.5 m) of stickup above grade. A reinforced concrete cap was then cast three ft (0.9 m) off the ground, embedding the piles 2 ft (0.6 m) into the concrete.

The purpose of the ground clearance was to eliminate any soil-cap interaction effects and to facilitate load transfer interpretation; all loads applied to the cap were resisted by the piles only.

To provide the vertical reaction, four rock anchors were installed. The anchors were grouted into the underlying limestone formation. After installation each anchor was proof-loaded to twice its design capacity. The rock anchors were connected to the reaction frame by bridge cables with open sockets at each end. Vertical loads were applied to the pile caps by jacks reacting against the frame.

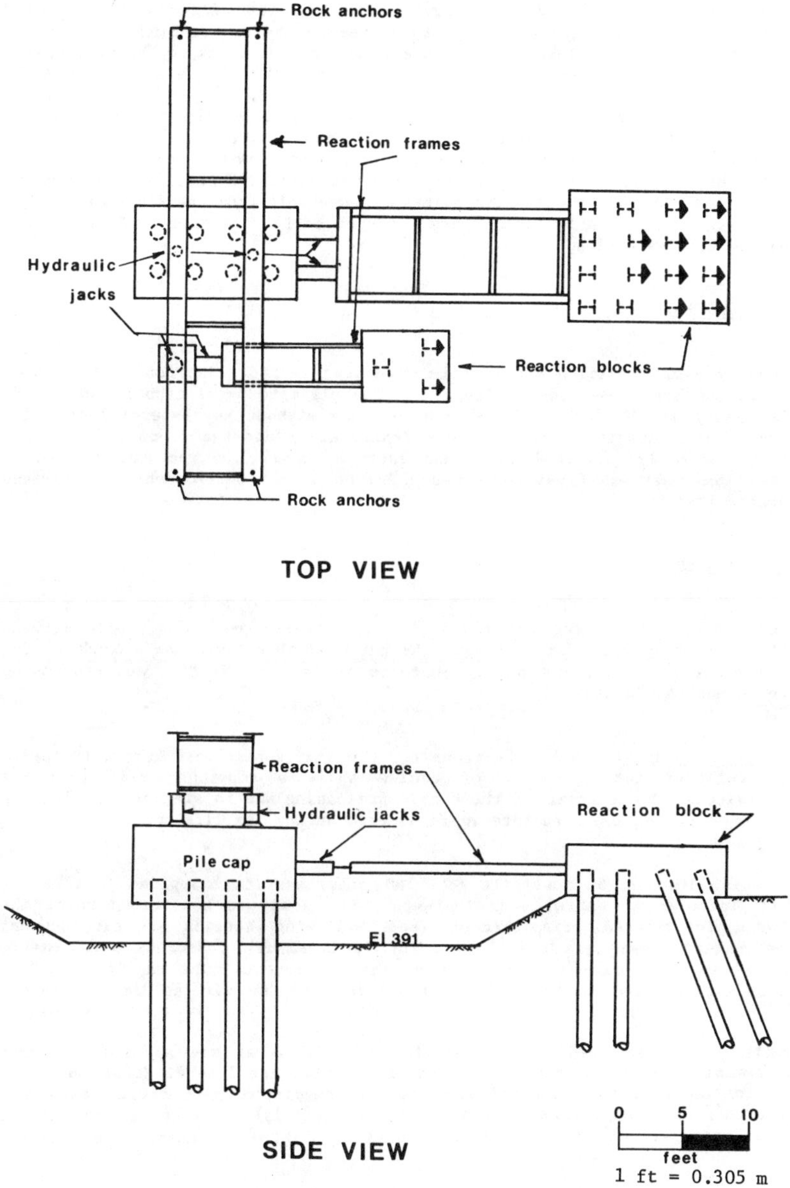

Figure 6 – LOAD TEST SETUP

Pile-supported concrete blocks provided the reaction for the horizontally acting jacks. A horizontal frame, supported on timber cribbing, was used to support the jacks and span the distance from reaction block to the pile cap.

The jacks applying the vertical load were placed on the cap so that the resultant acted at the geometric center of the as-built pile group. Because the piles deviated from the planned location, this was not necessarily at the center of the pile cap. The lateral load was applied 3 feet (0.9 m) from the top of the cap and at the third points transversely from the sides of the cap (see Fig. 6). All jacks were independently actuated and monitored continuously using load cells.

DESCRIPTION OF PILE GROUPS

Three pile groups and a single pile were load tested. The configurations of the groups are presented in Figure 7. The ultimate axial capacity of one 8-pile group and a single pile were determined without any lateral load. The lateral load capcities of the other 8-pile group and the 12-pile group were determined under combined axial and lateral loads. On the single pile an axial load test was first performed and then a test under combined axial and lateral load was done.

TEST PROCEDURES

The sequence of testing had three phases: cyclic preloading, pile driving effects testing and load testing. The purpose, objectives and procedures for each phase are described below. Methods of data acquisition and processing are summarized in Table 1.

Cyclic-Preloading. Each pile group and the single pile were laterally loaded and unloaded for a number of cycles while a constant axial load was maintained. The purpose of the cyclic preloading was to simulate the loading history a navigation structure would undergo during its lifetime.

An axial load of 30 tons/pile (267 kN/pile) was first applied in three or four increments. After each increment, the load was maintained until the change in vertical displacement (see following section on displacement measurements) was less than 0.01 in/hr (0.25 mm/hr). Various measurements were then made prior to applying the next increment. The magnitude of the loads was selected on the basis of an analysis of the existing dam structure.

Next, a lateral load of 6 tons/pile (53 kN/pile) was applied in three increments. Each increment was maintained until the lateral displacement in the loading direction (see following section on displacement measurements) was less than 0.01 in/hr (0.25 mm/hr). After the initial application, the lateral load was reduced to 0.6 tons/pile (5.3 kN/pile) until the lateral

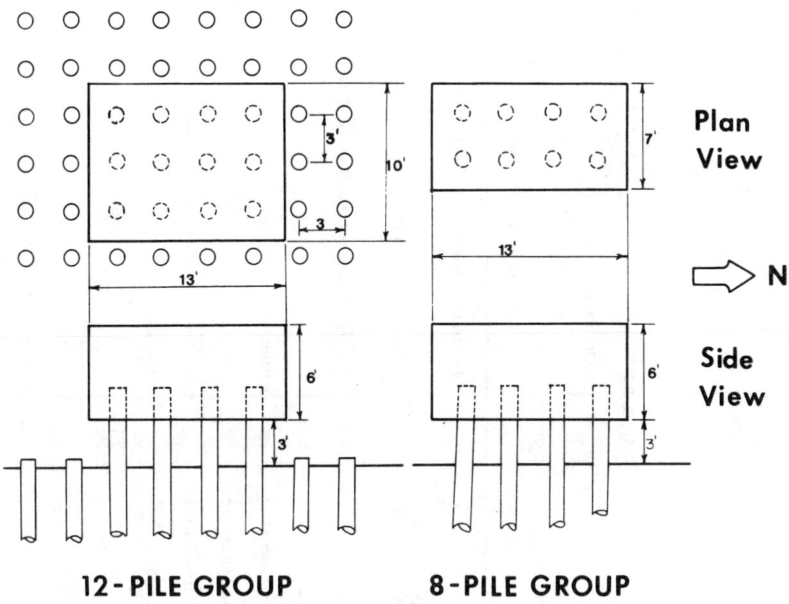

1 ft = 0.305 m

Figure 7 — PILE GROUP CONFIGURATIONS

TABLE 1

LOAD TEST MEASUREMENTS

Measurement	Equipment	Data Acquisition Method	Data Reduction Method
Applied load	Strain gage load cells	Voltmeter/Mini-computer system	Direct readout/computer
Vertical movement	Linear potentiometers/ dial gages/precise level	Mini-computer system/Manual/Manual	Computer/Manual/Manual
Horizontal movement	Linear potentiometers/ dial gages/optical survey	Mini-computer system/Manual/Manual	Computer/Manual/Manual
Pile cap tilt	Tiltmeter	Manual	Manual
Lateral pile deformation	Slope inclinometer	Manual	Manual
Lateral ground deformation	Slope inclinometer	Manual	Manual
Axial pile deformation	Strain gages/telltales	Mini-computer system/Manual	Mini-computer system/Manual
Temperature	Thermocouples	Manual	Manual

displacement rate criteria had been met. The lateral load was then alternated to 6 tons/pile (53 kN/pile) and 0.6 ton/pile (5.3 kN/pile) until the lateral displacement of the pile cap under the full lateral load reached an essentially constant value.

Pile Driving Effects Testing. This phase of the testing involved measuring the response of the pile groups to nearby pile driving. During this phase, the pile groups had sustained axial (30 tons/pile (267 kN/pile)) and lateral (6 tons/pile (53 kN/pile)) loads which were maintained during the entire phase.

The procedures used in this phase of the testing program will not be presented here. However, the effects on the monoiths will be briefly discussed as they pertain to the load tests.

Nearby pile driving caused lateral displacement and settlement of the single pile and the pile groups. As the pile cap was rigidly connected to the group piles, the lateral displacement of all pile heads was essentially the same. However, the pile driving caused unequal settlement of the piles. Those closest to the pile driving settled more resulting in a tilt of the pile cap. The differential settlement ranged from 0.4 to 0.8 inches (10 to 20 mm) and caused a tilt of 0.2 to 0.3 degrees. Thus the pile groups started with some distortions at the start of the load tests.

Axial Load Tests. Axial load tests to failure were made of the single pile and one eight-pile group. Only compressive loads were applied; no tension tests were performed. The procedure was as follows:

1. Make initial readings and measurements with no load on the pile or pile group.

2. Apply the load in increments of 30 tons/pile (267 kN/pile), i.e. 240 tons (2.1 MN) for the eight-pile group and 30 tons (267 kN) for the single pile. Maintain load until the settlement rate is less than 0.01 in/hr (0.25 cm/hr).

3. Make readings and measurements.

4. Increase the applied load in 30 tons/pile increments, repeating steps (2) and (3) until failure occurred.

   The single pile was considered as having failed when the pile tip movement was greater than 1.5 inches (38 cm) and the planned load increment could not be resisted by the pile.

For the eight-pile group, failure was taken as when the monolith had tilted so much that the continued stability of load frame system was questionable. A differential settlement of one corner of the pile cap of 1.2 inches (3.0 cm) was observed.

5. Removal of the load and final measurements and readings were taken.

Combined Load Tests. In the combined load tests, an axial load was first applied to the pile group and then the lateral load was increased in increments until failure occurred. The procedure used in the performance of the tests was as follows:

1. Make initial readings and measurements with no load on the pile or pile group.

2. Apply an axial compression load of 30 tons/pile (267 kN). Maintain load until settlement of monolith is less than 0.01 in/hr (0.25 cm/hr).

3. Make readings and measurements.

4. While maintaining axial load, apply a lateral load of 6 tons/pile (53 kN/pile). Maintain load until lateral displacement of pile cap is less then 0.01 in/hr (0.25 cm/hr).

5. Repeat step (3).

6. Increase lateral load by 3 tons/pile (27 kN) and maintain until lateral displacement of pile cap is less than 0.01 in/hr (0.25 mm/hr). Repeat step (3) and apply next load increment.

7. Repeat step (6) until failure occurs. Failure was defined as structural failure of the piles or connection with pile cap, or continued displacement of the monolith under constant load.

8. Remove lateral load and repeat step (3).

9. Remove axial load and repeat step (3).

DISPLACEMENT MEASUREMENTS

The determination of the pile cap displacements was done considering the three translation and three rotational degrees of freedom. Displacements at each corner of the pile were measured in the vertical and two horizontal directions, parallel and perpendicular to the lateral load direction. With these twelve measurements, the displacements and rotations were computed for the center of the bottom of the pile cap by considering it as a rigid body.

TEST RESULTS

Cylic Preloading. The results of this portion of the test program are summarized in Table 2. In general, the pile groups and single pile reached their maximum displacements about 10 cycles before the cyclic preloading was halted. Cyclic displacement amplitude between the minimum and maximum lateral loads was relatively constant after the first two cycles. During the early load cycles, displacement stabilization required 2 to 3 hours. For the late load cycles, stabilization occurred almost immediately after application of the load.

Axial Load Test. Axial load tests were performed on one eight-pile group and the single pile. The load-settlement curve and load-horizontal displacement curve are given in Figure 8. The maximum sustained load was 720 tons (6.4 MN) or an average of 90 tons/pile (800 kN/pile). From the strain gage data, individual pile loads ranged from 72 tons (640 kN) to 150 tons (1.3 MN). When the load was increased to 750 tons (6.7 MN), the pile cap continued to settle rapidly and unevenly, and showed no signs of stabilizing. The southeast corner of the pile cap settled the most and caused a precarious tilt in the loading system. Apparently that pile was unable to sustain the load. Upon unloading, there was no evidence of structural failure of any of the piles.

The load-settlement curve and load-horizontal displacement curve for the single pile are given in Figure 9. Under the maximum applied load, 110 tons (980 kN), the pile continued to slowly settle for six hours and did not stabilize. At that time, the test was halted and the pile had deflected significantly laterally.

Combined Load Tests. The results of lateral load tests under constant axial load are summarized in Table 3. Figures 10, 11 and 12 summarize lateral load-deflection data for the single pile, eight-pile group, and twelve-pile group, respectively.

In every case the maximum applied load resulted in horizontal displacement rates which approached but did not satisfy the stabilization criterion of no faster than 0.01 in. (0.25 mm) per hour after several hours. There were no obvious signs of structural failure of the piles, although cracking sounds were apparent at high load levels. Gaps of 1 to 2 in (2.5 to 5.0 cm)

TABLE 2

SUMMARY OF CYCLIC PRELOADING RESULTS

	12-pile group (3x4)	8-pile group 1 (2x4)	8-pile group 2 (2x4)	Single Pile
Maximum axial load, tons (kN)	360 (3203)	240 (2135)	240 (2135)	30 (267)
Maximum lateral load, tons (kN)	72 (641)	48 (427)	48 (427)	6 (53)
Settlement under maximum axial load only, inches (cm)	0.32 (0.81)	0.29 (0.74)	0.30 (0.76)	0.28 (0.71)
Lateral displacement with first application of maximum lateral load, inches (cm)	0.74 (1.88)	0.92 (2.34)	0.70 (1.78)	0.78 (1.98)
Lateral displacement at end of cyclic preloading with maximum lateral load, inches (cm)	1.26 (3.20)	1.07 (2.71)	1.05 (2.67)	0.91 (2.31)
Lateral displacement at end of cyclic preloading with minimum lateral load, inches (cm)	0.68 (1.73)	0.58 (1.47)	0.52 (1.32)	0.77 (1.96)
Total number of load cycles	40	29	40	23
Residual horizontal displacement after unloading, inches (cm)	0.53 (1.35)	0.49 (1.24)	0.45 (1.14)	0.42 (1.07)

Note: Lateral displacements measured 3 feet (0.9m) above ground surface.

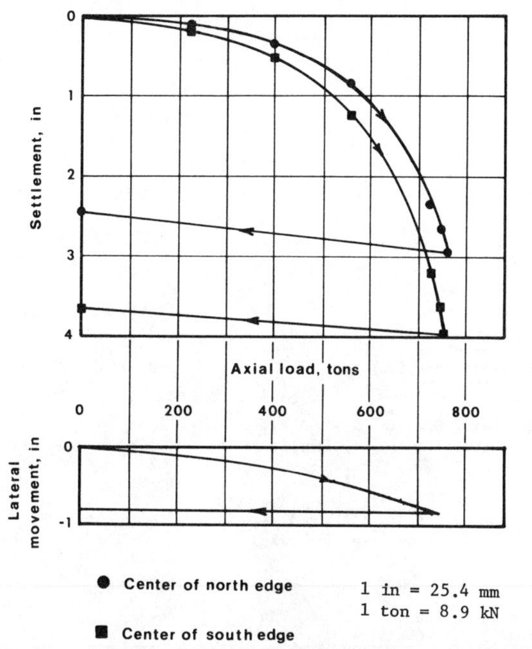

Figure 8 – LOAD-SETTLEMENT CURVE FOR PILE GROUP

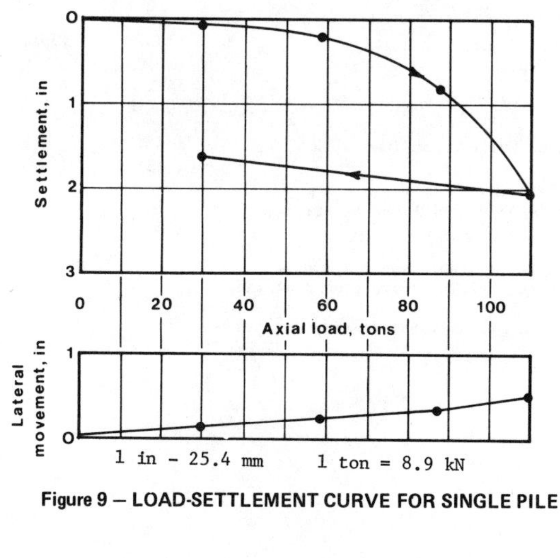

Figure 9 – LOAD-SETTLEMENT CURVE FOR SINGLE PILE

TABLE 3

SUMMARY OF COMBINED LOAD TEST RESULTS

	12-pile group (3x4)	8-pile group 1 (2x4)	Single Pile
Applied axial load, tons (kN)	360 (3203)	240 (2135)	30 (267)
Lateral load causing 0.25 in (6.3 mm) displacement, tons (kN); tons/pile, (kN/pile)	24 (214) 2 (18)	18 (160) 2.3 (20)	4.8 (43) 4.8 (43)
Lateral load causing 0.50 in (12.7 mm) displacement, tons (kN); tons/pile, (kN/pile)	44 (391) 3.7 (33)	35 (311) 4.4 (39)	8.2 (73) 8.2 (73)
Lateral load causing 1.0 in (25.4 mm) displacement, tons (kN); tons/pile, (kN/pile)	84 (747) 7 (62)	60 (534) 7.5 (67)	11.9 (106) 11.9 (106)
Lateral load causing 2.0 in (50.8 mm) displacement, tons (kN); tons/pile, (kN/pile)	144 (1281) 12 (107)	96 (854) 12 (107)	12.8 (114) 12.8 (114)
Maximum sustained load, tons (kN); tons/pile, (kN/pile)	172 (1530) 14.3 (127)	120 (1068) 15 (133)	15 (133) 15 (133)
Displacement at failure, in (mm)	5.7 (145)	5.2 (132)	2.9 (74)
Residual displacement after unloading, in (mm)	2.2 (56)	1.5 (38)	- -

Notes: 1. Lateral displacement measured 3 feet (0.9m) above ground surface

2. Tons/pile = Total load/No. of piles

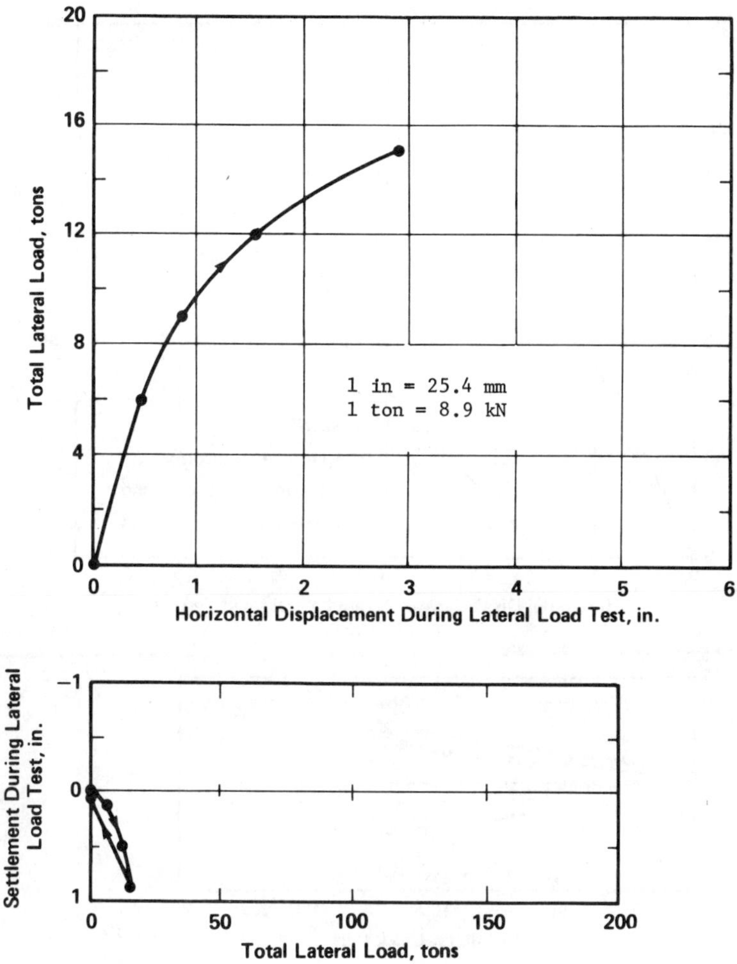

Notes:

1) Axial load during test was 30 tons
2) Horizontal displacement prior to load testing was 2.47 in.
3) Lateral load applied at 28 in. above ground surface
4) Data measured 33 in. above ground surface

Figure 10 – DISPLACEMENTS OF SINGLE PILE DURING LATERAL LOAD TEST

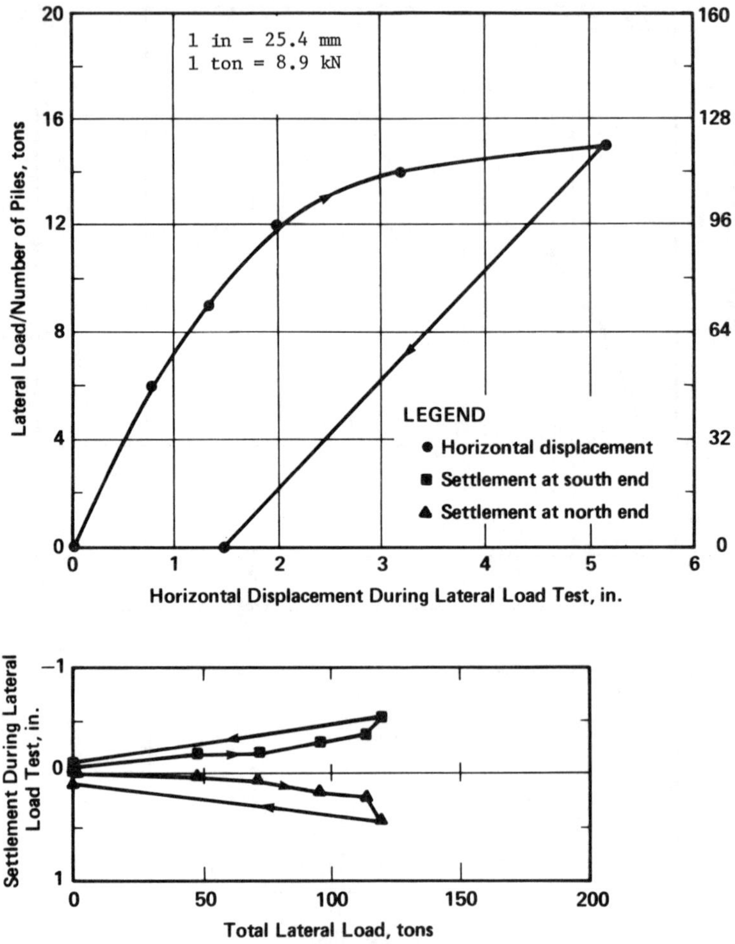

Notes:

1) Axial load during test was 240 tons
2) Horizontal displacement prior to load testing was 2.20 in.
3) Settlement prior to load testing was 0.95 in. at south end and 1.54 in. at north end
4) Residual displacement after complete unloading (axial and lateral)

Figure 11 – DISPLACEMENTS OF EIGHT-PILE GROUP CAP DURING LATERAL LOAD TEST

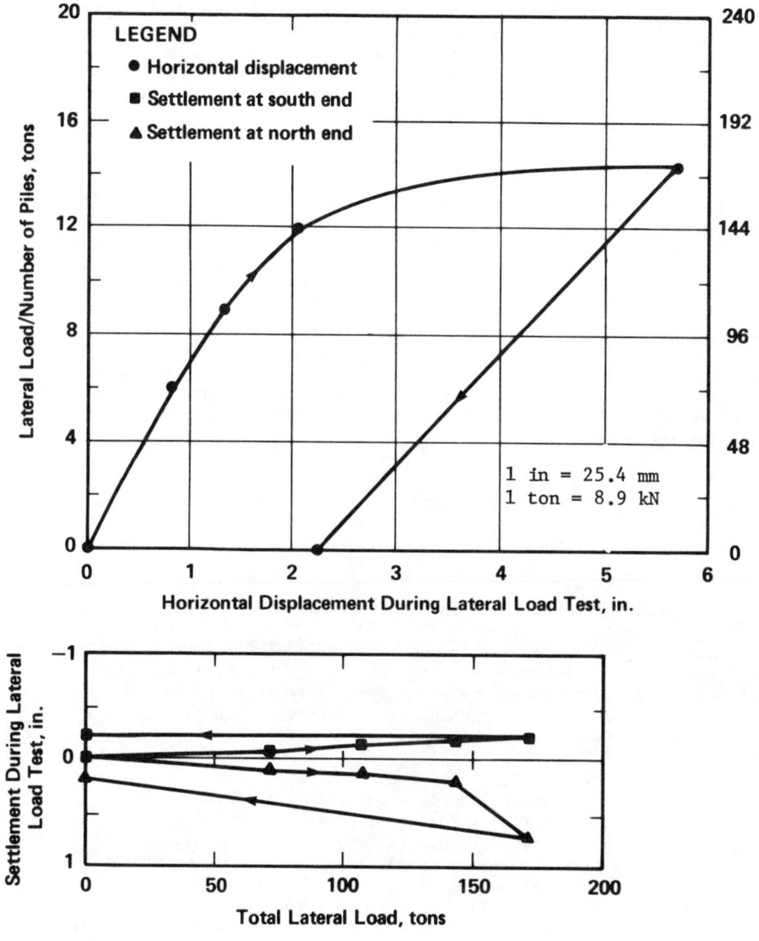

Notes:

1) Axial load during test was 360 tons
2) Horizontal displacement prior to load testing was 1.90 in.
3) Settlement prior to load testing was 1.11 in. at south end and 2.06 in. at north end
4) Residual displacement after complete unloading (axial and lateral)

Figure 12 – DISPLACEMENTS OF TWELVE-PILE GROUP CAP DURING LATERAL LOAD TEST

developed between the soil and the rear of the piles under the maximum applied loads. Soil deformations in front of the piles were obscured by water and snow cover. The ground remained thawed in the vicinity of the piles due to forced-air heater usage (warming the displacement measuring devices as well), and due to near surface maintenance of the ground water surface.

DISCUSSION OF TEST REULTS

The first comparison of importance is the axial capacity of single versus eight-pile group. It was noted that the single pile was "more efficient" than the group piles on average, although individual piles sustained between 65 percent and 130 percent of the single pile capacity under full group load. The timber piles in the group were jetted and driven from the south end toward the north end of the monolith (see Fig. 7) in sequence. Installation effects coupled with prior history (cyclic preloading and nearby pile driving) effects, no doubt influenced the observed response. Restrike measurements at the time of pile installation gave no indication of lower capacity piles at the south end of the cluster, suggesting that cyclic preloading, nearby pile driving, and group interaction effects under axial load were the major factors influencing the behavior.

In comparing single pile group response to lateral loads (under constant axial load) there are three details of the tests configuration deserving attention in terms of their influence on the test results.

1. The heads of the piles within each group were embedded 2 ft (0.6 m) into the cap, with 3 ft (0.9 m) "free span" between the base of the cap and the ground surface. This condition resulted in relative fixity at the base of the pile cap; the measured fixity moment resulted in a "zero crossing" of the bending moment (from negative at the cap to positive below the ground surface) near the ground surface.

2. The presence of axial load on the single pile involved an unknown degree of fixity as the pile translated horizontally relative to the (repositioned) axis of axial loading, causing a negative fixity moment. Combined with the positive moment induced by the position of the lateral thrust jack (approximately 28 in (71 cm) above the ground surface, the single pile cap rotated downward toward the direction of applied load, resulting in a varying, partially constrained head condition.

3. The configuration of the tests system for the twelve-pile monolith deviated from that shown in Fig. 6 in two important ways: the toe of the excavation slope was located 8 ft (2.4 m) from the pile cap, as opposed from 25 ft (7.6 m) or more in all other cases; and, one or two rows of non-structural piles were

driven surrounding the twelve-pile cluster to examine the influence of their installation and presence on monolith behavior during nearby pile driving.

Recognizing these conditions it is apparent that direct quanitative "efficiency" comparisons are not valid. In comparing the lateral load-deflection curves plotted as average load per pile, it is particularly interesting to note a remarkable agreement of these for the two pile groups. One would conclude that no apparent "group effects" were observed in the comparison. The presence of the excavation berm closer to the twelve-pile cluster, coupled with the effects of its surrounding rows of piles would tend to reduce deflection under a given load and to increase ultimate lateral capacity to some degree. Therefore, it is apparent that the larger pile cluster would have exhibited larger displacements and lower lateral load capacity per pile had these differences in test configuration been absent, an observation consistent with observations of most other investigators.

This paper summarizes the gross behavior observed in load tests on one single pile and three pile groups. Analyses and evaluations of the load transfer mechanisms that prevail are incomplete at the present time. Preliminary studies for one of the eight-pile groups indicate the quality of the remaining data (strain gages, inclinometers, etc.) is quite good (see Woodward-Clyde Consultants, 1979). This paper serves as an introduction to the static load testing aspects of a comprehensive field program, much of which remains to be evaluated at the present time.

APPENDIX - References

Gibbs, J. and Holtz, W. G. (1953), "Research on Determining the Density of Sands by Spoon Penetration Testing," Proc. 4th ICMSFE, Vol. 1, p. 35-39.

Woodward-Clyde Consultants (1979), "Foundation Investigation and Test Program, Locks and Dam No. 26,, Mississippi River, Alton, Illinois, Phase IV Report, Vol. III, Results and Interpretation of Pile Driving Effects Test Program," Contract No. DACW43-78-C-0005 for U.S. Army Engineer District, St. Louis.

Schmertmann, J. (1976) "An Updated Correlation Between Relative Density, D_r, and Fugro-type Electric Cone Bearing, q_c, "Contract Report DACW39-76-M-6646, Waterways Experiment Station, Vicksburg, MS.

EFFECTS OF PILE DRIVING SYSTEMS ON DRIVEABILITY AND CAPACITY OF CONCRETE PILES

by

CHRISTOPHER DAVID THOMPSON[1], M.ASCE, AND DAVID ELLIOT THOMPSON[2]

SUMMARY

The Case Method and CAPWAP wave equation analyses reliably predicted the ultimate bearing capacities of concrete piles driven to bearing in granular soils at two sites. The dynamic measurements indicate that it may be necessary to drive concrete piles to high driving resistances to develop the maximum capacity of the driving system. Additionally, beneficial reduction of tension was achieved by employing a hard capblock with a soft cushion.

INTRODUCTION

Extensive pile load testing programs were undertaken at the site of two projects close to Metropolitan Toronto in Southern Ontario, for the purpose of determining the most economical foundation types. On one project, six precast concrete piles were driven with two different hammers to various depths, in order to establish the minimum foundation costs per unit load, and to develop specifications for production pile driving. At the other site, steel H pipe and timber piles were installed in addition to four precast concrete piles. An open-end diesel hammer was used to drive the concrete piles, and the capblocks and cushions were changed to determine the effects of varying these components. This paper examines the results of the test programs with regards to the accuracy of the Case Method and CAPWAP wave equation analyses in estimating bearing capacities, and with regards to the effects of different hammers, capblocks and cushions on the driveability and induced stresses in the piles. Additionally, the use of the field measurements to predict pile damage is reviewed.

PROCEDURE

The testing at both of the sites included a geotechnical investigation, Pile Driving Analyser testing during driving, and conventional static load testing. The geotechnical investigations were carried out according to standard ASTM and CSA procedures to below the level of the deepest piles on each site. Standard penetration tests were undertaken and split-spoon samples obtained. The laboratory testing primarily involved classification of the soils, and moisture content and unit weight determinations.

The piles were 12-inch (305 mm) square precast reinforced concrete piles in segments of 25 and 40 feet (7.6 and 12.2 m) which were joined in the field by mechanical splices. Details of the piles and splices are shown in Figure 1.

[1]General Manager, The Trow Group Limited, Toronto, Ontario, Canada
[2]Manager of the Oshawa Office, The Trow Group Limited, Oshawa, Ontario, Canada

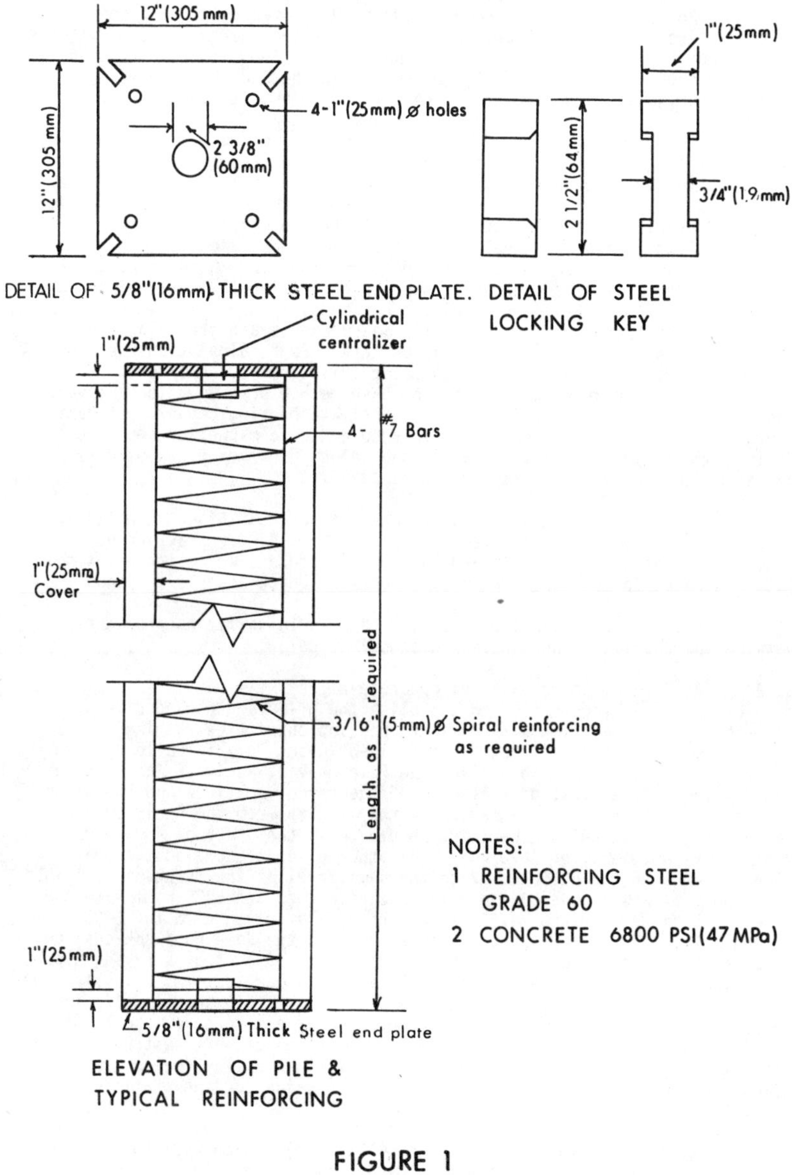

FIGURE 1

PRECAST CONCRETE PILE AND SPLICING DETAIL

All of the concrete piles were installed with either a Berminghammer B400 or B225 open-end diesel hammer. The B400 hammer has a rated energy of 46,000 ft-lbs (62 kJ) per blow, and the B225 a rated energy of 29,000 ft-lbs (39 kJ) per blow. The B400 has a piston weight of 5000 lbs (2268 kg) and a maximum piston stroke of 10 feet (3049 mm), and the B225 has a piston weight of 3000 lbs (1361 kg) and a maximum stroke of 9.75 feet (2973 mm).

The dynamic testing of the piles consisted of connecting equipment from the Pile Driving Analyser to the pile head and measuring strain and acceleration during driving. The stress wave speed of the piles was determined prior to pile driving by striking one end of each section with a sledge hammer and taking accelerometer measurements at the other end. From this data, the longitudinal wave velocity (wave speed) was calculated. The wave speeds were initially calculated from data recorded on a high speed oscillograph. They were subsequently checked by digitizing the data which had been stored on magnetic tape in the field. The forces in the piles were derived by means of elastic theory from the strain measurements and the velocities were obtained by integrating the acceleration data. Both calculations were performed automatically in the field computer (Pile Driving Analyser). In addition, the signals were used for the Case Method wave equation analysis (Goble et al, 1975) in the Analyser, to estimate bearing capacities, maximum energies, etc. The field signals were recorded on a magnetic tape by using an instrumentation F.M. tape recorder and selected hammer blows were re-analysed by the CAPWAP computer program (Rausche, 1975).

Static load testing was undertaken as per ASTM Standard D 1143-74. Five of the tests (three at Site A and two at Site B) were carried out according to Section 4.2 'Standard Loading Procedure', while three tests at Site A were performed according to Section 4.7 'Quick Load Test Method'.

There were some differences in the pile installation procedures between the two sites, and these are described below.

SITE A - At Site A, three pairs of concrete piles (Pile Nos. 6-1, 6-2, 7-1, 7-2, 3-2 and 3-3) were driven to establish the pile lengths and driving criteria to produce piles with ultimate bearing capacities ranging from 240 to 400 kips (1070 to 1780 kN) as defined in the Ontario Building Code. In order to achieve this objective, the Pile Driving Analyser was used during the driving of the first pile at each of the three locations and the piles were driven with the B400 hammer until the results from the Analyser indicated that their bearing capacities should be in the order of 400, 300 or 240 kips (1780, 1335, or 1070 kN). The other pile at each location was then driven with the B225 hammer to the same level as the first pile at that location. It was impossible to drive the pile with the B225 hammer to the same level as the 400-kip (1780 kN) capacity pile driven with the B400 hammer. During all of the pile driving for this project, the same capblock-cushion combination was employed. The capblock consisted of a steel striking block, 2-1/2 inches (64 mm) thick for the B225, and 5 inches (127 mm) thick for the B400, on four alternating plates of 1/2-inch (13 mm) aluminum and 1/2-inch (13 mm) rubber belting. The cushion consisted of six to seven sheets of 3/4-inch (19 mm) thick plywood. All of the test piles were restruck on the day following initial driving, to establish whether soil freeze or relaxation had occurred and to provide dynamic predictions of the static bearing capacity.

SITE B - At Site B, two piles (Pile Nos. TP3, TP4), which were to be load tested statically, were driven with a B400 hammer. In both cases, the capblock was the same as that used on Site A, and the cushion consisted of five sheets of 3/4-inch (19 mm) thick plywood. The first test pile was driven to the same depth as an adjacent 'H' pile which had been driven to a predicted, and subsequently statically tested, ultimate bearing capacity of approximately 800 kips (3560 kN) at a depth of 114 feet (34.8 m). The second concrete test pile was stopped at 54 feet (16.5 m) depth to determine whether an adequate bearing capacity was available at a higher level. Ultimate bearing capacities of 650 kips (2890 kN) and 470 kips (2090 kN) were predicted on the deep and shallow piles respectively by the Pile Driving Analyser at the end of initial driving.

Two other piles (Pile Nos. E1, E2), which were not load tested, were driven to examine the effects on the driveability of altering the capblock and cushion combinations. The capblocks and cushions employed for Pile E2 are listed in Table 1, and effects of changes in the capblocks and cushions and other aspects are shown in Figure 2.

TEST RESULTS

SITE A - Site A is a completed sanitary landfill disposal area situated on a ravine slope. The site is slightly higher in the centre than around the edges. At the test pile locations, the ground surface elevation ranged from 532.4 to 539.4 feet (162.3 to 164.5 m) geodetic, with the natural soil surface level varying from 497 to 515 feet (151.5 to 157.0 m) geodetic. The deepest and shallowest fill were at the location of Test Piles 7-1 and 2, and Test Piles 6-1 and 2, respectively.

The natural soils are interlayered alluvial deposits over glacial tills. The alluvial deposits, present above Elevations 480 to 492 feet (146.3 to 150 m), 42 to 54 feet (12.8 to 16.5 m) depth, are interbedded sands, silts and clays with some gravel. Their density ranges from compact to very dense. The underlying till is very dense silty sand with some gravel. The groundwater table was at approximate Elevation 505 feet (154.0 m) during the investigation, with perched groundwater at Elevation 514 feet (156.7 m) at some locations. The soil profile for Piles 7-1 and 2 is shown on Figure 3, which also shows typical profiles of driving resistance, maximum force at pile head, and ultimate bearing capacity predicted by the Pile Driving Analyser.

The piles were driven to founding in the alluvial deposits at tip Elevations of 511 to 484 feet (155.9 to 147.6 m), i.e. 28 to 49 feet (8.5 to 14.9 m) depth. They were stopped when the predicted ultimate bearing capacity on the Pile Driving Analyser ranged from 150 kips (670 kN) for Pile 6-1 to 370 kips (1645 kN) for Pile 7-2. The piles were restruck on the following day and the bearing capacity predictions had increased to a range of 205 kips (925 kN) for 6-1 to 385 kips (1715 kN) for 7-2. Details of these results are shown in Table 2.

The driving resistance of the piles was greatly different for the B225 and B400 hammers. The B400 drove the piles easily and the final driving resistances were four to five blows per inch (25 mm) with accumulative blows ranging from 500 to 1000. However, the B225 had difficulty in driving the piles and could not achieve the same penetration for Pile 7-1 as the B400 had for 7-2. The final driving resistances for this hammer were 17 to 39 blows per inch (25 mm) with accumulative blows of 1100 to 1700. Driving profiles for Piles 7-1 and 2 are shown in Figure 3.

TABLE 1

CAPBLOCK AND CUSHION COMBINATIONS FOR TEST PILE E1, SITE B

Depth feet (m)		CAPBLOCK		CUSHION		COMBINED STIFFNESS kip/in (kN/mm)
FROM	TO	NUMBER AND MATERIAL IN LAYERS	STIFFNESS kip/in (kN/mm)	NUMBER AND MATERIAL IN LAYERS	STIFFNESS kip/in (kN/mm)	
35 (10.7)	53 (16.2)	4 Aluminum 3 Rubber	5000 (876)	5 Plywood	2000 (350)	1400 (245)
53 (16.2)	82 (25.0)	4 Aluminum 3 Rubber	5000 (876)	5 New Plywood	2000 (350)	1400 (245)
82 (25.0)	93 (28.4)	4 Aluminum 3 Rubber	5000 (876)	4 Aluminum 3 Rubber	5000 (876)	2500 (438)
93 (28.4)	102 (31.1)	4 Aluminum 3 Rubber	5000 (876)	5 New Plywood	2000 (350)	1400 (245)
102 (31.4)	123 (37.5)	4 Aluminum 3 Rubber	5000 (876)	4 Aluminum 3 Rubber	5000 (876)	2500 (438)
123 (37.5)	124.7 (38.0)	4 Aluminum 3 Rubber	5000 (876)	6 New Marine Plywood	1600 (280)	1200 (210)
124.7 (38.0)	126.9 (38.7)	4 Aluminum 3 Rubber	5000 (876)	1 Aluminum 3 Rubber	5000 (876)	2500 (438)
126.9 (38.7)	127.3 (38.8)	2 Micarta 1 Aluminum	32000 (5604)	1 Aluminum 8 Rubber	1900 (333)	1800 (315)

NOTE: The nominal thickness of each Plywood layer: 3/4-inch (19 mm)
Rubber layer: 3/4-inch (19 mm)
Aluminum layer: 1/2-inch (13 mm)
Micarta layer: 2 -inch (51 mm)

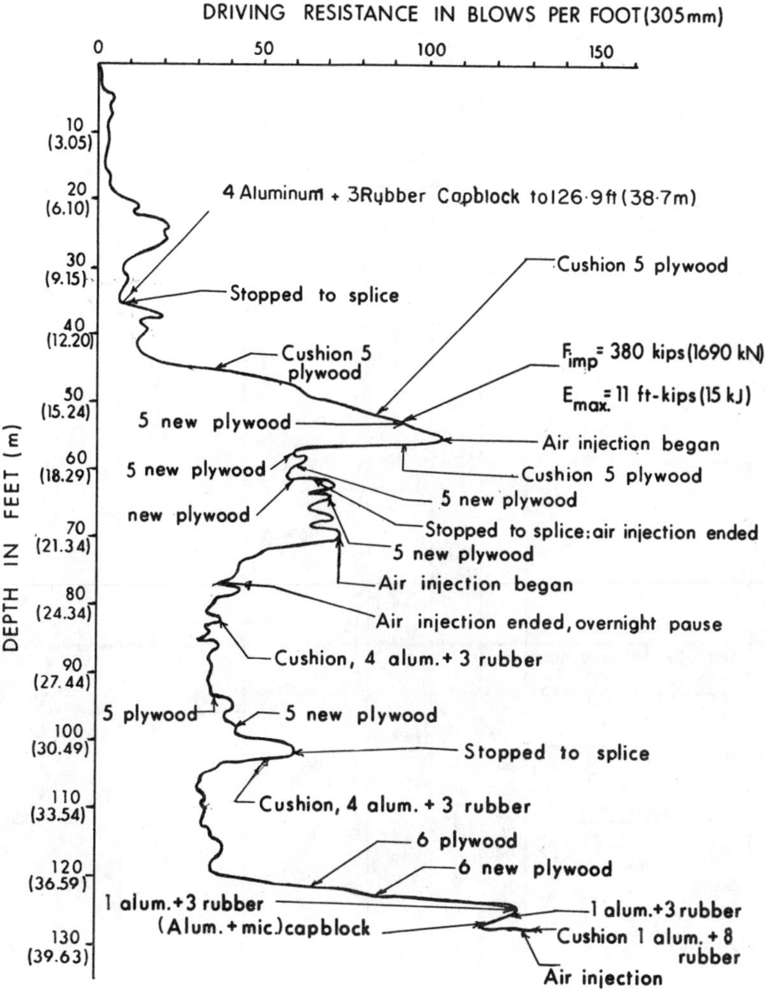

NOTE: DETAILS OF CAPBLOCK AND CUSHION PRESENTED IN TABLE 2

FIGURE 2

DRIVING RESISTANCE PROFILE FOR PILE E1, SITE B

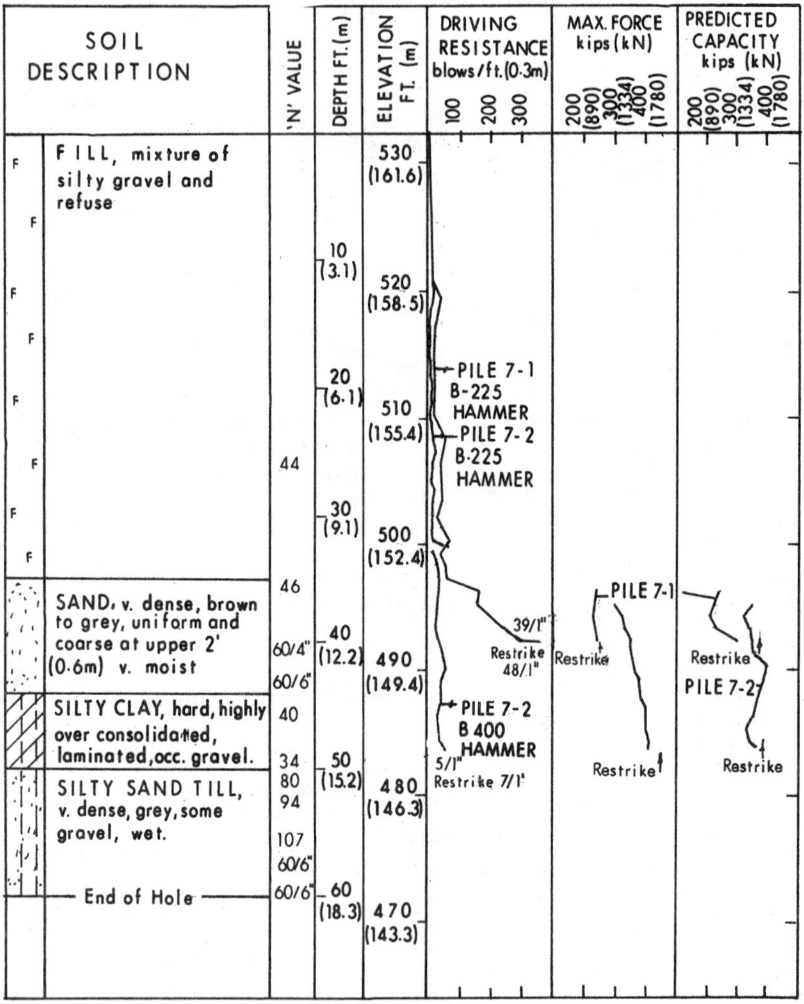

FIGURE 3

GEOTECHNICAL, PILE DRIVING RESISTANCE, MAXIMUM FORCE AND PREDICTED BEARING CAPACITY PROFILES FOR PILES 7-1, 7-2 SITE 'A'

TABLE 2

DRIVING STRESSES AND ULTIMATE BEARING CAPACITY

PILE			HAMMER		EQUIVALENT DRIVING RESISTANCE	MAXIMUM IMPACT STRESS AT PILE HEAD			ULTIMATE BEARING CAPACITY kips (kN)		
	LENGTH feet (m)						% OF COMPRESSIVE STRENGTH OF CONCRETE				
NO.	DRIVEN	EMBEDDED	MODEL	RAM RISE feet (m)	bl/in / 25 mm	ksi (MPa)	DESIGN STRENGTH	TESTED STRENGTH	ANALYSER PREDICTION	CAPWAP PREDICTION	LOAD TEST BY DAVISSON
3-2	60.0 (18.3)	37.4 (11.4)	B400	6 (1.8)	5	2.78 (19.2)	41	36	280 (1260)	--	--
				10 (3)	6	2.99 (20.6)	44	39	325 (1450)	310 (1390)	305 (1370)
3-3	50.0 (15.2)	36.7 (11.2)	B225	4 (1.2)	34	2.00 (13.8)	29	26	270 (1210)	--	--
				6 (1.8)	50	1.54 (10.6)	23	20	290 (1300)	260 (1160)	360 (1600)
6-1	35.0 (10.7)	27.9 (18.5)	B225	4 (1.2)	17	1.96 (13.5)	29	25	150 (660)	--	--
				4 (1.2)	12	1.73 (11.9)	25	22	205 (920)	--	235 (1040)
6-2	35.0 (10.7)	27.6 (18.4)	B400	6 (1.8)	4	2.90 (20.0)	43	38	235 (1040)	--	--
				6 (1.8)	4	3.45 (23.8)	51	45	300 (1340)	--	255 (1140)
7-1	65.0 (19.8)	41.0 (12.5)	B225	6 (1.8)	39	1.93 (13.3)	28	25	305 (1360)	--	--
				6.5 (2)	48	2.06 (14.2)	30	27	380 (1690)	280 (1250)	400 (1780)

TABLE 2 CONTINUED

TABLE 2 (CONTINUED)

DRIVING STRESSES AND ULTIMATE BEARING CAPACITIES

PILE			HAMMER		EQUIVALENT DRIVING RESISTANCE	MAXIMUM IMPACT STRESS AT PILE HEAD			ULTIMATE BEARING CAPACITY kips (kN)		
							% OF COMPRESSIVE STRENGTH OF CONCRETE				
NO.	LENGTH feet (m) DRIVEN	LENGTH feet (m) EMBEDDED	MODEL	RAM RISE feet (m)	bl/in / 25 mm	ksi (MPa)	DESIGN STRENGTH	TESTED STRENGTH	ANALYSER PREDICTION	CAPWAP PREDICTION	LOAD TEST BY DAVISSON
7-2	65.0 (19.8)	49.5 (15.1)	B400	8 (2.4)	5	2.78 (19.2)	41	36	370 (1020)	--	--
				6 (1.8)	7	3.13 (21.6)	46	41	385 (1710)	345 (1540)	350 (1560)
TP3	145.0 (442)	114.2 (34.8)	B400	8.5-9.0 (2.6-2.7)	5	4.00 (27.6)	59	42	650 (2190)	--	--
		114.2 (34.8)		8.5-9.0 (2.6-2.7)	12	2.78 (19.2)	41	39	485 (2150)	520 (2320)	500 (2220)
		114.3 (34.8)		8.5-9.0 (2.6-2.7)	12	3.22 (22.2)	47	34	570 (2540)	--	--
TP4	65.0 (19.8)	54.2 (16.5)	B400	9.0-9.5 (2.7-2.9)	5	5.15 (35.5)	76	54	470 (2090)	--	--
				9.0-9.5 (2.7-2.9)	6	4.97 (34.3)	73	53	470 (2090)	--	500 (2220)
		54.5 (16.6)		9.0-9.5 (2.7-2.9)	12	4.15 (28.6)	61	44			

NOTE: 1. Top line of values for each pile obtained at end of initial driving and second line at the beginning of restrike.
2. Bottom line for TP3 and TP4 gives values obtained at the end of redriving.
3. Ram Rise on restrike approximate for Piles 3-2, 3-3, 6-1, 6-2, 7-1 and 7-2 as only 3 or 4 blows undertaken.

There were tendencies for the maximum impact force at the top of the pile to increase with depth, reaching maxima of approximately 300 kips (1335 kN) with the B225 hammer and 400 kips (1780 kN) with the B400 hammer. These represent 25 to 40 per cent of the tested compressive strength of the concrete in the piles.

The piles were load tested and the results are presented in Figures 4 and 5 and Table 2. The ultimate bearing capacity of the piles, established by means of the Davisson Method, ranged from 235 to 400 kips (1045 to 1780 kN). They varied from the Pile Driving Analyser predictions by between 4 to 19 per cent with a standard deviation of ±14 per cent. The CAPWAP analysis was performed on Test Piles 7-1, 7-2, 3-2 and 3-3. The CAPWAP was considerably more accurate for the B400 at 1 per cent from the test results of Piles 7-2 and 3-2, than the 30 to 28 per cent errors for Piles 7-1 and 3-3, which were driven by the B225, as shown in Table 2.

SITE B - Site B was in a flat location which was grass covered. The piles were driven fairly close to each other, with very little differences in ground surface elevation at 638 feet (194.5 m) geodetic.

There were some differences in the soil stratigraphy between different areas of the overall site for the project. The typical profile at the location of the test piles is indicated in Figure 6, together with the pile driving resistances. This figure shows that there is 49 feet (14.9 m) of soft or loose to hard or compact clayey silt to silty sand before soil suitable for high capacity piles is present. The soils below this level are tills and range from between compact and dense clayey silt to loose to compact sandy silt above 106 feet (32.3 m) depth. Standard penetration tests N-values are consistently in the range of 45 to 80 blows per foot (305 mm) at the upper levels of the till.

Below 106 feet (32.3 m) depth the soils are very dense with N-values of 68 to 201 blows per foot (305 mm). The textural classification of this stratum ranges from clayey silt to silty sand, and it is layered at some levels. The boreholes were terminated at 131 feet (39.9 m) depth. The groundwater table was close to and slightly above the ground surface indicating minor artesian conditions.

The pile driving records for the concrete piles are shown in Figure 6. Piles TP3 and 4 were driven with the same capblock-cushion combination and show more variations in driving resistance at the same levels than would be expected for the lateral changes in soil conditions, which were the only apparent differences effecting driving. Piles E1 and E2 show a greater range of variation in driving resistance. However, there was little difference in the overall driveability, even though the combined capblock-cushion stiffness ranged from 1,200 to 2,500 kips/inch (210 to 438 kN/mm). Generally, there was little difference in the force and velocity traces for all cushions employed from those presented as dashed lines in Figure 7, except when the capblock-cushion combination was changed during driving of Pile E2 to an aluminum and micarta sandwich and a cushion of nine sheets of 3/4-inch (19 mm) thick rubber belting having an overall stiffness of 1550 kips/inch (270 kN/mm). This change resulted in the lower maximum force and velocity and extended impact illustrated by the solid lines in Figure 7. Also, when the capblock-cushion was returned to the more normally employed combination of a rubber capblock and plywood cushion with a similar overall stiffness of 1360 kips/inch (238 kN/mm), the driving resistance increased sharply, together with the impact force and ram stroke.

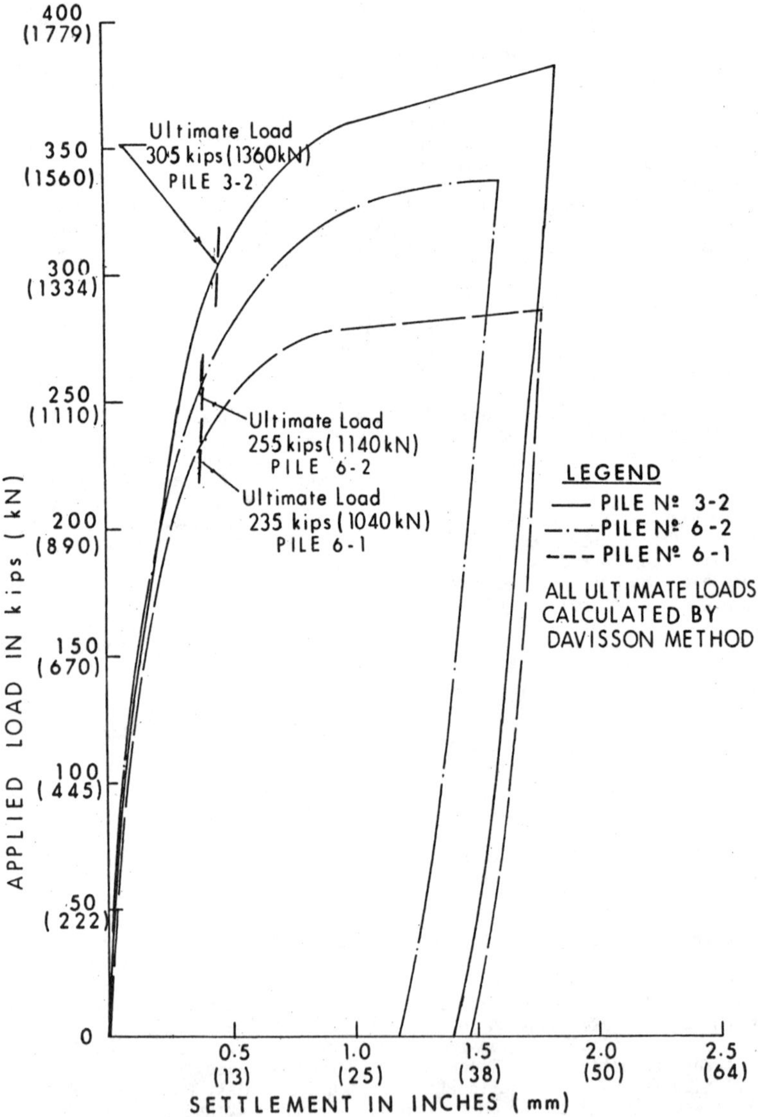

FIGURE 4
QUICK PILE LOAD TESTS, SITE 'A'

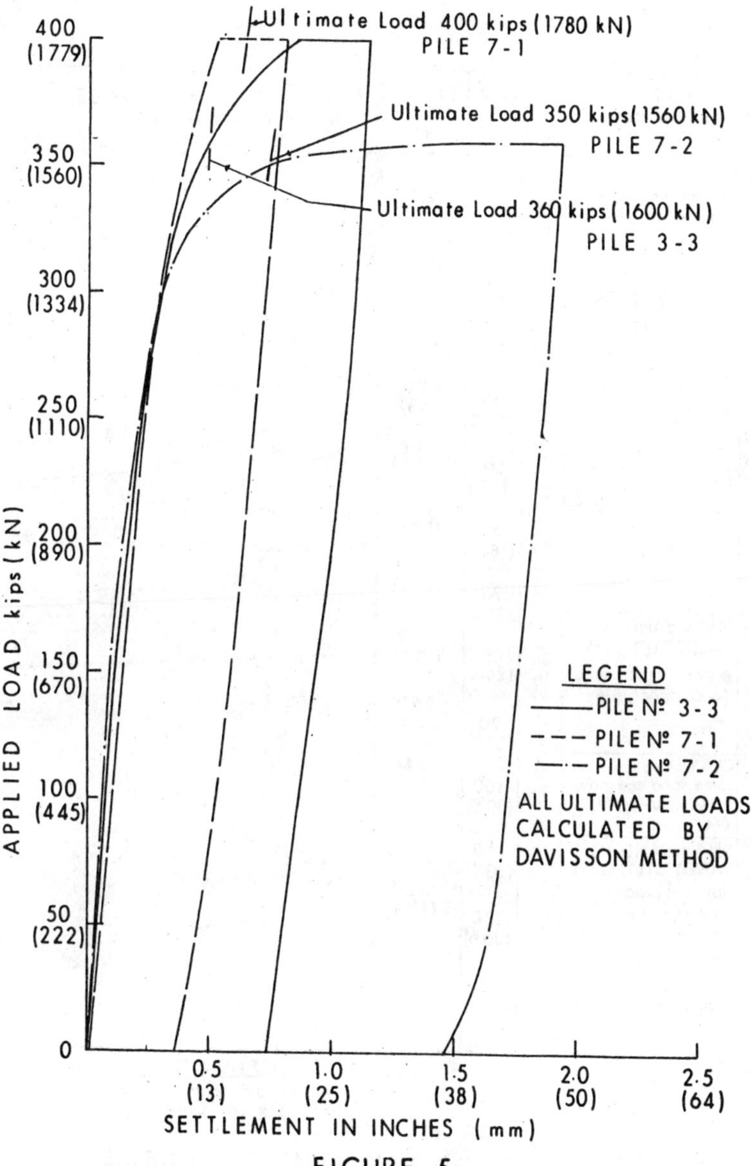

FIGURE 5
STANDARD PILE LOAD TESTS, SITE 'A'

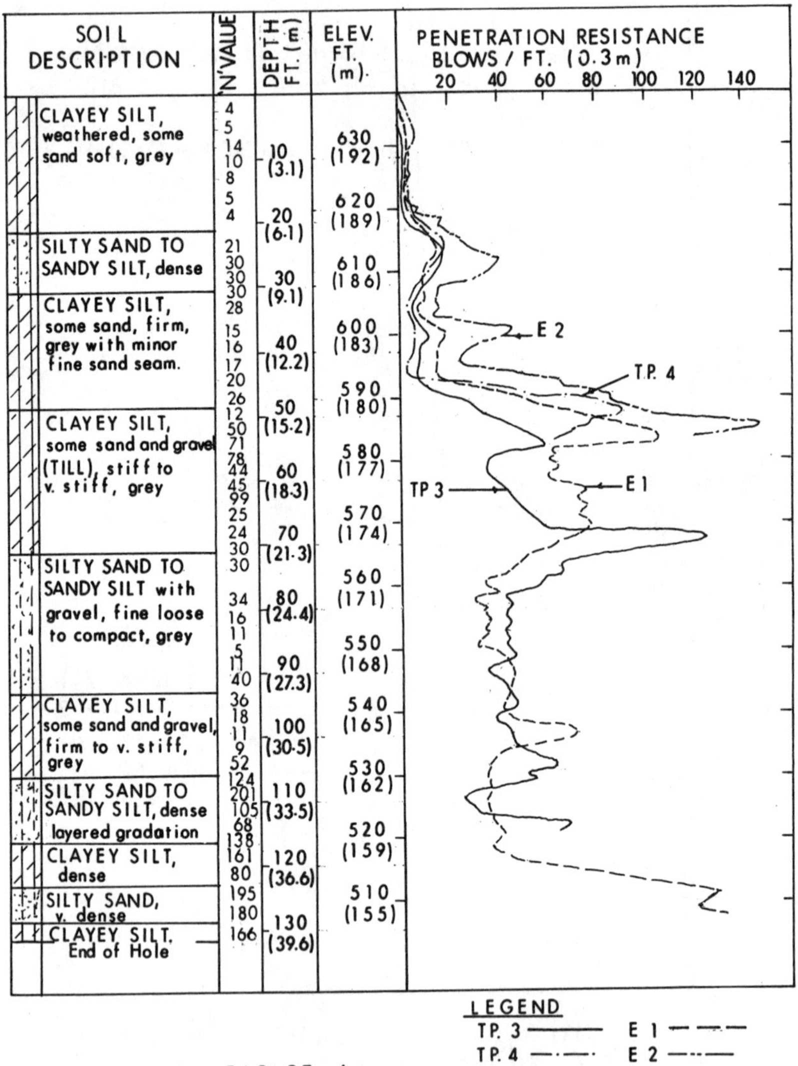

FIGURE 6
GEOTECHNICAL AND PILE DRIVING RESISTANCE PROFILES, SITE B

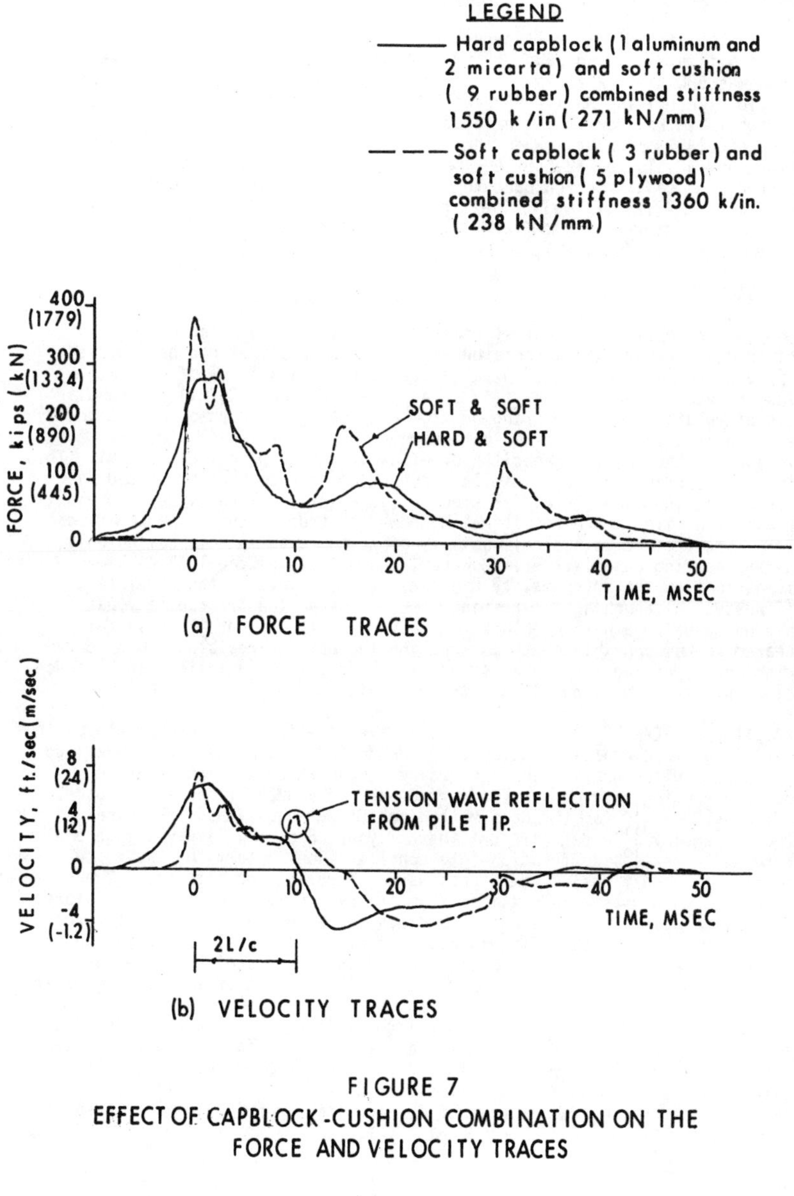

FIGURE 7
EFFECT OF CAPBLOCK-CUSHION COMBINATION ON THE FORCE AND VELOCITY TRACES

The impact force during driving varied significantly and reached a maximum of 600 kips (2670 kN) at 82 feet (25.0 m) depth in Pile E1, and 740 kips (3290 kN) at the end of driving TP4.

The static load test curves for Piles TP3 and 4 are shown in Figure 8. The ultimate bearing capacity of both piles according to the Davisson Method was 500 kips (2225 kN). The ultimate capacities initially predicted by field Pile Driving Analyser results were 720 kips (3205 kN) for TP3 and 460 kips (2045 kN) for TP4. These assumed stress wave speeds of 15,000 feet per second (4573 m/sec.). When the stress wave speed was checked by digitization, and a value of 12,330 feet per second (3759 m/sec.) was determined, the predicted capacities changed to 485 and 470 kips (2160 to 2090 kN) for Piles TP3 and 4 respectively, i.e. the error in prediction for TP3 was reduced from 44 to 3 per cent. The CAPWAP predicted that the ultimate capacity of TP3 was 520 kips (2310 kN), an error of 4 per cent.

DISCUSSION

The results of the test programs at Sites A and B indicated that dynamic measurements and associated wave equation analyses can be employed to provide reliable estimates of ultimate bearing capacities of concrete piles, driven to resistances of 4 to 40 blows per inch (25 mm) in granular soils of relative densities ranging from compact to very dense. Additionally, useful information was obtained on the relationship of impact stresses at the pile top and the bearing capacity for piles driven to practical refusal. This information indicated that a similar proportionality to that presented by Thompson and Thompson (1978) for steel piles exists at lower stress levels for concrete piles. The results also showed that differences in hammers can have a large effect on the driveability of piles, even in soil having only moderate bearing capabilities. However, the effects of capblock-cushion combinations on the driveability did not clearly relate to their relative stiffnesses, although an interesting observation on the driveability was achieved by using a hard capblock with a very soft cushion. This system appeared to be beneficial both in reducing the driving resistance and in reducing and extending the impact stress in the piles. The latter can be beneficial when driving in conditions where tension may occur in the piling.

PREDICTION OF BEARING CAPACITY - The two wave equation procedures employed in relation to field dynamic testing of piles during driving have been examined for concrete piles bearing in fine grained granular soils with relative densities ranging from compact to very dense. The predictions are compared in Figures 9 and 10 for the Case Method and CAPWAP equations. The Case Method is employed in the Pile Driving Analyser, and is a fairly simple procedure which takes side friction into consideration in terms of the force and velocity conditions as the stress wave returns to the pile top. The CAPWAP procedure uses data obtained from the Pile Driving Analyser and stored on magnetic tapes. The relevant parameters are adjusted to model the force traces from the velocity. The distribution of the external forces is one of the parameters varied in this modelling, and in these cases indicated proportions of side friction varying from 20 to 80 per cent of the total capacity.

The Case Method Analysis correlates well with load tests (Figure 9) with all eight test results within 19 per cent of the static load test. Two of the worst predictions were on Piles 3-3 and 6-1 and were both lower than static load test results. These piles were driven with the smaller B225 hammer and had driving resistances of 17 blows per inch (25 mm) or greater. The remaining pile which was driven with the B225 was No. 7-1. The Case Method

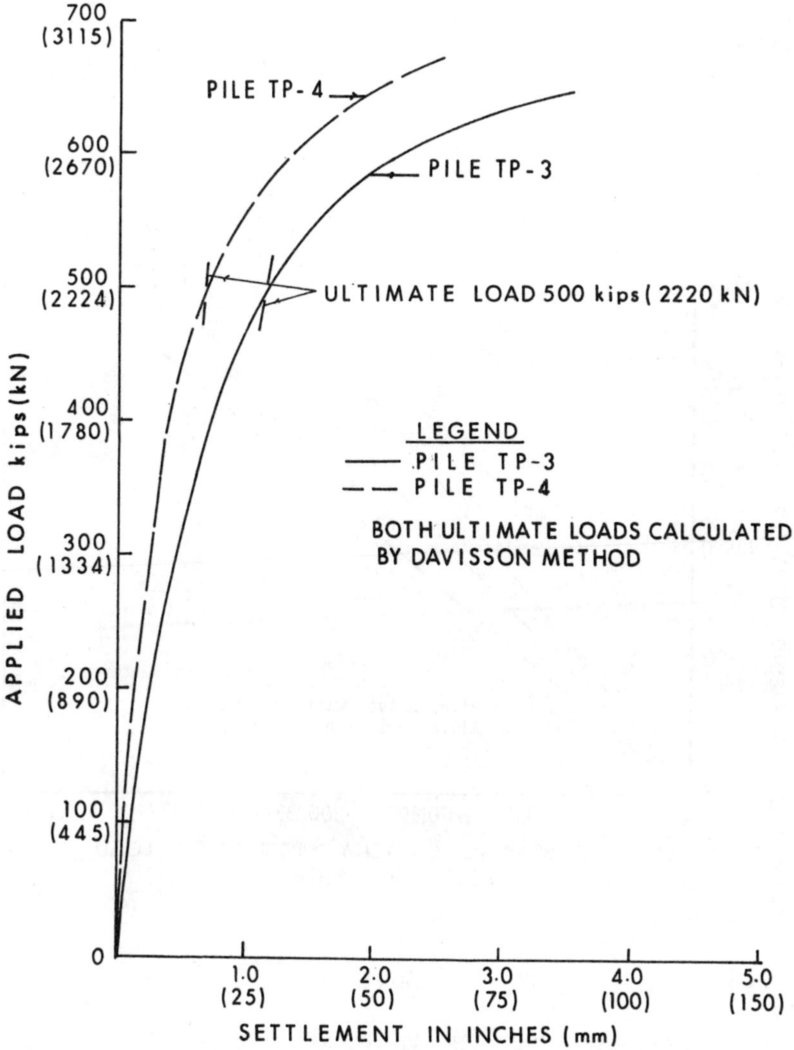

FIGURE 8
PILE LOAD TESTS, SITE 'B'

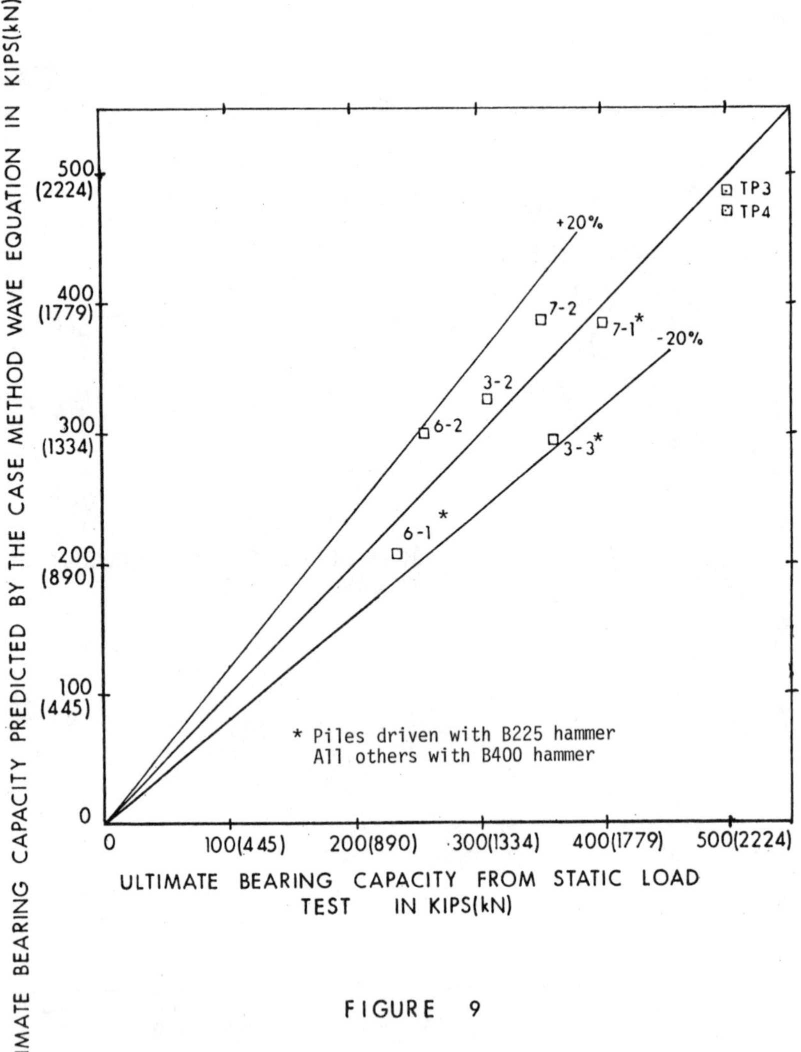

FIGURE 9

COMPARISON OF ULTIMATE BEARING CAPACITIES PREDICTED BY THE CASE METHOD WITH STATIC LOAD TEST RESULTS

estimate for this pile was also lower than static tests. These results indicate that the insufficient force was developed by the B225 in the conditions at Site A to dynamically prove the ultimate capacity. This is a problem which has been observed elsewhere with different types of piles and in different conditions. However, it is concluded that the Case Method wave equation presented accurate estimates of ultimate bearing capacities in fine grained granular soils with a wide variety of support conditions, but that it is more accurate when the pile is being moved by the hammer-capblock-cushion combination.

One potential source of inaccuracy in the Case Method revealed by the testing is the stress wave speed. The force calibration is a function of the stress wave speed. The selection interval for the second part of the calculation in this procedure is established by dividing twice the pile length by the stress wave velocity, and is therefore potentially critical. The wave speed of individual pile sections is determined experimentally in the field. The actual signal from the section end can be masked, if there are a number of minor flaws, such as hairline cracks, in the pile. This occurred for Pile TP3, Site B, when a stress wave speed of 15,000 feet per second (4578 m/sec.) was assumed in the field, instead of the value of 12,330 feet per second (3759 m/sec.) which was subsequently determined. This value resulted in an error of 44 per cent, rather than the 3 per cent which was achieved with the corrected wave speed. It is therefore clear that the wave speed can be critical to accurate bearing capacity predictions with the Case Method analysis.

The CAPWAP analysis appears to be more critically affected by the driving resistance or the capability of the hammer to develop the bearing capacity than is the Case Method, as illustrated by Figure 10. Piles 3-2, 7-2 and TP3 were driven with the B400 to driving resistances of 5 to 12 blows per inch (25 mm) and had a maximum difference from static tests of 4 per cent. Piles 3-1 and 7-1 were, however, driven with the smaller B225 to driving resistances of 34 to 39 blows per inch (25 mm) and were low by 28 to 30 per cent. This may indicate a trend for the CAPWAP procedure to underestimate capacities at high driving resistances. However, it is concluded that the CAPWAP analysis was very accurate, when the pile was being moved by the hammer, and could be correlated well with static load test results at higher driving resistances.

IMPACT STRESSES - The impact stresses at the top of the piles were examined to determine whether the relationship observed by Thompson and Thompson (1978), mainly on steel piles, exists for concrete piles. This relationship indicated that piles driven to practical refusal in granular soils have an ultimate static bearing pressure equal to 1.2 times the impact stress at the pile top at the end of initial driving. The results of the current testing on piles with final driving resistances of 10 blows per inch (25 mm) or greater are presented in Figure 11. These results show more scatter than for the steel piles and more correlation with a multiplication factor of unity from impact stresses to bearing pressure with a lower bound at 80 per cent of this value. It is interesting to note however that Piles 3-3 and 7-1 with the highest driving resistances were closer to the relationship for steel piles. This may indicate that practical refusal for concrete piles is at greater driving resistances than for steel piles. This would mean that the bearing capacity of concrete piles tends to increase significantly as the driving resistances increase to 20 to 30 blows per inch (25 mm), whereas the steel piles effectively reach close to their maximum capacities at driving resistances of approximately 10 blows per inch (25 mm). This trend is confirmed by the Case Method and CAPWAP bearing capacity predictions which were conservative at high driving resistances but more accurate when the piles were moving

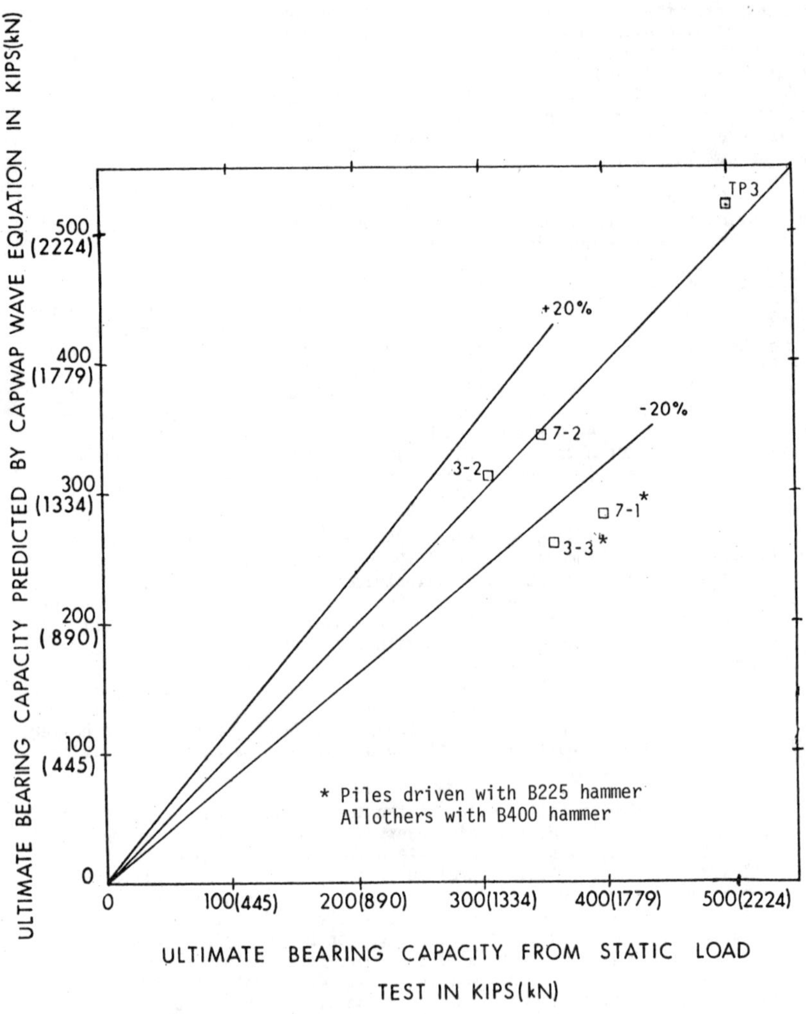

FIGURE 10

COMPARISON OF ULTIMATE BEARING CAPACITIES PREDICTED BY CAPWAP WITH STATIC LOAD TEST RESULTS

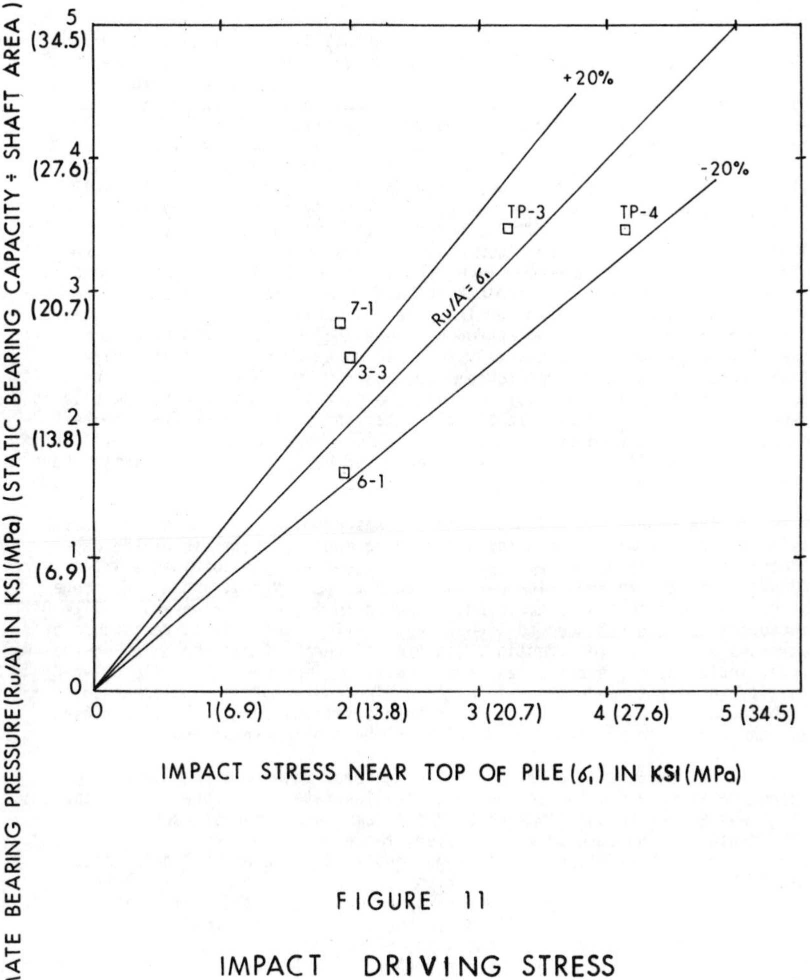

FIGURE 11

IMPACT DRIVING STRESS
VERSUS
ULTIMATE BEARING PRESSURE

appreciably. It may therefore be concluded that it is necessary to drive concrete piles to high penetration resistances in the order of 30 to 40 blows per inch (25 mm) to develop the maximum capacity of the hammer-capblock-cushion-pile-soil system.

EFFECTS OF HAMMER CAPACITY ON THE PILE DRIVEABILITY - At Site A, the driveability of piles installed with different sizes of hammers was examined. It was clear that the B400 hammer, with a rated energy of 46,000 ft-lbs per blow (62 kJ), was much more effective and economical than the B225 at a rated energy of 29,000 ft-lbs (39 kJ). The total magnitude of the impact force developed by the hammers was not greatly different being a maximum of 300 kips (1335 kN) for the B225 and 400 kips (1780 kN) for the B400. However, it can be observed that the lower force was not sufficient to effectively fail the soil, while the higher value was probably controlled by the maximum that the soil could support. The net result on the driveability was to lower the driving resistance from 30 to 40 blows per inch (25 mm) for the B225 to between 4 and 5 blows per inch (25 mm) for the B400. This indicates that matching of the hammer and capblock-cushion combination with the pile and soil had a significant effect on the pile driveability. As the compressive strength of the pile was not approached in either case, it can be observed that the large hammer was unlikely to cause damage problems at this site. It is concluded that damage problems due to overstressing the pile are primarily governed by the soils capacity to resist penetration. As long as the pile is penetrating the ground at a reasonable rate, the piles are unlikely to be damaged in compression by using a large hammer. However, little benefit in terms of driveability is likely to be achieved by increasing the hammer above the size required to easily fail the soil.

The capblock-cushion combination was examined at Site B. At this site, a B400 hammer was used for all driving and the stiffness of the capblock-cushion combination was varied significantly. The combined stiffness ranged from 1200 to 2500 kips per inch (210 to 438 kN/mm), and a wide range of capblock-cushions were employed, as shown in Table 1. However, at Site B these changes did not seem to significantly affect the driving resistance or stresses with a single exception. In general terms of all the piles on the site, including the steel piles, there was some improvement in pile driveability in easy to moderate driving conditions with softer capblock-cushion combinations. However, aspects other than stiffness seem to be significant and these may include internal damping in the capblock-cushion.

An interesting exception to the general trend was the use of a 'hard' capblock and a 'soft' cushion, as illustrated in Figure 7. The capblock was a micarta and aluminum sandwich, similar to the standard Delmag D22 capblock. The cushion was, however, two aluminum plates with nine 3/4-inch (19 mm) layers of rubber belting, resulting in an overall calculated capblock-cushion stiffness of 1550 kips per inch (271 kN/mm). While the overall stiffness was very similar to a number of the other combinations used, the hard capblock and soft cushion had the effects of flattening and extending the impact force and velocity traces as well as decreasing the driving resistance. This is a desirable effect for concrete piles as the reflected tension stress wave was eliminated and consequently the potential of damage in easy driving conditions would be substantially lessened.

PILE DAMAGE - During driving of Piles TP3 and E1, Site B, there was concern that they had been damaged because of irregularities in the force and velocity trace from the Pile Driving Analyser. However, the driving resistance profiles in Figure 6 did not show any changes, which would be attributed to damage. Also, changes in driving behaviour, which might indicate damage, were not observed during driving. Conversely, the driving resistance of both piles was substantially less than on the adjacent test 'H' pile which had similar impedance characteristics and had driven to the same level. Additionally, TP3 failed at 500 kips (2225 kN), as compared with 800 kips (3560 kN) for the test 'H' pile. There was some feeling that these differences were caused by a reduction in bearing capacity of the concrete piles due to artesian pressures at depth or by the comparatively large displacement of the soil by the concrete piles as compared with the 'H' pile. While these potential reasons cannot be completely eliminated without excavation of the pile, damage is indicated both by changes in the force and velocity traces shown for Pile TP3 in Figure 12 and by subsequent CAPWAP analysis. Figure 12 shows the gradual development of a 'free end' at 40 feet (12.2 m) below the gauges in the 105 feet (32.0 m) driving length at that time. This is indicated by the increase in velocity and corresponding decrease in force at 40 feet in Figure 12 (c). This depth coincides with the top splice in the pile. Additionally, the CAPWAP analysis shows the best match with the Analyser force and velocity traces if a low stiffness zone is inserted at the same location in the pile. Such a low stiffness zone would be caused by pile damage. Consequently, it is considered that damage has probably occurred approximately 40 feet (21.3 m) below the pile top, which could not be detected from the pile driving records or observations. It must therefore be concluded that damage can occur to concrete piles during driving which cannot be detected visually by an experienced engineer. This makes the testing of random piles with the Pile Driving Analyser a valuable checking procedure in cases where there could be any doubt about damage.

CONCLUSIONS

The dynamic testing of 12-inch (305 mm) square precast concrete piles at Sites A and B where the founding soils were compact to very dense fine grained granular soils provided some valuable information on dynamic methods of estimating bearing capacities and on the effects of hammers, capblocks and cushions on the driveability of piles.

(1) The Case Method and CAPWAP wave equations provided accurate estimates of ultimate bearing capacities for the eight test piles, as shown in Figures 9 and 10. Both tend to be conservative for piles driven to practical refusal. Stress wave speed data can be critical in accurately estimating the bearing capacity.

(2) It was observed that for five test piles driven to penetration resistances of more than ten blows per inch (25 mm), the ultimate bearing pressure was approximately equal to the final impact stress at the pile top during final driving. This is slightly less bearing pressure per unit impact stress than presented by Thompson and Thompson (1978), although piles driven to 30 to 40 blows per inch (25 mm), met the relationship indicated in that paper. Consequently, it is concluded that it may be necessary to drive concrete piles to high driving resistances to develop the maximum capacity of the hammer-capblock-cushion-pile-soil system.

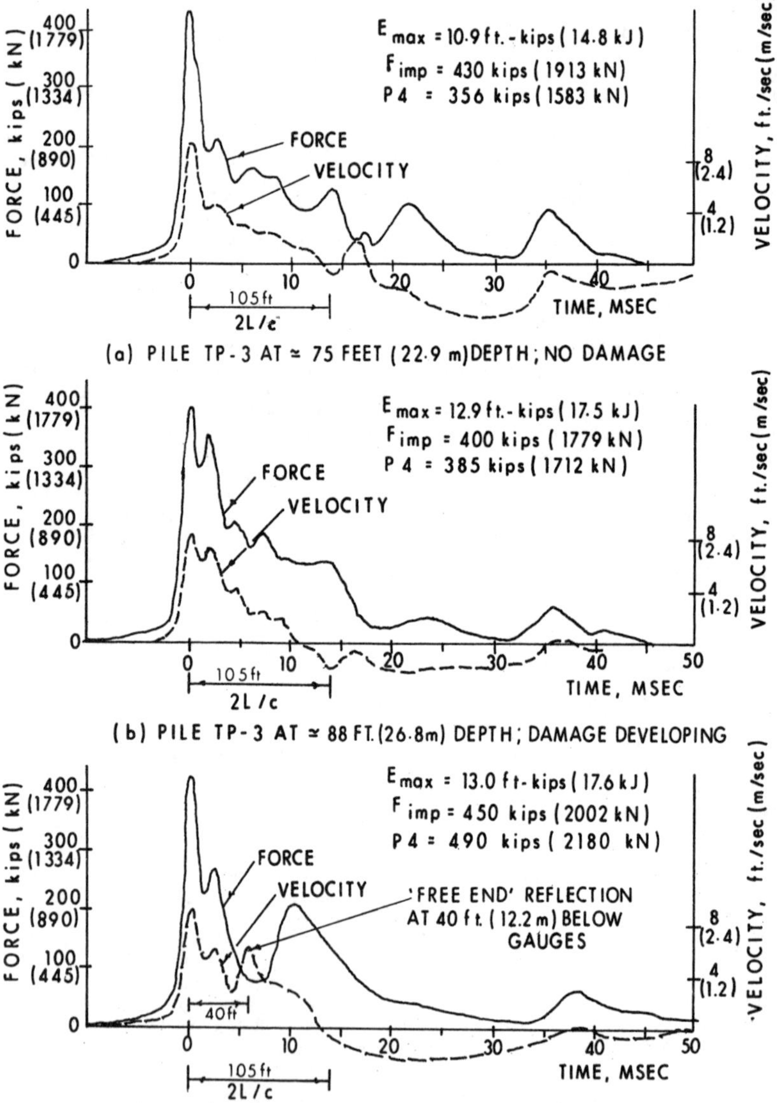

(a) PILE TP-3 AT ≃ 75 FEET (22.9 m) DEPTH; NO DAMAGE

(b) PILE TP-3 AT ≃ 88 FT. (26.8 m) DEPTH; DAMAGE DEVELOPING

(c) PILE TP 3 AT ≃ 101 FT. (30.8 m) PILE DAMAGE

FIGURE 12
FORCE AND VELOCITY TRACES SHOWING DEVELOPMENT OF PILE DAMAGE, PILE TP3

(3) It is important that the pile hammer should have sufficient rated energy to fail the soil during pile driving. However, increasing the hammer size above a certain limiting capability for a given hammer-soil combination is unlikely to significantly improve the driveability. On the other hand, it is unlikely to damage the pile in compression, as long as the pile is penetrating and there is a sufficient thickness of capblock-cushion.

(4) At Site B, a 'hard' capblock with a 'soft' cushion was observed to decrease the driving resistance, and flatten and extend the impact from the hammer, compared with other capblock-cushion combinations with similar stiffnesses. This capblock-cushion eliminated the tension effects at the pile head during moderate driving which would be beneficial in reducing damage to concrete piling.

(5) Pile damage was indicated by the Pile Driving Analyser and subsequent CAPWAP work. This could not be observed from visual observations of pile driving or from the pile driving records.

REFERENCES

Goble, G. G., Rausche, F., Likens, G., 'The Case Method', paper presented at the Fourth Case Pile Seminar, March 1975, Case Western Reserve University, Cleveland, Ohio.

Rausche, F., 'The Classical Wave Equation Approach and the Case Pile Wave Analysis', paper presented at the Fourth Case Pile Seminar, March 1975, Case Western University, Cleveland, Ohio.

Thompson, C. D., Thompson, D. E., 'The Influence of Driving Stresses on the Development of High Pile Capacities'. Proceedings of A.S.T.M. Symposium on Behaviour of Deep Foundations, Boston, Massachusetts, June 1978.

CASE HISTORY: CONCRETE-FILLED STEEL PIPE PILES

David E. Thompson[1], M. ASCE

and

John W. Critchfield[2], A.M. ASCE

INTRODUCTION

This case history documents six field load tests performed in connection with pile foundations installed for a major building project on the Boston waterfront. The project includes two 40-story apartment towers and an adjacent parking garage, supported by 90 ton (800 kN) design capacity concrete-filled steel pipe piles. Three of the six test piles failed to develop twice the design capacity, as required by the local building code.

The following sections describe site conditions, test pile installation and load test procedures. Load test data are then analyzed and conclusions are drawn relative to the cause of the observed failures.

SITE CONDITIONS

The relative locations of the two apartment towers and the garage are shown on Figure 1. Prior to construction, the ground surface was level at about El. +15 ft. (4.6 m) above Boston City Base Datum (El. 0.00 ft. BCB = El. 5.65 ft. Mean Sea Level). The area of Tower No. 2 was excavated to about El. +2 ft. (0.6 m) and a gravel working pad was installed prior to pile driving. No excavation was made at Tower No. 1 or at the garage prior to test pile installation.

---

[1]Principal, Haley & Aldrich, Inc., Cambridge MA

[2]Staff Engineer, Haley & Aldrich, Inc., Cambridge MA

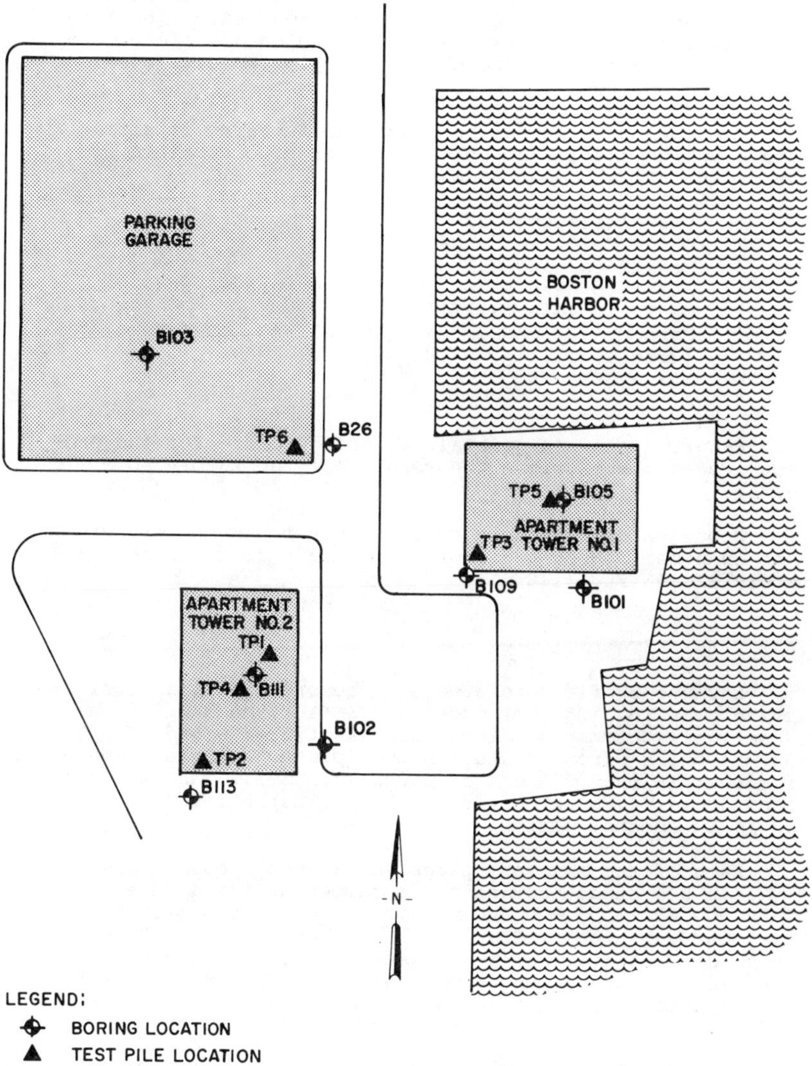

FIG. 1 - LOCATION PLAN

Logs were available from over 90 borings made within and adjacent to the project area. Figure 1 shows the locations of a few representative borings which are closest to the test piles and from which soil samples were selected for laboratory testing.

The soil profile can be described in terms of six strata, which are listed below in order of occurrence from the ground surface downward:

1. Miscellaneous FILL
2. ORGANIC SILT and PEAT
3. Medium to hard CLAY
4. Stiff to hard SANDY CLAY, trace gravel and boulders (Glacial Till)
5. Hard CLAYEY SAND, trace gravel and boulders (Glacial Till)
6. Shale BEDROCK

Figure 2 shows a generalized soil profile for the project area and summarizes the results of laboratory soil tests.

Groundwater levels in borings were found to vary with tide changes in Boston Harbor. Mean tide level is approximately El. +6 ft. (1.8 m).

TEST PILE INSTALLATION

Each test pile consisted of a concrete filled steel pipe having a 14 in. (356 mm) outside diameter and a 0.312 in. (8 mm) wall thickness. Each pipe was fitted with a 1.5 in. (38 mm) thick circular steel bottom plate having a diameter of either 14 in. (356 mm) or 14.75 (375 mm). Test piles 1, 2 and 3 had 14.75 in. bottom plates. Test piles 4, 5 and 6 had 14 in. bottom plates.

Each of the six test piles was driven closed-end with a Vulcan 010 single-acting steam hammer having a rated energy of 32,500 ft.-lb. (44000 J) and using a 8 in. (203 mm) thick oak cap block. The hammer was handled by a 60 ton (535 kN) capacity crawler crane with 85 ft. (25.9 m) long fixed leads. Where required, a spud was used to penetrate obstructions in the surficial fill materials. Driving records for each test pile are shown on Figures 3 through 8.

Following driving and just prior to concreting, each pipe was inspected by lowering an outdoor floodlight to the

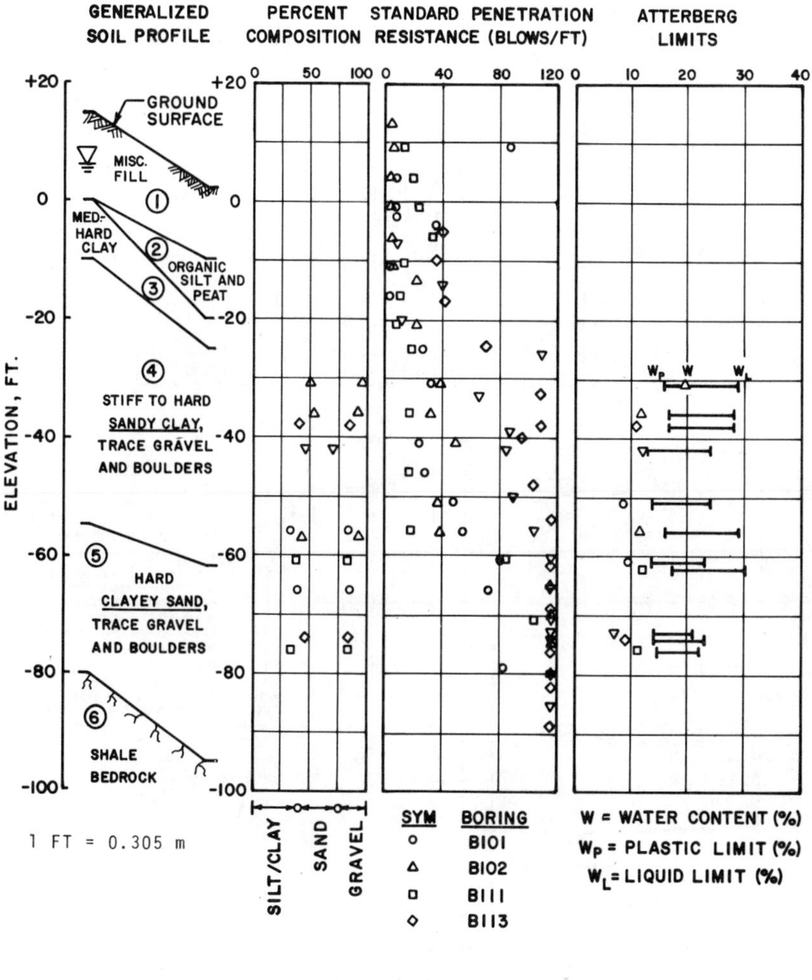

FIG. 2 - GENERALIZED SOIL PROFILE AND LABORATORY TEST DATA

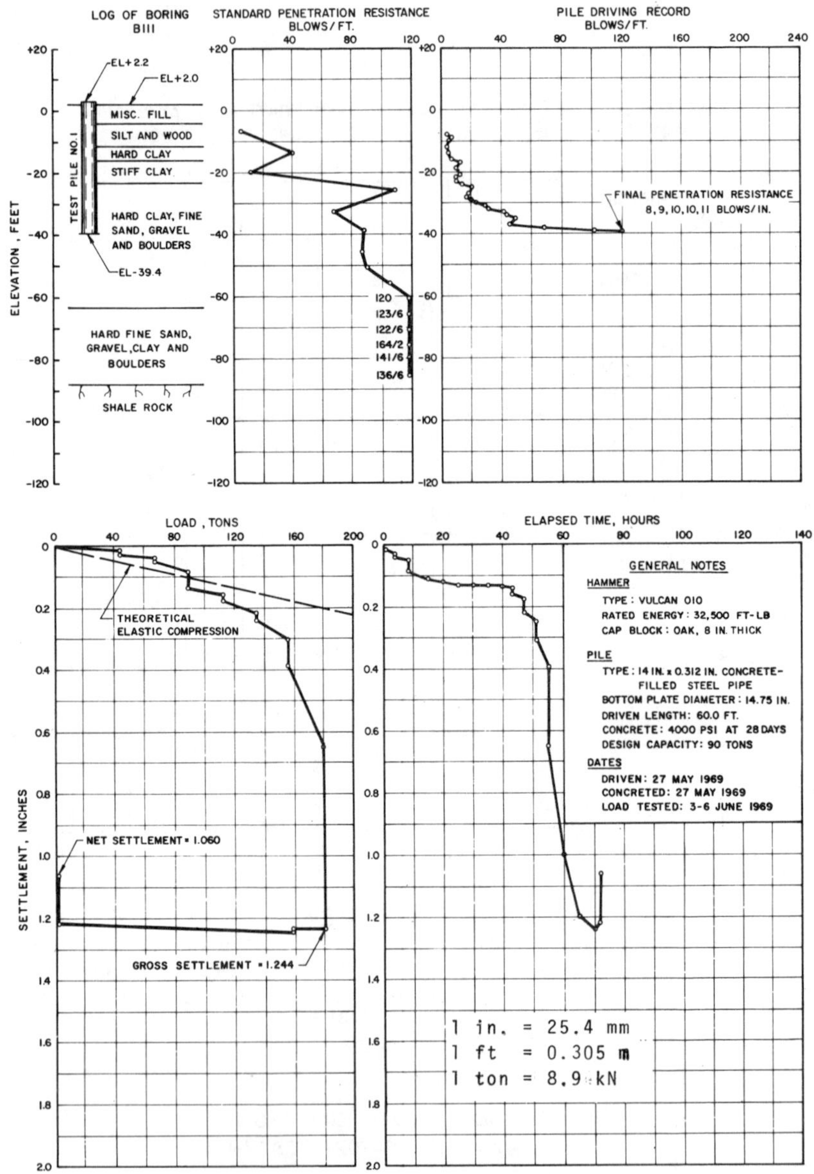

FIG. 3 - TEST PILE NO. 1

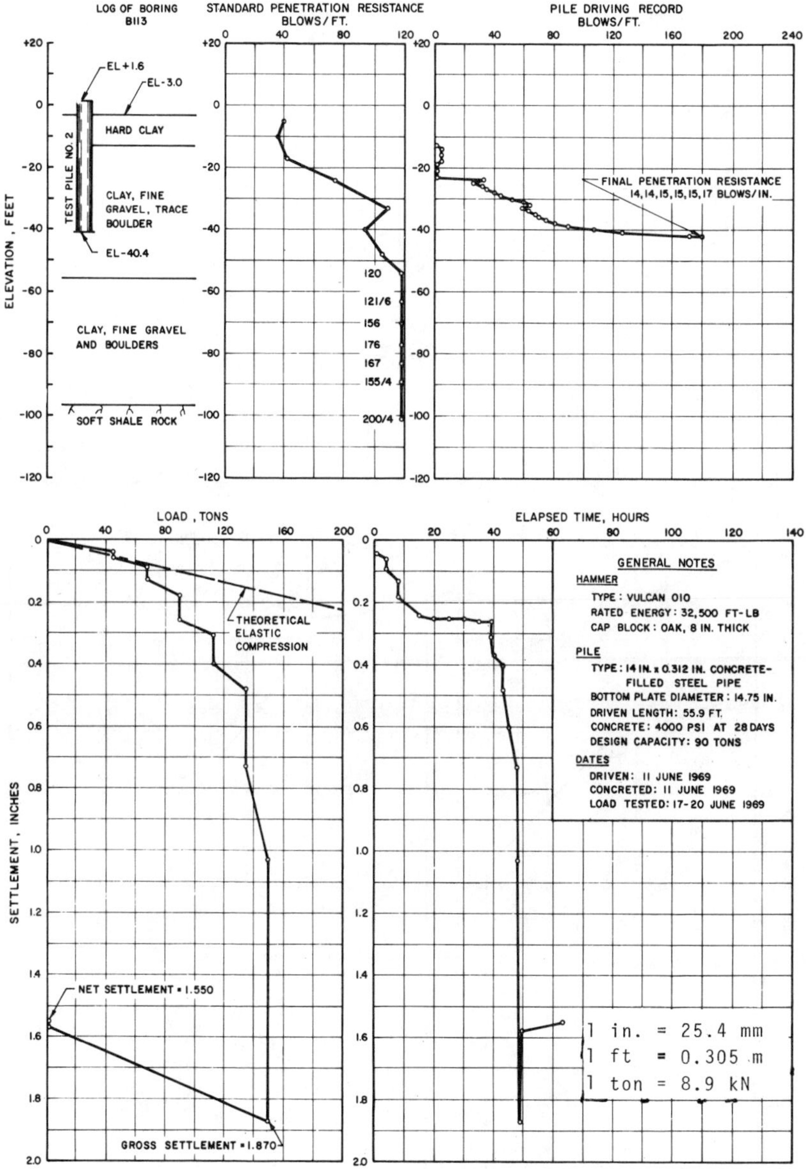

FIG. 4 - TEST PILE NO. 2

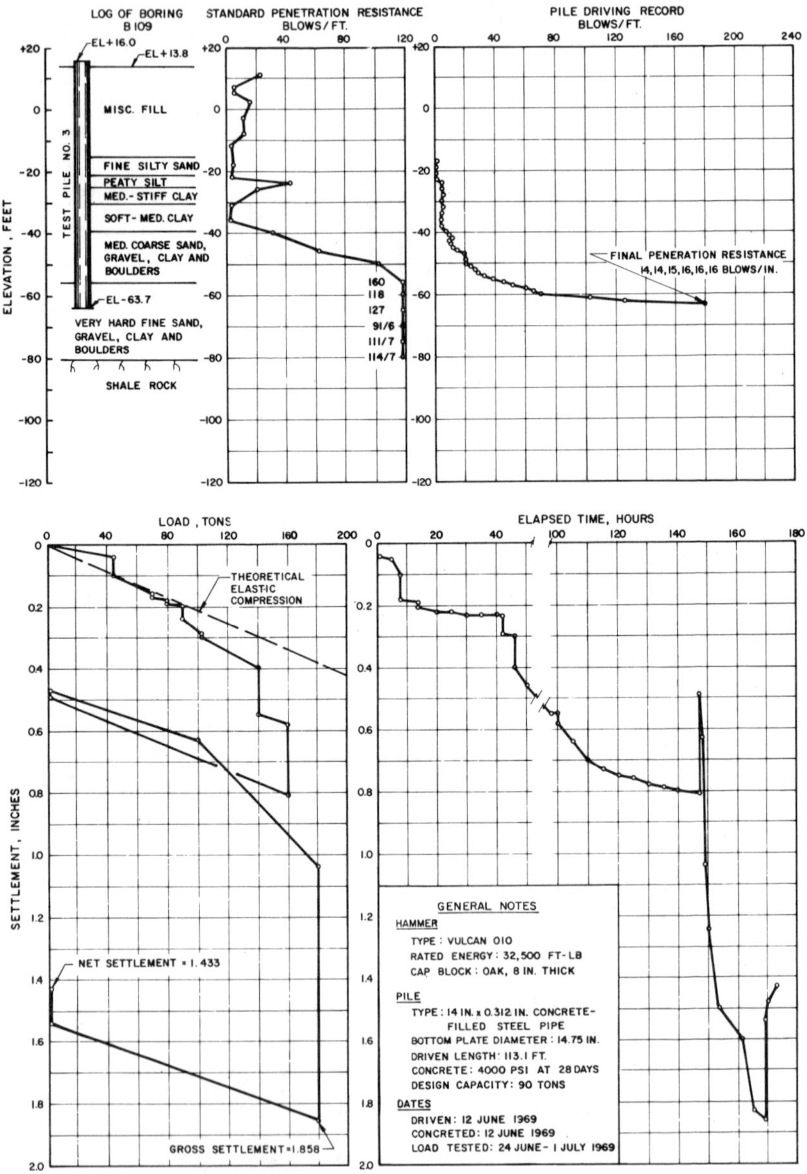

1 in. = 25.4 mm 1 ft = 0.305 m 1 ton = 8.9 kN

FIG. 5 - TEST PILE NO. 3

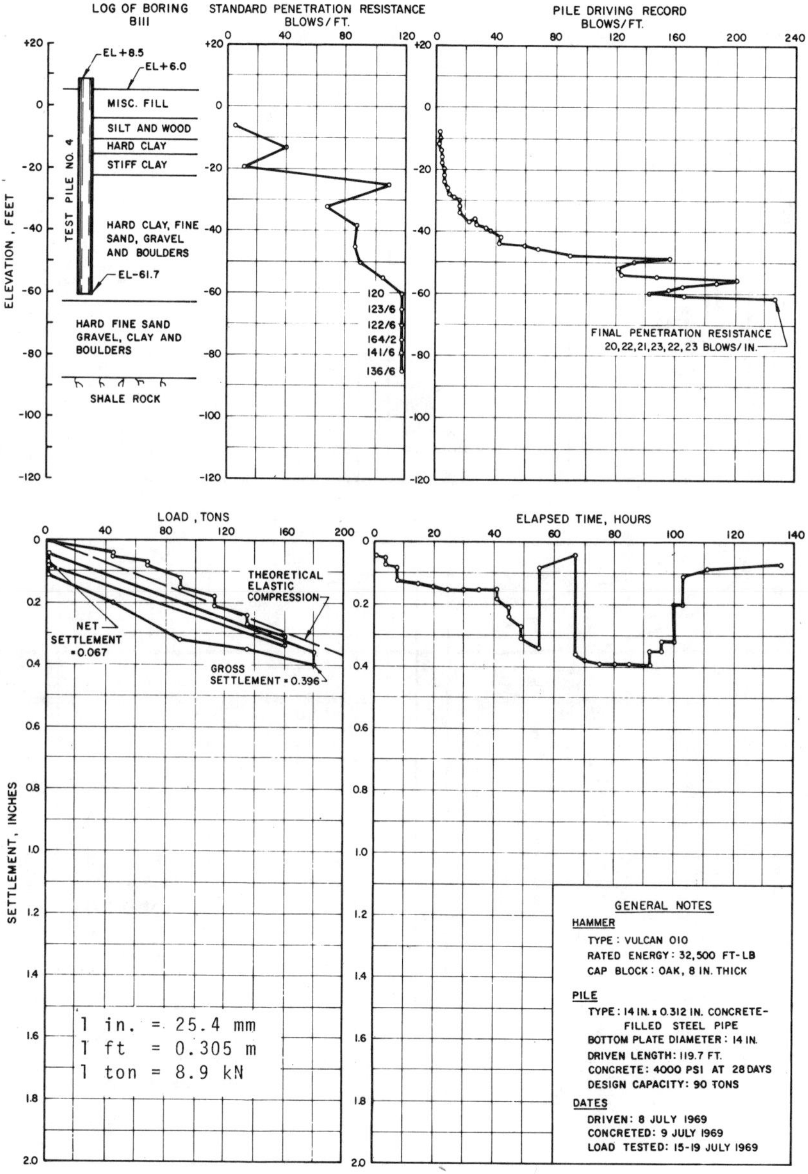

FIG. 6 – TEST PILE NO. 4

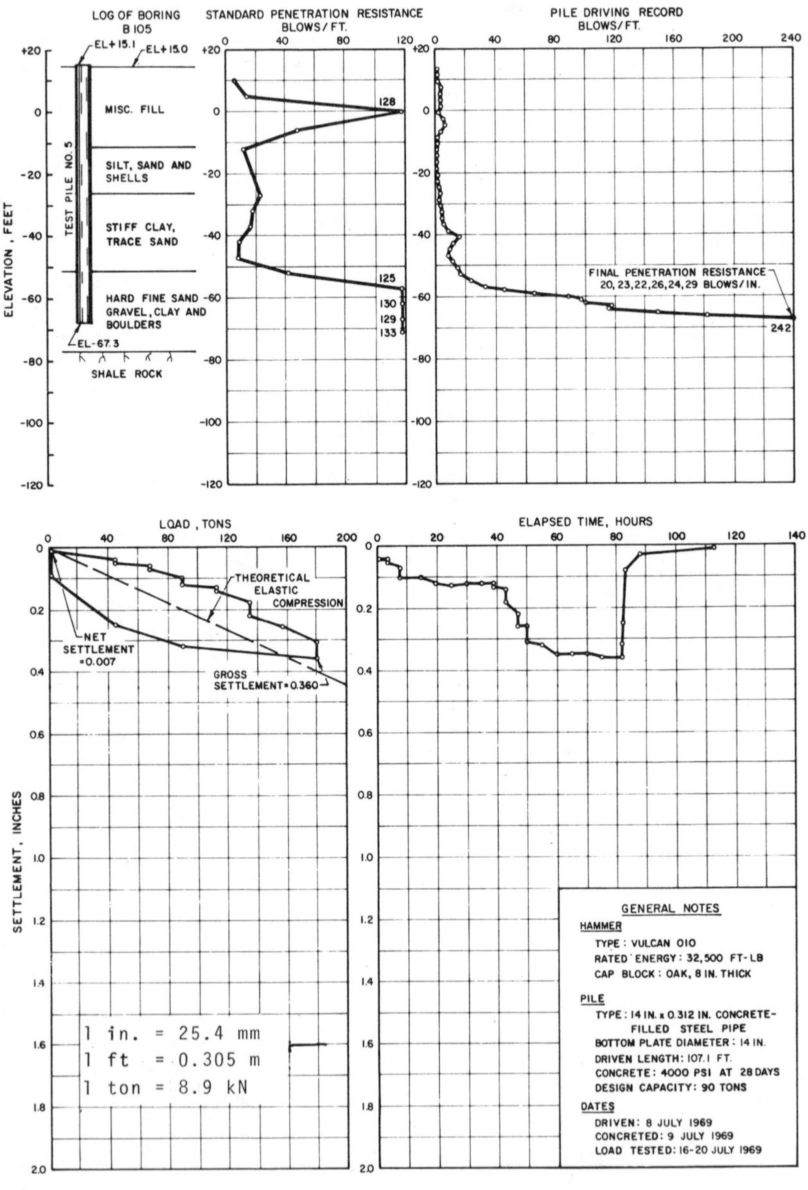

FIG. 7 – TEST PILE NO. 5

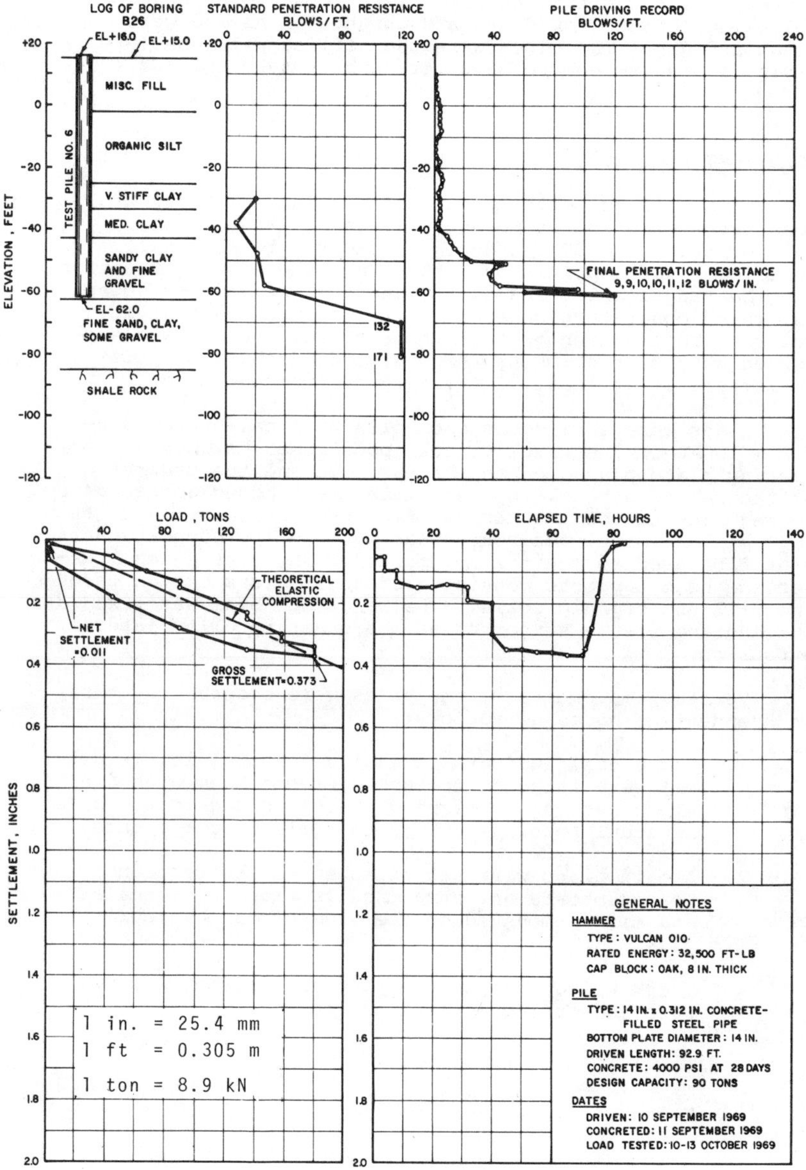

FIG. 8 – TEST PILE NO. 6

bottom of the shaft. The pipes were found to be reasonably plumb, free from dents and constrictions and were clean and dry. The pipes were then filled with concrete having a minimum compressive strength of 4000 psi (27,600 kN/m^2) at 28 days.

LOAD TEST PROCEDURE

In accordance with local building code requirements, load tests were set up to apply at least 180 tons (1600 kN), or twice the 90-ton design capacity, to each pile. Test loads were applied by jacking against a steel reaction beam bearing against the underside of a mat of steel beams on which concrete blocks were stacked as a dead weight. Prior to application of the jacking load, the mat was supported on timber cribbing which straddled the test piles at a distance of at least 15 ft. (4.6 m) from the pile.

Loads were applied to each pile by a calibrated 300-ton (2670 kN) hydraulic jack, positioned between a steel bearing plate placed on top of the pile and the underside of the reaction beam. Spacer plates were added as necessary between the top of the ram and the reaction beam. A 20 ft. (6.1 m) long steel reference beam was used to support instrumentation used to monitor movements of the test piles. Reference beam supports consisted of angle irons driven through the fill and organic soils. The reference beam was fixed at one end and freely supported at the other to allow lateral expansion.

Vertical movements of the piles were observed by three independent methods, as follows:

1. Three dial indicators reading in 0.001 in. (0.025 mm) were attached to the reference beam with dial stems resting on bearing plates welded to the pile at equal spacings around the pile circumference.

2. A thin piano wire was attached to the reference beam supports and stretched in front of a scale and mirror mounted on the side of the hydraulic jack.

3. Check elevations on the pile and reference system were made using a level and rod.

LOAD TEST RESULTS

The results of each pile load test are presented on Figures 3 through 8. Each figure includes the following data:

1. A log of the boring located nearest the test pile with a plot of Standard Penetration Resistance with depth.

2. The position of the test pile within the soil profile after driving.

3. The pile driving record.

4. Load-settlement and time-settlement plots from the pile load test.

As shown by the load test data, Test Piles 1, 2 and 3 could not support the required test load. Test Piles 4, 5 and 6 were loaded to 180 tons (1600 kN) without excessive settlement.

ANALYSIS

Installation and load test data for the six test piles are summarized in Table I. It is apparent from these data that the success or failure of the load tests was related to the type of bottom plate used. Test Piles 1, 2 and 3, fitted with oversize bottom plates, all failed. Test Piles 4, 5 and 6, which had flush bottom plates, each successfully supported the required test load without excessive settlement.

The effect of the bottom plate on the performance of the test piles can be illustrated in two ways:

1. Wave equation analyses can be used to compare the theoretical static resistance available at the time of driving to the capacity developed during load testing. Wave equation solutions and ultimate capacities estimated from the load-settlement curves are plotted on Figure 9.

   The ultimate capacities of Test Piles 1 and 3 were about the same as the theoretical static resistance predicted by wave equation analyses. Test Pile 2 developed significantly less capacity during the load test than was predicted, which suggests that

TABLE I

DATA SUMMARY

HAMMER DATA

TYPE:	VULCAN 010
RAM WEIGHT:	10,000 LBS.
RATED ENERGY:	32,500 FT.-LB.
RATED SPEED:	50 BLOWS/MIN.
CAP BLOCK:	OAK, 8-IN. THICK
ASSUMED EFFICIENCY:	70 PERCENT

METRIC EQUIVALENTS

1 IN. =	25.4 mm
1 FT. =	0.305 m
1 LB. =	0.453 kg
1 TON =	8.9kN
1 PSI =	6.9 kN/m^2

PILE DATA

TYPE:	CONCRETE-FILLED STEEL PIPE
SIZE:	14-INCH OUTSIDE DIAMETER, 0.312-INCH WALL THICKNESS
CONCRETE:	4000 PSI AT 28 DAYS
DESIGN CAPACITY:	90 TONS

LOAD TEST DATA

	TEST PILE NO.					
	1	2	3	4	5	6
PILE BOTTOM PLATE DIAMETER, IN.	14.75	14.75	14.75	14.0	14.0	14.0
PILE LENGTH WHEN DRIVEN, FT.	60.0	55.9	113.1	119.7	107.1	92.9
PILE LENGTH WHEN LOAD TESTED, FT.	41.6	42.0	79.7	70.2	82.4	78.0
PILE TIP ELEVATION, FT.	-39.4	-40.4	-63.7	-61.7	-67.3	-62.0
FINAL PENETRATION RESISTANCE, BLOWS/IN.	10	15	15	22	23	10
LOAD TEST DURATION, HRS.	72	63	74	136	113	84
MAXIMUM TEST LOAD, TONS	180	150	180	180	180	180
GROSS SETTLEMENT, INCHES	1.244	1.870	1.858	0.396	0.360	0.373
NET SETTLEMENT, INCHES	1.060	1.550	1.433	0.067	0.007	0.011
ULTIMATE CAPACITY, TONS	170	110	140	>180	>180	>180
ULTIMATE DYNAMIC RESISTANCE, TONS	160	170	150	160	160	140

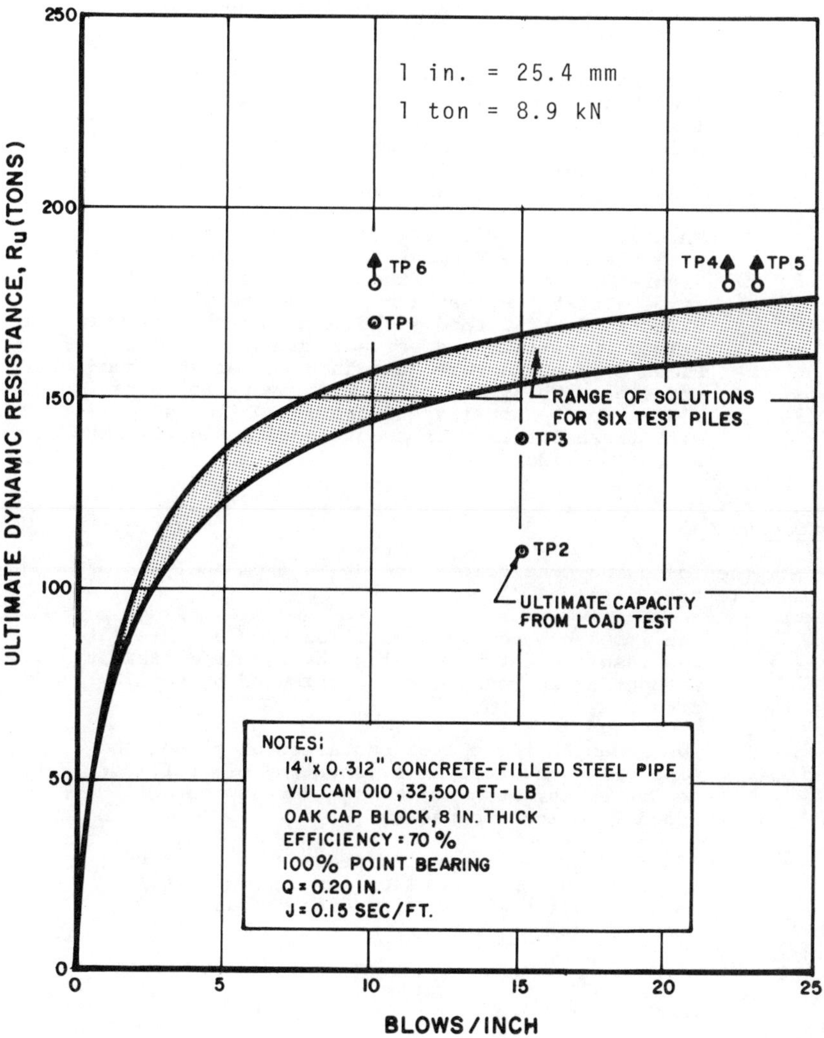

FIG. 9 - WAVE EQUATION ANALYSES

some relaxation occurred. In contrast, Test Piles 4, 5 and 6 each had capacities well above the theoretical static resistance present at the time of driving. That is, the piles with flush bottom plates developed additional capacity from soil freeze.

2. A difference in the behavior of the piles during load testing can also be seen by comparing the load-settlement plots to the theoretical elastic pile compression, assuming that the tip is fixed. Since no tell tales were installed, the actual elastic pile compression was not measured.

   Theoretical elastic pile compression is plotted with the load-settlement data for each pile on Figures 3 through 8. Examination of these data reveals that the measured settlements of Test Piles 1, 2 and 3 tend to be more than the elastic pile compression, whereas settlements of Test Piles 4, 5 and 6 were less than the elastic compression. Thus, the piles with oversize bottom plates were primarily end-bearing while the piles with flush bottom plates developed a significant amount of side friction.

CONCLUSIONS

The following conclusions are drawn from analysis of the pile test data:

1. The impedance of the pipe piles used at this site was insufficient to develop the required capacity without added resistance contributed by soil freeze.

2. Oversized bottom plates installed on closed-end pipe piles resulted in overcutting of stiff cohesive soils, and limited the development of added frictional capacity.

3. Added capacity from soil freeze should not be anticipated for pipe piles installed in stiff cohesive soils if oversize closure plates are used at the tip.

PREDRILLING: MINIMIZES THE UNDESIRABLE EFFECTS OF PILE INSTALLATION?

by

Richard P. Ray,[1] A.M. ASCE, Edward J. Ulrich, Jr.,[2] M.ASCE and James Malinak,[3] M.ASCE

Drilling pilot holes to facilitate pile installation and minimize the undesirable effects of pile driving has been a standard practice of pile installation in the overconsolidated sediments along the Texas Gulf Coast for at least two decades. Predrilling provides a means of installing displacement piles to greater penetrations and higher capacities without a proportionate increase in driving effort or soil heave during driving. However, the predrilling method may have an adverse effect on adjacent structures. Wet rotary predrilling methods with high circulation pressures may hydraulically fracture the soil formation, resulting in heave of nearby structures and a reduced capacity of inplace pile supported foundations.

This presentation provides field performance data on the effects of predrilling for the installation of piles at the Houston Lighting & Power Company's W. A. Parish Generating Station in Fort Bend County, Texas. As background information, the project, soil conditions, pile installation procedures and the hydraulic fracture concept are briefly described.

Observations of adjacent foundations during pile installation are reviewed to assess independently the effects of predrilling methods and pile driving. Results of a pile retapping program are included to augment the pile installation observation data.

Project Description

Houston Lighting & Power Company chose the W. A. Parish Generating Station as the site for two new 665 MW coal powered generating plants to help meet the Houston Area power demands. A vicinity map showing the site location is given in Fig. 1. Each plant consists of a turbine generator unit, boiler, precipitator and 600-ft high smoke stack extending in a line 760 ft long. The two plants have a combined width of over 600 ft. Column loads range upward of 2500 tons, and each smoke stack has a total weight of about 4000 tons.

About 8000 piles were needed to support the facilities. The piles, step tapered and mandrel driven, were installed through predrilled holes to provide a nominal working load of 100 tons. Embedded pile length ranged from about 75 to 90 ft.

1 ft = 0.305 m

1. Project Engineer, McClelland Engineers, Inc., Houston, Texas

2. Project Manager, McClelland Engineers, Inc., Houston, Texas

3. Lead Engineer, Houston Lighting & Power Company, Houston, Texas

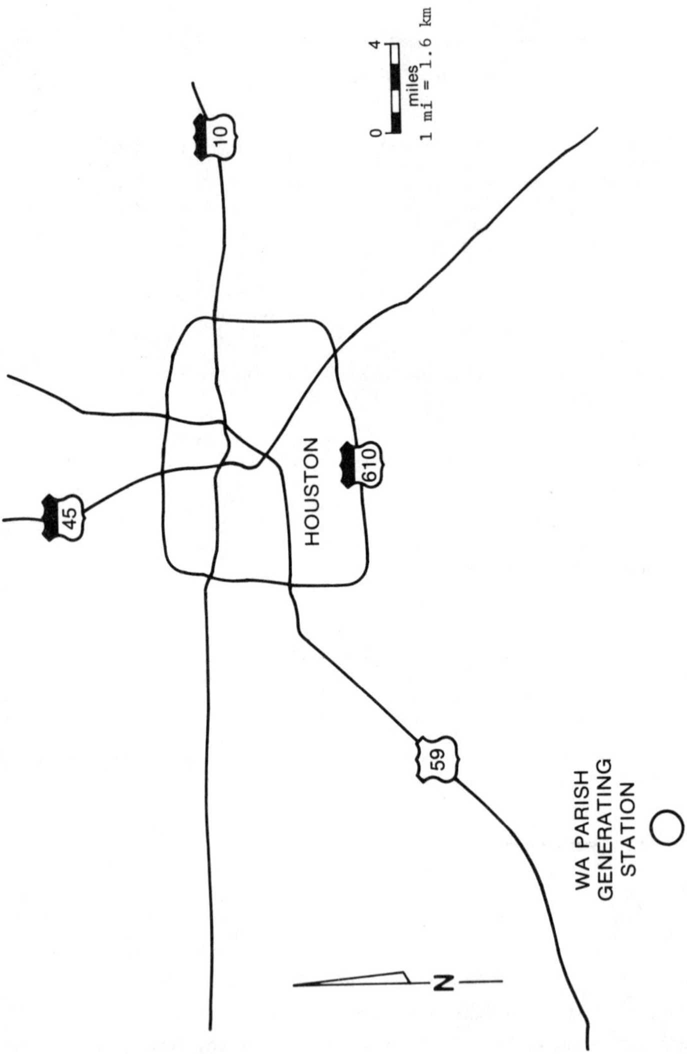

Fig. 1 VICINITY MAP

General Subsurface Conditions

The site is located in the West Gulf Coastal Plain geologic province and is underlain by overconsolidated clays that are interspersed with sand and silt layers. These soils are typical of the Beaumont clay formation which is a deltaic deposit of Pleistocene geologic age.

Generalized soil conditions and physical properties are presented in Fig. 2 along with typical pile make-up and installation characteristics. The uppermost soil horizon consists of firm to stiff sandy clay. Plasticity characteristics of this stratum are somewhat variable although liquid limits are often above 50. A zone of clayey silt and silty fine sand ranging from 4 to 13 ft in thickness underlies the upper clay. The granular zone is underlain by very stiff slickensided, fissured clay to about 50-ft depth. Below 50-ft depth, the clay is interbedded with sand and silt layers to about 85-ft depth. Dense fine to medium sand with scattered gravel exists to depths of about 125 ft. Very stiff slickensided clay with medium-dense to dense sand layers are found below 125 ft.

The depth to groundwater is at the surface of the upper silt zone, about 10 ft below the ground surface.

Pile Installation

Pile cutoff, at El 60, was typically 10 ft below the relatively flat ground surface. Pile penetration extended 4 to 15 ft into the major sand stratum encountered at about El -16, depending on the driving resistance and sand character. The piles were step-taper shells each 12 ft in length with shell sizes ranging from the No. 6 (16-3/8-in. OD) to the No. 0, (10-3/8-in. OD). The tip end of the pile was closed with a 1-in. thick plate, flush welded at the tip. The piles were driven with an internal steel mandrel that engaged drive rings located at each change in shell diameter and extended to the pile tip. The mandrel was 96 ft long, weighed 10 tons, and was equipped with a 1-ton aluminum-micarta driving block and shield. After driving, the shells were filled with concrete up to pile cut-off and the excess shell above cut-off was removed. Pile makeup and typical driving characteristics are compared to the generalized subsurface conditions in Fig. 2.

The step-taper piles were installed following a three-phase sequence:

(1) Wet rotary predrill to El +10 with a 10-in. fish-tailed bit attached to a string of 6-in. hollow stem drill pipe. Water pump discharge pressure was 200 psig. Drilling time was limited to 3 minutes.

(2) Pile placed in pilot hole immediately after drilling and driven to grade with a differential steam hammer rated at 24,450 ft-lb per blow at a rate of 100 bpm.

(3) The shells were backfilled with concrete after pile driving operations advanced 30 ft beyond.

1 ft = 0.305 m 1 in. = 25.4 mm 1 ton = 907 kg 1 psi = 6.9 kPa

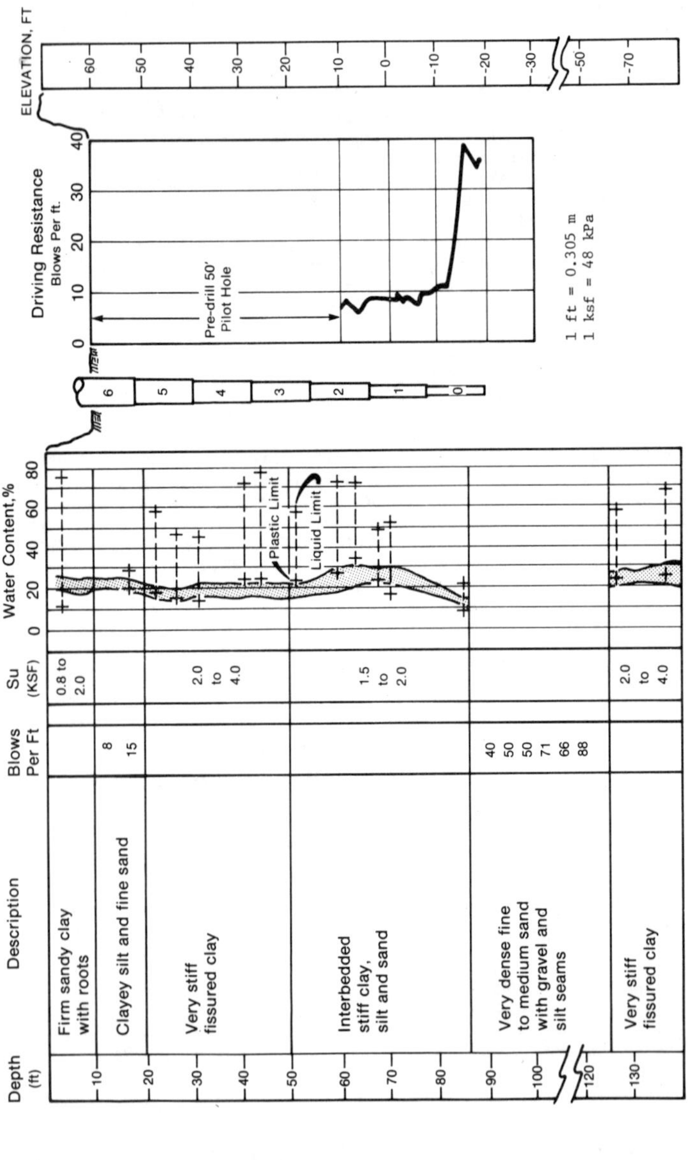

Fig. 2 TYPICAL SOIL CONDITIONS AND PILE INSTALLATION DATA

Installation Observations

Reported observations that nearby pile caps were moving upward during predrill operations resulted in a monitoring program to assess individually the influence of the wet rotary pilot hole operations and the influence of pile driving with respect to potential movement of adjacent foundations. Movements of existing pile foundations were observed during the installation of two 20-pile groups at the locations shown in Fig. 3. The principal observation efforts were focused on the small pile cap located near each of the 20-pile groups.

South Area Movements. Monitoring in the south area consisted of vertical and horizontal measurements of the columns on the adjacent pile caps during drilling and driving operations. The vertical measurements were obtained by continuous sighting with a surveyor's level on targets during pile installation to minimize effects of instrument error.

Results of the observations made on the existing pile foundation are presented in Fig. 4. No movement of the adjacent foundations was observed during driving; however, movement to 1 in. in the upward and lateral directions were observed due to the predrilling operations for the 20 piles. A 0.1-in. extension of the corrugated shell on Pile S1 and upward foundation movement of 0.1 in. were observed during the installation of a test hole to 60-ft depth (El 0). Observed movement occurred when drillwater circulation ceased at El 20, in the interbedded clay, sand and silt stratum.

North Area Movements. A variety of installation modifications were performed to provide an installation technique that was both compatible to the load test conditions and would eliminate excessive movement on adjacent structures. Ground surfacé heave markers were installed at locations shown in Fig. 5 to augment measurements made on the existing foundations. The installation modifications consisted of:

(1) Drilling a 60-ft deep relief hole prior to installation.

(2) Reducing the predrilling depth from 60 ft to 40 and 35 ft.

(3) Predrilling with a hollow stem continuous flight auger, nominally 10-in. diameter, to 45-ft depth and lubricating the piles with water both prior to and during driving.

(4) Drilling the pilot hole to 50-ft depth with a 10-in. diameter bit attached to a 4-in. diameter hollow stem drill pipe to increase the return annulus. Water pressures at the pump were near 200 psig.

(5) Predrilling to 60-ft depth using the continuous flight auger but with water discharged at the drill bit. Water pressure at the pump was about 40 psig.

1 in. = 25.4 mm 1 ft = 0.305 m 1 psi = 6.9 kPa

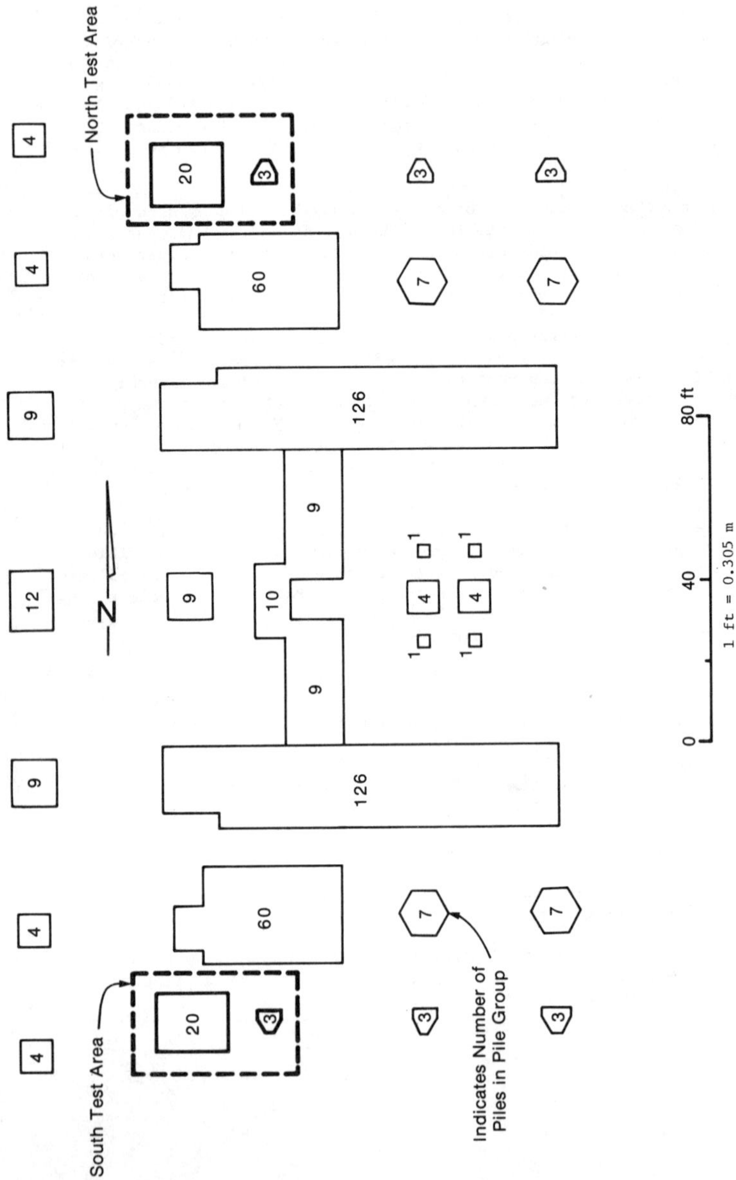

Fig. 3 PLAN OF BOILER FOUNDATION
NOTE PILE TEST AREAS

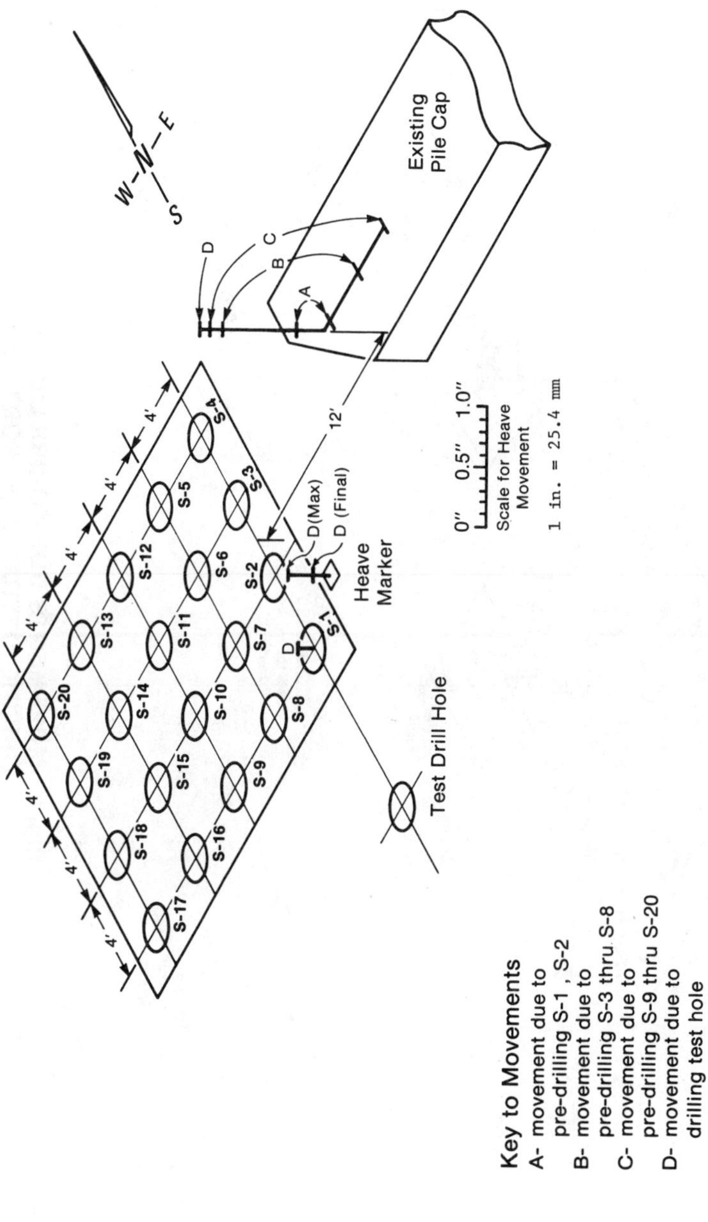

Fig. 4 MOVEMENTS DURING DRILLING, SOUTH AREA

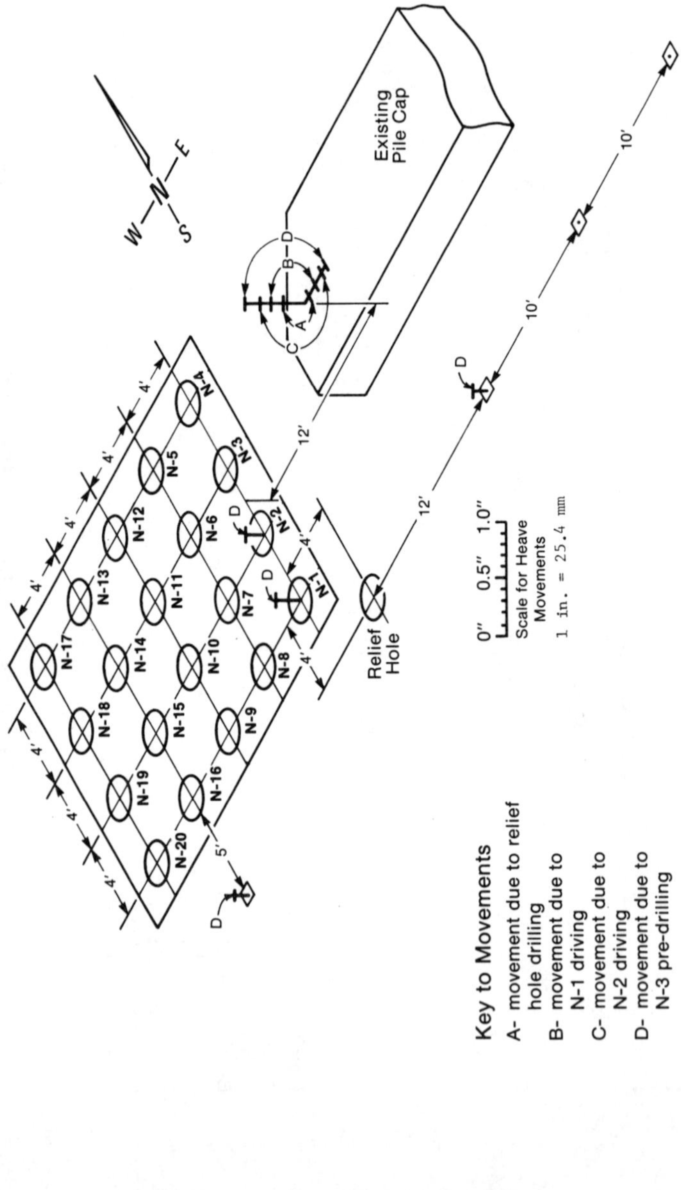

Fig. 5 MOVEMENTS DUE TO DRILLING AND DRIVING OPERATIONS, NORTH AREA

These modifications produced the results shown below and on Fig. 5:

Predrill Hole Length, Ft	Observed Foundation Movement, Inches Drilling	Driving	Drilling & Driving
60 (Relief Hole)	0.1		0.1
40 (N-1)	0 to 0.05	0.1	0.1 to .15
45 (N-2)	0	0.1	0.1
60 (N-3)	0.1	0	0.1
60 (Wet Auger)	nil	nil	nil

As shown by the observations, the relief hole did not accomplish its task, but rather added to the problem. A reduction in predrilling depth reduced drilling movement but added movement during driving. Dry augering eliminated the heave associated with drilling but heave during driving became a problem. This also affected installation by (a) inhibiting an acceptable correlation between test piles and production piles, and (b) maximizing the risk of pile damage during driving. The increased return annulus of the 4-in. drill stem did not aid in reducing hydraulic pressures at the bit as evidenced by the drilling heave. The "wet auger" method with reduced water pressure showed the most favorable results and was adopted.

Cause of Movement

The measurement data presented provide evidence that the predrilling process was the major contributor to foundation movement during pile installation. We believe the foundation movement is attributed to the phenomenon of hydraulic fracturing, i.e., the drill water fractured the soil formation when circulation was lost at El 20 and pump discharge pressures remained at 200 psig. In addition, water had seeped up to the surface along the soil-pile interface of previously driven piles.

Hydraulic Fracturing

As explained by Hubert and Willis (1957), the mechanics of hydraulic fracturing are as follows: the radial and tangential stresses in a borehole wall are normally compressive (Fig. 6). If the borehole is injected with a fluid to a pressure, P, the pressure will create tensile forces on the soil elements near the borehole, as in an expanding cylinder. If the fluid pressure is great enough to overcome the in situ compressive stresses and the tensile strength of the soil (or rock), fracturing will occur. The orientation and propagation of the fracture is influenced by the initial state of stress in the soil. Since the fracture will follow a path of least resistance, the preferred plane of cracking will be perpendicular to the direction of least stress. This implies that in soils with a lateral earth pressure coefficient, K_o, less than unity, cracks will form along a vertical

1 ft = 0.305 m   1 in. = 25.4 mm   1 psi = 6.9 kPa

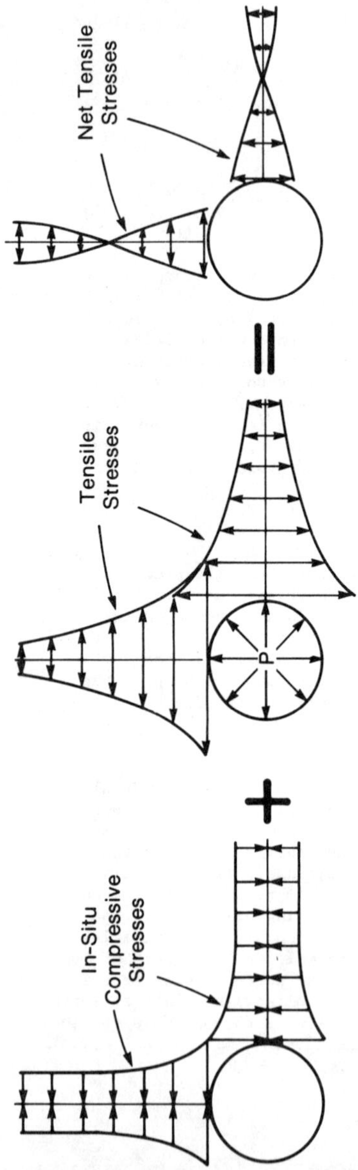

Fig. 6 STRESSES IN BOREHOLE DUE TO INJECTION PRESSURE "P"

plane; in soils with K_o greater than unity, cracks will form along a horizontal plane (Fig. 7). In order to propagate a fracture, the pressure in the borehole needs to be slightly above that required to keep the initial crack open. Due to soil nonhomogeniety, however, it is difficult to predict the specific propagation and orientation characteristics of the fracture.

The study of hydraulic fracturing in soils has been directed toward earth dam applications and as a method to determine in situ earth pressures. Massarsch and Broms (1978) have studied the effects of pore pressure dissipation and rapid reconsolidation due to fracturing during pile driving operations.

At this site, the predrilling data indicated the fracturing probably took place in the interbedded stratum encountered at about El 20. The fracture's influence was observed at the surface as far away as 20 ft from the borehole.

Pile Retapping Program

A pile retapping program was engaged to (1) identify the existing piles which had experienced upward movement due to new pile installation and (2) reseat those which had experienced excessive upward movement. Since the piles to be retapped were concreted, wave equation analyses were performed to investigate the compatibility of the new hammer-pile-soil system. The retap pile cushion assembly and wave equation analyses results are shown in Fig. 8 and 9, respectively. Interpretation of the wave equation results indicated the hammer system would be capable of restarting the pile without imposing excessive concrete stresses. To confirm our interpretations, piles that were known to be unaffected by new pile installation were retapped initially and termed "control piles". The observed butt behavior was then used as a guide to evaluate other retapped piles.

The retapping program included 232 piles. Three typical modes of butt movement were observed and are described below:

(1) Unaffected piles: These behaved like the control piles. Blow count versus butt movement curves were relatively linear with a flat slope.

(2) Moderately affected piles: Blow count/butt movement curves showed a low initial resistance that increased with penetration. Final slopes of the curves were comparable to the control piles.

(3) Critically affected piles: These piles exhibited retap curves similar to moderately affected ones up to a point where driving resistance then decreased. The cause of such behavior may have been structural damage due to uplift or retapping, and replacement piles were installed adjacent to these. The three modes are illustrated graphically in Fig. 10. Based on the retap curves, piles 40 ft distant were affected by a hydraulic fracture location.

---

1 ft = 0.305 m

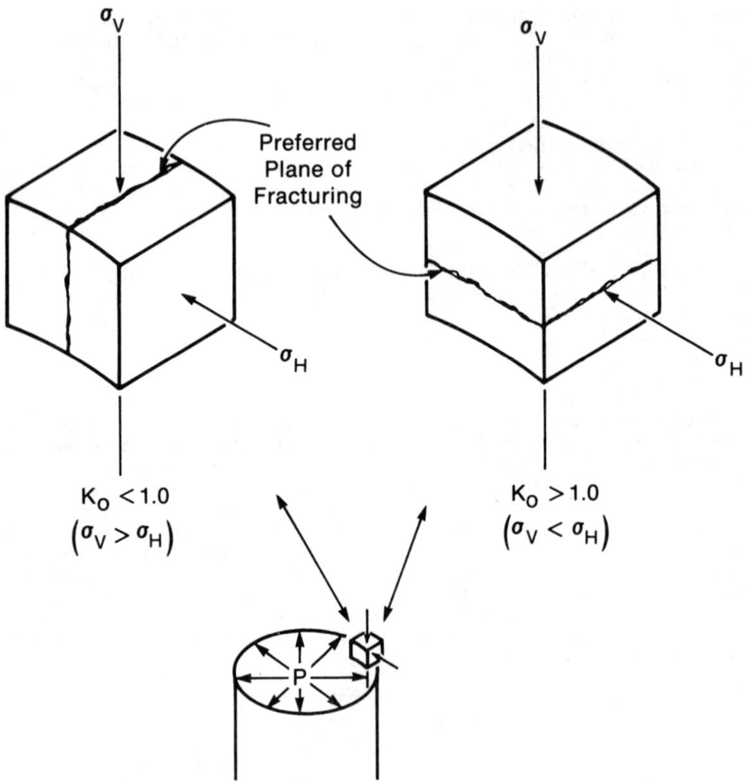

Fig. 7 PREFERRED PLANES OF FRACTURE AND INITIAL STATE OF STRESS

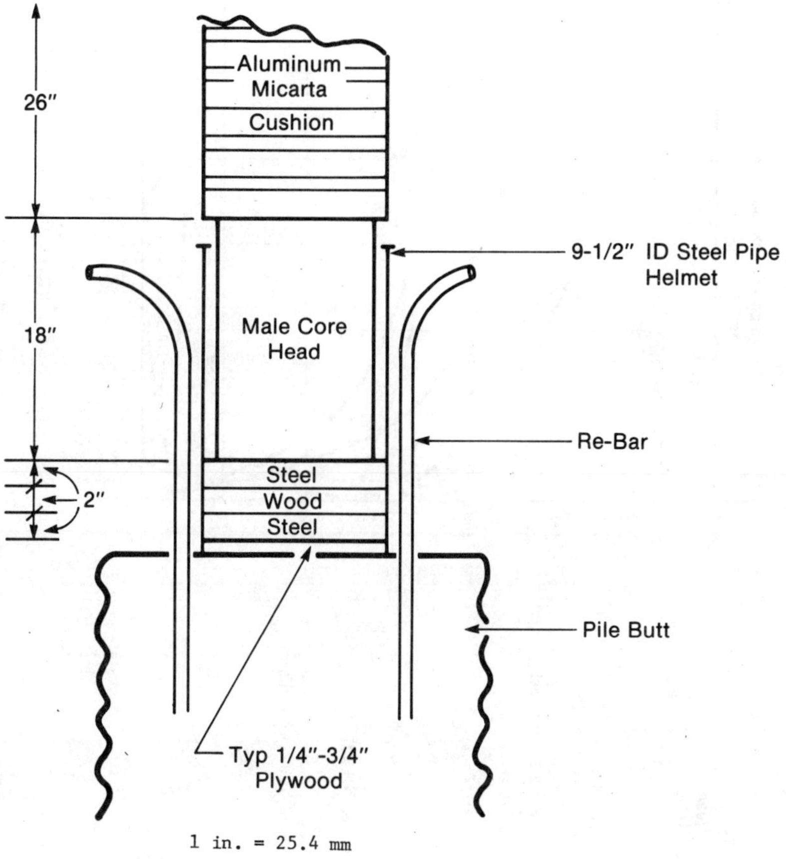

Fig. 8 TYPICAL RETAP PILE CUSHION ASSEMBLY
Horizontal Scale: 1″ = 7″ (Approx.)

Hammer Properties

Type : Differentaial acting steam
Rated Energy = 24,450 ft-lbs/Blow
Efficiency = 80%
Wt. of Ram = 8,000 lb.
Wt. of Pile Cap = 1,500 lb.
Spring Constants = Aluminum Micarta
Capblock E = 6.95 × 10^6 psi
2-in-thick Douglas Fir Plywood
Cushion E = 3.2 × 10^6 psi

Pile Parameters

Case No. :	1	2	3
Pile Length, ft :	81	81	84
Concrete Comp. Strength, psi :	6000	6000	6000
Concrete Elastic Modulus, psi :	4.4 x 10^6	4.4 x 10^6	4.4 x 10^6
Point Resistance, Percent :	50	0	50
Friction Distribution :	Uniform	Uniform	Uniform

Soil Parameters

Damping, J, sec/ft	Quake, Q, inches
Point = 0.15	Point = 0.10
Side = 0.05	Side = 0.10

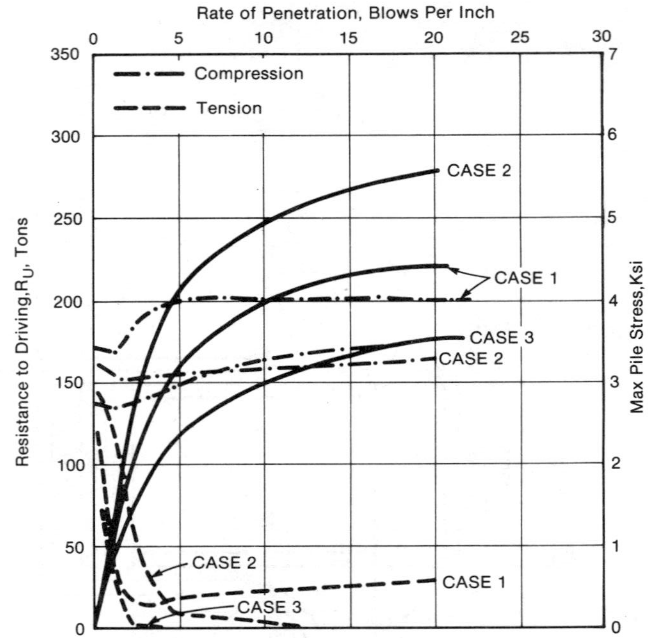

1 ft-lbf = 1.4 J
1 lbf/in^2 = 6.9 kPa
1 ksi = 6.9 MPa
1 ton force = 8.9 kN

1 lb = 0.45 kg
1 ft = 0.305 m
1 in. = 25.4 mm

Fig. 9 WAVE EQUATION PARAMETERS AND RESULTS

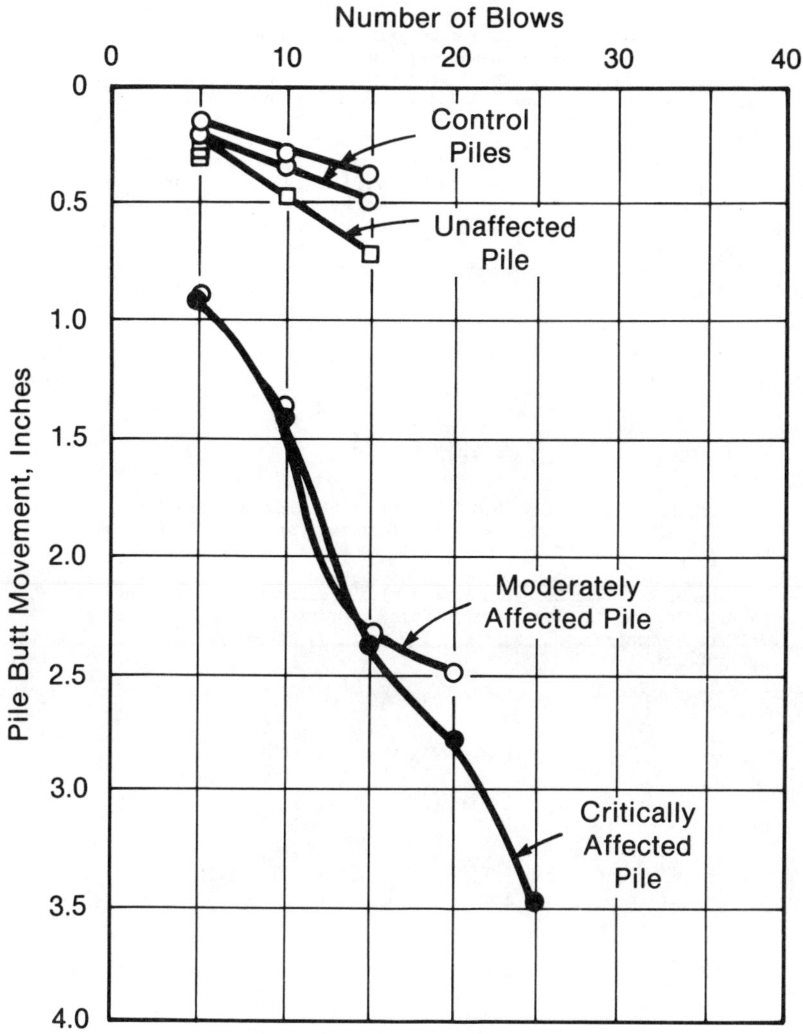

1 in. = 25.4 mm

Fig. 10 TYPICAL PILE RETAP CURVES

Conclusions

Due to observed movements of nearby pilecaps, a monitoring program was instituted to determine the cause and magnitude of these movements. The movements occurred during the predrilling operation and were on the order of one inch. Based on field observations, it was determined that high circulation pressures during predrilling was causing hydraulic fracture of an interbedded soil stratum, thereby resulting in heave of nearby pile groups. The predrilling technique was changed from a wet rotary method using 200 psi circulation pressure to a wet auger method with 40 psi circulation pressure. A pile retap program was performed in conjunction with pile-driving analyses to identify those piles which had experienced upward movement and reseat the affected piles.

---

1 psi = 6.9 kPa

REFERENCES

1 Hubbert, M.K. and Willis, D.G. (1957), "Mechanics of Hydraulic Fracturing," *Transactions, American Institute of Mining Engineers*, Vol. 210, pp. 153-168.

2 Massarsch, K.R. and Broms, B.B. (1978), "Fracturing of Soil Caused by Pile Driving in Clay," *Proceedings, 9th International Conference on Soil Mechanics and Foundation Engineering*, Tokyo, Vol. 1, pp. 197-200.

STRESS MEASUREMENTS FOR OFFSHORE PILE DRIVING

by Wendell R. Vines, A.M. ASCE[1]

and

Jack J. Amar[2]

INTRODUCTION

Pile driving can be one of the most time consuming and costly portions of offshore platform installation. The prediction of drivability is complicated by uncertainty about both the driving system performance and the driving resistance from the soil. The most costly result of such uncertainties is the occurrence of early refusal that could require remedial installation measures to advance the pile to an acceptable penetration. Field monitoring of driving system performance is being used more and more frequently to reduce the risk of costly refusal situations and to provide additional information for the evaluation of the need for remedial measures if refusal occurs.

When unexpected refusal was encountered on a platform in the Eugene Island Area in November, 1977, such monitoring services were requested by the operator. Two piles had refused after a delay for field welding the final add-on section, and documentation of the driving system performance was desired for an attempted re-drive of the piles. In addition, driving of the remaining add-on sections for other piles of the platform was monitored to avoid further delays due to questions about driving system performance.

The platform in question is an eight-pile drilling and production structure in the Eugene Island Area of the Gulf of Mexico. The platform is located approximately 80 miles (129 km) south of Morgan City, Louisiana and about 50 miles (80 km) off the Louisiana coast. The water depth at the platform site is 139 ft (42.3 m).

---

[1] Manager, Offshore Construction Engineering, TERA, Inc., Houston, Texas

[2] Director of Geotechnical Engineering, Soil Testing Services of Wisconsin, Inc.; formerly Project Engineer, McClelland Engineers, Inc., Houston, Texas

Soil Investigation

Soil conditions at the site were determined in May, 1977 by McClelland Engineers with a foundation boring to a penetration of 530 ft (162 m) below the mudline. The boring was drilled with a skid-mounted Failing 1500 rotary rig operated through a centerwell installed in the deck of a specially rigged supply boat. Samples of the soils encountered in the boring were obtained by a wire line sampler operated through the 3-1/2 in. (88.9 mm) ID drill pipe and driven with a 275-lb (1223 N) weight. A 2-1/8-in. (54 mm) ID liner sampler was used to obtain samples of soft clay and a 2-1/4 in. (57.2 mm) OD, 2-1/8-in. (54 mm) ID thin-walled tube sampler was used to obtain the remaining samples. The sampler weight was dropped approximately 5 ft (1.5 m) a sufficient number of times to secure the desired penetration of the samplers. The boring log is illustrated on Fig. 1.

The soil boring disclosed very soft to firm, highly plastic clay in Stratum I which extends to 92-ft (28 m) penetration. The undisturbed shear strength of these clays is in the range of 100 psf (0.478 N/cm^2) at the mudline and increases in a linear trend to 900 psf (4.309 N/cm^2) at the bottom of the stratum. Remolded shear strength is approximately one third to one half of the undisturbed shear strengths. Stratum II consists of stiff, highly plastic clay and extends to 200-ft (61 m) penetration. Shear strengths of undisturbed soil samples range from 1000 psf (4.788 N/cm^2) near the top of the stratum to 1600 psf (7.661 N/cm^2) at 200-ft (61 m) penetration. Remolded strengths of the clays in Stratum II are less than half the undisturbed strengths. Stratum III is a granular layer from 200-ft (61 m) to 330 ft (101 m) penetration that grades from a silty fine sand to a silt with increasing depth. The sampler resistance in this zone indicates a medium-dense to dense formation. Stratum IV, encountered at 330-ft (101 m) penetration and extending to 380-ft (116 m) penetration, consists of dense clayey silt with interbedded layers of stiff clay to about 350-ft (107 m) penetration. Stratum V is a very stiff layer of moderately plastic clay to a penetration of 410 ft (125 m). Shear strengths of the clays in this layer are quite variable; an interpreted

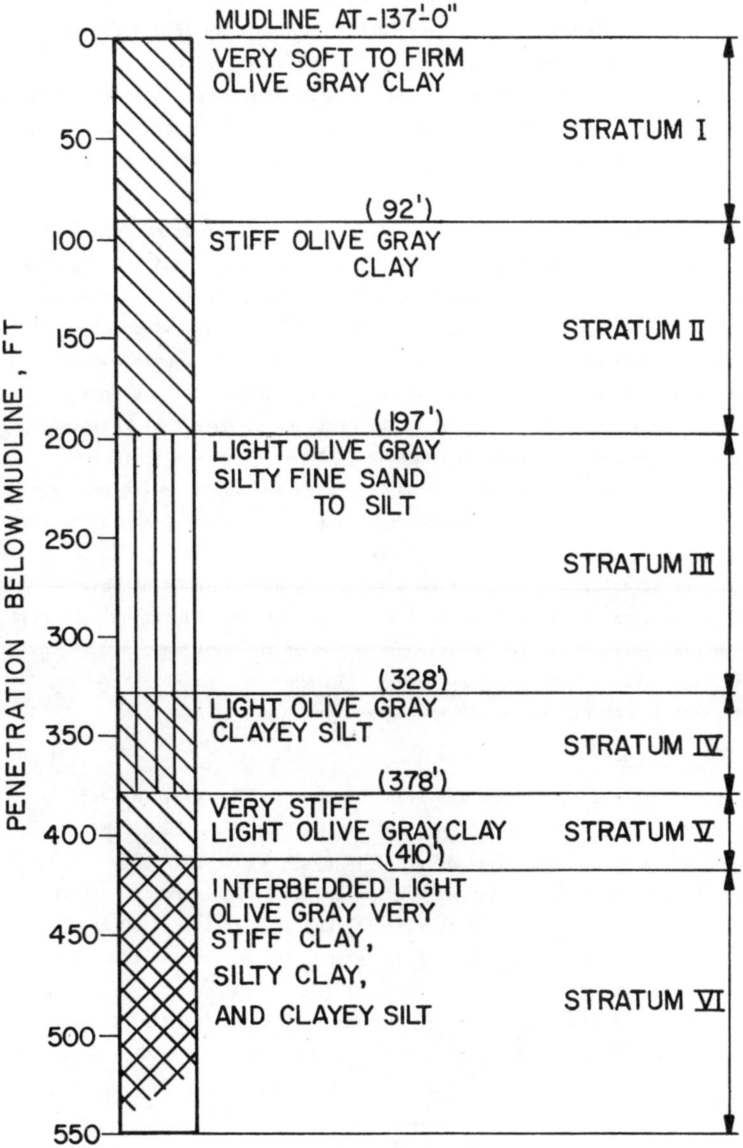

FIG. 1 BORING LOG

shear strength value of 3000 psf (14.36 N/cm^2) was selected for use in computing axial pile capacity. A very stiff to hard clay of moderate to high plasticity interlayered with silty clay and clayey silt is present below 410-ft (125 m) penetration and extends to the maximum explored depth of 530 ft (162 m) below the mudline. Shear strength of the clays in this layer ranged from about 1500 psf (7.18 N/cm^2) to almost 6000 psf (28.73 N/cm^2).

Eight steel pipe piles of 48-in. (1.22 m) diameter form the foundation for the platform. In addition to supporting the weight of the jacket, decks and equipment, the piles must resist lateral and axial forces produced by wind and wave action. This platform was designed to resist winds of up to 140 mph (225 kph) and wave heights of 56 ft (17.1 m). The environmental forces result in lateral or shear forces on the piles and also create an overturning moment on the structure. The overturning moment is resisted by compressive and tensile loads on the piles. In order to resist these forces, wall thicknesses of up to 2.00 in. (50.8 mm) and a design penetration of 410 ft (125 m) were selected for the piles. Figure 2 shows the pile wall thickness schedule.

As is typical for template-type offshore structures such as this one, the piles were too long to be installed in one section. Additional lengths of pile, called add-ons, are positioned on top of the pile in place in the jacket leg, field welded and then driven. This process is repeated until the desired length of pile has been driven into the seafloor.

Pile Installation

Installation of the piles was begun on October 27, 1977. The derrick barge used is 400 ft (121.9 m) long, 100 ft (30.5 m) wide, 28 ft (8.5 m) deep and has a normal operating draft of 14 ft (4.3 m). This size is fairly typical of the construction barges used in the Gulf of Mexico. Its revolving crane has a capacity of 800 tons (7,117 kN) at a radius of 60 ft (18.3 m) for lifts over the stern of the barge, and 600 tons (5,338 kN) at a radius of 65 ft (19.8 m) for revolving lifts.

Three hammer sizes were used in driving the 48-in. (1.22 m) diameter piles. Vulcan 340 and 060 steam hammers were used on early add-on sections.

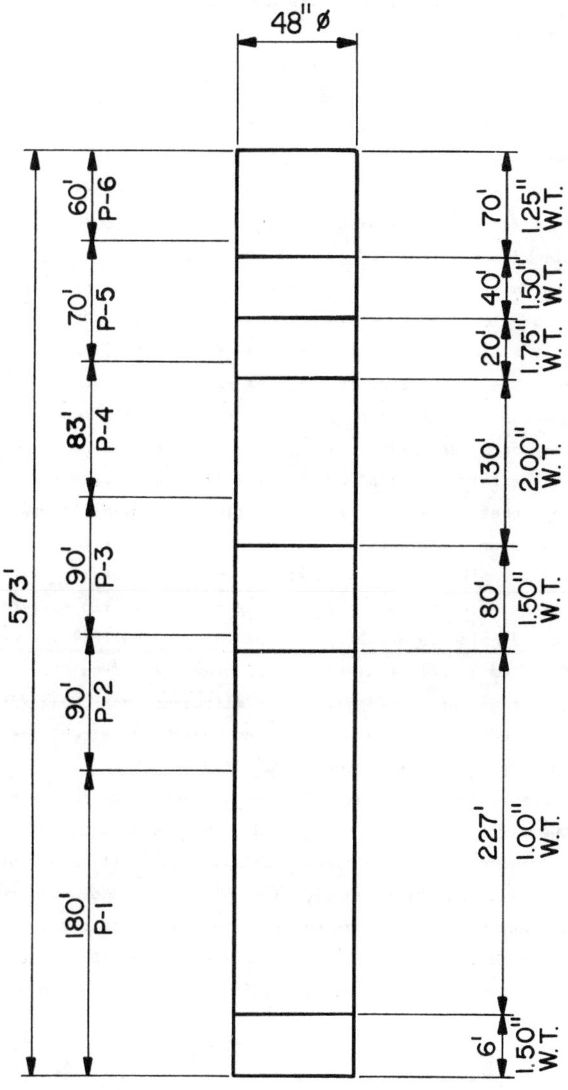

FIG. 2 PILE SCHEDULE

The final pile add-ons were driven with a Vulcan 3100 steam hammer which has a rated energy of 300,000 ft-lb (406,746 J). This hammer has a ram weight of 100,000 lb (45.4 Mg) and a nominal stroke of 3 ft (91.4 cm). A cushion made of Ascon discs (an asbestos-concrete material) and steel plates was used with the Vulcan 3100 hammer. Two Ascon discs and one steel plate made a layer of cushion material; seven to ten layers were used, resulting in a cushion thickness of about 6 in. (15.2 cm). A striker plate made of 6-in. (15.2 cm) steel was placed between the ram and the cushion. The pilecap used with this hammer weighed 45,900 lbs (20.8 Mg). The pilecap acts as an adapter to centralize the position of the hammer over the pile and to transmit the force of the hammer blow to the pile.

Figures 3, 4 and 5 show driving records for three of the piles for this platform. The plates show the driving resistance in terms of blows per foot as a function of penetration below the mudline. Also shown are any delays that occurred during driving. The driving records shown in Figs. 3 and 4 are fairly typical of the driving of the piles for this platform. Note that the driving records reflect the highly variable shear strengths of Stratum V and interbedding of Stratum VI. Figure 5 shows the driving record for pile B1, which first encountered refusal on November 8, 1977. The refusal occurred at a penetration of 362 ft (110 m) after a delay of 10 hours and 35 minutes during which the last add-on (P6) was welded onto the pile. The blow counts recorded for the previous add-on (P5) were somewhat higher than but not significantly different from those for P5 sections of the interior single-batter piles, which had already been driven to grade with no problems using the Vulcan 3100 hammer.

The driving records show the influence of setup time on the difficulty of driving the piles. On all three records, the blow counts increased after delays for welding of add-on sections. This is especially apparent for pile B4 and B1. Accurate quantitative estimates of increased driving resistance due to delays cannot be made since the early add-ons were not monitored and, therefore no information is available regarding the driving system performance before the setup occurred. However, it is apparent that higher blow

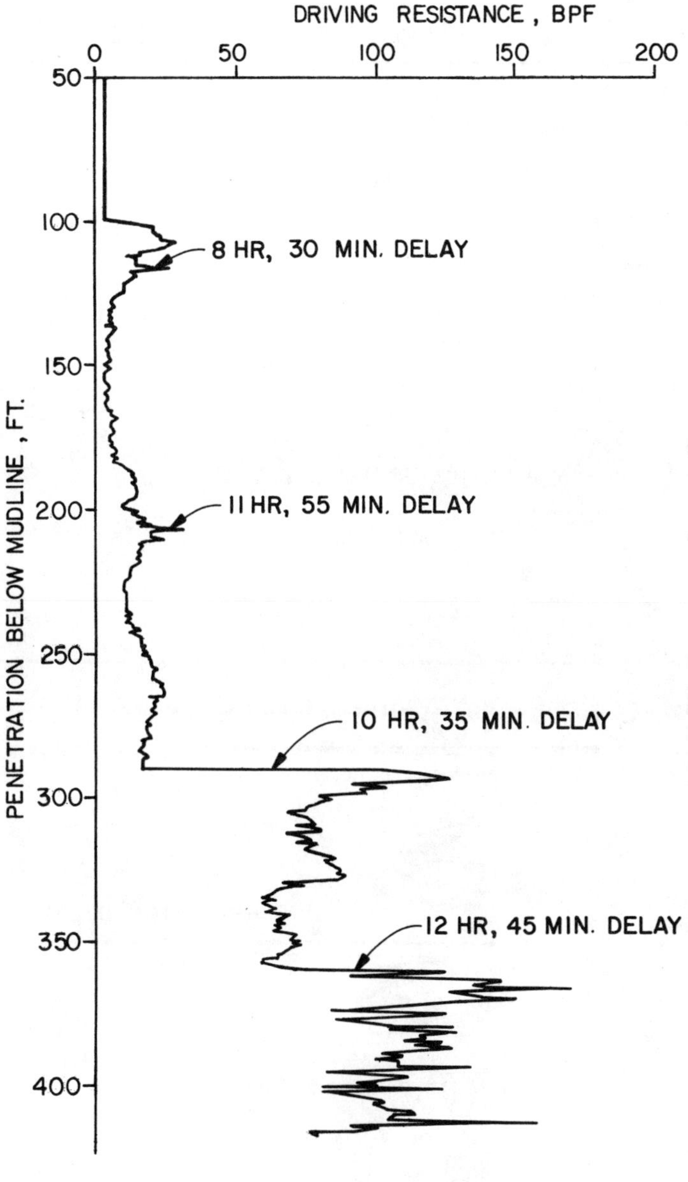

FIG. 3 DRIVING RECORD, PILE A4

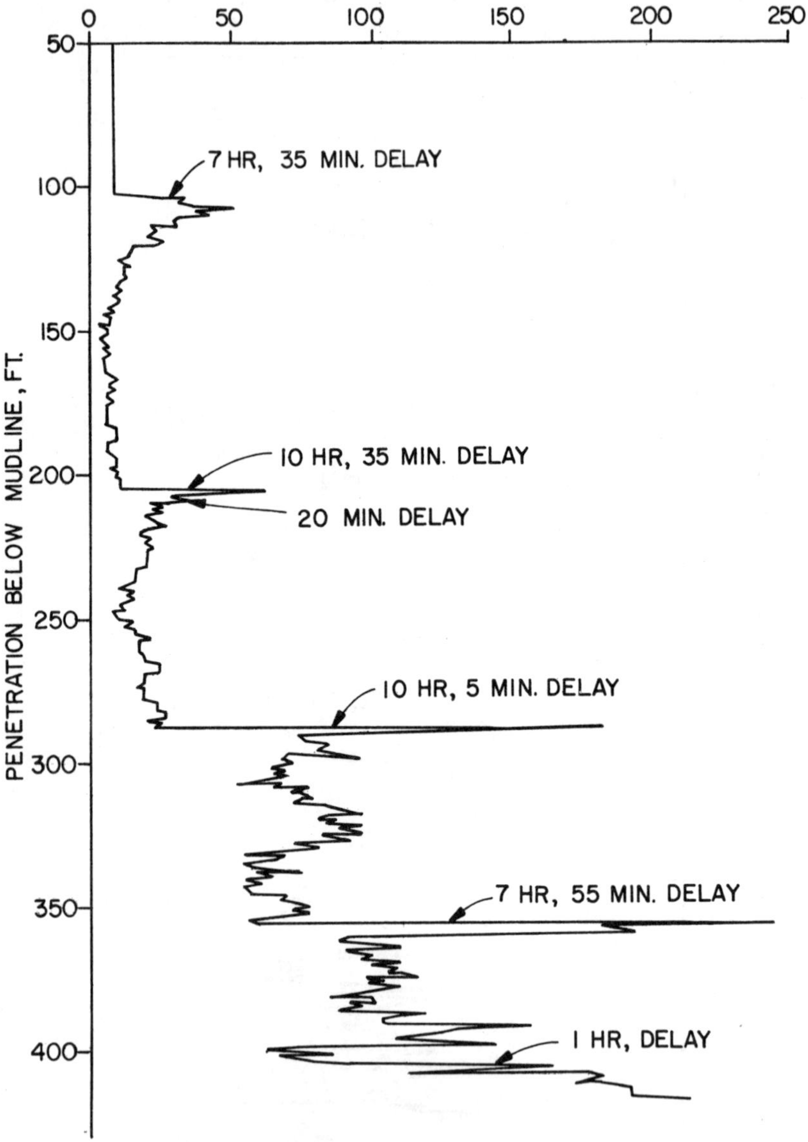

FIG. 4 DRIVING RECORD, PILE B4

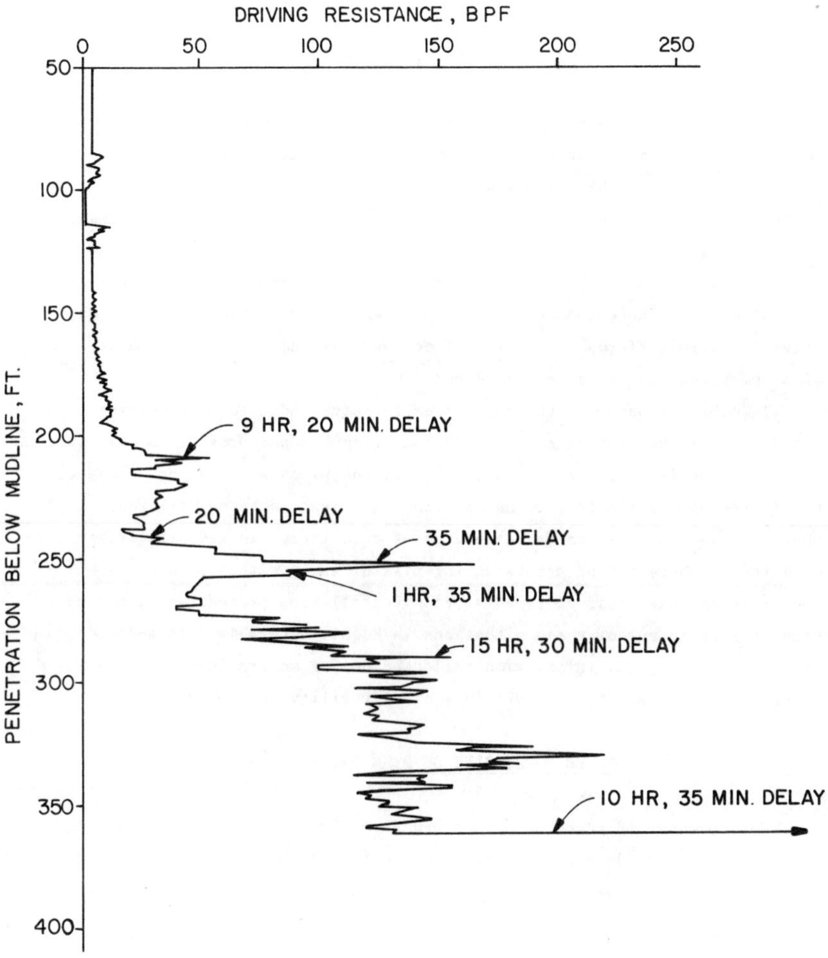

FIG. 5 DRIVING RECORD, PILE B1

counts, and in some cases much higher blow counts, were experienced after delays in driving. This suggests that significant setup occurred during those delays.

PURPOSE OF MONITORING

By monitoring the performance of a pile driving system, benefits are realized both on-site during pile driving and for future projects. In this situation, the main objective was to document the performance of the driving system during attempted restart of a refused pile. This eliminates questions regarding the working condition of the hammer or cushion when considering pile capacity, leaving only the soil stratigraphy and strength as variables. When no documentation of this type is available, the level of driving system performance and its effect on advancement of the pile are also questions that must be considered.

In other situations, the risk of early refusal due to poor driving system performance can be reduced by monitoring the performance to make sure all equipment is operating as expected. If the piles are designed close to the limits of drivability, if new driving equipment is being utilized, or if the installation location is in an area of relatively limited pile driving experience, there may be a substantial risk of early refusal. Because of the extremely high cost of remedial pile installation procedures, a reduction of risk in any of these situations is highly desirable. In addition to on-site benefits, the information collected during monitoring activities can be used to provide better predictions of drivability for future platforms.

FIELD MONITORING OF PILE DRIVING

Weldable strain gages were attached directly to the pile surface for use as sensors to detect the strain (or force) in the pile resulting from a hammer blow. Gages were attached to the refused pile in place in leg B1 about four feet (1.22 m) above the top of the jacket leg and about one foot (30.5 cm) below the field weld of the last add-on section (P6). This location allowed the gaging operation to be performed from the jacket grating. The gages on the other add-on sections were placed approximately 16 ft (4.9 m) below the top of the sections while the add-ons were located on the deck of the derrick barge. The gage locations for these piles were chosen to avoid non-uniform stress distributions which occur near the driven end of the pile. The gages were wired in a four-gage Wheatstone bridge arrangement to cancel temperature effects and to increase sensitivity of the measurements. A pair of gages were located on each end of a diameter of the pile (180° apart) to cancel beam bending effects. Protection from moisture and handling damage was provided by a vinyl mastic tape and an epoxy overcoat.

The acquisition of data during driving operations was accomplished using a set of specially-designed signal conditioning and recording equipment. The strain gages in place on the pile were connected to a signal conditioner which filters line noise using a phase-sensitive demodulator and then amplifies the signal from the strain gages. A mini-processor digitizes and stores the information in an internal memory. The signal is then displayed on a strip chart recorder and simultaneously recorded on magnetic tape.

Calibrations made immediately prior to the beginning of driving were used to interpret recorded blows. Timing marks recorded on the strip chart along with each force-time pulse allowed accurate analysis of the time-based characteristics of the impulse. An internal calibration resistor provided a signal of known magnitude which was used to calibrate the strip chart output of axial strain in the pile.

For the refused pile section approximately every fifth hammer blow was recorded during the attempted redrive on November 11, 1977. Since the attempt did not advance the pile, the second refused pile was not instrumented. For two remaining pile sections, which were driven on November 14, 1977, blows were recorded at each five-foot (1.5 m) interval of driving to obtain a good sampling of driving system performance. Additional blows were recorded for these piles at the beginning of driving to monitor the increase in hammer efficiency that occurred during the warm-up period of the driving system. Data could not be obtained for the last 10 ft (3 m) of driving on these sections, since the lead wire was cut as the gage location passed the top of the jacket leg.

INTERPRETATION OF MEASURED DATA

The data recorded consists of a set of time histories of strain at the gage locations. Strains are converted to forces with a calculated calibration value that assumes a Young's modulus of 30,000 ksi (2.07×10^7 N/cm^2) for steel and pile area based on nominal wall thicknesses. These impulses represent the axial force input from the driving system to that pile.

Estimates of hammer efficiency are obtained by matching the measured force-time pulse to a force-time history predicted by a wave equation analysis. There are three parameters in the wave equation simulation that can be varied to match the measured force-time pulse to the predicted pulse. Two of the parameters, stiffness and coefficient of restitution, are cushion properties and are related to time-based characteristics of the force-time pulse. The third parameter is ram velocity, which is a function of ram velocity at impact and which determines the level of axial stress in the pile for a given set of cushion parameters. Cushion properties affect the maximum axial stress level in the pile but the ram velocity has negligible influence on the characteristics of the curves related to cushion properties. The portion of the time history included in the measurements occurs prior to any reflections from below the mudline, except for very short piles, so that soil modeling parameters are not involved in matching the curves.

The estimated cushion properties can be used to predict the peak stress that would occur as a function of ram velocity at impact. From parameter studies, the peak dynamic axial stress for driving with a given set of cushion parameters is found to be directly proportional to ram velocity.

Ram velocity is related to efficiency by

$$V = (2ghe)^{1/2}$$

or,

$$e = (V/V_{100})^2$$

where V = ram velocity

g = acceleration of gravity

h = nominal stroke

e = hammer efficiency (fraction of rated energy)

V_{100} = $(2gh)^{1/2}$

Therefore, hammer efficiency as defined here is dependent only on the impact velocity of the ram. The quantity V_{100} is based on the nominal stroke of the hammer.

Since maximum dynamic axial pile stress is directly related to the impact velocity of the ram, hammer efficiency can be determined in terms of axial stress. If "f_{100}" is the maximum dynamic axial stress predicted by a wave equation analysis for a hammer at 100% efficiency and for the estimated cushion properties, and "f" is the measured maximum stress then, similar to above,

$$e = (f/f_{100})^2$$

In summary, cushion properties are first determined by matching time based characteristics of the measured data to predictions of these characteristics from wave equation analyses. Hammer efficiency is then estimated by using the estimated cushion properties to determine a predicted maximum axial pile stress which is compared to the measured maximum stress level produced by the hammer.

DRIVING SYSTEM PERFORMANCE

The first add-on section for which data were recorded had refused to move during earlier attempts at driving. There were three separate attempts made to drive that section on November 11, 1977 during which measurements were made. Progressively higher numbers of blows per minute, presumably indicating better steam quality, were noted for the three attempts, each of which consisted of approximately 300 blows. No advancement of the pile was observed. Two other pile sections were successfully driven to grade on November 14, 1977.

Cushion properties were determined for the steel plate and Ascon cushion used in driving the piles. The average value for stiffness was about 48,000 kips per in. (84.1 kN/cm) with values ranging from 30,000 to 85,000 kips per in. (52.5 to 148.9 kN/cm). The authors believe that the higher of the measured values of cushion stiffness occurred during periods of driving when much of the Ascon material had become pulverized, resulting in a loss of some of the cushioning effect of the material. The average coefficient of restitution was 0.72, with values ranging from 0.55 to 0.80.

Maximum axial stresses in the pile varied with hammer efficiency and cushion properties. The average maximum axial stress measured was 15.7 ksi (10.8 kN/cm^2), with a range from 9.0 to 20.0 ksi. (6.2 to 13.8 kN/cm^2).

The average efficiency of the Vulcan 3100 after warm-up was 38% for the driving of the three instrumented add-ons, with a range of individual values from 24 to 52%. The hammer appeared to be in normal operating condition: a 3-foot (0.91 m) stroke was measured and no steam leaks from the hammer or hoses were noted except in driving one section on November 14. However, the number of blows per minute noted for the attempted driving of the refused pile, which could be an indication of inadequate steam pressure or quantity, was 15 to 20 percent less than the manufacturer's published speed of operation. Other possible causes of low efficiency include improper timing of steam controls and friction in the hammer guides due to the batter of the piles.

As previously mentioned, re-driving of the refused pile was undertaken in three separate attempts. Because of the relatively short period of time of hammer operation during each attempt, it is possible that the hammer never warmed up. The results of the estimated cushion and hammer efficiency measurements indicate that the overall average hammer efficiency for these three attempts was 34%, as opposed to an average of 40% for the driving of the two piles on November 14. However, the last attempt at driving the refused pile showed an average efficiency of 44%, with a maximum of 49% that occurred near the end of the last attempt to re-start the pile, indicating that the hammer was probably working at least as well as on the other instrumented add-on sections.

COMPARISON OF CAPACITY REQUIREMENTS AND WAVE EQUATION PREDICTIONS

The need for remedial pile installation measures is one of the most difficult questions to resolve when dealing with a pile refusal situation. Remedial measures are very expensive, especially in an offshore setting, and they also present a possibility for actually reducing pile capacity. Therefore, remedial measures are to be avoided if the refused pile can be shown to have some reasonably acceptable factor of safety. Although the state-of-the-art soils models used in wave equation analyses are still relatively crude, such analyses provide some indications regarding the capacity of a refused pile. Such information is useful in evaluating the need for remedial measures when used in conjunction with the pile capacity predictions developed from soil borings.

The measured force-time pulses acquired in the field are used in a wave equation analysis to qualitatively help evaluate the piles' load carrying capability. This is accomplished by using the force-time pulse data as input to a wave equation computer analysis in place of using the hammer model in the program. Using force-time input eliminates variables associated with the driving system and leaves only the soil parameters for consideration. The wave equation analysis provides estimates of the permanent set at the pile tip as a function of driving resistance. The results are presented in the form of blows per foot versus driving resistance.

Besides providing the force-time data for the analysis, several soil parameters must be selected. These parameters include: (1) the elastic ground compression (quake) at the sides and tip of the pile; (2) the damping constants at the pile side and tip; and (3) the distribution of resistance during driving along the sides and tip at the penetrations to be analyzed. For this study, soil parameters were selected based on values recommended by the Texas Transportation Institute [4] A value of 0.10 in. (2.54 mm) was taken for soil elastic compression at the sides and tip of the pile while the damping was varied from 0.05 to 0.20 sec/ft (0.16 to 0.66 sec/m) along the pile side for the sands, silts, and clays shown in the soil borings. A value of 0.15 sec/ft (0.49 sec/m) was used for damping at the pile tip which was embedded in sandy silt. Resistance at the pile tip was assumed to be five percent of the total resistance and triangular distribution of soil resistance was assumed along the sides of the pile.

Results of the wave equation analysis for the force-time pulse measured in pile B1 at 362-ft (110 m) penetration where refusal occurred are presented in Fig. 6 in the form of driving resistance as a function of blow count. An estimate of driving resistance can be obtained by entering the graph with the measured blow count. This method has been used previously[2, 3] in the North Sea for estimating driving resistance. Since pile B1 was not restarted (very high blow count) the resistance was larger than could be overcome by the driving system and only a lower bound of driving resistance can be estimated. The analyses of the attempted driving of the refused pile show about 3600 kips (16×10^6 N) as a computed maximum driving resistance that can be overcome by this particular hammer-pile-soil system.

The predicted driving resistance from the wave equation analysis can also be compared to the estimated driving resistance based on soil properties from the boring at the site. In general, the resistance to continous driving in clay may be significantly less than the static capacity after setup occurs. Figure 7 shows the estimated range of driving resistance expected at the site. The lower bound represents a condition of continuous driving (reduced soil-pile adhesion in clays) whereas the upper bound curve is equal to the ultimate static pile capacity, taken to be representative of

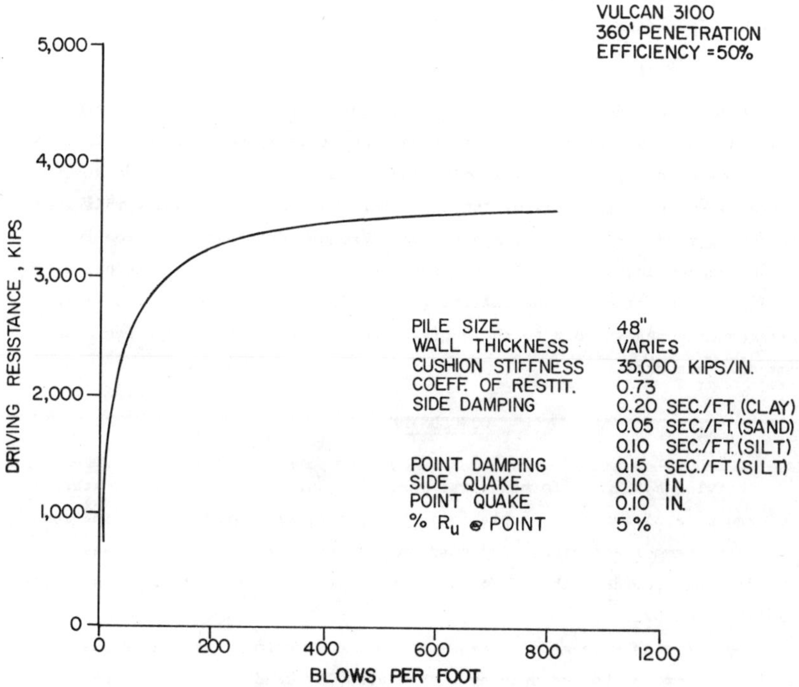

FIG. 6 DRIVING RESISTANCE FROM WAVE EQUATION ANALYSIS

full set-up of the soil. The ultimate static pile capacity was computed by the API RP 2A (April 1977) method.(1)

A comparison of resistance that the hammer-pile-soil system can overcome with the estimated driving resistance of the soil indicates that the soil resistance is within the range expected, and that at 362-ft (110 m) penetration with any significant set-up of the soil the hammer would not be able to restart the pile. These observations increase the level of confidence in predicted ultimate static resistance (pile capacity) developed from soil boring information.

Determination of Pile Adequacy

A decision was made to accept the refused piles at less than design penetration, based primarily on a comparison between the predicted ultimate static capacity of the refused piles and the design pile loads from a detailed structural analysis of the platform. The factor of safety indicated by that comparison was 1.39, about seven percent less than the normally recommended value of 1.5. The increased confidence in static pile capacity predictions considering the qualitative results of the wave equation analyses contributed support to the decision to accept the refused piles without remedial measures.

CONCLUSIONS

Driving system performance was monitored during attempts to re-start a refused pile of an offshore platform. The monitoring provided information regarding hammer efficiency and cushion properties for use in analyses of the driving records. This information was useful in evaluating the need for remedial measures.

Comparisons were made between the computed maximum resistance that could be overcome by the driving system actually used and the driving resistance estimated from soil borings. These comparisons indicate that at the penetration where refusal occurred, significant soil set-up would be expected to create more driving resistance than could be overcome by the hammer. Therefore, the driving experience for this platform supports the predictions of driving resistance based on soil boring information. The comparisons provide more confidence in the predicted ultimate static resistance (pile capacity) expected once complete soil setup is attained.

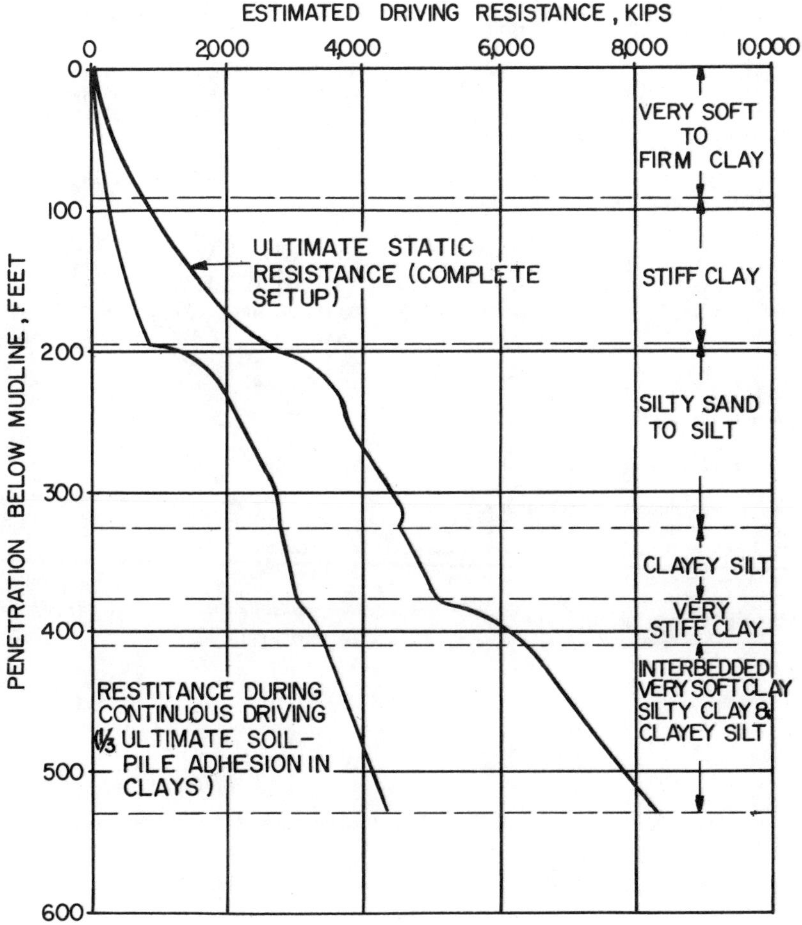

FIG. 7 ESTIMATED DRIVING RESISTANCE

APPENDIX I - REFERENCES

[1] API Recommended Practice for Planning, Designing and Constructing Fixed Offshore Platforms, Ninth Edition, American Petroleum Institute, Washington, D.C., April, 1977.

[2] Fox, D. A., V. J. R. Sutton and Y. Oksuzler, "North Sea platform piling--development of the Forties Field piles from West Sole and Nigg Bay Experience and Tests," Design and Construction of Offshore Structures, Institution of Civil Engineers, London, 1976.

[3] Hirsch, T. J., "Design, Driving and Evaluation of Long Offshore Piling", Ocean Engineering, November 15, 1975.

[4] Hirsch, T. J., Larry Carr and L. L. Lowery, Jr., Pile Driving Analysis Wave Equation User's Manual, TTI Program, Vol. II, FHWA Report No. IP-76-13.2, U.S. Department of Transportation, April 1976, p. C-8.

APPENDIX II - NOTATION

The following symbols are used in this paper:

e = Hammer efficiency
f = Measured maximum stress
f_{100} = Predicted maximum stress at 100% efficiency
g = Acceleration of gravity
h = Nominal stroke of hammer
V = Ram impact velocity
v_{100} = Nominal impact velocity of ram at 100% efficiency

HIGH CAPACITY PILE FOUNDATIONS FOR

CHINA STEEL CORPORATION INTEGRATED STEEL MILL

W. C. Wan [1], Paul Engeling [2], M ASCE and Edith A. Hacker [3], AM ASCE

INTRODUCTION

China Steel Corporation's integrated steel mill, located in Kaohsiung, is one of ten major infrastructure and industrial development projects which were undertaken by the Republic of China in the 1970's. The total project is divided into two phases each of which has two stages. Phase I, Stage 1 of the steel mill, which provides a design annual production capacity of 1.5 million metric tons, was completed in 1977 ahead of schedule. China Steel has been successful both in producing excellent quality steel in large quantities and in marketing the product. Phase I, Stage 2 construction has been accelerated and Phase II will probably follow in the near future. When completed, the China Steel mill will have an annual production capacity in excess of 8 million metric tons.

Retired Servicemen's Engineering Agency (RSEA) was and is the prime contractor for Phase I, Stages 1 and 2 construction, respectively. Raymond International Builders, Inc. supplied technical advisers, specialized pile driving equipment, and Raymond Step-Taper piles for both Stages 1 and 2 of Phase I. For Stage 1, more than 22,000 Step-Taper piles were installed, starting in 1974 and finishing in 1975. For Stage 2, 18,260 Step-Taper piles were installed starting in July 1979 and finishing in February 1980, a period of eight months. Co-operation between China Steel, RSEA and Raymond on this very large and important project has been excellent.

The China Steel mill site was chosen directly adjacent to the Kaohsiung harbor for ready access to ocean shipping, land transportation, electric power, labor and other support facilities. In the early planning stages, it was recognized that the soil conditions at the site were relatively weak and that the foundations for many of the facilities in the mill would have to be designed to carry

---

1. Deputy Chief Engineer, China Engineering Consultants, Inc., on special assignment to China Steel Corporation as Director, Civil Construction Division, Taiwan.

2. Assistant Vice President and Chief Foundation Engineer, Raymond Technical Facilities Inc., Houston, Texas.

3. Foundation Engineer, Raymond Technical Facilities Inc., Houston, Texas.

very heavy loads with minimum differential settlements. The coke ovens would present especially critical problems in this area. Extensive subsoil investigations showed that shallow foundations were feasible for light, one-story structures, but that deep foundations would be absolutely essential for the very heavy design loads in the mill itself. The construction schedule demanded that the foundations be completed quickly and efficiently and the very high concentrated loads required piles with very high capacity. China Steel engineers were aware that Raymond Step-Taper piles had been used to support the heavy facilities at the Talin Power Station just across the harbor from the steel mill site. The piles for that project had been installed rapidly and efficiently, and long-term settlement measurements proved that both total and differential settlements were minimal for such large structures. These were substantial considerations in the selection of Step Taper piles for the China Steel Project.

An extensile pile load testing program was initiated at the China Steel site several months before foundation construction was started. Several types of piles, including Raymond Step-Taper piles were driven and load tested. The Step-Taper pile was chosen based on its excellent performance and speed of installation.

SITE CONDITIONS

The China Steel plant site is located directly adjacent to the Kaohsiung harbor and has an area of 480 hectares (1186 acres). The site is located on a broad, flat plain which is bordered on the south by limestone hills. Most of the site was formerly cultivated farmland used to raise sugar cane and other crops. The part of the site directly adjacent to the harbor was devoted to commercial fish ponds.

The soil conditions at the site are totally unsuitable for heavily loaded shallow foundations. All heavy structures must be supported on pile or pier foundations. Lighter structures, such as two to three-story office buildings may be supported on compensated floating foundations. The soil conditions are very uniform across the entire site. Over 200 borings have been made on the site, and all of the borings show very similar conditions as illustrated in Figure 6. Soft to stiff silty clay topsoil, with some fill inclusions is found in the upper 12 to 18 feet (3.7 to 5.5 m) at all the borehole locations. Beneath the topsoil lies 4 to 8 feet (1.2 to 2.4 m) of soft to medium silty clay in all of the borings. For approximately 40 feet (12 m) beneath the clay lies a loose to medium dense silty sand layer. An occasional clay or silty clay seam may be found in this layer between depths of 28 and 35 feet (8.5 and 10.7 m) at some of the borings. Beneath the loose to medium dense silty sand lies a medium dense to dense silty sand to the terminal depth of the borings at about 165 feet (50 m). An occasional silty clay or silt seam was found between depths of 74 and 78 feet (22.6 and 23.8 m). As described above, there are more or less random variations in the locations of silt and clay layers and lenses, but the main layers, topsoil, clay, loose to medium dense sand, and finally medium dense to dense sand, are very consistent all across the site.

PILE FOUNDATION PERFORMANCE

Pile design loads were mostly 110 metric tons (1080 kN) for 96-ft (29.3 m) piles and 135 metric tons (1320 kN) for 102-ft (31.1 m) piles. There were also 84-ft (25.6 m) piles with 100 metric ton (980 kN) design loads and 72-ft (22 m) piles with 80 metric ton (780 kN) design loads. Extensive pile load testing programs were conducted on working piles both in Stages 1 and 2. A total of 31 Step-Taper piles were load tested. Pile load test capacities were very high, ranging up to 405 metric tons (3970 kN). Based on the excellent load test results, the design load for many of the piles was increased from 110 metric tons to 135 metric tons (1080 to 1320 kN).

General performance of the foundations over a period of four years in operation has been excellent. Total settlements have been measured in two plant areas, and these measurements have shown that differential settlements are minimal. Figure 1 shows settlement measurements for Battery I of the Stage 1 coke ovens and Figure 2 shows settlements of Cowper 11 of the blast furnace of Stage 1. Battery 1 had a maximum total settlement of only 36 mm (1.4 in.) and a maximum differential settlement of only 9 mm (0.4 in.) in four years. Cowper 11 had a maximum total settlement of 32 mm (1.3 in.) and a maximum differential settlement of only 6 mm (0.2 in.).

The ratios of the measured long-term settlements of several pile group foundations to the single pile settlement have been plotted as a function of pile foundation width in Figure 3. The group settlements were taken as the average of settlement measurements at the corners and/or centers of the pile caps. The single pile settlement values were taken as an average of the pile load test settlement at design load for all load test piles in the general area of the foundation being considered.

Skempton (see p. 540, Reference 3) and Vesić (4) have both published curves for estimating pile group settlement from single pile settlement data. Figure 3 shows Skempton's and Vesic's curves along with all data points obtained from the Stage 1 coke oven and blast furnace settlement observations. In almost all cases, actual pile group settlements were less than predicted by either Skempton's or Vesic's procedures.

PILE LOAD TESTS

The results of load tests on 32 Step-Taper piles were studied for this paper, 17 from Stage 1 and 15 from Stage 2. The load-settlement curves for 21 of the test piles were extrapolated to establish a failure load, since only 10 of the piles were tested to failure. The extrapolations were conservative and may underestimate the actual failure load by as much as 15 percent.

The criteria used to establish the failure load is discussed below and on page 214 of <u>Foundation Engineering</u> (2). The scales for the load-settlement curve are established by plotting the elastic compression of the pile (which is proportional to the load) at an angle of approximately 20 degrees below the load axis.

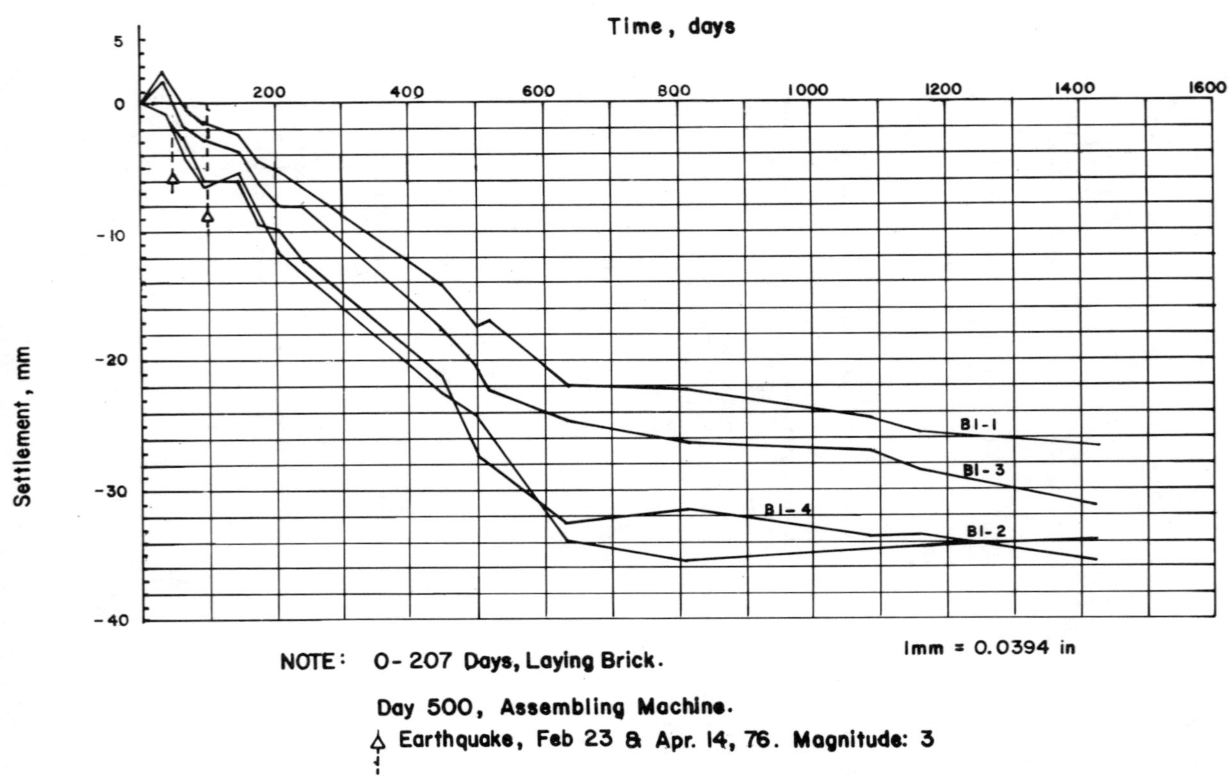

FIG. 1, OBSERVED SETTLEMENTS BATTERY I COKE OVENS, STAGE 1.

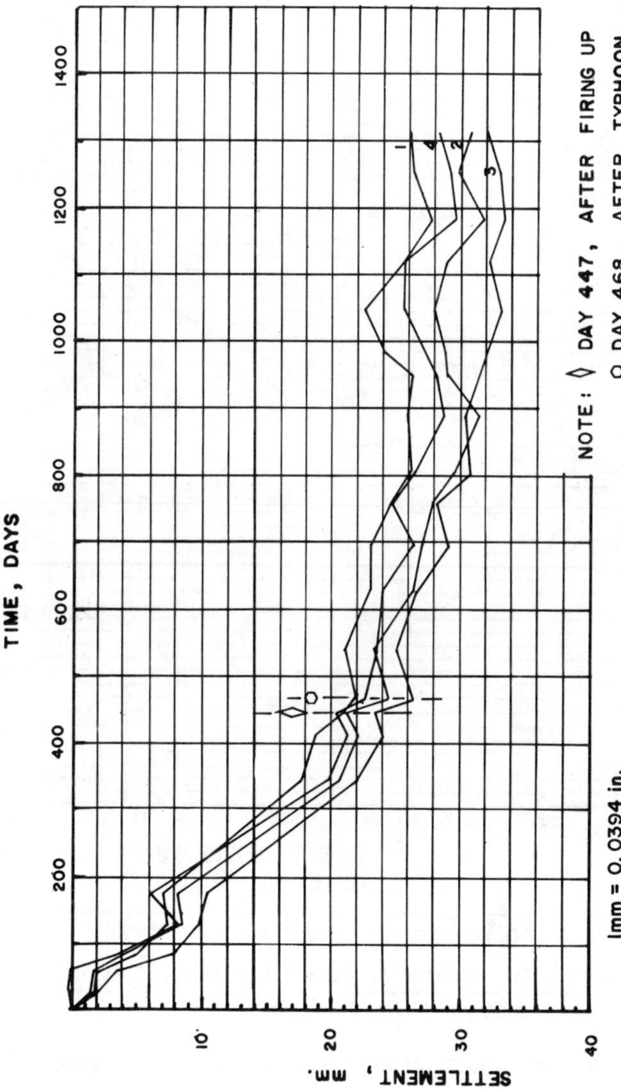

FIG 2, OBSERVED SETTLEMENTS BLAST FURNACE STAGE 1

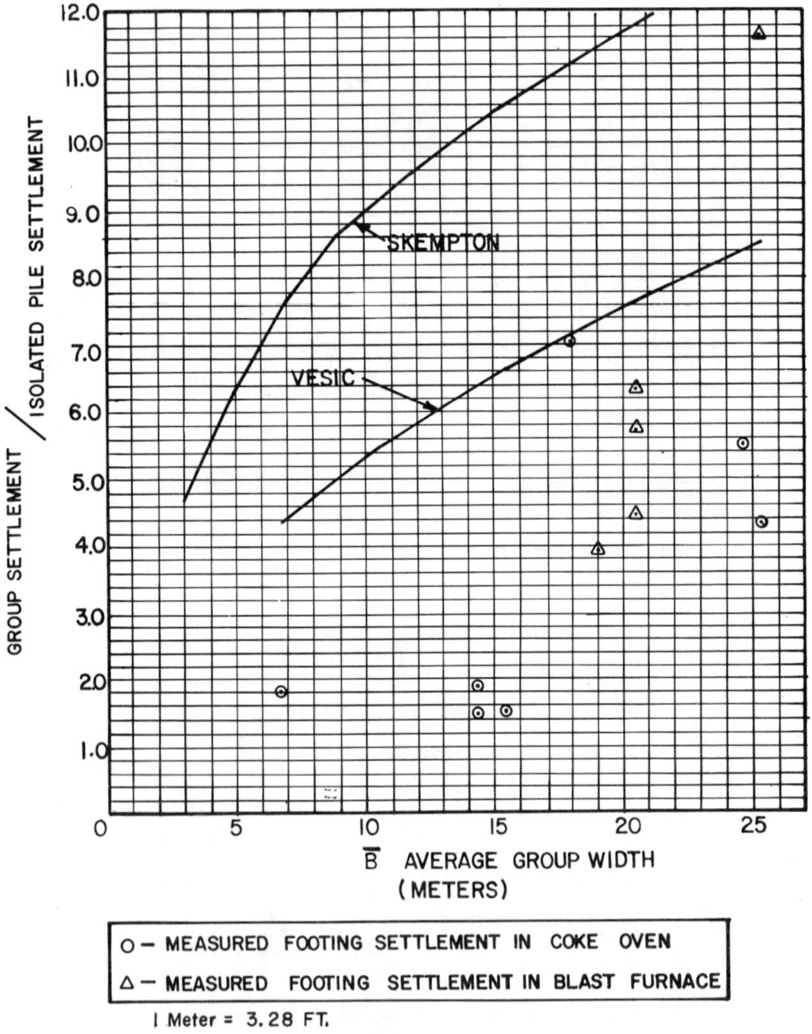

FIG, 3, MEASURED VERSUS CALCULATED SETTLEMENTS

The elastic compression of the pile is expressed by the equation S = PL/AE where S is the elastic compression settlement, P is the load, L is the pile length, A is the cross-sectional area of the pile, and E is the modulus of elasticity of the concrete. Since the Step-Taper pile has varying cross-sectional areas, the elastic compression was calculated for each section and the sum of the compressions for all the sections was taken as the elastic compresion of the pile. A second line, called the Davisson offset line, is drawn beneath and parallel to the elastic compression line. The second line intercepts the settlement axis at a settlement of 0.15 + 0.1d inches where d is the diameter of the pile measured in feet. Only those settlements, recorded at one hour or less after the load was applied, were plotted to determine failure load. The ultimate failure load is taken as the load at which the load-settlement curve for the test pile intercepts the offset line. Figures 4 and 5 show load-settlement curves for eight of the 32 load tests which were conducted on Step-Taper piles. Elastic compression lines and Davisson's offset lines are shown on the plots.

STATIC PILE CAPACITY ANALYSIS

The static capacity analyses for this project were made based on the boring or borings closest to the test pile considered.

Soil Strength Parameters

Standard penetration tests were made at 5-ft (1.5 m) intervals in most of the borings. The N-values from these tests were corrected for the effects of overburden pressure with the correction factors contained in Fig. 19.6, p. 312 of Foundation Engineering (2). The average corrected N-values for each stratum were then related to the angle of internal friction, ϕ using Fig. 19.5, p. 310 of Reference 2. The N-values in the soft silty clay layer were related to the undrained shear strength using Table 45.2, p. 347 of Soil Mechanics in Engineering Practice (3).

Calculation Procedure

Static pile capacity calculations were performed for both Stage 1 and Stage 2 test piles generally using the procedures proposed by Nordlund (1). An example of the detailed calculation procedure used is illustrated on Figure 6 and Table I. Certain modifications were made to the procedure as discussed below. Pile capacity in cohesive soils was calculated using the alpha method as discussed in Foundation Engineering (2) pp. 282-283.

The α factors used in the analysis ranged between 1.0 and 0.75 since the undrained shear strengths were low to medium, ranging between 200 and 1000 psf (9.6 and 47.9 kPa) (5). The adhesion, C_a, was then estimated to be the product of the alpha factor and the undrained shear strength of the clay. The frictional capacity of the topsoil was neglected since its contribution to total pile capacity was small and since it was jetted during pile installation.

The skin friction developed in the sand and silty sand layers was estimated using Nordlund's procedure (1) which is discussed and illustrated in detail below. The main advantage of Nordlund's procedure is that it considers the effects of pile displacement, volume and taper on pile capacity, which are not considered

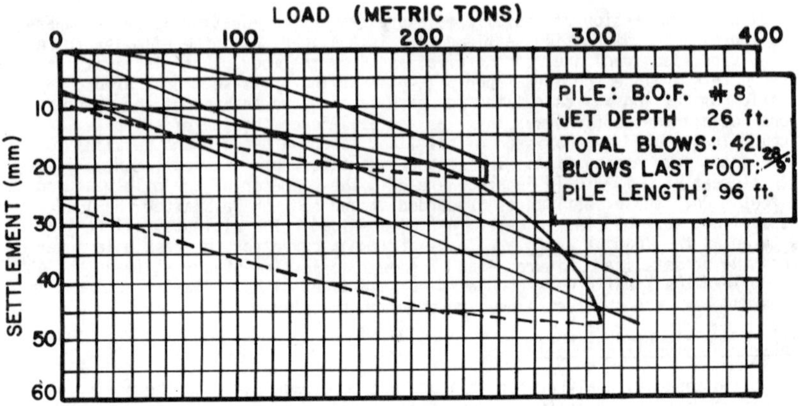

1mm = 0.0394 in.
1 Metric Ton = 9.8 kN

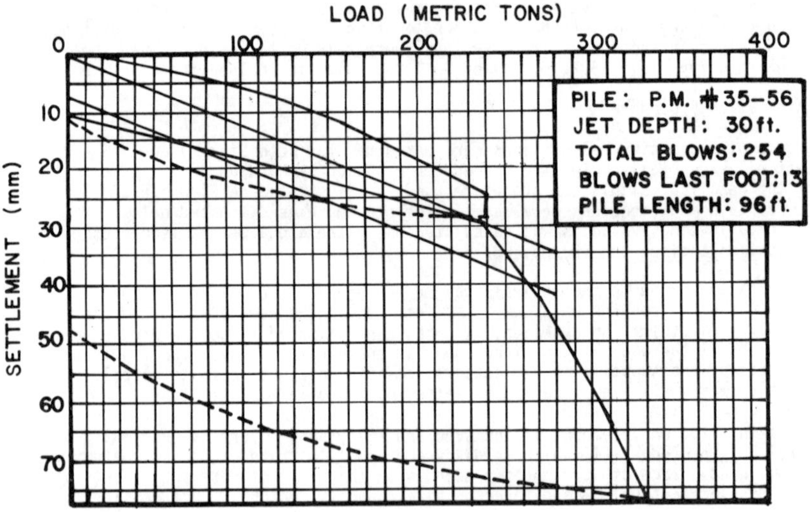

FIG. 4, TYPICAL LOAD-SETTLEMENT CURVES
PHASE 1 STAGE 1

(CONT'D)

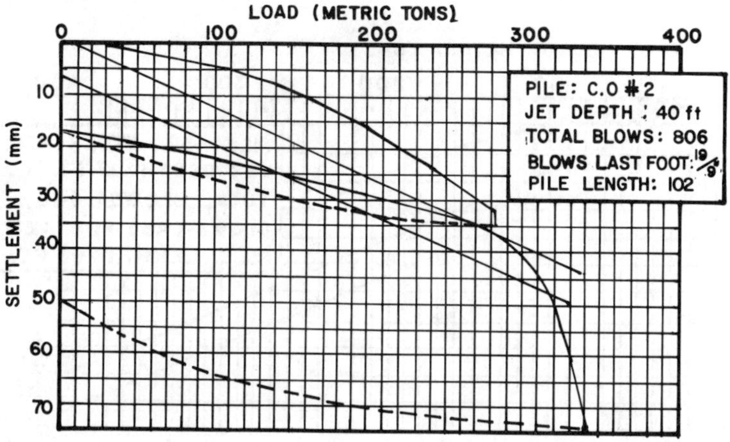

1 mm =0.0394 in.
1 Metric Ton = 9.8 kN

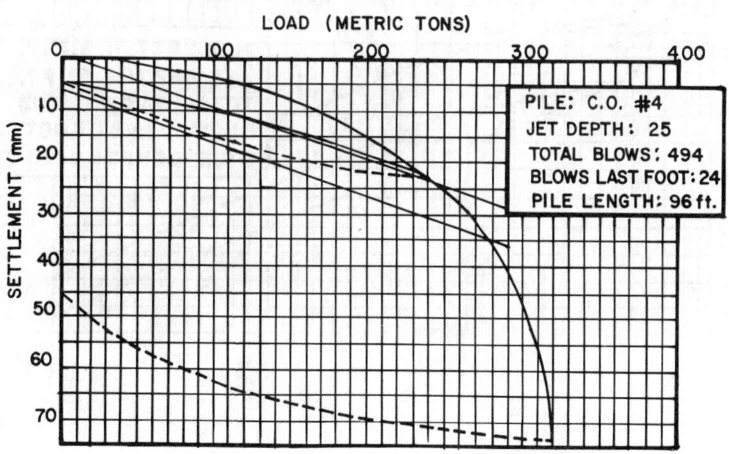

FIG. 4, (CONT'D) TYPICAL LOAD-SETTLEMENT CURVES
PHASE 1 STAGE 1

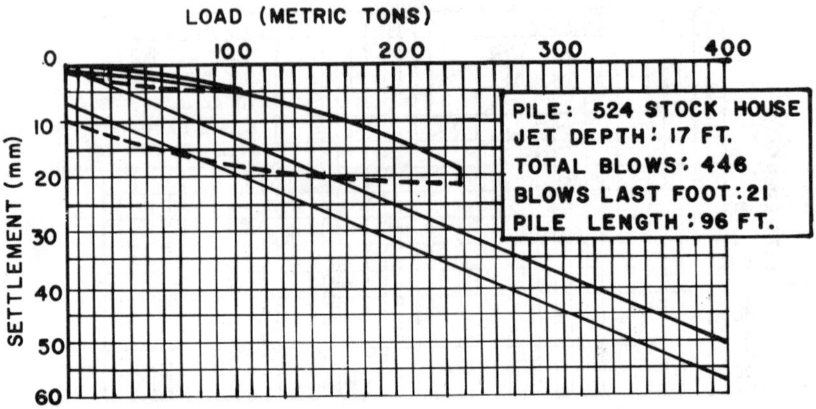

1 mm = 0.0349 in
1 Metric Ton = 9.8 kN

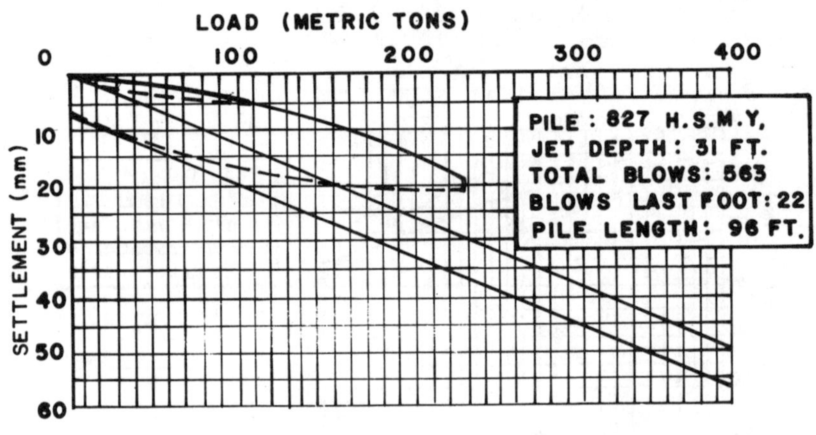

FIG. 5, TYPICAL LOAD-SETTLEMENT CURVES
PHASE I, STAGE 2

(CONT'D)

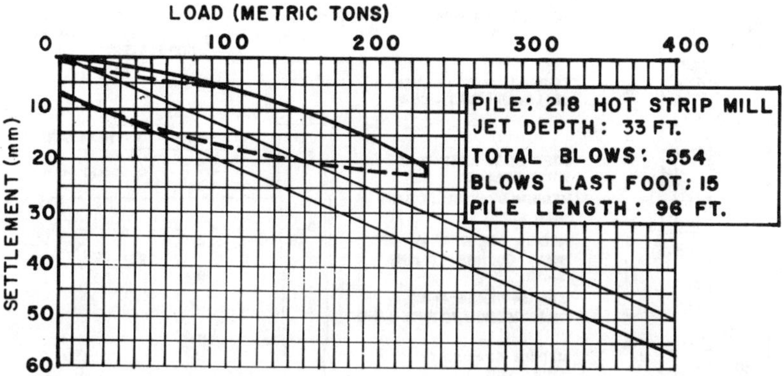

1 mm = 0.0394 in
1 Metric Ton = 9.8 kN

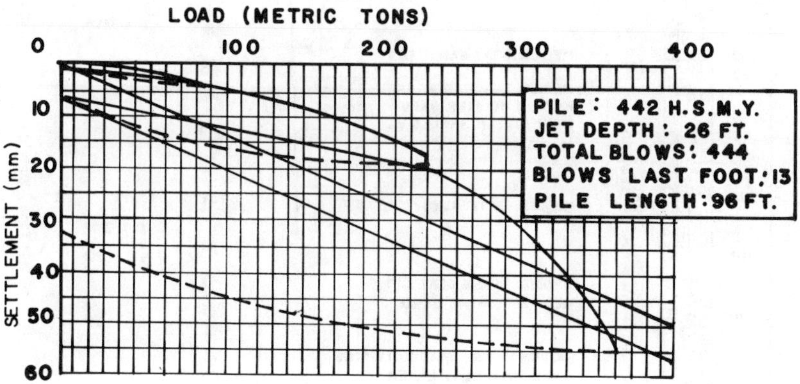

FIG. 5, (CONT'D) TYPICAL LOAD-SETTLEMENT CURVES
PHASE I, STAGE 2

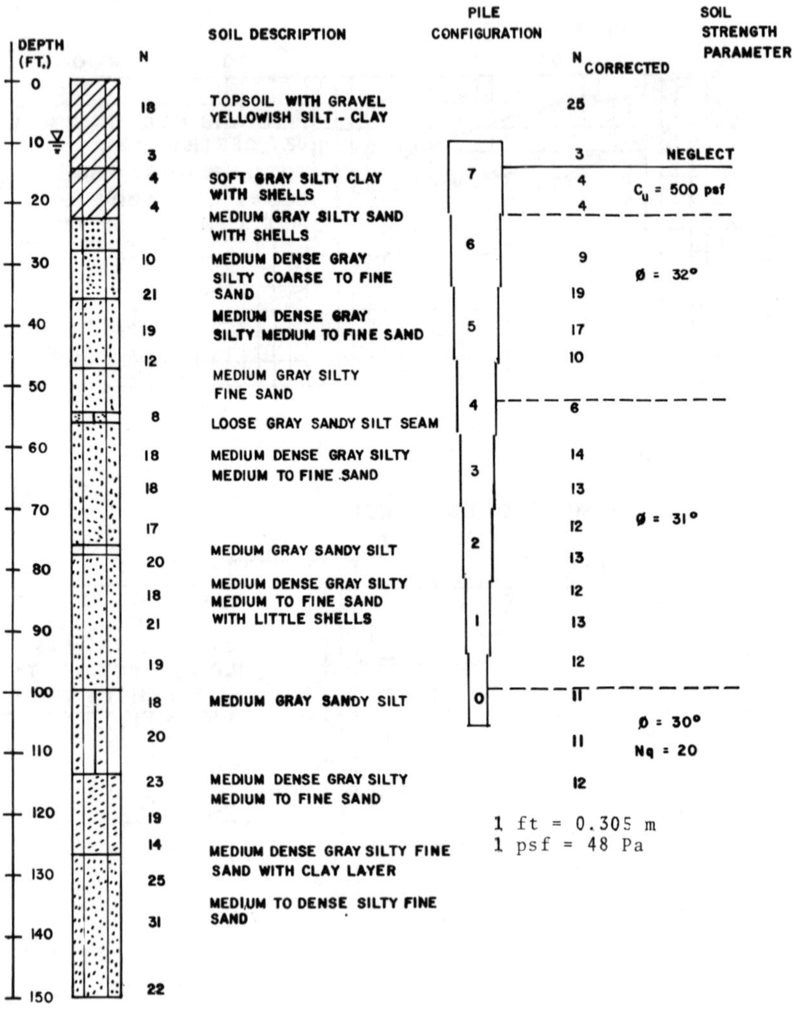

FIG. 6, SOIL AND PILE DATA FOR TYPICAL STATIC CAPACITY CALCULATION

Table I

Summary of Static Capacity Calculations
Pile 410 Stage 2

Depth ft	Test Pile	Soil Parameters a	v ft³/ft	δ/φ b	δ deg	sin δ	$\bar{p}_d$ c ksf	K_δ d	C_k e	C_a f ksf	f g ksf	Surface Area ft²	Adhesion or Friction kips
0 – 4													
4 – 13	7	Cu = 0.5 ksf, α = 1.0	--	--	--	--	--	--	--	0.50	--	40.7	20.4
13 – 24	6		1.5	1.02	32.64	0.5394	1.11		1.0	--	1.14	47.1	53.7
24 – 31	5	φ = 32°	1.3	0.97	31.04	0.5156	1.65	1.9		--	1.62	28.2	45.7
31 – 36							2.01			--	1.97	20.1	39.6
36 – 44	4		1.1	0.92	29.44	0.4915	2.40		0.96	--	2.15	30.1	64.7
44 – 48					28.52	0.4775	2.76			--	2.13	15.1	32.2
48 – 60	3		1.0	0.91	28.21	0.4727	3.24		0.94	--	2.42	42.0	101.6
60 – 72	2	φ = 31°	0.8	0.85	26.35	0.4439	3.96	1.68	0.91	--	2.69 h	38.9	97.3
72 – 84	1		0.7	0.82	25.42	0.4293	4.68		0.89	--	3.00 h	35.7	89.3
84 – 90	0		0.6	0.77	23.87	0.4047	5.22		0.88	--	3.12 h	16.5	41,2
90 – 96		φ = 30°			23.10	0.3923	5.58	1.45		--	2.79 h	16.5	41.3
96		Nq = 20					5.76						

Total Adhesion and Friction = 627 kips
End bearing = $q_pA_p = p_dN_qA_p$ = (5.76)(20)(0.6) = 69 kips
Total capacity P_u = 696 kips = 316 m-ton

Notes:
a. Alpha factor according to Reference 5
b. From Fig. 9 Reference 1
c. Effective pressure at mid-points except tip
d. From Fig. 8 Reference 1
e. From Fig. 7 Reference 1
f. $C_a = \alpha C_u$
g. $f = C_k K_\delta \bar{p}_d \sin\delta$
h. Use limiting value skin friction = 2.5 ksf

in most static analysis procedures. The displaced soil volume, v, per linear foot for the applicable Step-Taper section was related to the ratio of the angle of friction on the surface of sliding, δ, to the angle of internal friction, ϕ, using Fig. 9 (curve "d") in Reference 1 for Step-Taper piles. By multiplying this ratio by ϕ, δ was obtained. The taper of the pile, ω, was calculated from the pile dimensions. Knowing δ, ϕ and ω the ratio, K_δ, of the resultant of the effective normal and shear stresses on the incipient failure plane to the effective overburden pressure was interpolated from the various curves in Fig. 8 of Reference 1. The correction factor, C_k, for K_δ was taken from Fig. 7 of Reference 1. The effective overburden pressure p_d, was calculated assuming a dry unit weight of 120 pcf (1920 kg/m^3) and a submerged unit weight of 60 pcf (960 kg/m^3) below the water table. Between 5 and 10 feet (1.5 and 3 m) of topsoil was excavated before driving, and its weight was not included in the overburden pressure calculations. The water table was assumed to be located at the ground surface after the topsoil was excavated. The unit skin friction was then calculated for each soil layer or pile section by the following equation:

$$f = C_k K_\delta \bar{p}_d \sin \delta$$

where

- f = unit skin friction
- C_k = correction factor for K_δ
- K_δ = ratio of the resultant of the effective normal and shear stresses on the incipient plane of failure to the effective overburden pressure
- $\bar{p}_d$ = average effective overburden pressure.
- δ = angle of friction between pile and soil.

A limiting unit skin friction of 2500 psf (119.7 kPa) was used for Stage 2 data when the calculated unit friction exceeded this value. In Stage 1 data, the 2500 psf (119.7 kPa) limit was used for sands with ϕ of 32° or greater and a 2000 psf (95.8 kPa) limit was used for ϕ of less than 32°. The unit end bearing resistance, q_p, of the pile was estimated as the product of N_q times the effective overburden pressure at the pile tip. N_q was obtained from correlations with the corrected N-values in Fig. 19.5, p. 310 of Foundation Engineering (2).

The ultimate static bearing capacity, P_u, of the pile was then calculated by multiplying the unit adhesion, unit skin friction, and unit end bearing by the pile surface area over which each acts and then adding the products. This is represented by the following expression.

$$P_u = q_p A_p + \Sigma f_i A_{si} + \Sigma C_{aj} A_{sj}$$

where

P_u = ultimate pile bearing capacity
q_p = unit end bearing resistance
A_p = pile tip area
f_i = unit skin friction for soil layer i
A_{si} = pile surface area within soil layer i
C_{aj} = unit soil-pile adhesion for soil layer j
A_{sj} = pile surface area within soil layer j

COMPARISON OF TESTED CAPACITY WITH STATIC ANALYSIS

The calculated ultimate static capacities of the piles and the tested pile capacities have been compared in Figures 7 and 8 for Stage 1 and Stage 2 data respectively. For Stage 1, the predicted capacities were within 25 percent of the tested capacities for 94 percent of the test piles and within 20 percent for 88 percent of the piles. For Stage 2, the predicted capacities were within 25 per cent of the tested capacities for 87 percent of the test piles and within 20 percent for 80 percent of the piles. In general there was good agreement between predicted and tested capacities; in almost all cases, the actual tested pile capacity was higher than the capacity predicted by static pile capacity analysis.

The methods of correlating standard penetration tests with soil strength parameters are conventionally accepted. The approach used herein to estimate the frictional capacity of the piles in cohesionless soils is considered by some engineers as somewhat unconservative for most pile types. However, this approach has shown reasonable and generally conservative results when used with Step-Taper piles. The actual failure loads were determined in a conservative manner because several load-settlement curves had to be extrapolated to determine failure loads since few piles were loaded close to failure. The comparison of the predicted and tested capacities indicate that the approach used to calculate the predicted capacity was definitely conservative for these soil conditions and these piles.

WAVE EQUATION ANALYSIS

The wave equation analysis of pile driving predicts the soil load resistance that the hammer-pile system can overcome. The analysis predicts the ultimate load capacity of the pile at the time of driving versus penetration resistance in blows per inch. This ultimate capacity may or may not agree with the load test results depending on whether soil freeze or relaxation occurs after the pile is driven. A wave equation analysis was prepared for each pile length and hammer energy used. Knowing the final driving resistance of the pile, the capacity of the pile at the time of driving can be predicted. Figures 9 and 10 show a graphical representation of the wave equation prediction of pile capacity at the time of driving and the difference between this capacity and the capacity determined by tests sometime later. The capacities predicted by the wave equation analysis ranged from 16 to 77 percent of the actual capacities determined by load tests with an overall average of 48 percent. Although the individual pile capacities predicted by the wave equation varied considerably,

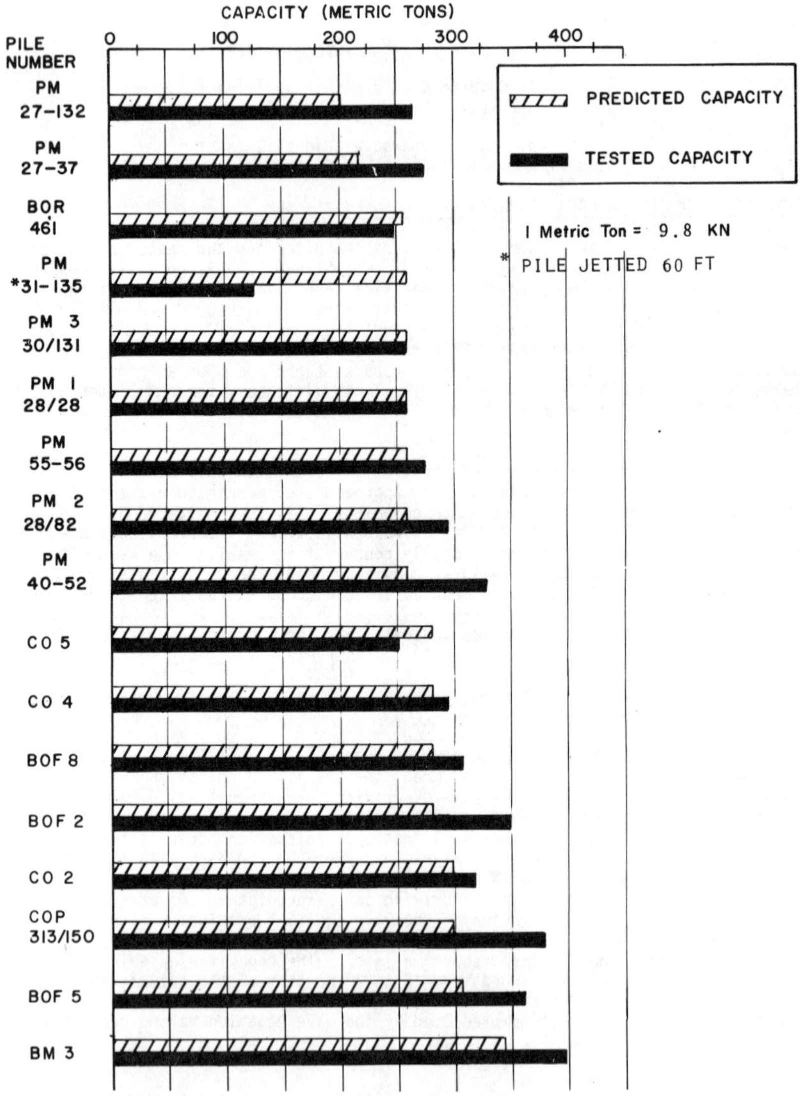

FIG. 7, PREDICTED STATIC ANALYSIS CAPACITY VS TESTED CAPACITY PHASE I, STAGE 1

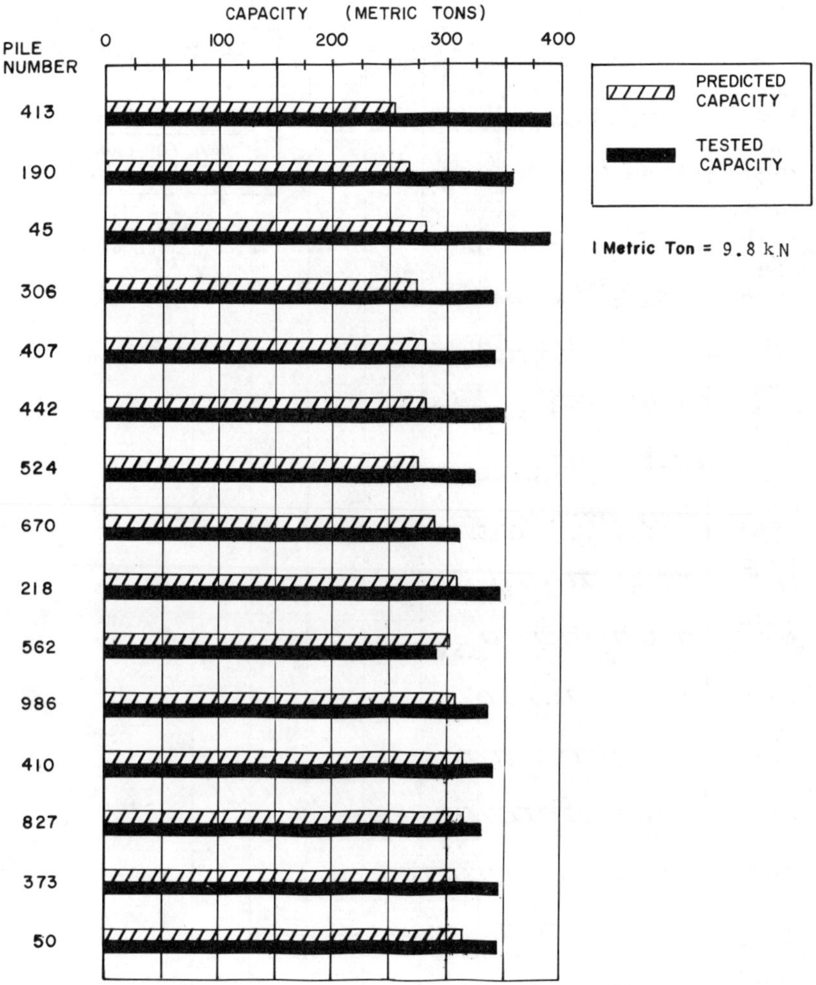

FIG. 8, PREDICTED STATIC ANALYSIS CAPACITY VS TESTED CAPACITY

PHASE I, STAGE 2

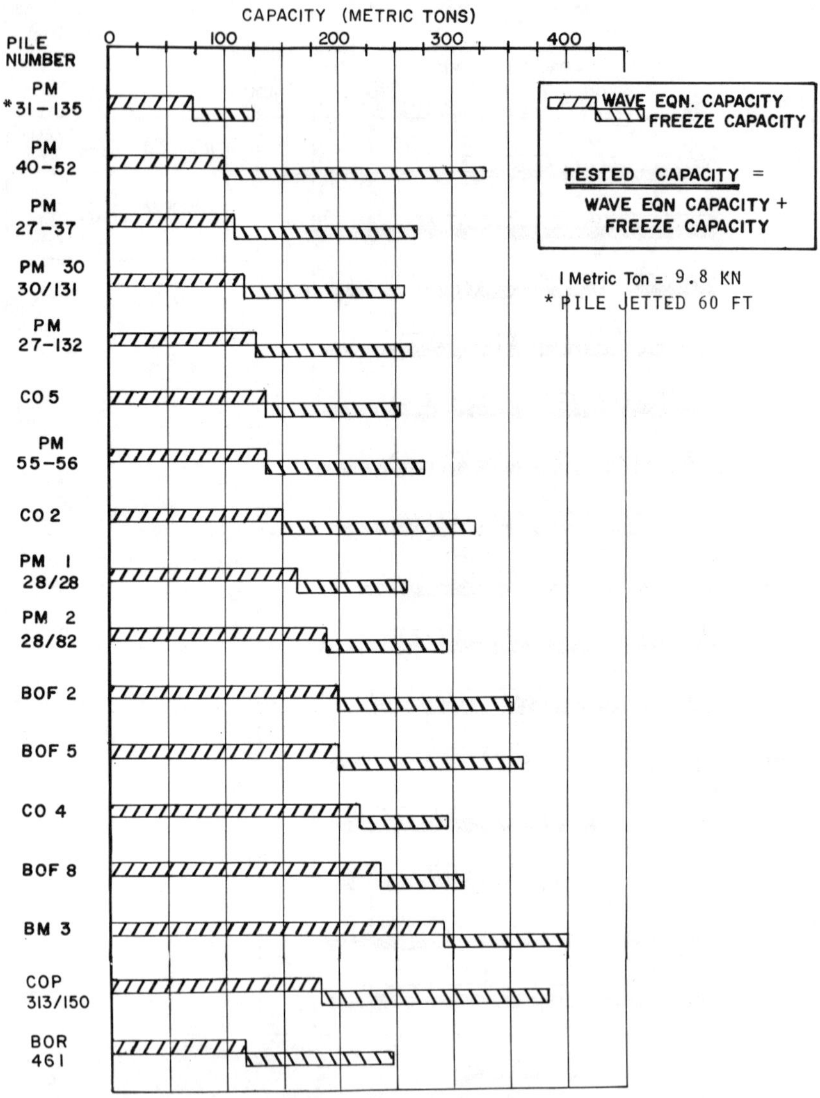

FIG. 9, CAPACITY PREDICTED BY THE WAVE EQUATION ANALYSIS AND THE FREEZE CAPACITY, PHASE I, STAGE 1

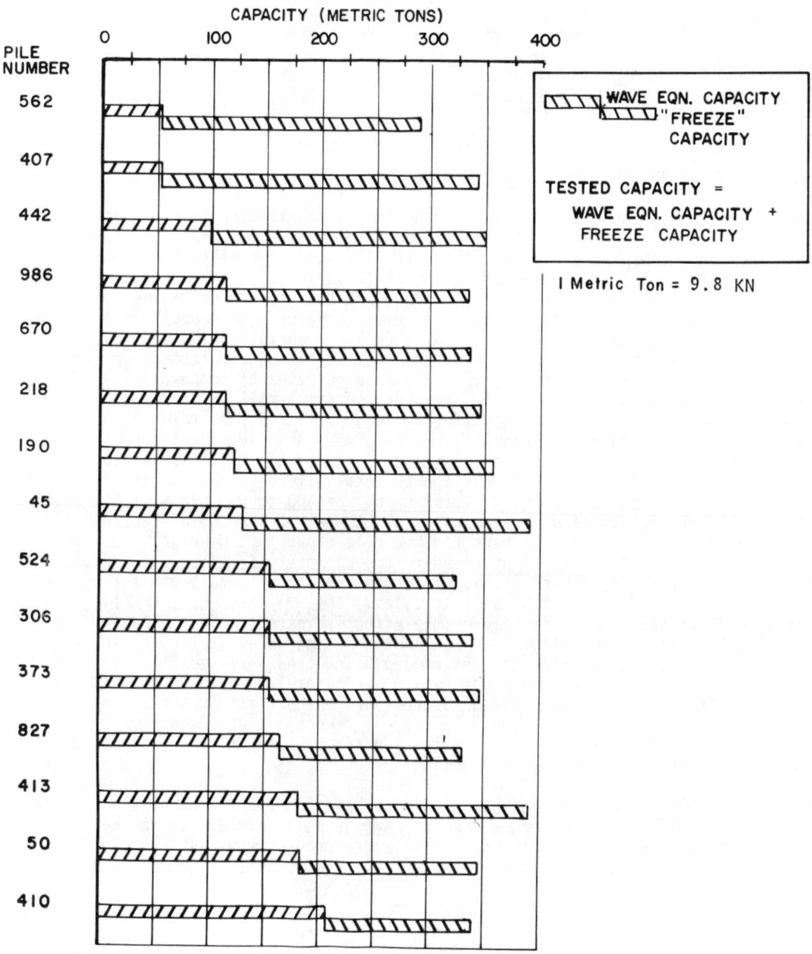

FIG. 10, CAPACITY PREDICTED BY THE WAVE EQUATION ANALYSIS AND FREEZE CAPACITY, PHASE I, STAGE 2

the tested capacities of these piles were quite uniform. Figures 11 and 12 are plots of ultimate pile load test capacity as a function of total hammer blows and hammer blows on the last foot of penetration respectively. It is quite apparent that pile capacity is not dependent on driving resistance. Soil freeze is an overwhelming factor in establishing pile capacity. Because of the great effect of soil freeze, pile capacity cannot be established by a driving resistance criterion, even with the use of wave equation analyses. As discussed below, it appears that jetting helped increase the excess pore water pressures normally associated with pile driving in these type soils which would further invalidate a driving resistance criterion. Fortunately, the soil conditions are sufficiently uniform all across the site so that pile penetration criteria can be used.

PILE INSTALLATION CONSIDERATIONS

Two of the main advantages of the Raymond Step-Taper pile for installation in the soil conditions which exist in the Kaohsiung area, are speed of installation and very high pile capacity. The very fine sands and silty sands which exist on this site lend themselves very well to an installation procedure which involves water jetting to ease pile driving such that the piles can be readily driven to the deep penetrations which are necessary for high capacity. Although excessive jetting did reduce friction pile capacity on this site, controlled jetting, used to ease pile penetration, did not have much effect on pile capacity. Figure 13 is a plot of pile capacity as a function of depth of water jetting for all 32 test piles from all areas of the steel-mill site. Figure 13 shows that for a depth of water jetting up to approximately 35 to 40 feet (10.5 m to 12 m), there is essentially no reduction of pile capacity caused by water jetting for the pile lengths installed in these soils. To establish some type of pile driving control in addition to total pile penetration, jetting depth was continually adjusted so that a total number of 400 to 500 blows of the pile driving hammer was achieved on each pile. If these had been point bearing piles, deeper water jetting would not have reduced pile capacity. Deep jetting reduced the friction capacity of these piles which are basically friction piles. The fine sand and silty fine sand at Kaohsiung are ideal soil conditions for water-jet assisted installation of Step-Taper piles. The jet injects water into the sands under pressure and this reduces the effective stresses in the sand. When the pile is inserted and allowed to run in the jetted hole, it further forces the water into the sand below the jetted depth and creates excess pore water pressures. These excess water pressures, which are maintained by pile driving, allow the piles to penetrate deeply in the dense sands which start at a depth of about 74 feet (23 m) below original ground surface. Deep penetration into the dense sands produces high pile capacity. Such deep penetration in such dense sands would be impossible without water jetting.

The excess pore water pressures begin to dissipate immediately when pile driving ceases. This is the phenomenon which is known as soil freeze. In the soil conditions which exist in Kaohsiung, soil freeze occurs very rapidly. Complete dissipation of excess pore water pressure takes place within a few hours after pile driving ceases.

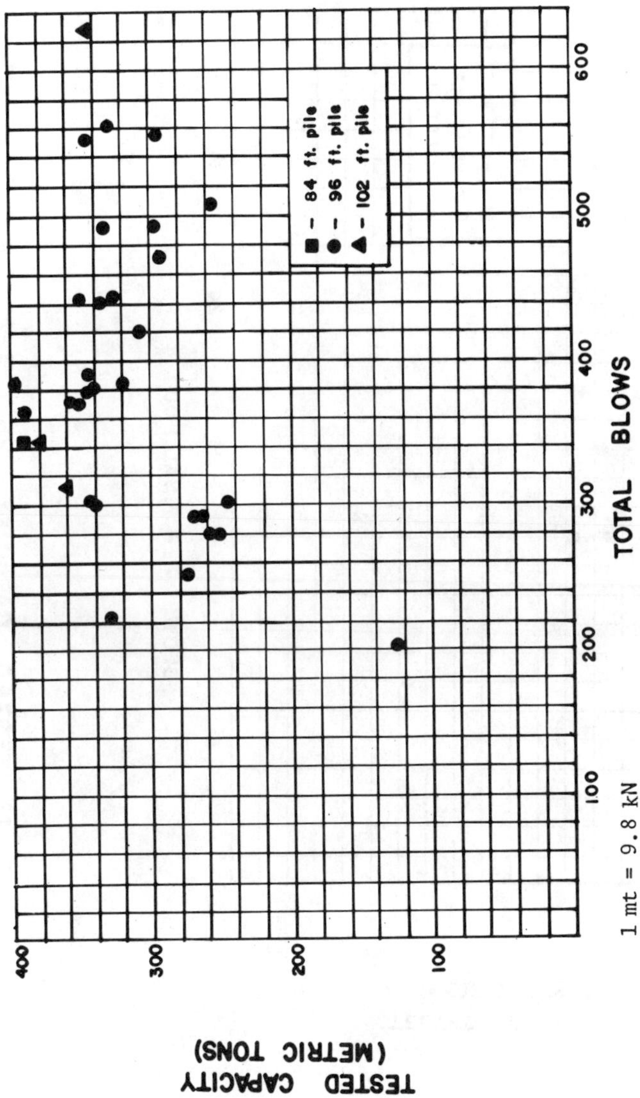

1 mt = 9.8 kN

FIG. 11, TESTED CAPACITY VS. TOTAL NUMBER OF BLOWS

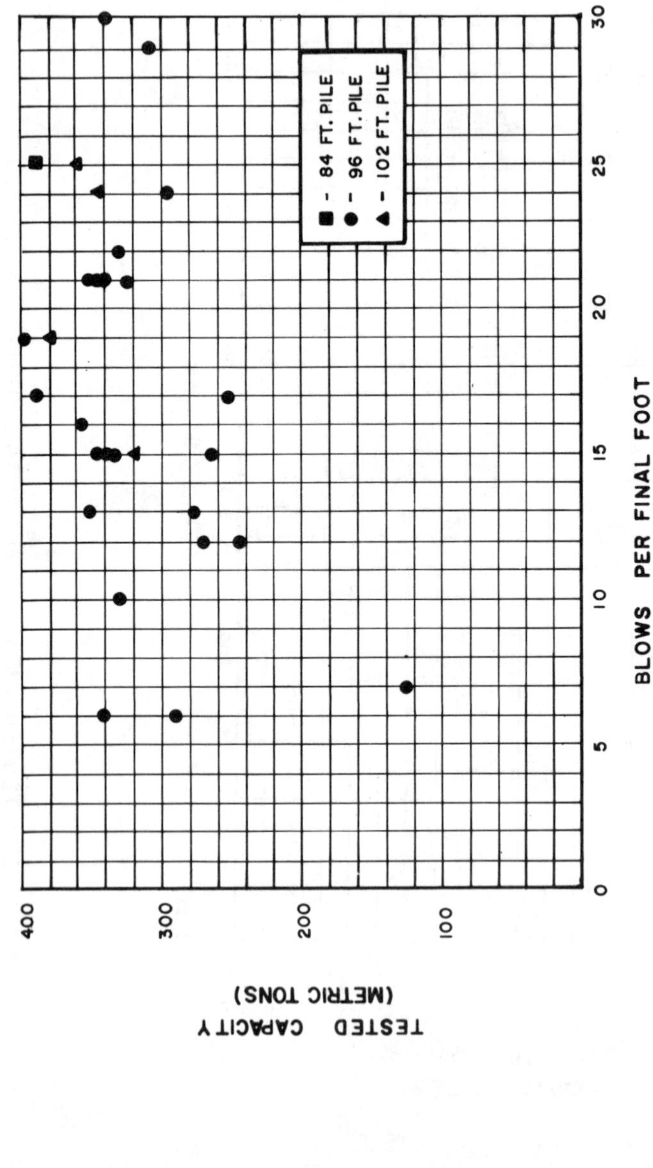

1 MT = 9.8 kN

FIG. 12, **TESTED CAPACITY VS. FINAL DRIVING RESISTANCE**

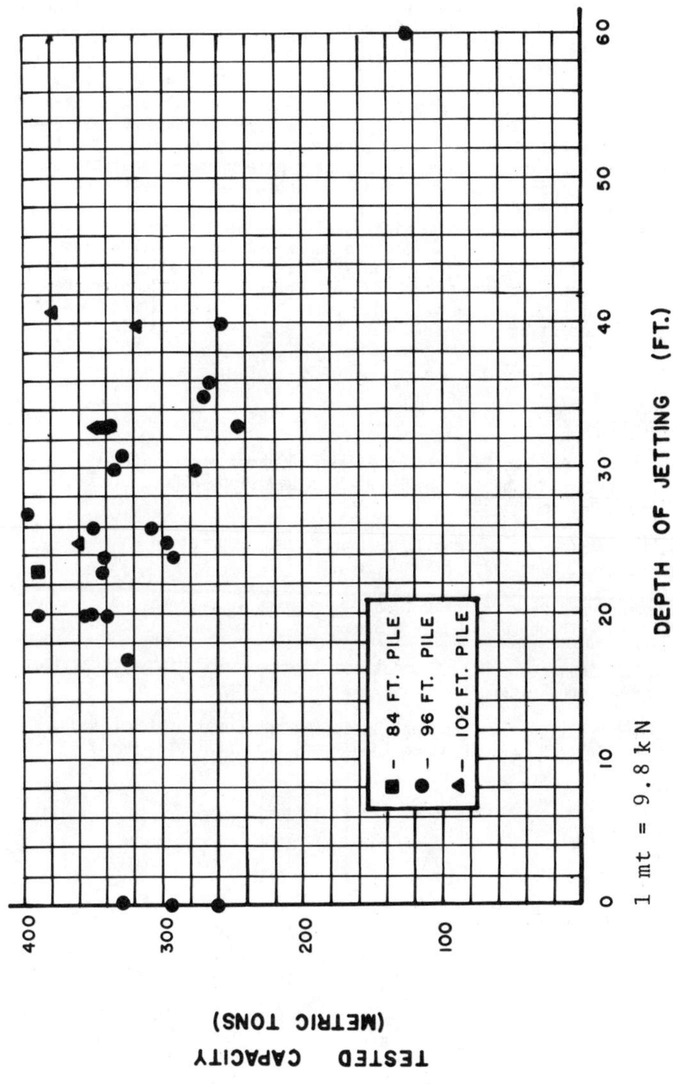

FIG. 13, TESTED CAPACITY VS. DEPTH OF JETTING

SUMMARY AND CONCLUSIONS

High capacity pile foundations were provided for China Steel Corporation's integrated steel mill in Kaohsiung, Taiwan by co-operation between China Steel, RSEA and Raymond. More than 40,000 Raymond Step-Taper piles were driven for design loads up to 135 metric tons (1320 kN). Load test capacities ranging up to 405 metric tons (3970 kN) were achieved by driving the piles deep into the lower dense sands at the site. Controlled water jetting was necessary to achieve the deep pile penetrations. Wave equation analysis, or any other driving criterion, is not applicable at this site because of the overwhelming effect of soil freeze. Pile penetration criteria along with load testing proved to be the best method of assuring pile load capacity. Nordlund's static pile capacity analysis procedure proved to be relatively accurate and was almost always somewhat conservative. Long term performance of these pile foundations have proven to be totally acceptable as measured in the highly settlement-sensitive coke oven structures of Phase I, Stage 1, as well as in the blast furnace structures. Pile driving was completed very rapidly and efficiently, thus allowing plant construction to be completed ahead of schedule. Phase I, Stage 1, plant operation has proven to be highly successful and profitable, and CSC looks forward to the successful completion of Phase I very soon and Phase II within the next few years.

APPENDIX I - REFERENCES

1. Nordlund, R. L. "Bearing Capacity of Piles in Cohesionless Soils", Journal of the Soil Mechanics and Foundation Division, ASCE, Vol. 89, No. SM3, May 1963, pp. 1-35.

2. Peck, R. B., Hanson, W. E., and Thornburn, T. H., Foundation Engineering, 2nd Edition, John Wiley & Sons, Inc., 1973

3. Terzaghi, Karl and Peck, Ralph B., Soil Mechanics In Engineering Practice, 2nd Edition, John Wiley & Sons, Inc., 1967.

4. Vesić, A. S., "Design of Pile Foundations," Synthesis of Highway Practice, 42, Transportation Research Board, 1977.

5. "Recommended Practice for Planning, Designing and Constructing Fixed Offshore Platforms," API RP2A American Petroleum Institute, March 1979.

APPENDIX II - NOTATION

A_p = pile tip area

A_s = pile surface area

C_a = unit soil-pile adhesion

C_k = correction factor for K_δ when $\delta \neq \phi$

C_u = undrained shear strength of soil

f = unit skin friction

K_δ = ratio of resultant of effective normal and shear stress on incipient failure plane at a point, to effective overburden pressure at that point.

N_q = bearing capacity factor

P_d = effective overburden pressure

P_u = ultimate bearing capacity

q_p = unit end-bearing resistance

v = volume of displaced soil per unit length of pile

α = reduction factor for C_u

δ = friction angle on surface of sliding

ϕ = angle of internal friction - soil

ω = taper of pile expressed in degrees

Subject Index

Author Index